Edited by
Alberto Fernandez-Nieves,
Hans M. Wyss, Johan Mattsson,
and David A. Weitz

Microgel Suspensions

Related Titles

Tadros, T. F. (ed.)

Topics in Colloid and Interface Science

Volume 1: Self-Organized Surfactant Structures

2011

ISBN: 978-3-527-31990-9

Tadros, T. F.

Rheology of Dispersions

Principles and Applications

2010

ISBN: 978-3-527-32003-5

Tadros, T. F. (ed.)

Topics in Colloid and Interface Science

Volume 2

2008

ISBN: 978-3-527-31991-6

Tadros, T. F. (ed.)

Colloids and Interface Science Series

6 Volume Set

2008

ISBN: 978-3-527-31461-4

Schramm, L. L.

Dictionary of Nanotechnology, Colloid and Interface Science

2008

ISBN: 978-3-527-32203-9

Platikanov, Dimo / Exerowa, Dotchi (eds.)

Highlights in Colloid Science

2008

ISBN: 978-3-527-32037-0

Edited by
Alberto Fernandez-Nieves, Hans M. Wyss,
Johan Mattsson, and David A. Weitz

Microgel Suspensions

Fundamentals and Applications

WILEY-VCH Verlag GmbH & Co. KGaA

The Editors

Dr. Alberto Fernandez-Nieves
Georgia Inst. of Technology
School of Physics
Atlanta, GA 30332
USA

Dr. Hans M. Wyss
Eindhoven University of Technology
ICMS & Mech. Engineering
5612AZ Eindhoven
The Netherlands

Dr. Johan Mattsson
University of Leeds
School of Physics and Astronomy
Leeds, LS2 9JT
UK

Prof. David A. Weitz
Harvard University
Dept. of Physics
Pierce Hall, 29 Oxford Street
Cambridge, MA 02138
USA

All books published by **Wiley-VCH** are carefully produced. Nevertheless, authors, editors, and publisher do not warrant the information contained in these books, including this book, to be free of errors. Readers are advised to keep in mind that statements, data, illustrations, procedural details or other items may inadvertently be inaccurate.

Library of Congress Card No.: applied for

British Library Cataloguing-in-Publication Data
A catalogue record for this book is available from the British Library.

Bibliographic information published by the Deutsche Nationalbibliothek
The Deutsche Nationalbibliothek lists this publication in the Deutsche Nationalbibliografie; detailed bibliographic data are available on the Internet at http://dnb.d-nb.de.

© 2011 WILEY-VCH Verlag & Co. KGaA, Boschstr. 12, 69469 Weinheim, Germany

All rights reserved (including those of translation into other languages). No part of this book may be reproduced in any form – by photoprinting, microfilm, or any other means – nor transmitted or translated into a machine language without written permission from the publishers. Registered names, trademarks, etc. used in this book, even when not specifically marked as such, are not to be considered unprotected by law.

Cover Grafik-Design Schulz, Fußgönheim
Typesetting Thomson Digital, Noida, India
Printing and Binding Fabulous Printers Pte Ltd

Printed in Singapore
Printed on acid-free paper

ISBN: 978-3-527-32158-2

Contents

Preface XIII
List of Contributors XVII

Part One Synthesis 1

1 Microgels and Their Synthesis: An Introduction 3
Robert Pelton and Todd Hoare
1.1   Introduction 3
1.1.1 Defining Microgels 3
1.1.1.1 The Generic Microgel: Structure and Characterization 4
1.1.2 Microgels Are Special 6
1.1.3 The Microgel Landscape 7
1.2   Microgel Synthesis 8
1.2.1 Introduction 8
1.2.2 Approach 1: Microgels Formed by Homogeneous Nucleation 10
1.2.2.1 Emulsion Polymerization and Surfactant-Free Emulsion Polymerization 11
1.2.2.2 Homogeneous Nucleation of Microgels from Linear Polymers 14
1.2.2.3 Core–Shell Microgels 14
1.2.3 Approach 2: Microgels from Emulsification 15
1.2.4 Approach 3: Microgels by Polymer Complexation 16
1.2.5 Exotic Microgels 18
1.2.6 Summary 19
1.3   Particle Derivatization 19
1.3.1 Chemical Coupling to Microgels 19
1.3.2 Microgel Decross-Linking 19
1.3.3 Charged Microgels from Nonionic Precursors 20
1.3.4 Nanoparticle-Filled Gels 21
1.4   Microgel Purification and Storage 22
1.4.1 Microgel Characterization 22
1.5   Conclusions 25
       References 25

2	**Polymerization Kinetics of Microgel Particles** 33	
	Abdelhamid Elaissari and Ali Reza Mahdavian	
2.1	Introduction 33	
2.2	Polymerization Processes 36	
2.3	Kinetics of Polymerization Reaction 37	
2.3.1	The Influence of Initiators 39	
2.3.2	The Effect of the Cross-Linking Agent 41	
2.3.3	The Effect of Functional Monomers 42	
2.3.4	Kinetic Aspects of Microgel Formation 45	
2.4	Conclusions 49	
	References 49	
3	**New Functional Microgels from Microfluidics** 53	
	Jin-Woong Kim and Liang-Yin Chu	
3.1	Introduction 53	
3.2	Monodisperse Thermosensitive Microgels Fabricated in a PDMS Microfluidic Chip 54	
3.3	Monodisperse Thermosensitive Microgels Fabricated in a Capillary Microfluidic Device 57	
3.4	Monodisperse Thermosensitive Microgels with Tunable Volume-Phase Transition Kinetics 62	
3.5	Monodisperse Thermosensitive Microgels with Core–Shell Structures Containing Functional Materials 63	
3.6	Monodisperse Thermosensitive Microgels with Multiphase Complex Structures 65	
3.7	Conclusions 68	
	References 68	

Part Two Physical Properties of Microgel Particles 71

4	**Swelling Thermodynamics of Microgel Particles** 73	
	Benjamin Sierra-Martin, Juan Jose Lietor-Santos, Antonio Fernandez-Barbero, Toan T. Nguyen, and Alberto Fernandez-Nieves	
4.1	Introduction 73	
4.2	Swelling Thermodynamics 76	
4.2.1	Polymer/Solvent Mixing 76	
4.2.2	Rubber Elasticity 79	
4.2.3	Ionic Effects 81	
4.2.3.1	Ideal Gas Contribution 81	
4.2.3.2	Electrostatic Energy of a Homogeneously Charged Microgel 83	
4.2.3.3	Contribution from Counterion Correlations 84	
4.2.3.4	Effect of a Slightly Inhomogeneous Fixed-Charge Distribution 86	
4.2.4	Equilibrium: Equation of State 88	
4.3	Theory Versus Experiment 94	
4.3.1	Role of Flory Solubility Parameter 94	

4.3.2	Influence of Cross-Linking Density	97
4.3.3	Effect of Charge Density	98
4.3.4	Salt Effects	100
4.3.5	Effect of Added Polymer	102
4.3.6	Cononsolvency: Swelling in Solvent Mixtures	104
4.3.7	Surfactant Effects	106
4.4	Additional Aspects	108
4.4.1	Elastic Moduli	108
4.4.2	Brief Remarks on Swelling Kinetics	108
4.5	Summary	110
	References	113

5 Determination of Microgel Structure by Small-Angle Neutron Scattering 117

Walter Richtering, Ingo Berndt, and Jan Skov Pedersen

5.1	Introduction	117
5.2	Form Factor of Microgels	118
5.3	Core–Shell Particles	129
5.2	Summary	131
	References	131

6 Interactions and Colloid Stability of Microgel Particles 133

Brian Vincent and Brian Saunders

6.1	Theoretical Background	133
6.1.1	Introduction	133
6.1.2	Van der Waals Interactions	136
6.1.3	Electrostatic Interactions	137
6.1.4	Depletion Interactions	139
6.1.5	Criteria for Dispersion Stability	140
6.2	Experimental Studies	141
6.2.1	Temperature- and Electrolyte-Induced Aggregation	141
6.2.2	Depletion-Induced Aggregation	151
6.2.3	Heteroaggregation	153
6.2.4	Probing Interactions between Microgel Particles	154
	References	160

Part Three Phase Behavior and Dynamics of Microgel Suspensions 163

7 Structure and Thermodynamics of Ionic Microgels 165

Christos N. Likos

7.1	Introduction	165
7.2	Effective Interparticle Potentials	167
7.3	The Fluid Phase of Ionic Microgels	174
7.4	Genetic Algorithms for the Crystal Structures	178
7.5	Phase Behavior	182

7.6	Summary and Concluding Remarks *189*	
	References *191*	

8	**Elasticity of Soft Particles and Colloids Near the Jamming Threshold** *195*	
	Matthieu Wyart	
8.1	Introduction *195*	
8.2	Structure and Mechanical Stability *196*	
8.3	Elastic Moduli *199*	
8.3.1	Force Balance and Contact Deformation Operators *199*	
8.3.2	Energy Expansion and Virtual Force Field *200*	
8.3.3	Elastic Moduli *202*	
8.4	Summary and Conclusion *204*	
	References *205*	

9	**Crystallization of Microgel Spheres** *207*	
	Zhibing Hu *207*	
9.1	Introduction *207*	
9.2	Synthesis and Characterization of PNIPAM Microgels *208*	
9.3	Phase Behavior of Dispersions of PNIPAM Microgels at Room Temperature *210*	
9.3.1	Characterization of Different Phases Using UV–Visible Spectroscopy *212*	
9.4	Temperature- and Polymer Concentration-Dependent Phases of the PNIPAM Microgel Dispersions *213*	
9.5	Theoretical Investigation of Phase Behavior *215*	
9.6	Phase Diagram in Terms of Volume Fraction *217*	
9.7	The Interparticle Potential *219*	
9.8	Annealing and Aging Effects *220*	
9.9	Kinetics of Crystallization *222*	
9.10	Crystallization Along a Single Direction *224*	
9.11	Summary and Outlook *225*	
	References *226*	

10	**Melting and Geometric Frustration in Temperature-Sensitive Colloids** *229*	
	Ahmed M. Alsayed, Yilong Han, and Arjun G. Yodh	
10.1	Introduction *229*	
10.2	The Experimental System *232*	
10.2.1	Synthesis of NIPA Microgel Particles *232*	
10.2.2	Microscopy and Temperature Control *235*	
10.2.3	Characterization: Dynamic Light Scattering *238*	
10.2.4	Characterization: Video Microscopy Measurement of Interparticle Potentials *240*	
10.3	"First" Melting in Bulk (3D) Colloidal Crystals *242*	

10.3.1	Background	*242*
10.3.2	Sample Preparation and Imaging	*244*
10.3.3	Positional Fluctuations and the Lindemann Parameter	*246*
10.3.4	Bulk Melting	*247*
10.3.5	"First" Melting Near Grain Boundaries	*248*
10.3.6	"First" Melting Near Dislocations	*250*
10.3.7	Positional and Angular Fluctuations Near Defects	*251*
10.3.8	Summary	*253*
10.4	Melting in Two Dimensions: The Hexatic Phase	*254*
10.4.1	Theoretical Background	*254*
10.4.2	Experimental Background	*257*
10.4.3	2D Samples	*258*
10.4.4	Data Analysis	*259*
10.4.5	The Hexatic Phase and Other Features of the Phase Diagram	*264*
10.4.6	The Order of the Phase Transitions	*265*
10.4.7	Summary	*265*
10.5	Geometric Frustration in Colloidal "Antiferromagnets"	*266*
10.5.1	Background	*266*
10.5.2	The Experimental System	*269*
10.5.3	Antiferromagnetic Order	*270*
10.5.4	Stripes and the Zigzagging Ground State	*270*
10.5.5	Dynamics	*271*
10.5.6	Summary	*273*
10.6	Future	*273*
	References	*274*

Part Four Mechanical Properties *283*

11 Yielding, Flow, and Slip in Microgel Suspensions: From Microstructure to Macroscopic Rheology *285*
Michel Cloitre

11.1	Introduction	*285*
11.2	Advanced Techniques for Microgel Rheology	*286*
11.2.1	Macroscopic Shear Rheology	*286*
11.2.2	DWS-Based Microrheology	*288*
11.2.3	Real Space Particle-Tracking Techniques	*290*
11.3	Near-Equilibrium Properties and Linear Rheology of Microgel Suspensions	*291*
11.3.1	Dilute Regime and Paste Formation	*291*
11.3.2	Linear Viscoelasticity of Microgel Pastes	*293*
11.3.3	Elastic Properties of Concentrated Microgel Pastes	*294*
11.4	Yielding, Flow, and Aging	*297*
11.4.1	Yielding	*297*
11.4.2	Flow of Microgel Pastes	*299*
11.4.3	Slow Dynamics and Aging of Microgel Pastes	*301*

11.5	Slip and Flow of Microgel Suspensions Near Confining Surfaces	*303*
11.5.1	Wall Slip *303*	
11.5.2	Direct Measurements of Slip Velocity *304*	
11.5.3	Elastohydrodynamic Lubrication as the Origin of Wall Slip *306*	
11.6	Outlook *307*	
	References *307*	

12 Mechanics of Single Microgel Particles *311*
Hans M. Wyss, Johan Mattsson, Thomas Franke, Alberto Fernandez-Nieves, and David A. Weitz

12.1	Compressive Measurements by Variation of the Osmotic Pressure *312*
12.2	Capillary Micromechanics: Full Mechanical Behavior of a Single Microgel Particle *315*
12.3	Discussion: Effects of Particle Softness on Suspension Rheology *322*
12.4	Microgels as Model Glasses: Soft Particles Make Strong Glasses *323*
12.5	Analogy to Emulsions and Foams *324*
	References *324*

13 Rheology of Industrially Relevant Microgels *327*
Jason R. Stokes

13.1	Introduction *327*
13.2	Flow Behavior *328*
13.2.1	Influence of Phase Volume and Concentration *329*
13.2.2	Shear Rheology of Concentrated Microgel Suspensions *332*
13.2.3	Linear Viscoelasticity of Concentrated Microgel Suspensions *335*
13.3	Microgel Suspension Rheology in Applications *338*
13.3.1	Coating Formulations *339*
13.3.2	Biomedical, Pharmaceutical, Personal Care, and Cosmetic Products *342*
13.3.3	Biopolymer Microgels for Food and Other Applications *345*
13.3.3.1	Starch Microgels *345*
13.3.3.2	Biopolymer Microgels and Particle Anisotropy *347*
13.4	Outlook *350*
	References *351*

Part Five Applications *355*

14 Exploiting the Optical Properties of Microgels and Hydrogels as Microlenses and Photonic Crystals in Sensing Applications *357*
L. Andrew Lyon, Grant R. Hendrickson, Zhiyong Meng, and Ashlee N. St. John Iyer

14.1	Introduction *357*
14.2	Responsive Microgel and Hydrogel-Based Lenses *358*
14.3	Photonic Crystals *362*
14.4	Other Responsive Systems *368*

14.5	Summary *372*	
	References *372*	
15	**Microgels in Drug Delivery** *375*	
	Martin Malmsten	
15.1	Introduction *375*	
15.2	Polymer Gels *376*	
15.3	Polymer Microgels *378*	
15.3.1	Temperature Triggering of Microgels *380*	
15.3.2	Electrostatic Triggering of Microgels *382*	
15.3.3	Triggering of Microgels by Specific Metabolites *385*	
15.3.4	Microgel Triggering by External Fields *385*	
15.3.5	Microgel Triggering by Degradation *389*	
15.4	Polymer Microcapsules *393*	
15.4.1	Microcapsule Triggering *395*	
15.5	Swelling, Loading, and Release Kinetics *399*	
15.6	Outlook *402*	
	References *403*	
16	**Microgels for Oil Recovery** *407*	
	Yuxing Ben, Ian Robb, Peng Tonmukayakul, and Qiang Wang	
16.1	Introduction *407*	
16.2	Microgels Used in Oil Recovery *410*	
16.2.1	Guar *410*	
16.2.1.1	Gel Formation *411*	
16.2.2	Rheology of Guar Gels and its Relation to Proppant Transport *415*	
16.2.2.1	Proppant Transport *416*	
16.2.3	Xanthan *416*	
16.2.4	Gels for Gravel Packing *418*	
16.2.5	Gels for Fluid Loss Control *419*	
16.2.6	Gels for Pills *420*	
16.3	Concluding Remarks *421*	
	References *421*	
17	**Applications of Biopolymer Microgels** *423*	
	Eugene Pashkovski	
17.1	Introduction *423*	
17.2	Origin, Production, and Molecular Properties of Xanthan Gum *425*	
17.3	Characterization of Xanthan and CMC Microgels *429*	
17.4	Rheology of Silica Suspensions in Xanthan Microgel Pastes *436*	
17.5	Aging of Concentrated Xanthan Suspensions *442*	
17.6	Conclusions *446*	
	References *447*	

Index *451*

Preface

Microgels are suspensions of colloidal-scale particles each comprised of a cross-linked polymer network. The intrinsic viscoelasticity of these particles leads to a fascinating and rich range of suspension properties. Microgels can be made responsive to external stimuli including pH, temperature, magnetic and electric fields, flow, or osmotic pressure. The kinetics of the response can be well controlled and properties such as the timescale of swelling or the microgel elasticity can be well tuned. Thus, by adjusting the properties of the microgel particles during synthesis, bulk suspension properties including rheology, dynamics, and structure can be finely tuned. Recent advances in synthesis methods have also led to a wide range of sophisticated morphologies of microgels, including core–shell structures, interpenetrating polymer networks, microgels containing nanoparticles, Janus particles, and functionalized microgels designed for sensor or pharmaceutical delivery applications. Moreover, microgels can be made biodegradable or biocompatible, which makes them valuable in many biomedical applications.

This tunable nature of the physical properties of microgels leads to a surprisingly rich suspension-phase behavior; as a consequence, these suspensions exhibit a wide range of material properties. Microgels are thus increasingly used in a wide variety of industrial applications, where they can provide new functionalities and be used as additives to control the properties of composite materials. This versatility also means that microgels constitute an ideal model system for fundamental studies of physical processes such as crystallization and glass formation, where the colloidal size offers insight that is often difficult to achieve for atomic or molecular materials. Microgels have thus become a highly active topic of research both for applied and fundamental studies.

In this book, we attempt to give an overview of the fascinating topic of microgels, starting from their synthesis, continuing with their physical properties, phase behavior, dynamics, and mechanical behavior, and finishing with discussions about some key applications that exploit the features of these materials. The book is organized in the following parts:

1) **Synthesis**: Different methods for synthesizing microgels are described. Pelton and Hoare (Chapter 1) give a comprehensive overview of the synthesis of different microgel systems, Elaissari and Mahdavian (Chapter 2) discuss the

polymerization kinetics during synthesis, and Kim and Chu (Chapter 3) describe novel microgel synthesis routes based on microfluidic devices, which offer fascinating possibilities for precise control of the morphology of individual particles.

2) **Physical properties of microgel particles**: Sierra-Martin et al. (Chapter 4) discuss the thermodynamics of microgel particles and their suspensions. Richtering et al. (Chapter 5) describe how small-angle neutron-scattering techniques can be used to determine the structure both of individual microgels and of suspensions, while Vincent and Saunders (Chapter 6) describe the interactions between microgel particles in suspension and their effects on dispersion stability.

3) **Phase behavior and dynamics of microgel suspensions**: Microgel particles can exhibit widely varying elastic properties, charge, and morphology, which in turn affect the structure, dynamics, and phase behavior of the suspensions. Likos (Chapter 7) gives a detailed account of the structure and thermodynamics observed in ionic microgels, where the presence of charge yields a rich phenomenology. Wyart (Chapter 8) describes how packed assemblies of soft elastic particles behave near jamming and discusses connections to different glass-forming systems. Hu (Chapter 9) discusses how crystallization takes place in microgel systems where the specific properties of the microgels often yield distinctly different behavior from that of hard-sphere colloids. Finally, Alsayed et al. (Chapter 10) describe an intriguing and unexpected use of microgels as model systems for studying geometrically frustrated systems that exhibit properties reminiscent of those observed in magnetic systems.

4) **Mechanical properties**: The mechanical properties of microgel suspensions is essential for many of their applications, and the tunable nature of the individual microgel particles provides a powerful means to tailor their bulk rheological properties. Cloitre (Chapter 11) describes the fascinating linear and nonlinear rheological properties of microgel suspensions, showing that variations in softness and interactions can lead to distinct changes in their behavior. To fully understand how the rheological suspension behavior corresponds to the properties of individual microgel particles, it is essential to directly measure the mechanics of single particles. This is the topic of the chapter by Wyss et al. (Chapter 12) who describe techniques for determining the full elastic properties of single microgel particles. Applications that use microgels to control mechanical properties are numerous. Based on industrially relevant systems, Stokes (Chapter 13) discusses how microgel properties can be exploited to control the mechanical behavior.

5) **Applications**: The last four chapters give examples of the important use of microgels for applications. Lyon et al. (Chapter 14) describe how the optical properties of microgels can be used in sensors and optical systems. Malmsten (Chapter 15) discusses the wide use of microgel systems in drug delivery applications. Ben et al. (Chapter 16) describe the use of microgels in advanced oil recovery techniques, and finally Pashkovski (Chapter 17) discusses the

important use of biologically derived microgels in many personal care, pharmaceutical, and cosmetic products.

Microgels will remain a rich topic of study in the future as many aspects of their behavior remain poorly understood. Moreover, the synthesis of new microgel-based materials with novel behavior is a highly active research area. Thus, this book does not offer a complete description of microgels and their properties; instead, the contributions serve both to provide a background for researchers and engineers working with these fascinating materials and to provide a source of inspiration for future work.

October 2010

Alberto Fernandez-Nieves, Hans M. Wyss
Johan Mattsson, and David A. Weitz

List of Contributors

Ahmed M. Alsayed
University of Pennsylvania
Department of Physics and Astronomy
209 South 33rd Street
Philadelphia
PA 19104
USA

and

CNRS/UPENN/Rhodia
Complex Assemblies of Soft Matter
UMI 3254
Bristol
PA 19007
USA

Ingo Berndt
RWTH Aachen University
Institute of Physical Chemistry
Landoltweg 2
52056 Aachen
Germany

Yuxing Ben
Halliburton Technology Center
2600 South 2nd Street
Duncan
OK 73536-0470
USA

Liang-Yin Chu
Sichuan University
School of Chemical Engineering
Chengdu
Sichuan 610065
China

Michel Cloitre
ESPCI-Paris Tech
Soft Matter and Chemistry
10 rue Vauquelin
75 005 Paris
France

Abdelhamid Elaissari
Université Lyon 1
CNRS, UMR 5007
Laboratoire d'Automatique et de Génie
des Procédés
Villeurbanne
69622 Lyon
France

Thomas Franke
University of Augsburg
Department of Physics
Augsburg
Germany

Microgel Suspensions: Fundamentals and Applications
Edited by Alberto Fernandez-Nieves, Hans M. Wyss, Johan Mattsson, and David A. Weitz
Copyright © 2011 WILEY-VCH Verlag GmbH & Co. KGaA, Weinheim
ISBN: 978-3-527-32158-2

Antonio Fernandez-Barbero
University of Almeria
Department of Applied Physics
Almeria 04120
Spain

Alberto Fernandez-Nieves
Georgia Institute of Technology
School of Physics
Atlanta
GA 30332-0430
USA

Yilong Han
Hong Kong University of Science and Technology
Department of Physics
Clear Water Bay
Kowloon
Hong Kong

Grant R. Hendrickson
Georgia Institute of Technology
School of Chemistry and Biochemistry
Atlanta
GA 30332-0400
USA

and

Parker H. Petit Institute for Bioengineering and Bioscience
Atlanta
GA 30332
USA

Todd Hoare
McMaster University
Department of Chemical Engineering
JHE 374
1280 Main St. W
Hamilton
Ontario
Canada L8S 4L7

Zhibing Hu
University of North Texas
Department of Physics
Denton
TX 76203
USA

Jin-Woong Kim
Amore-Pacific Co.
R&D Center
314-1, Bora-dong
Giheung-gu
Yongin-si
Gyeonggi-do 449-729
Korea

Juan Jose Lietor-Santos
Georgia Institute of Technology
School of Physics
Atlanta
GA 30332-0430
USA

Christos N. Likos
Faculty of Phyiscs
University of Vienna
Boltzmanngasse 5
1090 Vienna
Austria

L. Andrew Lyon
Georgia Institute of Technology
School of Chemistry and Biochemistry
Atlanta
GA 30332-0400
USA

Parker H. Petit Institute for Bioengineering and Bioscience
Atlanta
GA 30332
USA

Ali Reza Mahdavian
Iran Polymer and Petrochemical Institute
Polymer Science Department
P.O. Box 14965-115
Tehran, 14967
Iran

Martin Malmsten
Uppsala University
Department of Pharmacy
P.O. Box 580
751 23 Uppsala
Sweden

Johan Mattsson
Chalmers University of Technology
Department of Applied Physics
Göteborg
Sweden

and

University of Leeds
School of Physics and Astronomy
Leeds, LS2 9JT
UK

Zhiyong Meng
Georgia Institute of Technology
School of Chemistry and Biochemistry
Atlanta
GA 30332-0400
USA

and

Parker H. Petit Institute for Bioengineering and Bioscience
Atlanta
GA 30332
USA

Toan T. Nguyen
Georgia Institute of Technology
School of Physics
Atlanta
GA 30332-0430
USA

Eugene Pashkovski
Unilever R&D
Trumbull
CT 06 611
USA

Jan Skov Pedersen
University of Aarhus
Department of Chemistry and iNANO
Interdisciplinary Nanoscience Center
Langelandsgade 140
800 Aarhus C
Denmark

Robert Pelton
McMaster University
Department of Chemical Engineering
JHE-136
Hamilton
Ontario
Canada L8S 4L7

Walter Richtering
RWTH Aachen University
Institute of Physical Chemistry
Landoltweg 2
52056 Aachen
Germany

Ian Robb
Halliburton Technology Center
2600 South 2nd Street
Duncan
OK 73536-0470
USA

Brian Saunders
The University of Manchester
School of Materials
Manchester M1 7HS
UK

Benjamin Sierra-Martin
Georgia Institute of Technology
School of Physics
Atlanta
GA 30332-0430
USA

Ashlee N. St. John Iyer
Georgia Institute of Technology
School of Chemistry and Biochemistry
Atlanta
GA 30332-0400
USA

Parker H. Petit Institute for
Bioengineering and Bioscience
Atlanta
GA 30332
USA

Jason R. Stokes
Unilever Corporate Research
Colworth Science Park
Sharnbrook
Bedfordshire MK44 1LQ
UK

The University of Queensland
School of Chemical Engineering
Brisbane
Queensland 4072
Australia

Peng Tonmukayakul
Halliburton Technology Center
2600 South 2nd Street
Duncan
OK 73536-0470
USA

Brian Vincent
University of Bristol
School of Chemistry
Bristol BS8 1TS
UK

Qiang Wang
Halliburton Technology Center
2600 South 2nd Street
Duncan
OK 73536-0470
USA

David A. Weitz
Harvard University
Department of Physics & School of
Engineering and Applied Science
(SEAS)
Cambridge
MA 02138
USA

Matthieu Wyart
Dept. of Physics and Center for Soft
Matter Research
New York University
2-4 Washington Place
New York
NY 10003
USA

Princeton University
Lewis-Sigler Institute
Princeton
NJ 08544-1014
USA

Hans M. Wyss
Eindhoven University of Technology
ICMS & Mech. Engineering
5612AZ Eindhoven
The Netherlands

Arjun G. Yodh
University of Pennsylvania
Department of Physics and Astronomy
209 South 33rd Street
Philadelphia
PA 19104
USA

Part One
Synthesis

1
Microgels and Their Synthesis: An Introduction
Robert Pelton and Todd Hoare

1.1
Introduction

This chapter introduces microgels and overviews their preparation, cleaning, and characterization. Some aspects of microgel derivatization and storage will also be summarized.

1.1.1
Defining Microgels

In a 1949 publication entitled "Microgel, a new macromolecule," Baker coined the term "microgel" to describe cross-linked polybutadiene latex particles [1]. The word "micro" referred to the size of the gel particles, which might now be termed "nano" since the diameters of his gels were less than 1000 nm. The "gel" part of Baker's microgel referred to the ability of the particles to swell in organic solvents. Baker's work emphasized that microgels consisted of very high molecular weight polymer networks. In other words, each gel particle was an individual polymer molecule. A revised definition of microgels is now given, followed by an introduction to the unique characteristics of microgels.

We define a microgel as *a colloidal dispersion of gel particles*. Implicit in this definition are three criteria:

1) Microgels fall within the particle size range of 10–1000 nm, typical of colloidal particles [2].
2) Microgels are dispersed in a solvent.
3) Microgels are swollen by the solvent.

Our definition encompasses a wide range of microgel materials. At one extreme are Baker's cross-linked latex particles in swelling solvent; at the other extreme, many biological cells satisfy the above definition. However, with the possible exception of Pollack's book [3], few links have been made between biological cells and microgels. Thus, the discussion of microgel preparation in this chapter will be

Microgel Suspensions: Fundamentals and Applications
Edited by Alberto Fernandez-Nieves, Hans M. Wyss, Johan Mattsson, and David A. Weitz
Copyright © 2011 WILEY-VCH Verlag GmbH & Co. KGaA, Weinheim
ISBN: 978-3-527-32158-2

restricted to synthetic microgels based on polymers of both petrochemical and biological origin.

Both surfactant and polymeric micelles also fit the above three criteria. However, these species are not usually called microgels. In the case of surfactant micelles, individual micelles have a finite lifetime with rapidly exchanging surfactant monomers whereas microgels have a static composition. At the other extreme, aqueous block copolymer micelles based on long hydrophobic blocks can be long-lived; however, the hydrophobic cores tend not to swell very much with water. Thus, stable block copolymer micelles are more akin to latex than to microgels. These considerations lead to a fourth criterion for defining microgels.

4) Microgels have stable structures. Either covalent or strong physical forces stabilize the polymer network. On the other hand, like any colloidal dispersion, microgel particles can aggregate (flocculate or coagulate) as described in Chapter 6.

Finally, Baker suggests that each microgel particle is composed of one polymer molecule [1]. Although this is true for many microgels prepared by vinyl polymerization, we propose that this requirement is too restrictive. For example, we will describe microgels prepared by polyelectrolyte complex formation giving particles containing many polymer chains.

Of course, attempts to define a class of materials can be problematic at the boundaries. Compare, for example, a polystyrene-core (water-insoluble) poly(N-isopropylacrylamide) (PNIPAM)-shell (water swellable) microgel with a polystyrene latex bearing a monolayer of surface PNIPAM. Are they both microgels? Where is the boundary between microgels and nanogels? Does a microgel have to be swollen? Some speak of the latex-to-microgel transition [4] – is this useful? We leave these questions for others.

1.1.1.1 The Generic Microgel: Structure and Characterization

To facilitate our discussion of microgel preparation strategies, it is useful to consider the generic microgel. Figure 1.1 shows a schematic representation of the generic microgel at three distance scales – the data are based on a PNIPAM microgel [5]. The suspension consists of a dispersion of uniform gel particles $\sim$ 500 nm in diameter. The microgel particles are present as distinct particles undergoing Brownian motion. To the naked eye, highly swollen microgel suspensions are nearly transparent, whereas slightly swollen gel suspensions are milky white like a conventional latex dispersion. Highly swollen particles have a refractive index close to that of water. Thus, swollen microgels scatter little light compared to dispersions of unswollen organic polymers, such as polystyrene.

In general, microgels are very resistant to aggregation (i.e., colloidally stable) because the surfaces often bear electrical charges and dangling surface chains (hairs). High colloidal stability is further illustrated by the ability of freeze-dried [6] or precipitated (ultracentrifuged) microgels to spontaneously redisperse in water. This is unusual. For example, dried or coagulated polystyrene latex is virtually impossible to redisperse completely in water. In general, microgels tend to be more colloidally

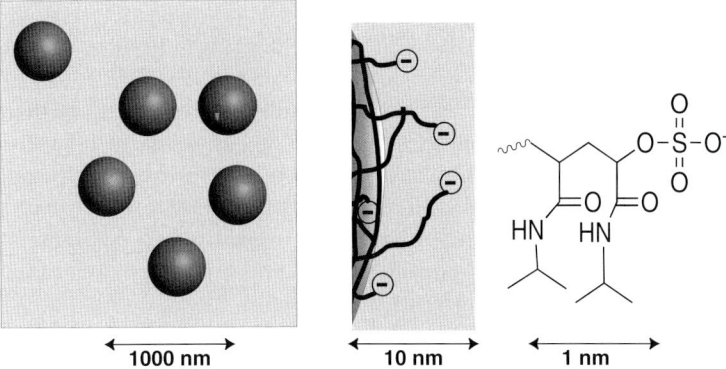

Figure 1.1 The essential features of a microgel: a water (solvent) swollen polymer network present as a colloidal dispersion.

stable in the swollen form where van der Waals attraction is diminished and surface hairs can sterically stabilize the microgel particles.

Individual microgel particles are usually spherical and consist of a water swollen cross-linked polymer network. Figure 1.2 shows a transmission electron micrograph of the first PNIPAM microgel [7]. The dark circles are disks arising from the dehydration of an ordered layer of spheres.

We will learn from the following sections and in other chapters that although gels with a uniform particle distribution are quite common, it is very difficult to prepare microgels with a uniform distribution of cross-links or bound charge throughout the volume of individual gel particles.

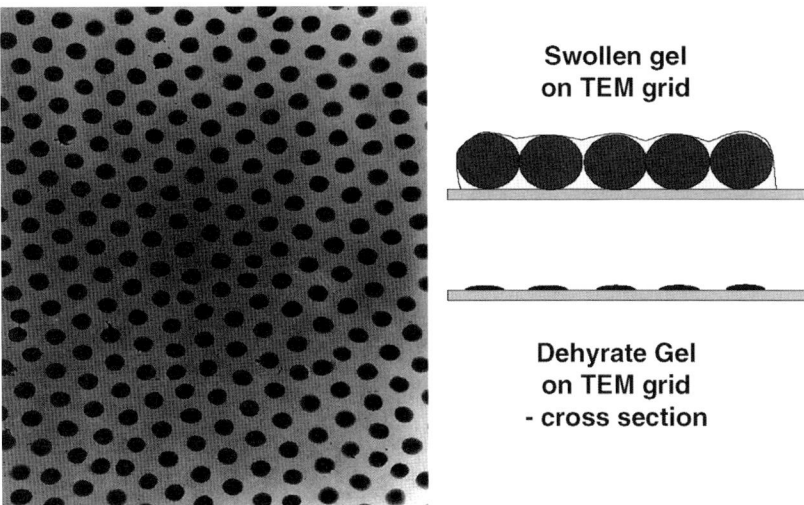

Figure 1.2 Transmission electron micrograph of the first PNIPAM microgel.

Typical water contents for microgels vary from 10 to 90 wt% depending upon the detailed chemistry of the microgel dispersion. Microgel swelling is described in detail in Chapter 4. Many publications give swelling ratios that are derived from particle size measurements under two different solvency conditions. In contrast, relatively few papers give microgel molecular weight because it is surprisingly difficult to measure the average dry mass per microgel particle. A consequence of this difficulty is that many publications give neither the water content nor the number concentration of microgel dispersions. Microgel molecular weight can be measured by (1) the measurement of the size of the microgels under low swelling conditions and assuming a water content – for example, two water molecules per NIPAM moiety [8]; (2) packing the microgels into colloidal crystals and estimating the degree of swelling from the wet and dry masses of the colloidal crystal [5]; (3) calculating the effective particle volume fraction from viscosity measurements [9], which when coupled to the swollen diameter and dry solids content gives number concentration and molecular weight.

Microgel number concentration can be measured directly by single-particle counting using flow cytometry [10] or manually with a hemocytometer. Indirect methods are usually based on measurement of the dry solids content and the microgel molecular weight.

The schematic representation of the generic microgel in Figure 1.1 shows the presence of short polymer chains extending from the gel surface. The presence of these chains was postulated in the first PNIPAM microgel publication in order to explain the exceptionally high colloidal stability in concentrated electrolyte. Surface chains would provide steric stabilization [7]. In general, the surface topology of microgels has been poorly described in the literature. In most cases, we know neither the length distributions nor the density of surface chains. We are likely to know more about the surface chains when (i) living radical techniques are used to grow surface chains on existing particles [11–13]; (ii) monomers such as vinyl acetic acid are used, which act as chain transfer agents [14]; or (iii) macromonomers are used to decorate the microgel surface with polyethylene glycol chains [15]. An elegant example from Kawaguchi's group involved using living radical polymerization to grow PNIPAM hairs on a core particle [16].

The generic microgel in Figure 1.1 has negative charge groups covalently bound to the polymer network. Virtually all microgels are electrically charged – it is difficult to prepare a nonionic aqueous microgel. The main sources of the electrically charged groups are ionic free radical initiators and/or ionic monomers copolymerized into the polymer network.

1.1.2
Microgels Are Special

The existence of hundreds of scientific publications, patents, and this book suggests that microgels are important. Interest in microgels comes from their unique blend of properties combining useful aspects of conventional macrogels with useful properties of colloidal dispersions.

Microgels share a number of properties with macrogels. Most importantly, both macrogels and microgels swell with water (or solvent) to an extent controlled by the cross-link density, the polymer/water compatibility, and the presence of electrical charges. Microgel swelling properties are described in Chapter 4. Perhaps one of the biggest driving forces for microgel research is that, like macrogels [17], microgels can be "intelligent" or "responsive," meaning their degree of swelling can be controlled by temperature, pH, magnetic fields, light, and specific solutes such as glucose [18–21]. Controllable swelling has been applied to demonstrate the uptake and release of solutes, including drugs [22], proteins, and surfactants [23, 24].

The colloidal nature of microgels gives them significant advantages over macrogels. These include, in decreasing order of importance, the following:

1) Microgel suspensions are free-flowing liquids unless highly concentrated. Indeed, their flow properties depend upon the volume fraction of swollen particles and are approximately independent of cross-link density, whereas macrogels flow only at very low levels of cross-linking near the gel point.
2) Microgels respond very rapidly to environmental changes. The very high surface to volume ratios facilitate mass transport to and from the microgels.
3) Exotic microgel morphologies can be used to fine-tune properties. For example, there is no macrogel equivalent of the wide range of core–shell particle architectures.
4) Colloidal science techniques including electrophoresis, dynamic light scattering, and small-angle light scattering provide structural information not usually available for macrogels.
5) Microgels can be assembled into useful larger objects such as 2D assemblies at the air–water [25, 26] and oil–water interfaces [27–29]. Examples of 3D structures are colloidal crystals giving environmentally sensitive optical properties [30, 31] and layer-by-layer assemblies [32–42].

1.1.3
The Microgel Landscape

The microgel field is rapidly evolving with ever increasing complexity. However, some generalizations can be made to help create perspective. There are two microgel worlds that are virtually exclusive – commercial microgels and academic microgels. The commercial gels have been used in large quantities since the 1960s. Two common classes of commercial gels are nonaqueous and alkali swellable microgels. Nonaqueous microgels are described in the paints and coatings patent literature [43]. Alkali swellable microgels are based on cross-linked acrylic acid latexes that swell when the pH is raised. These are widely used in formulated products to control rheological properties [44–46].

The academic microgel literature has exploded in the last decade and we can generalize to emphasize major trends. First, most scientific publications employ "homemade" instead of commercial microgels. Second, most of the publications involve aqueous microgels. Finally, most of the aqueous microgel studies describe microgel particles on the basis of PNIPAM, which is readily polymerized into linear water-soluble polymers [47], microgels [7], or macrogels [48]. PNIPAM derivatives have received much attention because the microgels are very uniform and the swelling properties are temperature sensitive [7]. The organic chemistry of PNIPAM and the other major microgel platform polymers will be summarized in another section later on.

1.2
Microgel Synthesis

1.2.1
Introduction

The goals of microgel synthesis include controlling the particle size distribution, the colloidal stability, and the distribution of specific functional groups such as cross-linker, charged groups, or reactive centers for further chemical derivatization. There are three possible starting points for microgel preparation:

1) **From monomer.** This is the most common approach and is described in the most detail here. Table 1.1 lists many of the vinyl monomers that have been used to prepare microgels. Monofunctional monomers are nonionic, cationic, or anionic. Of the bifunctional cross-linking monomers N,N-methylenebisacrylamide (MBA) is the most widely used. Polyethylene glycol dimethacrylate is an attractive choice for acrylate-based microgels given that it offers the additional flexibility of varying the length of the PEG chain between the cross-link points. The cross-linker solubility can influence microgel properties [49].
2) **From polymer.** Aqueous polymer solutions can be emulsified in oil and chemically cross-linked. Another route to microgels based on existing polymers is the formation of colloidal polyelectrolyte complexes by mixing oppositely charged polymers in dilute solution.
3) **From macrogels.** It is possible to mechanically grind a macrogel to form microgels. There are very few reports of this in the literature [50]. We tried grinding polyvinylamine (PVAm) macrogels and obtained large, irregularly shaped microgels [51].

It is convenient to divide the diverse range of microgel preparation strategies into three classifications based on the particle formation mechanism – those formed by *homogeneous nucleation*, those formed by *emulsification*, and those formed by *complexation*. Homogeneous nucleation refers to those preparations in which microgel particles are generated from initially homogeneous (or nearly so) solutions. Emulsification refers to those methods where aqueous droplets of a "pregel" solution are formed in an oil or brine phase and, in the second step, the droplets are polymerized and/or cross-linked into a microgel. Finally, microgels can be prepared by mixing two dilute, water-soluble polymers that form complexes in water.

Table 1.1 Vinyl monomers used to prepare microgels.

Monofunctional nonionic

Acrylamide (AM)	N-Ethylacrylamide [126]	N-Ethyl methacrylamide [49]
N-Isopropylacrylamide (NIPAM) [7]	N-Vinylformamide (NVF) [51]	N-Vinyl caprolactam [127]
Vinylpyrrolidone [127]		

Anionic monofunctional

4-Vinylphenylboronic acid [128]	Phenylboronic acid methacrylamide [129]	[130]
Acrylic acid	Methacrylic acid	Fumaric acid [118]
Maleic acid [118]	Vinyl acetic acid [14]	

Cationic monofunctional

| Allylamine [131] | Diallyldimethyl ammonium chloride (DADMAC) [132] | [133] |

(Continued)

Table 1.1 (Continued)

[134]	2-(Dimethylamino)ethyl methacrylate [135]	1-Vinylimidazole [136]
N-3-Dimethylaminopropyl methacrylamide [137]	2-(Methacryloyloxy) ethyl trimethyl ammonium chloride [74]	

Bifunctional nonionic cross-linker

1,3-Divinylimidazolid-2-one (BVU) [51]	N,N'-Methylenebisacrylamide [7]	N,N'-(1,2-Dihydroxyethylene) bisacrylamide [81]
1,4-Butanediol diacrylate [46]	1,3-Butanediol dimethacrylate [49]	Tetraethylene glycol dimethacrylate [49]
1,4-Butanediol dimethacrylate [49]	N,N'-Bis(acryloyl)cystamine [91, 92]	Polyethylene glycol dimethacrylate [138]

1.2.2
Approach 1: Microgels Formed by Homogeneous Nucleation

In homogeneous nucleation, a solution of soluble monomer, including some type of cross-linking agent, is fed into the reactor and microgel particles form over the course of polymerization. A key requirement for discrete microgel particle formation is that the polymer formed must be insoluble under the polymerization conditions; monomers giving soluble polymers under the polymerization conditions will form a macrogel. For example, PNIPAM microgels readily form when the monomer is polymerized in water at 70 °C because PNIPAM is water insoluble at high temperature [47]. In contrast, acrylamide (see Table 1.1), a common monomer with a similar chemical structure to PNIPAM, gives a water-soluble polymer at all temperatures, so polyacrylamide microgels cannot be prepared by homogeneous polymerization in water. Polymerization of aqueous acrylamide solutions gives a macrogel.

Microgel preparations involving homogeneous nucleation include the following types of polymerizations: emulsion polymerization (EP), surfactant-free emulsion polymerization (SEP), and microgel formation from dilute polymer solution. Each of these is described in the following sections.

1.2.2.1 Emulsion Polymerization and Surfactant-Free Emulsion Polymerization

Emulsion polymerization is the primary process for preparation of commercial latex dispersions involving monomers of limited water solubility. Typically, the reactor is charged with water, surfactant, monomer, and a water-soluble free radical initiator. The monomer is initially present as a suspension of large monomer drops, whereas at the end of the polymerization the polymer is present as surfactant-stabilized latex particles, typically about 100 nm in diameter. The locus of polymerization is in the aqueous phase and the growing latex particles – the monomer droplets serve as a reservoir replenishing the dissolved monomer in the aqueous phase. The theoretical basis of emulsion polymerization has been investigated extensively – the major conclusions are well described in virtually every polymer textbook, and more details are given in specialized works such as Gilbert's [52].

In the mid-1970s, there was much activity in the academic community around a variation of emulsion polymerization called surfactant-free emulsion polymerization [53]. For example, with this method monodisperse polystyrene latexes can be prepared simply with water, styrene monomer, and potassium persulfate initiator. Upon heating at ∼60 °C under nitrogen, the persulfate decomposes into sulfate radicals that initiate styrene polymerization. Sulfate groups terminating polystyrene chains end up at the water/polystyrene interface, conferring electrostatic stabilization and preventing aggregation.

The first PNIPAM microgel was prepared with a variation of the polystyrene SEP recipe in which styrene was replaced with NIPAM and a little MBA was included to prevent microgels from dissolving when the temperature was lowered at the end of the polymerization [7]. Figure 1.3 shows the mechanism of PNIPAM microgel formation. Sulfate radicals generated in solution initiate the homogeneous polymerization of NIPAM and MBA. However, the insolubility of the PNIPAM network under polymerization conditions causes the growing polymer chain to phase separate, forming precursor particles that are not colloidally stable. As the aggregated precursor particles coalesce, the charged chain ends tend to concentrate at the particle/water interface. Therefore, as the aggregates grow, the surface charge density increases until a point is reached where the growing particle is colloidally stable with respect to similar sized or larger particles. These first formed stable particles are called primary particles. To achieve a monodisperse product, the primary particles

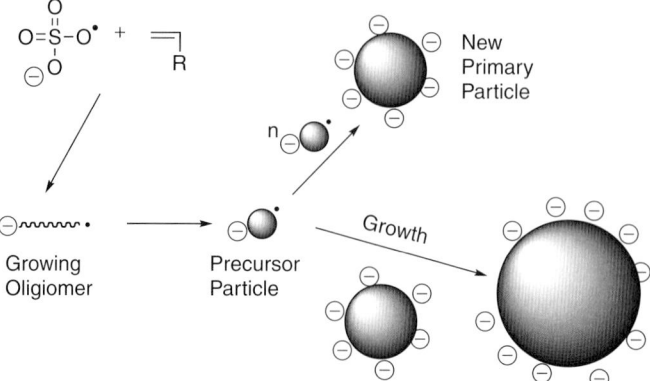

Figure 1.3 Microgel formation by surfactant-free emulsion polymerization. Initially, unstable precursor particles aggregate to form new primary particles. At the end of nucleation stage, all new precursor particles are captured by existing stable particles.

must be formed at low monomer conversion. In later stages of polymerization, all newly formed precursor particles deposit onto existing stable microgels contributing to particle growth.

There are few variables in the above PNIPAM microgel SEP, thus it is difficult to obtain a wide range of average microgel diameters. Using a surfactant, such as sodium dodecyl sulfate (SDS) influences microgel particle nucleation and thus the final size [5, 54]. Figure 1.4 shows that microgel diameter decreases with SDS

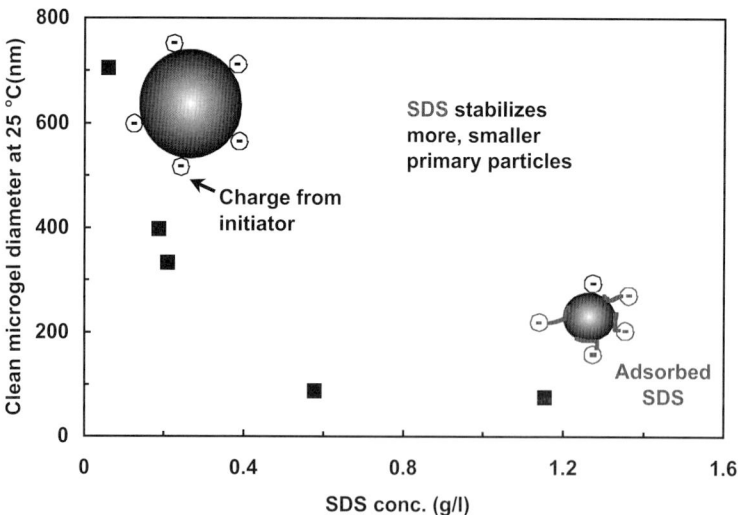

Figure 1.4 The influence of sodium dodecyl sulfate on the size of the resulting PNIPAM microgels. Data from Ref. [5].

concentration by a factor of 7. The role of the SDS is to stabilize the primary particles so that they are smaller than those prepared without SDS. The smaller the primary particles, the higher the total number of primary particles that are initially formed, resulting in smaller microgels for the same dose of monomer. Figure 1.4 illustrates the transition from SEP (i.e., no SDS) to EP (i.e., SDS above the critical micelle concentration). SDS addition also gave higher microgel yields and more uniform particles. Of course, it may be necessary to remove the surfactant after the preparation, depending upon the application. Standard approaches to microgel cleaning are described in a later section.

Herein we refer to the process shown in Figure 1.3 as surfactant-free emulsion polymerization because of the similarities with styrene SEP. However, there is some difficulty with this nomenclature. Emulsion polymerization applies to monomers with low water solubility, whereas virtually all vinyl monomers used to make microgels are water soluble (see Table 1.1). Therefore, many authors use the term "precipitation polymerization" to describe these microgel SEPs.

The majority of microgel recipes involve batch EP or SEP in which all of the monomer and initiators are added at the beginning. However, even the simplest PNIPAM microgel has two comonomers (NIPAM and MBA) and many of the most interesting microgels have been prepared with three or more monomers. The presence of more than one monomer type introduces complexity in any free radical copolymerization arising from the differences in monomer reactivity. For example, we showed many years ago that MBA polymerizes substantially faster than NIPAM in the PNIPAM microgel SEP [55]. Thus, the cross-linker density is higher in the first formed polymer than in the last. In other words, the microgel periphery will be less cross-linked and more swollen than the microgel core.

Recently, Hoare has employed kinetic modeling to predict the radial distribution of cross-links and carboxyl groups across a microgel particle [56]. The distributions are sensitive to the monomer chemistry and reactivity. For example, Figure 1.5 compares the distribution of carboxyl groups across an individual gel for PNIPAM microgels prepared using methacrylic acid or vinyl acetic acid as the carboxylic comonomer. Methacrylic acid polymerizes faster than PNIPAM; thus, the carboxyl groups are concentrated in the particle core. In contrast, vinyl acetic acid reacts more slowly and primarily by chain transfer instead of free radical propagation, resulting in the concentration of carboxyl groups at the end of hairs on the microgel surfaces. Chapter 2 gives a more detailed account of polymerization kinetics and Chapter 5 describes microgel structural characterization by neutron scattering.

In summary, most microgel recipes employ batch polymerizations and a few use semibatch strategies. Furthermore, there have been very few reaction engineering studies involving significant modeling of microgel formation. Finally, there have been some unusual variations of EP and SEP. Cao *et al.* reported microgel polymerizations in supercritical carbon dioxide [57]. Boyko *et al.* compared water and D_2O for the preparation of poly(*N*-vinylcaprolactam-*co*-*N*-vinylpyrrolidone) – heavy water was a poor solvent for microgels.

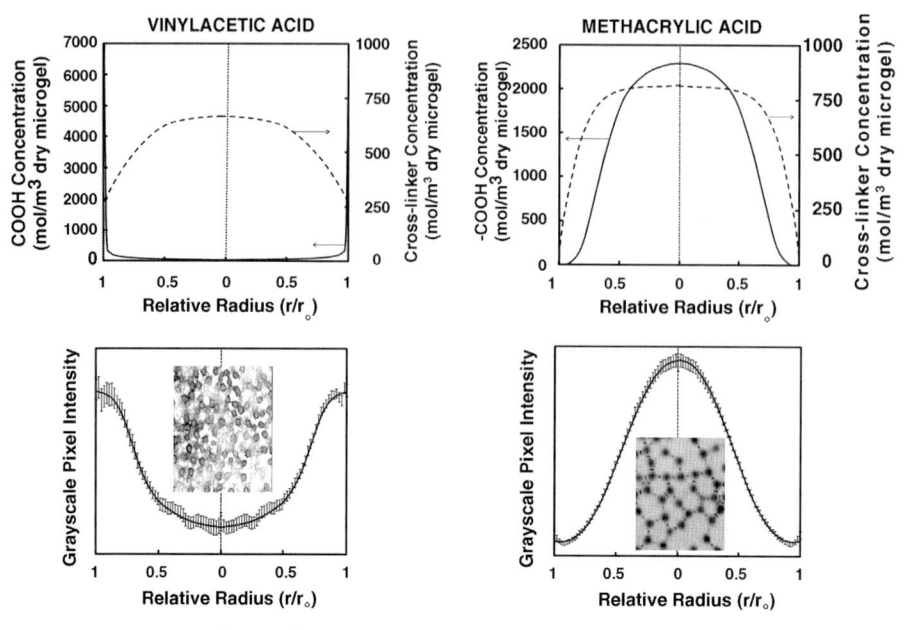

Figure 1.5 Distribution of cross-links and carboxyl groups for microgels prepared with vinyl acetic acid (VAA–NIPAM) and methacrylic acid (MAA–NIPAM). The top curves were computed whereas the bottom figures are experimental [56].

1.2.2.2 Homogeneous Nucleation of Microgels from Linear Polymers

There have been a few reports describing the conversion of linear polymer solutions to microgels [58–64]. In the case of PNIPAM, heating dilute linear polymer solutions above the VPTT gives slightly swollen, colloidally stable microgels [60]. To prevent microgels from dissolving on cooling, it is necessary to cross-link the gels. For example, Kuckling *et al.* used UV photocross-linking to stabilize phase-separated microgels [62].

A related approach is to prepare diblock copolymers that micellize [65] and cross-link the core. For example, Charleux's group reported microgels prepared by adding a little cross-linker during the nitroxide-mediated living radical polymerization of poly (acrylic acid-*b*-diethylacrylamide) under conditions in which the diethylacrylamide block phase separates [66]. Block copolymer micelles tend to be small, so this method will give relatively small microgels.

1.2.2.3 Core–Shell Microgels

Core–shell latex particles prepared by emulsion polymerization have been available for decades. The core particles are first prepared by conventional emulsion polymerization. In the second step, the first-stage particles are used as seeds for the second-stage shell polymerization. There are a number of challenges in the preparation of core–shell particles, including microgels. First, we must control nucleation in the first stage to generate uniform seeds, whereas in the second stage, nucleation must be prevented. That is, for uniform core–shell particles, all new stage-2 polymer chains must deposit on existing particles, avoiding secondary nucleation of stage-2 particles. Secondary nucleation of stage-2 polymer is a common problem easily identified with electron microscopy, which can reveal a population of small stage-2 particles coexisting with larger core–shell latex. Another challenge involves rearrangement of core–shell particles into more complex morphologies. For example, it is frequently difficult to prepare core–shell particles in which the shell is more hydrophobic than the core. There is a strong thermodynamic driving force to minimize interfacial energies by producing raspberry, stuffed olive, and other complex shapes [67]. To achieve nonequilibrium structures, it is necessary to freeze structures by cross-linking or working below T_g.

Core–shell microgels have been prepared since the earliest days of microgel development. We prepared the first polystyrene-core, PNIPAM-shell microgels in a two-step procedure – first preparing a polystyrene surfactant-free latex and then grafting PNIPAM onto the particles [68]. The key point is that the PNIPAM polymerization must be carried out at room temperature where PNIPAM is soluble and will not nucleate new particles. The state of the art in core–shell microgels is exemplified by a series of papers from Lyon's group, who prepared PNIPAM-core plus PNIPAM-*co*-acrylate shell gels and the inverse gels [69]. There have been more than 50 scientific publications on core–shell microgels from 1988 to 2008, with most appearing after 2000. This activity reflects the promise of core–shell architectures in controlled swelling, uptake, release, and sensing applications. On the negative side, most papers assume core–shell morphology without proving it.

1.2.3
Approach 2: Microgels from Emulsification

In this group of methods, an aqueous "pregel" solution is suspended in an oil or brine phase to give a water-in-oil emulsion – see Figure 1.6. The pregel can be either a monomer or a polymer solution. In the second gelation step, the emulsion droplets undergo a chemical reaction to gel each emulsion droplet. This type of polymerization is often called "inverse emulsion polymerization" or "miniemulsion polymerization" [70, 71]. A distinction between these two types of polymerizations is that miniemulsion recipes include a solute for the dispersed phase with ultralow solubility in the continuous phase to prevent Oswald ripening. For oil-in-water emulsions, the solute is a hydrophobic long-chain alkane, whereas for water-in-oil emulsions, salts provide this function.

Two cases for the gelation step are illustrated in Figure 1.6. In the homogeneous case, essentially a solution polymerization or cross-linking reaction occurs throughout the drop. An example of this case is Landfester's preparation of cross-linked 100 nm polyacrylic acid microgels using cyclohexane as the continuous phase and 50% water in the dispersed (monomer) phase [71].

The second case illustrated in Figure 1.6 occurs when reaction of the pregel causes new particles to nucleate within the emulsion droplet. A good example of this is the work of Dowding, Vincent, and Williams, who reported the evolution of emulsion

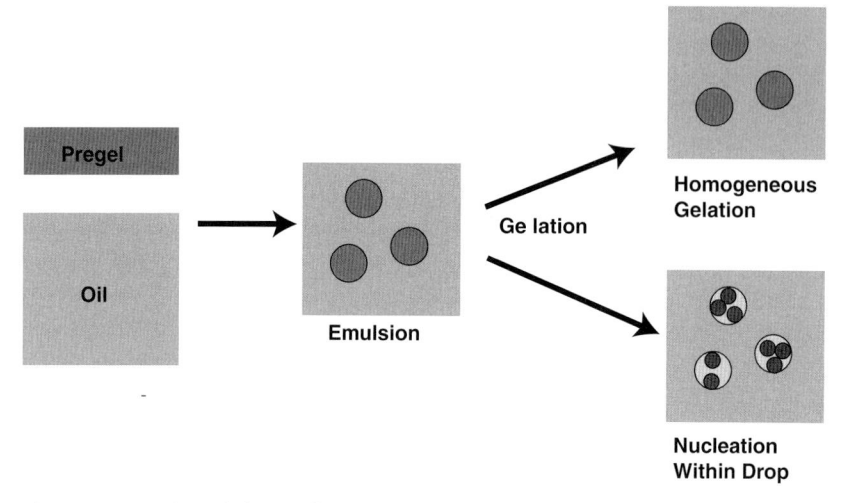

Figure 1.6 Pregel emulsification followed by gelation to give microgels.

droplet size for the inverse emulsion polymerization of PNIPAM [72]. They found that the produced microgels were much smaller than the emulsion droplets, suggesting more than one microgel particle formed per emulsion droplet.

Finally, emulsion can be prepared by conventional oil-in-water techniques [73] or using a particle-at-a-time microfluidic methods [74]. An interesting variation, described in a patent, involves forcing a gelling polymer solution through a membrane or packed bed to generate a microgel suspension [63].

1.2.4
Approach 3: Microgels by Polymer Complexation

A completely different approach to microgel synthesis involves the mixing of dilute solutions of oppositely charged polyelectrolytes to form colloidally dispersed, polyelectrolyte complexes [75, 76]. The principle is illustrated in Figure 1.7. The cooperative electrostatic attraction between oppositely charged chains gives stable polymer networks.

To achieve colloidally stable microgels, it is critical that one of the components is in excess to give charged microgels that are electrosterically stabilized. This is illustrated in a recent example from our laboratory where we determined the phase diagram for microgel formation when dilute, cationic polyvinylamine is mixed with anionic carboxymethyl cellulose (CMC) [77]. The phase diagram, reproduced in Figure 1.8, illustrates that stable microgels were obtained when either polymer was in excess. In contrast, stoichiometric mixtures gave macroscopic precipitates. The swelling of PVAm–CMC microgels is determined by the effective cross-link density, charge content, and polymer/water affinity. We have modeled the corresponding macrogel swelling by conventional gel theories [78].

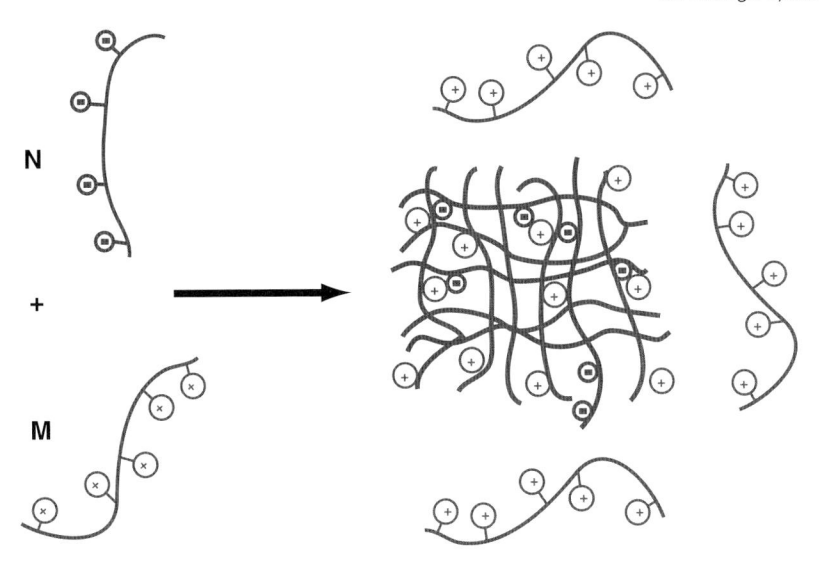

Figure 1.7 Polyelectrolyte complex formation, a route to microgels. In this case, excess cationic polymer gives cationic microgel.

Microgel preparation from polyelectrolyte complexation is attractive because it is relatively inexpensive and thus applicable for commodity applications such as strengthening paper [79]. However, the method has two serious deficiencies. First, it is difficult to prepare microgels with a narrow particle size distribution. Second, it is

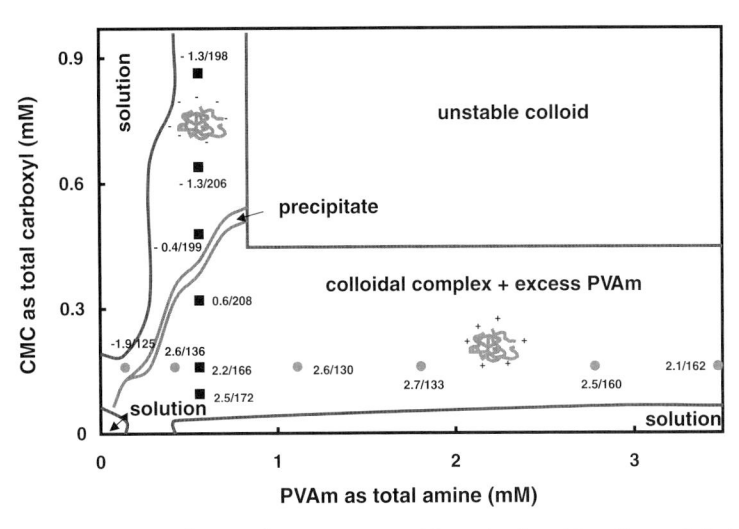

Figure 1.8 Phase diagram illustrating microgel formation by mixing dilute polyvinylamine with dilute carboxymethyl cellulose. The first number beside the points is the electrophoretic mobility ($10^{-8}\,m^2/(V\,s)$); the second number is the particle size (nm) [77].

difficult to prepare colloidally stable microgels without a significant content of excess soluble polymer that is difficult to separate.

1.2.5
Exotic Microgels

Finally, we mention some unusual microgel architectures. The colloid/nanoparticle literature contains many recent reports on the synthesis of Janus particles in which each particle has two areas (faces) with widely differing properties. The first report of Janus microgels we have found is from Kawaguchi's group (one of the microgel pioneers). In this approach, PNIPAM-co-acrylic microgels were prepared, assembled at the oil/water interface of an emulsion (i.e., a Pickering emulsion), and derivatized only on the "water side" by converting carboxyl groups to amines [29]. The resulting microgels had a carboxyl face and an amine face (see Figure 1.9). Other routes to Janus microgels have also been published; however, Kawaguchi's method has the advantage that it could be performed on a commercial scale to produce many particles.

Hollow colloidal particles are another example of unusual morphologies that have been around for many years in the latex and pigment industry. The usual approach is to prepare a core–shell particle and then dissolve away the core. There have been a few reports using this approach to prepare hollow microgels. Fu's group dissolved the silica core from silica-core PNIPAM-shell microgels [80] whereas Lyon's group dissolved a PNIPAM core cross-linked with N,N'-(1,2-dihydroxyethylene)bisacrylamide, which was oxidatively degraded [81].

Finally, Hu and coworkers were the first to report the preparation of interpenetrating network (IPN) microgels in which acrylic acid was polymerized inside PNIPAM microgels [82]. Similarly, polyacrylamide polyacrylic acid IPNs were prepared by inverse emulsion polymerization [83]. The IPN microgels were made in a two-step procedure. In the first step, polyacrylamide microgels were prepared by inverse emulsion polymerization using cyclohexane as the continuous phase. In the second step, an aqueous solution of acrylic acid plus cross-linker was added, emulsified, and reacted.

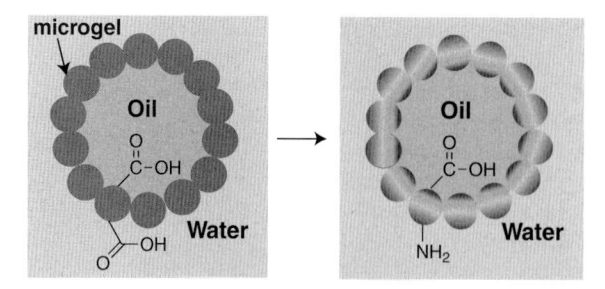

Figure 1.9 Janus particles from Pickering emulsions [29]. The oil phase protects one face of the microgels while the remaining microgel surface groups are converted from carboxyls to amines.

1.2.6
Summary

Most microgels are prepared from vinyl monomers using some form of free radical polymerization. Conventional emulsion polymerization is most commonly used for commercial microgels, whereas surfactant-free emulsion polymerization is frequently employed in preparations for academic investigations. Finally, none of the polymerization schemes is unique to microgels. Many of the same techniques are used to prepare latexes and other polymers.

1.3
Particle Derivatization

In this section, we summarize some important chemical transformations involving microgels. Many microgel applications, such as medical diagnostics [84], involve microgels that have functional molecules attached after the microgel is synthesized.

1.3.1
Chemical Coupling to Microgels

Most of the standard techniques for coupling (conjugating) small molecules, peptides, oligonucleotides, and proteins are applicable to microgels [85]. Indeed, microgels offer important advantages: first, microgels can be centrifuged and readily redispersed, which facilitates cleaning (more on this later); second, subtle changes can be followed by dynamic light scattering, which is sensitive to swelling, and microelectrophoresis, which is sensitive to surface charge; third, microgels are generally more colloidally stable than latexes and other nanosized support particles. The usual starting points for microgel derivatization are carboxyl or amine groups. Biotin [86], streptavidin [87], proteins [88], and oligonucleotides [89] have been conjugated to microgels. Carbodiimide-based coupling chemistries seem to be the most popular. However, most publications do not include good descriptions of either the amounts or the location of coupled material.

1.3.2
Microgel Decross-Linking

Microgel mechanical properties, swelling properties, and solute release properties are all sensitive to cross-link density. Thus, one can imagine applications where it is desirable to reduce microgel cross-link density in response to a stimulus. Macrogels, cross-linked with peptides, have been decomposed by enzymes [90]. Similarly, disulfide cross-linked macrogels have been decross-linked under reducing conditions [91, 92]. A good microgel example is the oxidative decomposition of N,N'-(1,2-dihydroxyethylene)bisacrylamide that Lyon exploited to prepare hollow microgels [81].

MBA (structure in Table 1.1), the most popular cross-linker, is hydrolytically robust. For example, Hoare showed that most of the cross-links survived 0.5 M NaOH during the hydrolysis of acrylamide moieties in a copolymer microgel. In contrast, the next section describes the preparation of polyvinylamine microgels in which MBA did hydrolyze in an environment rich with primary amine groups.

1.3.3
Charged Microgels from Nonionic Precursors

There are two important approaches for the conversion of a nonionic latex or microgel to charged, highly swollen microgel – alkali swellable latexes and polyvinylamine microgels from poly(N-vinylformamide). Alkali swellable latexes are one of the earliest types of aqueous microgels and they have been recently reviewed by Tam and Tam [93]. An example is the classic work of Wolfe and coworkers who reported the preparation and swelling characteristics of latex made by the emulsion polymerization of ethyl acrylate, methacrylic acid, and the cross-linker 1,4-butanediol diacrylate [46]. Raising the pH ionized the methacrylic acid groups that, in turn, caused the particles to swell.

The conversion of nonionic poly(N-vinylformamide) microgels to positively charged polyvinylamine microgels is another example. The chemistry is shown in Figure 1.10. The critical aspect of this scheme is the cross-linking monomer BVU (see Table 1.1). This monomer is attractive because, unlike MBA and EGBM, it is resistant to acid and alkaline hydrolysis. For example, we tried to prepare polyvinylamine microgels by hydrolysis of poly(N-vinylformamide) microgels cross-linked with MBA cross-linker. The MBA links did not survive the hydrolysis step, whereas BVU gels remained intact [51].

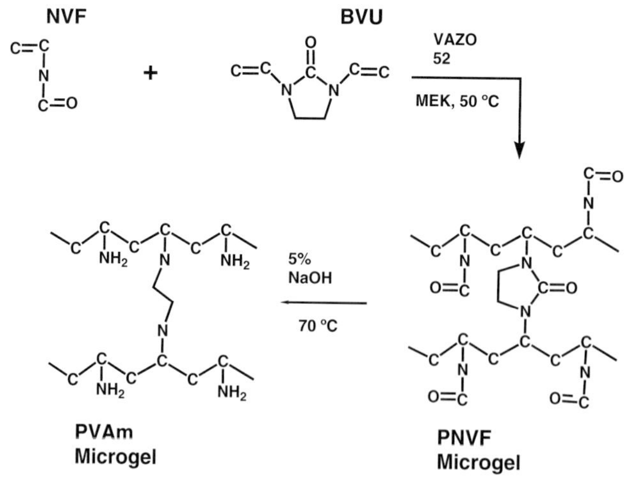

Figure 1.10 Chemistry of the poly(NVF-co-MBA) and poly(NVF-co-BVU) hydrolysis [51].

1.3.4
Nanoparticle-Filled Gels

A recent trend is to extend the utility of microgels by embedding nanoparticles. Gold and magnetic and iron oxide microgels have received the most attention. Gold-derivatized microgels offer some interesting properties. In some cases, gold adsorbs light, converts it to heat, and induces a phase transition [94]. In addition, gold-filled microgels can change color with swelling, reflecting the increase in the gold particle separation [95]. Gold-loaded microgels have been reported to be more colloidally stable than free gold sols [96]. Magnetic microgels are attractive for diagnostic applications because they are easily isolated from aqueous suspensions.

There are four common approaches to making microgels loaded with nanoparticles:

1) **Mixing of microgel and nanoparticle suspensions**. The criteria are that the nanoparticles must adhere to the microgel and that they must be small enough to penetrate the gel. Adhesion is usually driven by electrostatics; however, for charge-stabilized microgels, there is tendency to destabilize the gel by adsorbing oppositely charged particles. Also, we have found no examples showing good penetration of nanoparticles into microgels. Examples of this approach have been published for gold-loaded microgels [94, 97, 98] and for magnetic microgels [99].

2) **Growing nanoparticles in the presence of microgels**. A good general discussion of this approach has recently been published by Kumacheva's group [100]. Usually, ionic groups in the gel act as nucleation sites for nanoparticle formation. The facile transport of small molecules into the gel structure offers the possibility of high nanoparticle loading. Published examples include gold-loaded microgels [95, 96, 101–103] and magnetic gels [104, 105].

3) **Growing microgel in the presence of preformed nanoparticles acting as seeds**. The nanoparticle loading by this method is limited by the original nanoparticle concentration, which is usually low [99]. Gold-filled microgels have been prepared this way [106], as was one of the first magnetic microgels reported by Kondo *et al.* who prepared polystyrene particles loaded with magnetic particles and, in the second step, grafted PNIPAM onto the latex surface [107]. Silica-coated magnetic particles have also been coated with PNIPAM to give a core–shell microgel [108].

4) **Layer-by-layer and core–shell assembly**. Decher has promoted the generation of multilayer films on macroscopic surfaces by exposing them sequentially to solutions of cationic and anionic polyelectrolytes with interstage washing [109]. Microgels can replace one of the polyelectrolytes to give microgel layers on a surface [32]. These macrostructures from microgels are beyond the scope of this chapter. On the other hand, the same sequential adsorption approach can be used to coat microgels with nanoparticles. For example, Sauzedde and coworkers describe an elegant procedure in which anionic ~10 nm iron oxide particles are adsorbed onto cationic polystyrene-core PNIPAM-shell microgels [110, 111]. In the second step, a carboxylated PNIPAM encapsulating shell is coated to

encapsulate the iron oxide, giving an overall content of up to 23 wt% iron oxide. Others have employed similar approaches [41, 100, 108].

1.4
Microgel Purification and Storage

For most applications, microgels must be cleaned before characterization and use. The standard approach is serum replacement. Microgels are centrifuged, the supernatant is decanted, and the microgels are redispersed in clean electrolyte solution. This process is repeated a few times. An ultracentrifuge is required for highly swollen microgels that have a density close to that of water. For larger scale preparations, membrane-based separations will remove low molecular weight contaminants [112].

Polymer colloid purification and characterization was extensively studied in the 1970s. Wilkinson *et al.* published an excellent review of this body of work [113]. Wilkinson concluded, "A recommended cleaning procedure would involve the use of micro-filtration/serum exchange (with an acid wash to ensure complete exchange of cations for protons), steam stripping for removal of residual monomer (this will result in hydrolysis of groups, such as sulfate groups, and possibly loss of stability) and efficient sparging with pure nitrogen to remove carbon dioxide." Although this review focused on surfactant-free polystyrene latex and other hydrophobic latexes, these conclusions generally apply to microgels, except that microgels usually are not stream stripped, a technique for removing styrene, benzaldehyde, and other hydrophobic small molecules. Because microgel monomers are water soluble, they can be removed easily with serum replacement or dialysis.

Bacterial contamination is a significant problem with hydrophobic latex [114] and, although not discussed in the literature, the same is true for microgels. Commercial latexes employ bactericide. However, this is a contaminant for fundamental studies and may interfere with biomedical applications. The best alternative is freeze-drying (lyophilization) and low-temperature storage of the dry powder [6].

1.4.1
Microgel Characterization

Microgel characterization is a large topic. Specific techniques are discussed in other chapters and in a recent review [115]. The important properties include molecular weight, particle size, and charge content as functions of pH, temperature, and ionic environment.

Dynamic light scattering has become the standard technique for measuring particle size [116], as it is rapid, accurate, and reproducible. However, as emphasized by Ballauff, diameters measured by dynamic light scattering are sensitive to surface tails and thus are larger than corresponding sizes obtained by static light scattering, X-ray scattering, or neutron scattering [117]. Electron microscopy gives a qualitative indication of monodispersity, the extent of aggregation, and the presence of small particles from secondary nucleation.

Microgel charge content can be measured as a function of pH by a combination of conductometric and potentiometric titration [113]. Conductometric titration gives an accurate measure of the total titratable charge, whereas potentiometric titration gives the degree of ionization as a function of pH. To measure strong acid groups such as sulfate, the microgels must be ion-exchanged to replace potassium or sodium ions with protons. The shapes of the potentiometric curves are very sensitive to the charge distribution within the microgels. Figure 1.11 illustrates the behavior of the two extreme cases described in Figure 1.5, VAA–NIPAM microgels with isolated surface carboxyls and MAA-NIPAM gels with carboxyls concentrated in the cores [118]. The VAA–NIPAM ionization occurred over a low and relatively narrow pH, similar to dilute acetic acid. In contrast, the interactions of neighboring carboxyls extended the ionization curve for MAA-NIPAM over a wider and higher pH range. More detailed comparisons with other distributions and with isothermal titration calorimetry are given in Hoare's paper [118].

The charge content of linear, soluble polyelectrolytes is conveniently measured by the polyelectrolyte (colloid) titration. First proposed by Terayama [119], the titration exploits the fact that polyelectrolyte complex formation is often stoichiometric and irreversible. The original procedure involved end point detection with dyes, although commercial streaming current detectors are now popular [120]. There is only one recent publication describing microgel charge content measurement by the polyelectrolyte titration. High molecular weight titrants will only "see" the exterior microgel charges giving the opportunity to distinguish the interior from surface charges – similar approaches have been used to measure surface charges on porous wood pulp fibers [121].

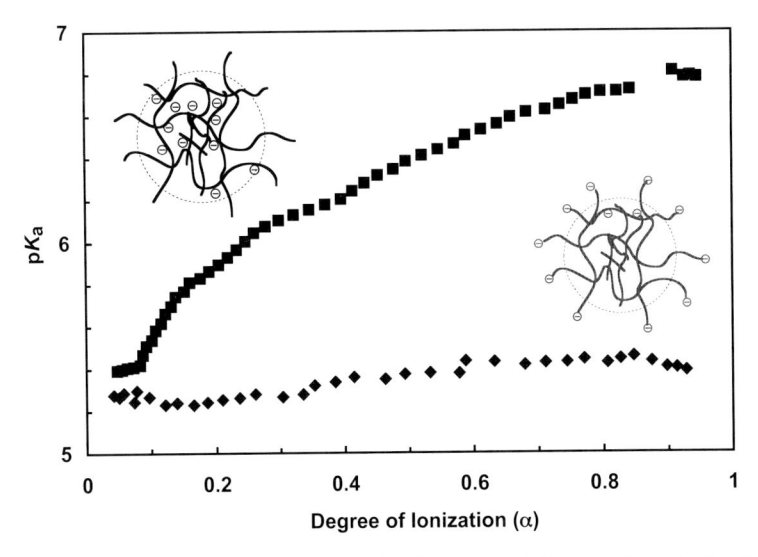

Figure 1.11 The carboxyl distribution within the microgel influences the ionization behavior as a function of pH [118]. Interactions between closely spaced carboxyl groups cause the pK_a to be a function of pH – the "polyelectrolyte effect."

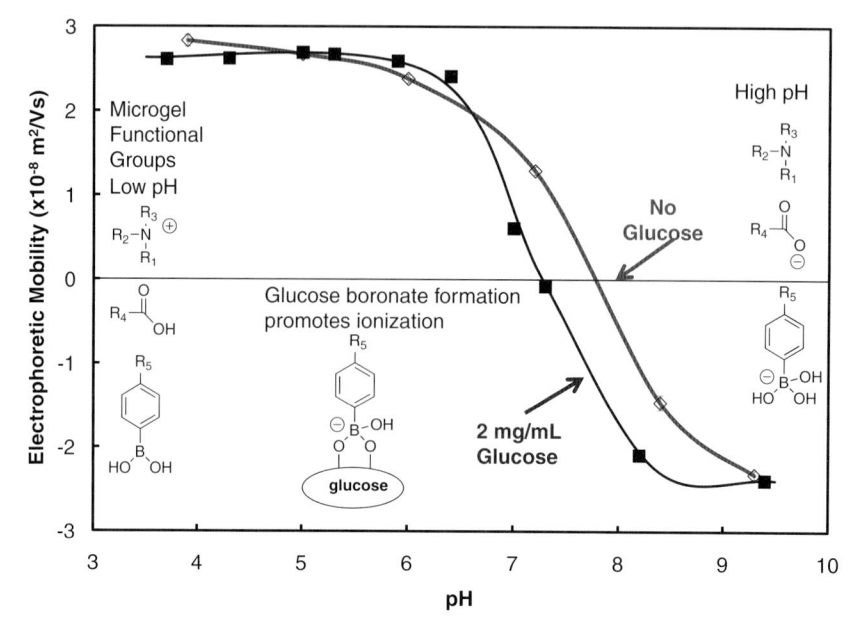

Figure 1.12 Electrophoretic mobility of a glucose-sensitive amphoteric microgel [21].

Finally, the electrophoretic mobility (particle velocity divided by electric field strength) of microgels is sensitive to the surface charge density [116, 122]. Figure 1.12 illustrates the utility of microelectrophoresis. The curves show mobility as a function of pH for an amphoteric latex bearing acrylic, phenyl boronic acid, and tertiary amino groups. At low pH the microgels have a net positive charge whereas at high pH they

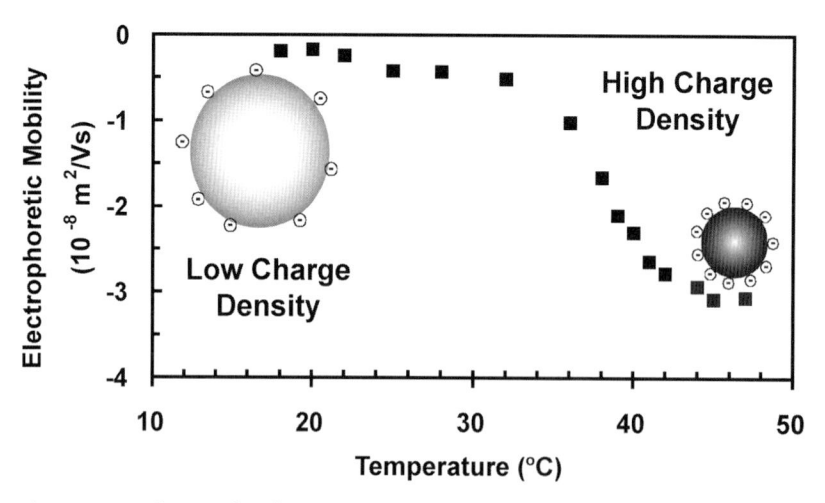

Figure 1.13 Influence of swelling on charge density and electrophoretic mobility [116].

are negative. The isoelectric point is sensitive to the presence of glucose because glucose-boronate ester shifts the boronate ionization equilibrium [21].

Microgel electrophoretic mobility is a sensitive function of swelling. High swelling gives low mobility because the effective surface charge density of swollen particles is low (see Figure 1.13). This behavior has been modeled by a number of authors [116, 123–125].

In summary, most colloid characterization techniques have been applied to microgels including scattering (light, X-ray, and neutron), rheology, potentiometric titration, electron microscopy, isothermal titration calorimetry, electrophoresis, and atomic force microscopy. In addition, many solution techniques such as NMR, time-resolved fluorescence, and FTIR have been applied to microgel characterization. A major challenge is the topochemical characterization within microgels. Although some progress has been made in this regard [115], the distribution of cross-links, electrical charges, and other functional groups within a particle remains difficult to quantify.

1.5
Conclusions

The first decade of the new millennium has produced much microgel research. From a synthetic perspective, the hot areas have been nanoparticle-filled microgels, microgels from living radical polymerizations, nonpolymerization routes for creating microgels from existing polymers, novel morphologies such as Janus particles, and many attempts to link microgel composition to efficacy in biological applications.

On the other hand, some important areas have received little attention. For example, there have been essentially no polymer reaction engineering approaches applied to achieve better composition control through kinetic modeling, process control, and getting away from batch polymerizations. Related issues that must be solved before large-scale commercial applications can be developed include high solids polymerizations, the elimination of residual monomers, and high-volume/throughput cleaning strategies.

From a scientific perspective, we lack good nucleation kinetic models to predict particle size and dispersity for particle formation by both polymerization and coacervation-based approaches. We also lack facile and accurate measurements of particle cross-linker and function group distributions within gels.

References

1 Baker, W.O. (1949) Microgel, a new macromolecule. *Ind. Eng. Chem.*, **41**, 511–520.
2 Hiemenz, P.C. and Rajagopalan, R. (eds) (1997) *Principles of Colloid and Surface Chemistry*, 3rd edn, revised and expanded, CRC Press.
3 Pollack, G.H. (2001) *Cells, Gels, and the Engines of Life*, Ebner and Sons, Seattle, WA.

4 Dupin, D., Rosselgong, J., Armes, S.P., and Routh, A.F. (2007) Swelling kinetics for a pH-induced latex-to-microgel transition. *Langmuir*, **23**, 4035–4041.

5 McPhee, W., Tam, K.C., and Pelton, R. (1993) Poly(N-isopropylacrylamide) lattices prepared with sodium dodecyl sulfate. *J. Colloid Interface Sci.*, **156**, 24–30.

6 Agbugba, C.B., Hendriksen, B.A., Chowdhry, B.Z., and Snowden, M.J. (1998) The redispersibility and physico-chemical properties of freeze-dried colloidal microgels. *Colloids Surf. A*, **137**, 155–164.

7 Pelton, R.H. and Chibante, P. (1986) Preparation of aqueous lattices with N-isopropylacrylamide. *Colloids Surf.*, **20**, 247–256.

8 Pelton, R. (2000) Temperature-sensitive aqueous microgels. *Adv. Colloid Interface Sci.*, **85**, 1–33.

9 Wolfe, M.S. and Scopazzi, C. (1989) Rheology of swellable microgel dispersions: influence of crosslink density. *J. Colloid Interface Sci.*, **133**, 265–277.

10 Bromberg, L.E. and Temchenko, M. (2003) Responsive microgel and methods related thereto. Pat. US 2003/0152623 Al, Aug. 14.

11 Okay, O. and Funke, W. (1990) Steric stabilization of reactive microgels from 1,4-divinylbenzene. *Makromol. Chem. Rapid Commun.*, **11**, 583–587.

12 Saito, R. and Ishizu, K. (1997) Flower type microgels. 1. Synthesis of the microgels. *Polymer*, **38**, 225–229.

13 Suzuki, D., Tsuji, S., and Kawaguchi, H. (2005) Development of anisotropic thermo-sensitive hairy particles using living radical graft polymerization. *Chem. Lett.*, **34**, 242–243.

14 Hoare, T. and Pelton, R. (2004) Highly pH and temperature responsive microgels functionalized with vinylacetic acid. *Macromolecules*, **37**, 2544–2550.

15 Kim, K.S., Cho, S.H., and Shin, J.S. (1995) Preparation and characterization of monodisperse polyacrylamide microgels. *Polym. J. (Tokyo)*, **27**, 508–514.

16 Tsuji, S. and Kawaguchi, H. (2004) Temperature-sensitive hairy particles prepared by living radical graft polymerization. *Langmuir*, **20**, 2449–2455.

17 Osada, Y. and Ross-Murphy, S.B. (1993) Intelligent gels. *Sci Am.*, **268**, 82–87.

18 Nolan, C.M., Gelbaum, L.T., and Lyon, L.A. (2006) 1H NMR investigation of thermally triggered insulin release from poly(N-isopropylacrylamide) microgels. *Biomacromolecules*, **7**, 2918–2922.

19 Zhang, Y.J., Guan, Y., and Zhou, S.Q. (2006) Synthesis and volume phase transitions of glucose-sensitive microgels. *Biomacromolecules*, **7**, 3196–3201.

20 Hoare, T. and Pelton, R. (2007) Engineering glucose swelling responses in poly(N-isopropylacrylamide)-based microgels. *Macromolecules*, **40**, 670–678.

21 Hoare, T. and Pelton, R. (2008) Charge-switching, amphoteric glucose-responsive microgels with physiological swelling activity. *Biomacromolecules*, **9**, 733–740.

22 Vinogradov, S.V. (2006) Colloidal microgels in drug delivery applications. *Curr. Pharm. Des.*, **12**, 4703–4712.

23 Kawaguchi, H., Kisara, K., Takahashi, T., Achiha, K. *et al.* (2000) Versatility of thermosensitive particles. *Macromol. Symp.*, **151**, 591–598.

24 Ballauff, M. and Lu, Y. (2007) "Smart" nanoparticles: preparation, characterization and applications. *Polymer*, **48**, 1815–1823.

25 Zhang, J. and Pelton, R. (1999) Poly(N-isopropylacrylamide) microgels at the air–water interface. *Langmuir*, **15**, 8032–8036.

26 Fujii, S., Ryan, A.J., and Armes, S.P. (2006) Long-range structural order, Moire patterns, and iridescence in latex-stabilized foams. *J. Am. Chem. Soc.*, **128**, 7882–7886.

27 Ngai, T., Auweter, H., and Behrens, S.H. (2006) Environmental responsiveness of microgel particles and particle-stabilized emulsions. *Macromolecules*, **39**, 8171–8177.

28 Lawrence, D.B., Cai, T., Hu, Z., Marquez, M., and Dinsmore, A.D. (2007) Temperature-responsive semipermeable

28 capsules composed of colloidal microgel spheres. *Langmuir*, **23**, 395–398.
29 Suzuki, D., Tsuji, S., and Kawaguchi, H. (2007) Janus microgels prepared by surfactant-free pickering emulsion-based modification and their self-assembly. *J. Am. Chem. Soc.*, **129**, 8088–8089.
30 Weissman, J.M., Sunkara, H.B., Tse, A.S., and Asher, S.A. (1996) Thermally switchable periodicities and diffraction from mesoscopically ordered materials. *Science*, **274**, 959–960.
31 Hu, Z.B., Lu, X.H., and Gao, J. (2001) Hydrogel opals. *Adv. Mater.*, **13**, 1708–1712.
32 Serpe, M.J., Jones, C.D., and Lyon, L.A. (2003) Layer-by-layer deposition of thermoresponsive microgel thin films. *Langmuir*, **19**, 8759–8764.
33 Nolan, C.M., Serpe, M.J., and Lyon, L.A. (2004) Thermally modulated insulin release from microgel thin films. *Biomacromolecules*, **5**, 1940–1946.
34 Serpe, M.J. and Lyon, L.A. (2004) Optical and acoustic studies of pH-dependent swelling in microgel thin films. *Chem. Mater.*, **16**, 4373–4380.
35 Serpe, M.J., Yarmey, K.A., Nolan, C.M., and Lyon, L.A. (2005) Doxorubicin uptake and release from microgel thin films. *Biomacromolecules*, **6**, 408–413.
36 Prevot, M., Dejugnat, C., Mohwald, H., and Sukhorukov, G.B. (2006) Behavior of temperature-sensitive PNIPAM confined in polyelectrolyte capsules. *ChemPhysChem*, **7**, 2497–2502.
37 De Geest, B.G., Dejugnat, C., Prevot, M., Sukhorukov, G.B. et al. (2007) Self-rupturing and hollow microcapsules prepared from bio-polyelectrolyte-coated microgels. *Adv. Funct. Mater.*, **17**, 531–537.
38 Glinel, K., Dejugnat, C., Prevot, M., Scholer, B. et al. (2007) Responsive polyelectrolyte multilayers. *Colloids Surf. A*, **303**, 3–13.
39 Kreft, O., Prevot, M., Mohwald, H., and Sukhorukov, G.B. (2007) Shell-in-shell microcapsules: a novel tool for integrated, spatially confined enzymatic reactions. *Angew. Chem., Int. Ed.*, **46**, 5605–5608.
40 Sorrell, C.D. and Lyon, L.A. (2007) Bimodal swelling responses in microgel thin films. *J. Phys. Chem. B*, **111**, 4060–4066.
41 Wong, J.E., Gaharwar, A.K., Muller-Schulte, D., Bahadur, D., and Richtering, W. (2007) Layer-by-layer assembly of a magnetic nanoparticle shell on a thermoresponsive microgel core. *J. Magn. Magn. Mater.*, **311**, 219–223.
42 Wong, J.E., Muller, C.B., Laschewsky, A., and Richtering, W. (2007) Direct evidence of layer-by-layer assembly of polyelectrolyte multilayers on soft and porous temperature-sensitive PNiPAM microgel using fluorescence correlation spectroscopy. *J. Phys. Chem. B*, **111**, 8527–8531.
43 Saatweber, D. and Vogt-Birnbrich, B. (1996) Microgels in organic coatings. *Prog. Org. Coat.*, **28**, 33–41.
44 Ketz, R.J., Prud'homme, R.K., and Graessley, W.W. (1988) Rheology of concentrated microgel solutions. *Rheol. Acta*, **27**, 531–539.
45 Barreiro-Iglesias, R., Alvarez-Lorenzo, C., and Concheiro, A. (2003) Poly(acrylic acid) microgels (carbopol® 934)/surfactant interactions in aqueous media. Part II. Ionic surfactants. *Int. J. Pharm.*, **258**, 179–191.
46 Rodriguez, B.E., Wolfe, M.S., and Fryd, M. (1994) Nonuniform swelling of alkali swellable microgels. *Macromolecules*, **27**, 6642–6647.
47 Heskins, M. and Guillet, J.E. (1968) Solution properties of poly(N-isopropylacrylamide). *J. Macromol. Sci. A: Chem.*, **2**, 1441–1455.
48 Tanaka, T. (1981) Gels. *Sci Am.*, **244**, 124–136. 138.
49 Hazot, P., Delair, T., Pichot, C., Chapel, J.P., and Elaissari, A. (2003) Poly(N-ethylmethacrylamide) thermally-sensitive microgel latexes: effect of the nature of the crosslinker on the polymerization kinetics and physicochemical properties. *C. R. Chim.*, **6**, 1417–1424.
50 Griffin, J.M., Robb, I., and Bismarck, A. (2007) Preparation and characterization of surfactant-free stimuli-sensitive microgel dispersions. *J. Appl. Polym. Sci.*, **104**, 1912–1919.

51 Miao, C., Chen, X., and Pelton, R. (2007) Adhesion of poly(vinylamine) microgels to wet cellulose. *Ind. Eng. Chem. Res.*, **46**, 6486–6493.

52 Gilbert, R.G. (1995) *Emulsion Polymerization: A Mechanistic Approach*, Academic Press, London.

53 Goodwin, J.W., Ottewill, R.H., and Pelton, R. (1979) Studies on the preparation and characterization of monodisperse polystyrene lattices. 5. Preparation of cationic lattices. *Colloid Polym. Sci.*, **257**, 61–69.

54 Andersson, M. and Maunu, S.L. (2006) Structural studies of poly(N-isopropylacrylamide) microgels: effect of SDS surfactant concentration in the microgel synthesis. *J. Polym. Sci. B: Polym. Phys.*, **44**, 3305–3314.

55 Wu, X., Pelton, R.H., Hamielec, A.E., Woods, D.R., and McPhee, W. (1994) The kinetics of poly(N-isopropylacrylamide) microgel latex formation. *Colloid Polym. Sci.*, **272**, 467–477.

56 Hoare, T. and McLean, D. (2006) Kinetic prediction of functional group distributions in thermosensitive microgels. *J. Phys. Chem. B*, **110**, 20327–20336.

57 Cao, L.Q., Chen, L.P., and Lai, W.C. (2007) Precipitation copolymerization of N-isopropylacrylamide and acrylic acid in supercritical carbon dioxide. *J. Polym. Sci. A: Polym. Chem.*, **45**, 955–962.

58 Frank, M. and Burchard, W. (1991) Microgels by intramolecular crosslinking of poly(allylamine) single chains. *Makromol. Chem. Rapid Commun.*, **12**, 645–652.

59 Brasch, U. and Burchard, W. (1996) Preparation and solution properties of microhydrogels from poly(vinyl alcohol). *Macromol. Chem. Phys.*, **197**, 223–235.

60 Chan, K., Pelton, R., and Zhang, J. (1999) On the formation of colloidally dispersed phase-separated poly(N-isopropylacrylamide). *Langmuir*, **15**, 4018–4020.

61 Fundueanu, G., Nastruzzi, C., Carpov, A., Desbrieres, J., and Rinaudo, M. (1999) Physico-chemical characterization of Ca-alginate microparticles produced with different methods. *Biomaterials*, **20**, 1427–1435.

62 Kuckling, D., Vo, C.D., and Wohlrab, S.E. (2002) Preparation of nanogels with temperature-responsive core and pH-responsive arms by photo-cross-linking. *Langmuir*, **18**, 4263–4269.

63 Chauveteau, G., Tabary, R., Renard, M., and Omari, A. (2003) Process for preparing microgels of controlled size. US Pat. 1086976, June 17.

64 Xia, X., Tang, S., Lu, X., and Hu, Z. (2003) Formation and volume phase transition of hydroxypropyl cellulose microgels in salt solution. *Macromolecules*, **36**, 3695–3698.

65 Shen, H.W. and Eisenberg, A. (2000) Block length dependence of morphological phase diagrams of the ternary system of PS-b-PAA/dioxane/H_2O. *Macromolecules*, **33**, 2561–2572.

66 Delaittre, G., Save, M., and Charleux, B. (2007) Nitroxide-mediated aqueous dispersion polymerization: from water-soluble macroalkoxyamine to thermosensitive nanogels. *Macromol. Rapid Commun.*, **28**, 1528–1533.

67 Sundberg, D.C. and Durant, Y.G. (2003) Latex particle morphology, fundamental aspects: a review. *Polym. React. Eng.*, **11**, 379–432.

68 Pelton, R.H. (1988) Polystyrene and polystyrene-butadiene latexes stabilized by poly(N-isopropylacrylamide). *J. Polym. Sci. Polym. Chem.*, **26**, 9–18.

69 Jones, C.D. and Lyon, L.A. (2000) Synthesis and characterization of multiresponsive core–shell microgels. *Macromolecules*, **33**, 8301–8306.

70 Sudol, E.D. and El-Aasser, M.S. (1997) Miniemulsion polymerization. In *Emulsion polymerization and emulsion polymers*, (eds. P.A. Lovell and M.S. El-Aasser). Chichester, p 699.

71 Landfester, K., Willert, M., and Antonietti, M. (2000) Preparation of polymer particles in nonaqueous direct and inverse miniemulsions. *Macromolecules*, **33**, 2370–2376.

72 Dowding, P.J., Vincent, B., and Williams, E. (2000) Preparation and swelling properties of poly(NIPAM) "minigel"

particles prepared by inverse suspension polymerization. *J. Colloid Interface Sci.*, **221**, 268–272.

73 Chen, L.W., Yang, B.Z., and Wu, M.L. (1997) Synthesis and kinetics of microgel in inverse emulsion polymerization of acrylamide. *Prog. Org. Coat.*, **31**, 393–399.

74 Kim, J.W., Utada, A.S., Fernandez-Nieves, A., Hu, Z.B., and Weitz, D.A. (2007) Fabrication of monodisperse gel shells and functional microgels in microfluidic devices. *Angew. Chem., Int. Ed.*, **46**, 1819–1822.

75 Fuoss, R.M. and Sadek, H. (1949) Mutual interaction of polyelectrolytes. *Science*, **110**, 552–554.

76 Michaels, A.S. (1965) Polyelectrolyte complexes. *Ind. Eng. Chem. Res.*, **57**, 32–40.

77 Feng, X., Pelton, R., Leduc, M., and Champ, S. (2007) Colloidal complexes from poly(vinyl amine) and carboxymethyl cellulose mixtures. *Langmuir*, **23**, 2970–2976.

78 Feng, X. and Pelton, R. (2007) Carboxymethyl cellulose:polyvinylamine complex hydrogel swelling. *Macromolecules*, **40**, 1624–1630.

79 Champ, S., Auweter, H., Leduc, M., and Noerenberg, R. (2004) Manufacture of aqueous polyelectrolyte complex dispersions for increasing wet strength of paper and paperboard. Germany Pat. WO 2004096895 A2, November 11.

80 Zha, L.S., Zhang, Y., Yang, W.L., and Fu, S.K. (2002) Monodisperse temperature-sensitive microcontainers. *Adv. Mater.*, **14**, 1090–1092.

81 Nayak, S., Gan, D.J., Serpe, M.J., and Lyon, L.A. (2005) Hollow thermoresponsive microgels. *Small*, **1**, 416–421.

82 Xia, X. and Hu, Z. (2004) Synthesis and light scattering study of microgels with interpenetrating polymer networks. *Langmuir*, **20**, 2094–2098.

83 Owens, D.E., Jian, Y.C., Fang, J.E., Slaughter, B.V. *et al.* (2007) Thermally responsive swelling properties of polyacrylamide/poly(acrylic acid) interpenetrating polymer network nanoparticles. *Macromolecules*, **40**, 7306–7310.

84 Pichot, C. (2004) Surface-functionalized latexes for biotechnological applications. *Curr. Opin. Colloid Interface Sci.*, **9**, 213–221.

85 Hermanson, G.T. (1996) *Bioconjugate Techniques*, Academic Press, San Diego, CA.

86 Nayak, S. and Lyon, L.A. (2004) Ligand-functionalized core/shell microgels with permselective shells. *Angew. Chem., Int. Ed.*, **43**, 6706–6709.

87 Su, S., Ali, M.M., Filipe, C.D.M., Li, Y., and Pelton, R. (2008) Microgel-based inks for paper-supported biosensing applications. *Biomacromolecules*, **9**, 935–941.

88 Hamerska-Dudra, A., Bryjak, J., and Trochimczuk, A.W. (2007) Immobilization of glucoamylase and trypsin on crosslinked thermosensitive carriers. *Enzyme Microb. Technol.*, **41**, 197–204.

89 Delair, T., Meunier, F., Elaissari, A., Charles, M.H., and Pichot, C. (1999) Amino-containing cationic latex–oligodeoxyribonucleotide conjugates: application to diagnostic test sensitivity enhancement. *Colloids Surf. A*, **153**, 341–353.

90 Plunkett, K.N., Berkowski, K.L., and Moore, J.S. (2005) Chymotrypsin responsive hydrogel: application of a disulfide exchange protocol for the preparation of methacrylamide containing peptides. *Biomacromolecules*, **6**, 632–637.

91 Plunkett, K.N., Kraft, M.L., Yu, Q., and Moore, J.S. (2003) Swelling kinetics of disulfide cross-linked microgels. *Macromolecules*, **36**, 3960–3966.

92 Bajomo, M., Steinke, J.H.G., and Bismarck, A. (2007) Inducing pH responsiveness via ultralow thiol content in polyacrylamide (micro)gels with labile crosslinks. *J. Phys. Chem. B*, **111**, 8655–8662.

93 Tan, B.H. and Tam, K.C. (2008) Review on the dynamics and micro-structure of pH-responsive nano-colloidal systems. *Adv. Colloid Interface Sci.*, **136**, 25–44.

94 Das, M., Sanson, N., Fava, D., and Kumacheva, E. (2007) Microgels loaded with gold nanorods: photothermally triggered volume transitions under physiological conditions. *Langmuir*, **23**, 196–201.

95 Suzuki, D. and Kawaguchi, H. (2006) Hybrid microgels with reversibly changeable multiple brilliant color. *Langmuir*, **22**, 3818–3822.

96 Pich, A., Karak, A., Lu, Y., Ghosh, A.K., and Adler, H.J.P. (2006) Tuneable catalytic properties of hybrid microgels containing gold nanoparticles. *J. Nanosci. Nanotechnol.*, **6**, 3763–3769.

97 Karg, M., Pastoriza-Santos, I., Perez-Juste, J., Hellweg, T., and Liz-Marzan, L.M. (2007) Nanorod-coated PNIPAM microgels: thermoresponsive optical properties. *Small*, **3**, 1222–1229.

98 Jones, C.D. and Lyon, L.A. (2003) Photothermal patterning of microgel/gold nanoparticle composite colloidal crystals. *J. Am. Chem. Soc.*, **125**, 460–465.

99 Rubio-Retama, J., Zafeiropoulos, N.E., Serafinelli, C., Rojas-Reyna, R. et al. (2007) Synthesis and characterization of thermosensitive PNIPAM microgels covered with superparamagnetic gamma-Fe_2O_3 nanoparticles. *Langmuir*, **23**, 10280–10285.

100 Zhang, J.G., Xu, S.Q., and Kumacheva, E. (2004) Polymer microgels: reactors for semiconductor, metal, and magnetic nanoparticles. *J. Am. Chem. Soc.*, **126**, 7908–7914.

101 Xu, J., Zeng, F., Wu, S.Z., Liu, X.X. et al. (2007) Gold nanoparticles bound on microgel particles and their application as an enzyme support. *Nanotechnology*, **18**, 265704.

102 Suzuki, D. and Kawaguchi, H. (2005) Gold nanoparticle localization at the core surface by using thermosensitive core–shell particles as a template. *Langmuir*, **21**, 12016–12024.

103 Suzuki, D., McGrath, J.G., Kawaguchi, H., and Lyon, L.A. (2007) Colloidal crystals of thermosensitive, core/shell hybrid microgels. *J. Phys. Chem. C*, **111**, 5667–5672.

104 Bhattacharya, S., Eckert, F., Boyko, V., and Pich, A. (2007) Temperature-, pH-, and magnetic-field-sensitive hybrid microgels. *Small*, **3**, 650–657.

105 Das, M., Zhang, H., and Kumacheva, E. (2006) Microgels: old materials with new applications. *Annu. Rev. Mater. Res.*, **36**, 117–142.

106 Singh, N. and Lyon, L.A. (2007) Au nanoparticle templated synthesis of pNIPAm nanogels. *Chem. Mater.*, **19**, 719–726.

107 Kondo, A., Kamura, H., and Higashitani, K. (1994) Development and application of thermosensitive magnetic immunomicrospheres for antibody purification. *Appl. Microbiol. Biotechnol.*, **41**, 99–105.

108 Deng, Y.H., Wang, C.C., Shen, X.Z., Yang, W.L. et al. (2005) Preparation, characterization, and application of multistimuli-responsive microspheres with fluorescence-labeled magnetic cores and thermoresponsive shells. *Chem. Eur. J.*, **11**, 6006–6013.

109 Decher, G. and Schlenoff, J.B. (2002) *Multilayer Thin Films: Sequential Assembly of Nanocomposite Materials*, Wiley-VCH Verlag GmbH, Weinheim.

110 Sauzedde, F., Elaissari, A., and Pichot, C. (1999) Hydrophilic magnetic polymer latexes. 2. Encapsulation of adsorbed iron oxide nanoparticles. *Colloid Polym. Sci.*, **277**, 1041–1050.

111 Sauzedde, F., Elaissari, A., and Pichot, C. (1999) Hydrophilic magnetic polymer latexes. 1. Adsorption of magnetic iron oxide nanoparticles onto various cationic latexes. *Colloid Polym. Sci.*, **277**, 846–855.

112 Labib, M.E. and Robertson, A.A. (1978) Application of a diafiltration technique in latex studies. *J. Colloid Interface Sci.*, **67**, 543–547.

113 Wilkinson, M.C., Hearn, J., and Steward, P.A. (1999) The cleaning of polymer colloids. *Adv. Colloid Interface Sci.*, **81**, 77–165.

114 Wilkinson, M.C., Sherwood, R., Hearn, J., and Goodall, A.R. (1979) Influence of residual styrene monomer, benzaldehyde, and microbial activity on the surface properties of polystyrene lattices. *Br. Polym. J.*, **11**, 1–6.

115 Hoare, T. and Pelton, R. (2008) Characterizing charge and crosslinker distributions in polyelectrolyte microgels. *Curr. Opin. Colloid Interface Sci.*, **13**, 413.

116 Pelton, R.H., Pelton, H.M., Morphesis, A., and Rowell, R.L. (1989) Particle sizes and electrophoretic mobilities of poly(N-isopropylacrylamide) latex. *Langmuir*, **5**, 816–818.

117 Ballauff, M. (2003) Nanoscopic polymer particles with a well-defined surface: synthesis, characterization, and properties. *Macromol. Chem. Phys.*, **204**, 220–234.

118 Hoare, T. and Pelton, R. (2006) Titrametric characterization of pH-induced phase transitions in functionalized microgels. *Langmuir*, **22**, 7342–7350.

119 Terayama, H. (1952) Method of colloid titration (a new titration between polymer ions). *J. Polym. Sci.*, **8**, 243–253.

120 Pelton, R., Cabane, B., Cui, Y., and Ketelson, H. (2007) Shapes of polyelectrolyte titration curves. 1. Well-behaved strong polyelectrolytes. *Anal. Chem.*, **79**, 8114–8117.

121 Borukhov, I., Andelman, D., Borrega, R., Cloitre, M. *et al.* (2000) Polyelectrolyte titration: theory and experiment. *J. Phys. Chem. B*, **104**, 11027–11034.

122 Hunter, R. (1981) *Zeta Potential in Colloid Science, Principles and Applications*, Academic Press, London.

123 Ohshima, H., Makino, K., Kato, T., Fujimoto, K. *et al.* (1993) Electrophoretic mobility of latex particles covered with temperature-sensitive hydrogel layers. *J. Colloid Interface Sci.*, **159**, 512–514.

124 Fernandez-Nieves, A., Fernandez-Barbero, A., de las Nieves, F.J., and Vincent, B. (2000) Motion of microgel particles under an external electric field. *J. Phys.: Condens. Matter*, **12**, 3605–3614.

125 Rasmusson, M., Vincent, B., and Marston, N. (2000) The electrophoresis of poly(N-isopropylacrylamide) microgel particles. *Colloid Polym. Sci.*, **278**, 253–258.

126 Lowe, J.S., Chowdhry, B.Z., Parsonage, J.R., and Snowden, M.J. (1998) The preparation and physico-chemical properties of poly(N-ethylacrylamide) microgels. *Polymer*, **39**, 1207–1212.

127 Boyko, V., Richter, S., Burchard, W., and Arndt, K.F. (2007) Chain dynamics in microgels: poly(N-vinylcaprolactam-*co*-N-vinylpyrrolidone) microgels as examples. *Langmuir*, **23**, 776–784.

128 Elmas, B., Onur, M.A., Senel, S., and Tuncel, A. (2002) Temperature controlled RNA isolation by N-isopropylacrylamide–vinylphenyl boronic acid copolymer latex. *Colloid Polym. Sci.*, **280**, 1137–1146.

129 Hazot, P., Delair, T., Elaissari, A., Chapel, J.P., and Pichot, C. (2002) Functionalization of poly(N-ethylmethacrylamide) thermosensitive particles by phenylboronic acid. *Colloid Polym. Sci.*, **280**, 637–646.

130 Garcia-Salinas, M.J., Romero-Cano, M.S., and de las Nieves, F.J. (2002) Colloidal stability of a temperature-sensitive poly(N-isopropylacrylamide/2-acrylamido-2-methylpropanesulphonic acid) microgel. *J. Colloid Interface Sci.*, **248**, 54–61.

131 Garcia, A., Marquez, M., Cai, T., Rosario, R. *et al.* (2007) Photo-, thermally, and pH-responsive microgels. *Langmuir*, **23**, 224–229.

132 Deng, Y.L. and Pelton, R. (1995) Synthesis and solution properties of poly(N-isopropylacrylamide-*co*-diallyldimethylammonium chloride). *Macromolecules*, **28**, 4617–4621.

133 Loxley, A. and Vincent, B. (1997) Equilibrium and kinetic aspects of the pH-dependent swelling of poly(2-vinylpyridine-*co*-styrene) microgels. *Colloid Polym. Sci.*, **275**, 1108–1114.

134 Ma, G.H. and Fukutomi, T. (1991) Studies on the preparation and characterization of poly(4-vinylpyridine) microgel. 1. Preparation with polymer emulsifier. *J. Appl. Polym. Sci.*, **43**, 1451–1457.

135 Amalvy, J.I., Wanless, E.J., Li, Y., Michailidou, V. *et al.* (2004) Synthesis and

characterization of novel pH-responsive microgels based on tertiary amine methacrylates. *Langmuir*, **20**, 8992–8999.

136 Ito, S., Ogawa, K., Suzuki, H., Wang, B. et al. (1999) Preparation of thermosensitive submicrometer gel particles with anionic and cationic charges. *Langmuir*, **15**, 4289–4294.

137 Eke, I., Elmas, B., Tuncel, M., and Tuncel, A. (2006) A new, highly stable cationic-thermosensitive microgel: uniform isopropylacrylamide–dimethylaminopropylmethacrylamide copolymer particles. *Colloids Surf. A*, **279**, 247–253.

138 Ma, X.M., Cui, Y.J., Zhao, X., Zheng, S.X., and Tang, X.Z. (2004) Different deswelling behavior of temperature-sensitive microgels of poly(N-isopropylacrylamide) crosslinked by polyethyleneglycol dimethacrylates. *J. Colloid Interface Sci.*, **276**, 53–59.

2
Polymerization Kinetics of Microgel Particles
Abdelhamid Elaissari and Ali Reza Mahdavian

2.1
Introduction

Over the last 50 years, polymeric colloidal dispersions have undergone significant development with increasingly diversified applications. This is particularly the case in the biomedical and pharmaceutical fields where the interest in well-defined colloidal systems explains the considerable efforts devoted to their research. This research is motivated by the possibilities to control the particle properties for targeted applications. Polymer-based particles such as latexes, microgels, or structured particles can be formulated by using a wide range of polymer assembly and polymerization processes in dispersed media. Knowledge of the polymerization kinetics, such as the individual polymerization conversion of the reactant as a function of time, the reactivity of the reactant, the decomposition kinetics of the used initiator, and the mechanisms involved in the formation of the particles, permits a good control over the colloidal properties of the microgel particles in terms of their size (from about 10 nm to several microns), their size distribution, their morphology, and their interfacial and swelling properties.

Polymeric materials in their dispersed state can be produced by different processing routes, such as classical emulsion polymerization, with or without surfactants, the formation of mini- and microemulsions, as well as dispersion and precipitation processes. Among the available particle systems, rigid polymer particles such as hydrophobic polystyrene-based latexes have been at the origin of many applications; for example, they are used in *in vitro* biomedical diagnostic tests. Using polymer-based particles as colloidal dispersions for biomedical (therapeutic and diagnostic) applications requires considering quite a large number of variables related to the macromolecular structure, internal morphology, surface properties, and the colloidal stability of the particles formed.

Microgels are colloidal gel particles that consist of chemically cross-linked three-dimensional polymer networks; these networks are able to dramatically shrink or swell by expelling or absorbing large amounts of water in response to external stimuli [1–4]. Thus, microgel particles can act like "nanosponges" and offer many

Microgel Suspensions: Fundamentals and Applications
Edited by Alberto Fernandez-Nieves, Hans M. Wyss, Johan Mattsson, and David A. Weitz
Copyright © 2011 WILEY-VCH Verlag GmbH & Co. KGaA, Weinheim
ISBN: 978-3-527-32158-2

potential applications in medicine, environmental science, and industry. The large change in size can be achieved, for example, either by modifying the pH, temperature, or ionic strength of the medium or by applying electric or magnetic fields; it is this response that makes microgels desirable for applications in drug delivery [5, 6], biosensing [7], diagnostics [8, 9], bioseparation [10], and optical devices [11–14]. In particular, biocompatible microgels are of great interest as drug delivery vehicles. Numerous targets for therapeutic drugs, such as tumors and inflamed tissues, exist under acidic conditions. This can be exploited using microgel-based drug delivery systems that release a drug directly at the target in response to the local pH conditions. Many protein-loaded microgels have already been synthesized and investigated for drug delivery applications.

The special properties of microgel particles are due to both the presence of covalent bonds between different polymer chains (i.e., cross-linking) and the presence of active functional groups. The latter allows, for example, the polymer network to retain water and microgels to exhibit interesting physical properties not seen with common polymer latex microspheres. Figure 2.1 shows the changes in hydrodynamic volume of lightly cross-linked poly(4-vinylpyridine)-silica (P4VP-SiO$_2$) nanocomposite microgel particles [15]. The shaded region in the figure indicates the pH range for which flocculation was observed. The midpoint of this region corresponds approximately to the isoelectric point, where the net charge of the particles is zero.

Thermosensitive microgel particles are relatively new colloidal systems with many potential applications due to the reversibility in microgel characteristics induced by temperature changes [1, 16–18]. To further expand their range of applicability, there

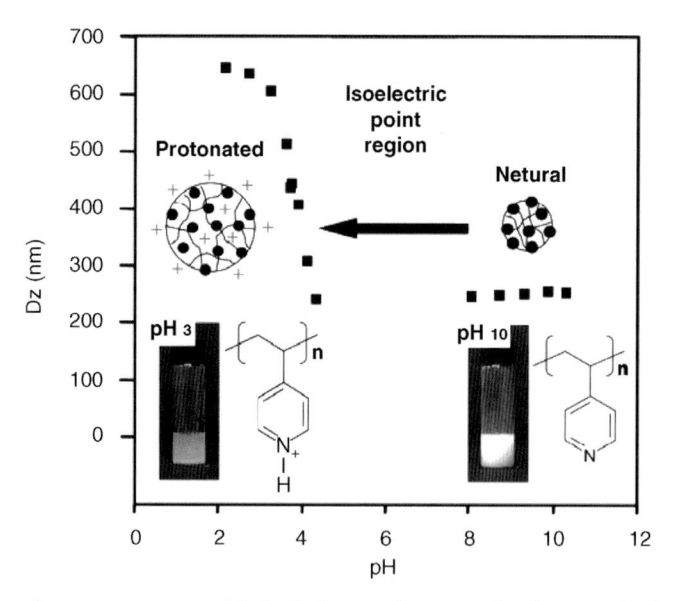

Figure 2.1 Variation of the hydrodynamic diameter with solution pH for lightly cross-linked P4VP-SiO$_2$ nanocomposite microgel particles. The photographs show the visual appearance of this nanocomposite microgel dispersion at pH 3 and pH 10, respectively [15].

have been efforts to generate microgels that are complexed with functionalized materials that impart additional desirable properties to microgels [19–23]. The added materials range from molecules to microparticles and are typically complexed with the gel matrix through specific interactions. The resulting microgels usually show a drastic decrease in their physical response to external stimuli compared to that of the original polymer networks [24–26]; this is an undesirable side effect since in most cases this responsiveness is exactly the property one wishes to exploit. In addition to functionality, the size distribution of a population of microgels is important; for instance, it is critical to provide a homogeneous distribution of microgels in surface formulations [27] and in controlling the release kinetics of encapsulates or adsorbents [28].

From the standpoint of performance and applicability, there is a need for methods to generate monodisperse microgels with a high sensitivity to external stimuli. The materials that are incorporated to add complementary functions should not affect the size and morphology of the formed particles. A flexible and straightforward method for generating such microgel particles is the use of capillary microfluidic techniques [29]. This enables a variation in the chemical composition, without affecting the control over the size of the microgel particles. A range of other novel synthesis methods for producing microgel particles have recently been developed. For instance, the use of membrane emulsification techniques allows preparation of monodisperse poly(acrylamide-co-acrylic acid) hydrogel microspheres [30, 31] with a typical particle size of 1–10 μm. Another recent example is the development of microemulsion polymerization techniques for the synthesis of thermally sensitive polyampholyte nanogels [32].

In this chapter, we will focus both on the direct synthesis of microgel particles in dispersed media and on the kinetics involved. The synthesis of microgel particles is performed mainly using dispersion radical polymerization techniques. These are the most commonly used and most thoroughly investigated processes; they have been studied regarding the polymerization kinetics, the partial conversion of the used monomers, and the chemical composition of the dispersed medium as a function of polymer conversion.

Understanding the kinetics of polymerization processes is of fundamental importance for efficiently generating polymeric products for a wide variety of applications. Kinetic studies of different polymerization reactions have therefore been an important topic since the 1920s when the classical definition of a macromolecule was introduced by Staudinger. However, the evaluation and analysis of reaction rates and molecular weight distributions for free radical polymerization using different synthesis routes are rather complicated; some simplifications are thus required to readily approach this topic. For preparation of gels or microgels, it is necessary to make a cross-linked polymer or highly branched high molecular weight network, where the molecular weight is of only minor importance. Hence, the study of molecular weight and its distribution will not be the scope of this chapter and we will instead focus on kinetic studies and rate equations that are used to describe the processes involved in gel formation.

Various kinds of microgels have been reported in the literature [33–36], but the most examined are microgels that are responsive to stimuli such as pH, salinity, and temperature. In the case of thermally sensitive microgel particles, systematic studies have been carried out on various monomers such as N-isopropylacrylamide (NIPAM), N-isopropylmethacrylamide (NIPMAM), and more recently N-ethylmethacrylamide (NEMAM). For the case where NIPMAM is the main monomer, a wide range of studies have been performed to evaluate the influence of experimental variables on both polymerization kinetics and the physiochemical properties of the latex. Introducing a cross-linking agent is necessary both to provide cohesion to the precipitated poly(NIPMAM) chains and to avoid the resolubilization of the particles when lowering the temperature below the volume-phase transition temperature (VPTT). In general, increasing the concentration of cross-linking agents leads to a decrease in the formation of water-soluble polymer (WSP) and to a reduction in the swelling capacity of the particle. This emphasizes the strong effect that cross-linking agents have both on the microgel particle nucleation step and on the growth process. It is, however, interesting to note that the increase in cross-linking agents such as methylene-bis-acrylamide (BA) in dispersion radical polymerization in water has little effect on the overall polymerization kinetics and on the final hydrodynamic particle size [37].

2.2
Polymerization Processes

Radical polymerization has been widely used in order to produce water-soluble polymers; the different processing routes are schematically illustrated in Figure 2.2.

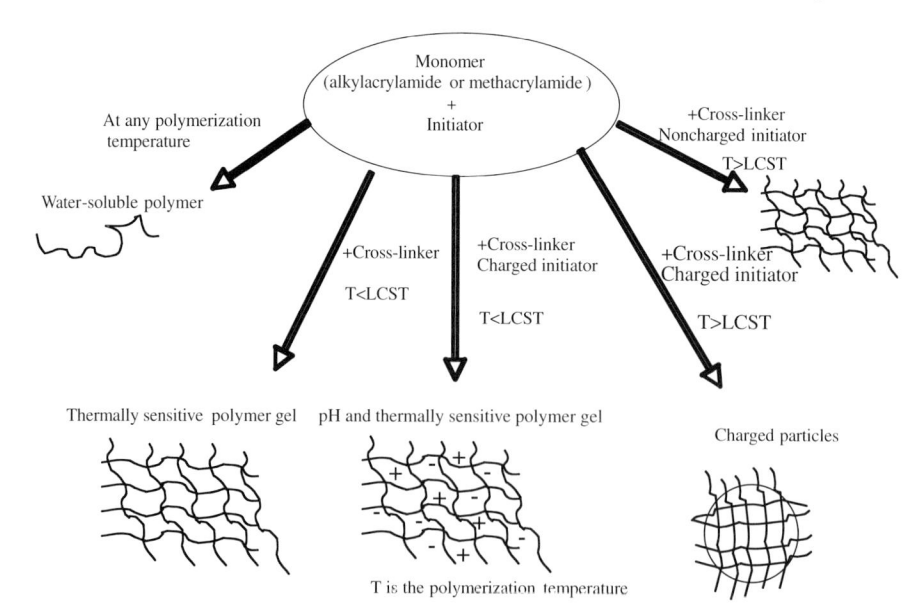

Figure 2.2 Illustration of the synthesis of thermally and pH-sensitive water-soluble polymer and gel-based polymer materials.

Without the addition of any cross-linking agent, radical polymerization for a water-soluble system leads to the formation of single chains that are not chemically cross-linked but which may form nonchemically cross-linked, the so-called physical gel, at very high concentrations. In the presence of charged or noncharged radical initiators, the use of cross-linking agents in polymerization leads to the formation of a macroscopic gel when the polymerization is performed below the LCST (lower critical solution temperature) of the corresponding homopolymer. However, when polymerization is performed in the presence of charged initiators above the LCST, colloidal microgel particles are obtained. Thus, polymerization conducted above the LCST is necessary in order to induce the precipitation and nucleation required for the formation of microgel particles. The most commonly used and studied polymerization processes are thus performed in the presence of a cross-linking agent above the LCST. Generally, charged initiators [38, 39], and in some cases charged comonomers, are used [40]; this leads directly to charged microgel dispersions. The formation of microgels sensitively depends on a range of parameters such as the chosen solvent, the concentration of monomer(s), the molar ratio of monomers/cross-linking agent, the polymerization temperature, the presence of a surfactant, salinity, pH, and the nature and concentration of the initiator. The selection of appropriate stabilizing agents, which prevent the aggregation of particles, is also of great importance.

2.3
Kinetics of Polymerization Reaction

The formation of particles in a polymerization reaction is mainly influenced by the conditions during the nucleation period, by the conditions during the growth period, and by the colloidal stability of the final particles. In dispersion polymerization, precipitation in the early stages of polymerization drives the nucleation process. To run any polymerization reaction, energy needs to be supplied to the system in order to induce the initiation state via initiator decomposition. In such radical dispersion polymerization, water-soluble monomers (i.e., acrylamide or methacrylamide derivatives) may be used. The most thoroughly studied initiators are potassium persulfate (KPS), 4,4'-azobis(4-cyanovaleric acid) (V501), and 2-2'-azobisamidinopropane dihydrochloride (V50). In the following sections, we will examine the influence on and the contribution of each of these reactants to the polymerization process.

For microgels based on acrylamide, N-alkylacrylamide, or N-alkylmethacrylamide, various systems have been investigated such as N-isopropylacrylamide/methylene bisacrylamide/water-soluble radical initiators (i.e., KPS and V50). The polymerization, in this case, can be conducted at temperatures below 60 °C and at low solid content (less than 5% wt in the presence of free-stabilizing agents), while usually larger solid fractions and higher temperatures are required. It has been pointed out that polymerization is rapid compared to classical polymerization in dispersed media such as emulsion and miniemulsion processes. For the latter systems, a total monomer conversion can often be achieved within less than 30 min.

Due to high reactivity of the used reactants (the reactivity of the cross-linker is often higher than the reactivity of the monomer), particle formation is fast. If no inhibitors are used, the nucleation step can often take place within less than 1 min. Such fast reactions make it difficult to experimentally follow the reaction kinetics in terms of the partial conversion of individual monomers. The first studies of the kinetics of these systems were reported by Pelton et al. [41] and then by Meunier et al. [40]; they succeeded in separately monitoring the concentrations of both the cross-linker and the monomer.

In dispersion polymerization of water-soluble monomers, the final microgel particles are generally submicron in size and exhibit a narrow size distribution. This can be attributed to high reactivity of the reactants used and to short duration of the nucleation step. Duracher et al. [37] have shown this by investigating the dependence of the hydrodynamic particle size on the polymerization conversion. The mass of one particle is given by the total polymer mass, m, divided by the number of particles, N_p. Assuming a homogeneous density of the particles, their radius (R_h) is given by

$$R_h \approx N_p^{-1/3} \cdot m^{1/3} \tag{2.1}$$

where the polymer mass, m, is equal to the total polymerization conversion, which can be determined using gas chromatography [40] or NMR techniques [37].

At constant number of particles, the particle size should increase linearly with the total mass m of polymer to the power of 1/3. Indeed, Duracher et al. found the diameter determined from SEM to increase linearly with the overall conversion to the power of 1/3 (i.e., $m^{1/3}$), as shown in Figure 2.3. This linear variation reflects the

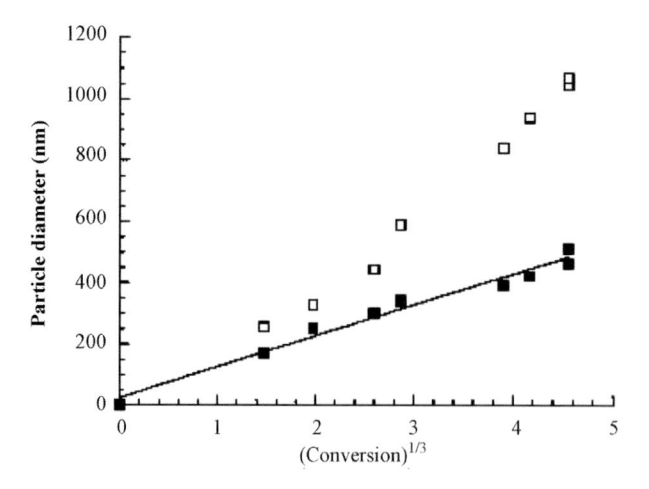

Figure 2.3 Hydrogel particle size versus conversion$^{1/3}$ (i.e., $m^{1/3}$) for N-alkylmethacrylamide-based microgels. The particle size in this study was investigated above and below the volume-phase transition temperature ($T_{VPT} = 45\,°C$). The hydrodynamic size, measured by light scattering (□), was compared with the size determined by SEM (■). Adapted from Duracher et al. [37].

constancy of the number of particles during the polymerization route. However, the hydrodynamic size, as determined from light scattering, did not show the same linear dependency as the SEM data. This difference highlights the fact that in the swollen state, the particles do not exhibit a uniform internal density. While in SEM the size is measured in a dried state, where the particle density is uniform, in light scattering the hydrodynamic size of the microgel particles was measured in the swollen state at room temperature. The deviation from a linear behavior shown in Figure 2.3 thus indicates that in the swollen state, the internal density of the particles is size dependent. Similar behavior has been reported by Kamijo et al. [42] for acrylamide–acrylic acid microgel particles prepared by precipitation polymerization carried out in isopropyl alcohol.

Transmission and scanning electron microscopy (SEM) are often used for characterization of the formed particles in different stages of reaction; they generally show that the particles obtained have a uniform size distribution, as shown in Figure 2.4 for the case of poly(NIPMAM) microgels.

It is evident from such SEM investigations of particle size versus polymerization time that the system shows a uniform size distribution already in the early stages of the reaction. This indicates both a very short duration of the nucleation phase and the absence of any further nucleation during the growth phase. The same behavior has been observed in the case of other systems such as NIPAM/BA/cationic initiator [37] and acrylamide/BA/methacrylic acid/AIBN in isopropanol medium [42].

2.3.1
The Influence of Initiators

The use of initiators is of paramount importance in radical polymerization reactions for production of particles. Numerous studies show that increasing the initiator concentration leads to a marked increase in the polymerization rate. An example is shown in Figure 2.5 where the polymer conversion is plotted as a function of reaction time for two different initiator concentrations.

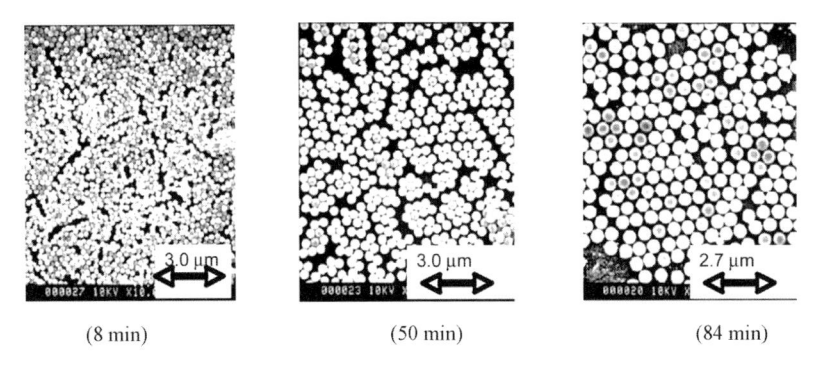

(8 min)   (50 min)   (84 min)

Figure 2.4 SEM images of N-alkylmethacrylamide-based microgels as a function of polymerization time [37].

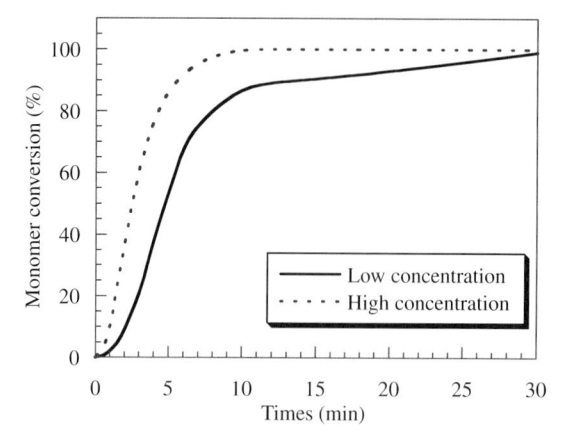

Figure 2.5 Monomer conversion versus polymerization time for two initiator concentrations. (Example for poly (NIPAM) particles: polymerization temperature $T = 70\,°C$, [NIPAM] = 48.51 mmol, [BA] = 3 mmol, [V50] = 0.1 (—) to 1 mmol (---) for total volume = 250 ml [44].

The dependence of the polymerization rate (R_p) on the initiator concentration ([I]) can be expressed by the following empirical relation [43]:

$$R_p \approx [I]^\alpha \tag{2.2}$$

where α is a positive exponent that depends on the particular system. The initiator usually decomposes and its concentration is thus not constant. The polymerization kinetics are therefore controlled not only by the initiator concentration itself but also by the initiator decomposition rate. The thermal decomposition rate of the initiator is related to the temperature of the medium as follows:

$$N_{dec} = [I](1 - e^{-k_d t}) \tag{2.3}$$

Here, k_d is the decomposition rate constant of the initiator at a given temperature, [I] is the initial concentration of initiator, and N_{dec} is the amount of decomposed initiator after a given time and at a given temperature.

The exponent α depends both on the system studied and on the polymerization conditions. Surprisingly, for dispersion polymerization of water-soluble monomers, this exponent is generally lower than that typically found in classical emulsion and miniemulsion polymerization. For instance, $\alpha \approx 0.18$ for NIPAM/BA/KPS at 70 °C and $\alpha \approx 0.4$ in the case of emulsion polymerization of styrene. This discrepancy is poorly understood and additional work is needed to clarify its origin.

The effect of initiator concentration and the influence of temperature on the polymerization reaction are clearly related to each other. An increase in the incubation temperature enhances the decomposition rate of the initiator and consequently increases the concentration of radicals in the media. Hence, the polymerization temperature affects the final molecular weight and its distribution. The instantaneous rate of polymerization R_p mainly depends on the propagation rate coefficient

(k_p), on the average number of radicals per particle ($\bar{n}$), and on the number of particles (N_p). These dependencies are well described by the following rate equation [45]:

$$R_p = \frac{k_p N_p [M]_p \bar{n}}{N_{av}} \qquad (2.4)$$

where $[M]_p$ and N_{av} are the concentration of monomer in the polymer particle and Avogadro's number, respectively.

2.3.2
The Effect of the Cross-Linking Agent

The cross-linking of polymer chains is of key importance in the production of microgel particles. While such cross-linking can be induced both via chemical and physical means, in most cases, such as in radical dispersion polymerization, the polymer chains are *chemically* cross-linked by the use of cross-linking agents. The selection of cross-linking agents has an important influence both on the formation of the particles and on their final properties. Due to the high reactivity of cross-linking agents compared to usual monomers, the addition of cross-linking agents leads to a slight increase in the overall polymerization rate. This has been reported by Meunier et al. [46], who have studied the effect of the concentration of methylene-bis-acrylamide on the kinetics of NIPAM polymerization, as shown in Figure 2.6.

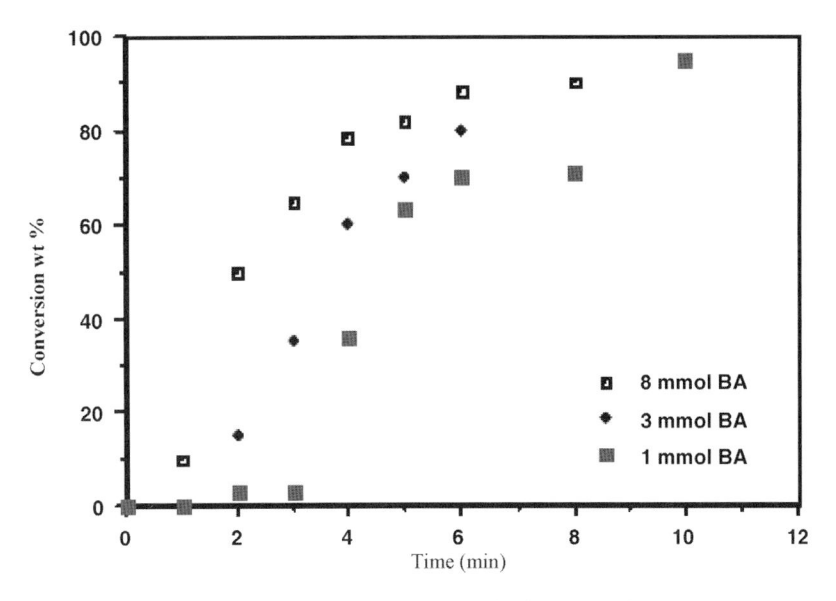

Figure 2.6 Effect of the concentration of methylene-bis-acrylamide on the batch surfactant-free kinetics of NIPAM polymerization in the presence of V50 (2,2'-azobis(2-methylpropionamidine) dihydrochloride). Polymerization temperature $T = 70\,°C$;, [NIPAM] = 48.51 mmol, [BA] = 1–8 mmol, [V50] = 0.3 mmol for total volume = 250 ml [46].

2 Polymerization Kinetics of Microgel Particles

The cross-linking agent also affects the physical properties of the final microgels. When high levels of cross-linking agent are used, the formed microgels are highly cross-linked and, consequently, the swelling ability and flexibility of such microgels are drastically reduced; this phenomenon has been clearly established in the case of thermally sensitive microgel particles [37, 47]. The effect has also recently been examined by Guillermo et al. [48] by using NMR-techniques; this yields relevant information about the heterogeneous internal structure of microgel particles. These measurements revealed the existence of a gradient in composition ranging from the core to the shell of the microgel. A large amount of cross-linker leads to the formation of a highly cross-linked core surrounded by a less cross-linked shell, as schematically shown in Figure 2.7 (right). For low amounts of cross-linker, the cross-linker concentration decays more gradually. The microgels can be thought of as consisting of three domains, as schematically shown in Figure 2.7 (left): a highly cross-linked core (almost hard core), a medium cross-linked shell, and a weakly cross-linked outer shell (hair-like shell).

2.3.3
The Effect of Functional Monomers

The functionalization of microgels is performed either directly during the particle formation stage or through an additional surface functionalization step. In the latter

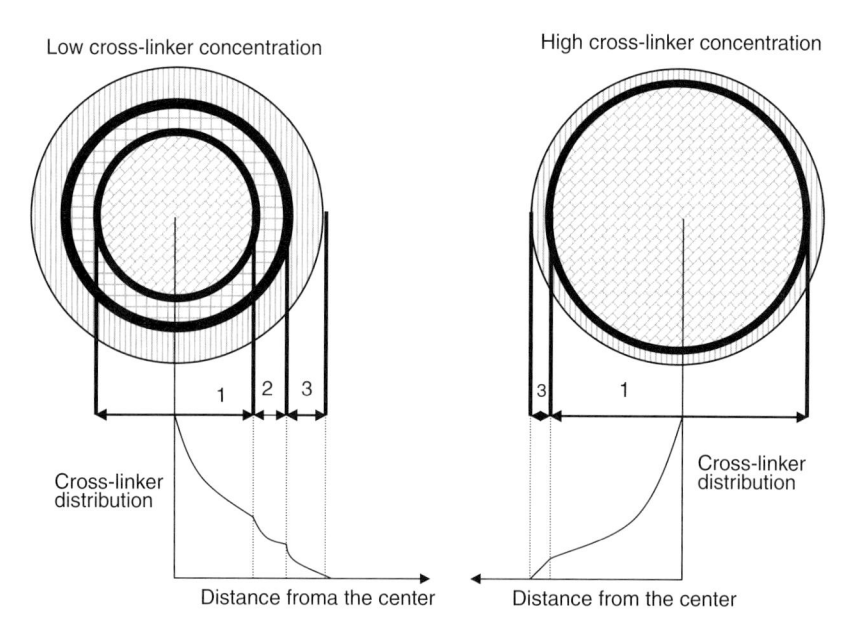

Figure 2.7 (Top) Schematic illustrations of microgel particles prepared via precipitation polymerization as a function of cross-linker concentration: (1) highly cross-linked part, (2) medium cross-linked shell, and (3) weakly cross-linked shell. (Bottom) Cross-linking distribution, as suggested by NMR studies [48].

case, the polymeric particles are used as seeds in a subsequent functionalization that is performed through modification reactions or through a seeded polymerization of functional monomers on the surface of the particles. The problem with such a process is related to the location of the functional groups. Generally, the reactive groups are not homogeneously distributed in the microgel structure. Due to this distribution, microgels are often pH and salinity sensitive provided that polar or ionic monomers are used. In precipitation polymerization, functionalization can be induced either by using water-soluble functional initiators or by using small amounts of highly water-soluble functional comonomers. The latter approach has been widely used in the synthesis of structured and functionalized microgels. Most studies on microgels have been conducted on thermally sensitive microgels. The functionalization of such microgels has been performed using acrylic acid, N-vinylbenzylimino diacetic acid [49], aminoethylmethacrylate hydrochloride [40], phenyl boronic acid [50–53], thiol-containing monomer [46], and cyanoacrylate [54, 55]. Generally, increasing the water-soluble functional comonomer concentration in batch polymerization leads to

- **A more rapid polymerization kinetics** (Figure 2.8). The concentration of functional monomers directly affects the formation of radicals and oligoradicals present in the polymerization medium. Consequently, the overall polymerization rate is rapid compared to the case without any functional monomers present.

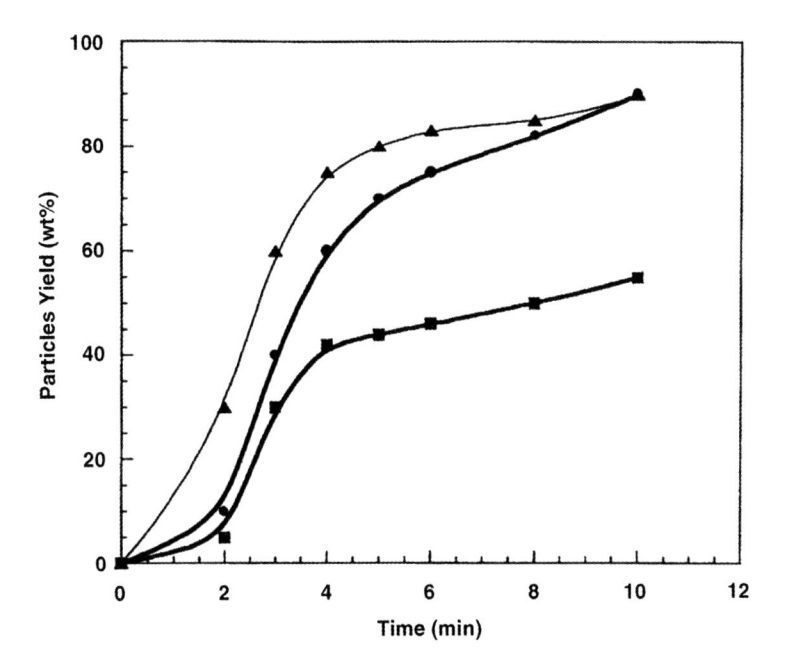

Figure 2.8 Effect of functional monomer on polymerization kinetics (particle yield). Batch NIPAM polymerization using vinyl benzyl isothiouronium chloride (VBIC) as functional monomer [46]. Polymerization temperature $T = 70\,°C$, [NIPAM] = 48.51 mmol, [BA] = 3 mmol, [V50] = 0.3, [VBIC] (■) for 0, (●) for 0.024 mmol, and (▲) for 0.12 mmol for total volume = 250 ml [44].

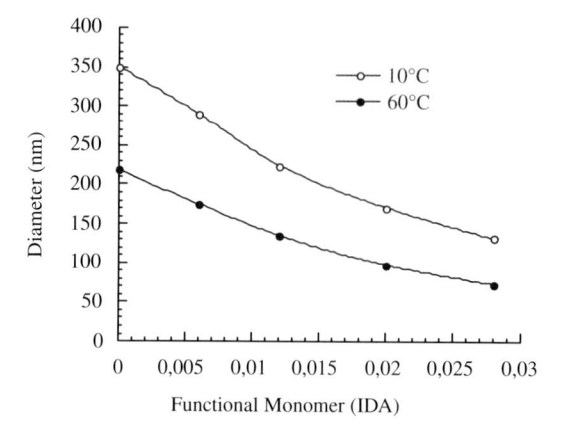

Figure 2.9 Effect of the functional comonomer N-vinyl benzyl imino diacetic acid (IDA in g) on the hydrodynamic particle diameter, as measured by quasielastic light scattering (QELS) at 10 and 60 °C [49]. Polymerization at 70 °C using 50 ml of water, 1 g of NIPMAM, 0.12 g of BA, 0.012 g of KPA, and IDC from 0 to 0.03 g. The particles size was measured after removal of the water-soluble polymer via a centrifugation process [49].

- **A smaller particle size** (Figure 2.9). The reduction in particle size for higher concentrations of functional monomer has been attributed to the enhancement of precursor formation and to the number of stable particles, which become the polymerization loci.
- **A higher percentage of water-soluble polymer in the continuous phase** (Figure 2.10). This effect is related to the good water solubility of oligomers containing functional monomers.

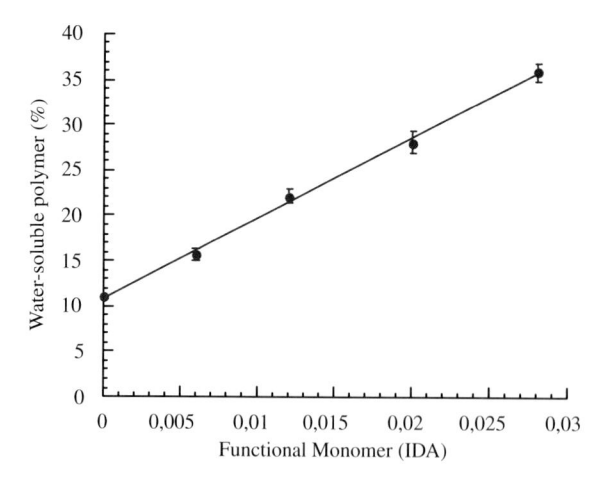

Figure 2.10 The influence of N-vinyl benzyl imino diacetic acid (in g) on water-soluble polymer formation [49]. The WSP amount was determined from the supernatant of the final latexes via a centrifugation process. Polymerization at 70 °C using 50 ml of water, 1 g of NIPMAM, 0.12 g of BA, 0.012 g of KPA, and IDC from 0 to 0.03 g [49].

2.3.4
Kinetic Aspects of Microgel Formation

In dispersion polymerization, the main reaction loci are the polymer particles. Due to their small size, typically smaller than 300 nm, they remain well dispersed in the water phase. The most well-known example of this polymerization route is the so-called radical segregation [56–58]. In order to understand this effect, let us consider a simplified model, the so-called "0-1-2" system [59]: in this model description, a radical "1" entering a particle without other radicals, "0", propagates until a second radical "2" enters the same particle. When the two radicals meet, a chemical bond is formed via a fast termination between the two radicals, thus creating again a particle without radicals. This process repeats itself indefinitely in every particle.

A good example of the relevance of segregation of different species inside polymer particles is polymer gelation [60–63]. In homogeneous systems, polymer gelation occurs both due to polymerization of single polymer chains and due to the union of these chains; if, during this process, some chains combine with chains of similar size, a fast geometric growth takes place, and extremely large polymer chains are soon formed. Typical examples of gelling systems are those exhibiting cross-linking or chain transfer to polymer followed by bimolecular termination [60, 64–66]. In both cases, the polymer chains form a macroscopically space-spanning network, a gel. On the other hand, when the same process occurs in emulsion polymerization, the gel growth is naturally limited by the fact that polymerization is inhomogeneous as it only takes place on the polymer particles; this effect is also known as the limited space effect [67]. Even though the chemical composition is identical to gel polymers formed in a homogeneous system, these chains have a finite size and thus do not form macroscopic gels, but instead lead to the formation of microgels.

The simple kinetic scheme shown in Table 2.1 is typical for radical polymerization kinetics, including polymer branching. It consists of radical entry into the polymer particle from the water phase (in emulsion polymerization), monomer propagation, bimolecular termination by radical combination, and chain transfer to polymer. According to the current description of gel formation [57, 58, 63, 68, 69], the

Table 2.1 Kinetic scheme of radical polymerization reactions and rate constants[a].

Definition	Reaction	Rate constant
Decomposition of initiator	$I \to 2R^0$	k_d (Rate-determining step)
Formation of active species	$R_e^0 + M \to RM^0$	k_i
Propagation	$P_n^0 + M \to P_{n+1}$	k_p
Termination:		
Recombination	$P_n^0 + P_m^0 \to P_{n+m}$	k_t
Disproportionation	$P_n^0 + P_m^0 \to P_n + P_m$	k_t
Chain transfer to polymer	$P_n^0 + P_m \to P_n + P_m^0$	k_{fp}

a) I: initiator; R^0: radical initiator; R_e^0: effective radical; M: monomer; P_i^0: polymeric radical with "i" repeating unit.

combination of a connecting event (bimolecular termination) with a branching event (transfer to polymer) is a necessary, although not sufficient, condition to obtain gelation.

Let us assume that the initiator I decomposes into two equally reactive primary radicals, with a rate constant k_d according to

$$-\frac{d[I]}{dt} = k_d[I] = \frac{1}{2}\frac{d[R^0]}{dt} \tag{2.5}$$

From Equation 2.5, it follows that the concentration of the initiator in the reaction medium decreases exponentially:

$$[I] = [I]_0 \cdot e^{-k_d t} \tag{2.6}$$

The rate of formation of the "effective" primary radicals R_e^0 can be written as

$$\frac{d[R_e^0]}{dt} = 2k_d f[I] \tag{2.7}$$

Here, f is the probability for a primary radical to react with a monomer M rather than to combine with another radical and form a "dead" reaction product.

Initiation involves consumption of these primary radicals by reaction with the monomer:

$$-\frac{d[R_e^0]}{dt} = k_i[R_e^0][M] \tag{2.8}$$

At steady state, the rates of formation and consumption of primary radicals have to be equal. Thus, by combining Equations 2.7 and 2.8, the steady-state concentration of effective primary radicals is obtained as

$$[R_e^0] = \frac{2k_d f[I]}{k_i[M]} \tag{2.9}$$

The propagation step follows from the rate of monomer consumption as

$$R_p = -\frac{d[M]}{dt} = k_p[P_n^0][M] \tag{2.10}$$

where we have neglected the monomer consumed in the initiation process, and where $[P_n^0]$ stands for the steady-state concentration of active sites during the growth process, irrespective of their size and assuming equal reactivity for each of them:

$$[P_n^0] = \sum_{i=1}^{\infty}[P_{ni}^0] \tag{2.11}$$

A termination reaction always involves two radicals, reacting either by recombination or by disproportionation. In the former case, a bond is formed between two growing radicals. In the latter case, a hydrogen atom is transferred from one macroradical to the other, whereby one of the polymer chains is saturated. Kinetically,

2.3 Kinetics of Polymerization Reaction

the two types of termination are identical.

$$-\frac{d[P_n^0]}{dt} = k_t[P_n^0]^2 \tag{2.12}$$

In order to obtain gelation in the particle, the rate of chain transfer to polymer must be at least equal or more than the rate of termination ($R_{fp} \geq R_t$). This ensures the formation of a gel instead of linear or weakly branched chains.

The rate of consumption of growing radicals through chain transfer to polymer is given by

$$-\frac{d[P_n^0]}{dt} = k_{fp}[P_n^0][P_m] \tag{2.13}$$

If the number of growing sites is assumed to stay constant (over small time intervals), the rates of initiation and of termination are to be set equal. It is worth noting that chain transfer to polymer (k_{fp}) does not affect the equilibrium concentration of growing sites. This results in the following steady-state equations:

$$k_t[P_n^0]^2 = 2k_d f[I] \tag{2.14a}$$

$$k_t[P_n^0]^2 = k_{fp}[P_n^0][P_m] \tag{2.14b}$$

where we have made the simplifying assumption that $R_{fp} = R_t$. Solving for $[P_n^0]$, we obtain

$$[P_n^0] = \left(\frac{2k_d f[I]}{k_t}\right)^{1/2} = \frac{k_{fp}}{k_t}[P_m] \tag{2.15}$$

$$[P_m] = \left(\frac{2k_d f[I]}{k_t}\right)^{1/2} \frac{k_t}{k_{fp}} = \left(\frac{k_t}{k_{fp}^2} R_i\right)^{1/2} \tag{2.16}$$

Introducing this expression into Equation 2.10 gives the general equation for the rate of polymerization:

$$R_p = -\frac{d[M]}{dt} = k_p \left(\frac{2k_d f[I]}{k_t}\right)^{1/2} [M] = \left(\frac{k_p^2}{k_t} R_i\right)^{1/2} [M] \tag{2.17a}$$

$$R_p = -\frac{d[M]}{dt} = k_p \frac{k_{fp}}{k_t}[P_m][M] = \left(\frac{k_p^2}{k_t} R_i\right)^{1/2} [M] \tag{2.17b}$$

It follows that the rate of propagation (which is the overall rate of monomer consumption) is first order with respect to the monomer concentration (proportional to [M]) and one-half order with respect to the initiator concentration (proportional to $[I]^{1/2}$).

It is reasonable to assume that the kinetics of the reaction obeys the simple assumption of $R_{fp} = R_t$ (Equations 2.17a and 2.17b). However, this assumption may be violated because different mechanisms are important at different stages of the reaction:

- At the early stages of polymerization, the kinetics of the reaction should be due to the conventional radical polymerization (Equation 2.17a).
- If the polymerization is carried out in the concentrated solution to higher conversions, the viscosity of the reaction medium increases as the reaction proceeds. This increase in viscosity does not affect the rate of decomposition of the initiator. But it will result in a drastic decrease in R_t because of inaccessibility of the reaction parameters (two radicals simultaneously). As a consequence, the reaction pathway will divert from the usual termination mechanism leading to chain transfer into polymer and to gelation. At this stage, it is reasonable that polymer chains are more accessible to the growing sites and thus (k_{fp}) will control the formation of polymer gels in the system.

The assumptions underlying these calculations are the following:

- k_p is independent of the length of the chain to which the growing sites are attached.
- The concentration of the active sites is constant, though the lifetime of each of them is extremely short. This implies that the initiator concentration stays constant or does not change significantly during the process. The validity of this hypothesis is limited to low conversions, more precisely for process durations that are short compared to the half-life time of the initiator.
- If this condition is not fulfilled, account has to be taken of the initiator consumption during the process, by introducing (Equation 2.6) into the rate equation (Equations 2.17a and 2.17b):

$$R_p = -\frac{d[M]}{dt} = k_p \left(\frac{2k_d f}{k_t}\right)^{1/2} [M]_t [I]_0^{1/2} e^{-(k_d t/2)} \tag{2.18}$$

To express the conversion as a function of time, Equations 2.17a, 2.17b have to be integrated. Calling [M$_0$] and [I$_0$] the initial monomer and initiator concentrations, one gets

–if [I] is held constant or does not vary over the period of time t

$$\ln \frac{[M]_0}{[M]} = k_p \left(\frac{2k_d f}{k_t}\right)^{1/2} [I]^{1/2} t \tag{2.19}$$

–if the initiator consumption during the process is taken into account

$$\ln \frac{[M]_0}{[M]} = 2k_p \left(\frac{2f}{k_d k_t}\right)^{1/2} [I]_0^{1/2} \left[1 - e^{-(k_d t/2)}\right] \tag{2.20}$$

Since the degree of conversion π is given by $([M]_0 - [M])/[M]_0$, Equations 2.19 and 2.20 are suited to express the degree of conversion versus time as

$$\ln \frac{[M]_0}{[M]} = -\ln(1-\pi) \tag{2.21}$$

As a result, the studies of polymerization kinetics in the microgel formation reveal a close similarity to the kinetics observed in conventional radical polymerization reactions. This may be attributed to the close similarities of the general mechanisms and reaction steps involved. Even though in microgel formation the chain transfer into polymer is predominant, it also plays an important role in the gel formation process.

2.4
Conclusions

This chapter covers the preparation of microgel particles using the most commonly used polymerization processes – dispersion and precipitation polymerization in dispersed media. The latter process has largely been explored in producing pH-, salinity-, and temperature-sensitive microgel particles. As a general tendency, when the polymerization is conducted using water-soluble reactants and radical initiation steps, the increase in temperature or initiator concentration leads to an enhancement both in terms of the polymerization kinetics and in terms of the overall final conversion. Increasing the amount of cross-linkers leads to a highly cross-linked microgel matrix and to low amounts of water-soluble polymer formation in the continuous phase. The copolymerization in the presence of functional monomers leads to the formation of charged microgel particles. The functionality distribution (i.e., functional groups) in the polymer matrix is related to the reactivity of each polymerizable reactant. In such a polymerization process, the driving force involved in the nucleation step is related to the nature of the solvent, which should induce the precipitation of polymer chains. This nucleation process is the key process in the polymerization mechanism, which largely determines the final properties of the microgel particles. Increasingly complex microgel systems are being explored, keeping in mind their biomedical applications such as in drug delivery or *in vitro* biomedical diagnostics. Further research on the mechanisms and kinetics of these polymerization processes will thus be necessary to elucidate the relevant mechanisms and control the properties of the final microgel particles.

References

1 Saunders, B.R. and Vincent, B. (1999) *Adv. Colloid Interface Sci.*, **80**, 1–25.
2 Nerapusri, V., Keddie, J.L., Vincent, B., and Bushnak, I.A. (2006) *Langmuir*, **22**, 5036–5041.
3 Bradley, M., Ramos, J., and Vincent, B. (2005) *Langmuir*, **21**, 1209–1215.
4 Qiu, Y. and Park, K. (2001) *Adv. Drug Deliv. Rev.*, **53**, 321–339.
5 Murthy, N., Xu, M., Schuck, S., Kunisawa, J., Shastri, N., and Frechet, J.M. (2003) *Proc. Natl. Acad. Sci. USA*, **100**, 4995–5000.
6 Langer, R. and Peppas, N.A. (2003) *AIChE J.*, **49**, 2990–3006.
7 Holtz, J.H. and Asher, S.A. (1997) *Nature*, **389**, 829–832.
8 Miyata, T., Jige, M., Nakaminami, T., and Uragami, T. (2006) *Proc. Natl. Acad. Sci. USA*, **103**, 1190–1193.
9 Olsen, K.G., Ross, D.J., and Tarlov, J. (2002) *Anal. Chem.*, **74**, 1436–1441.

10 Arai, F., Ng, C., Maruyama, H., El-shimy, H., and Fukuda, T. (2005) *Lab. Chip*, **5**, 1399–1403.
11 Hu, Z. and Huang, G. (2003) **115**, 4947–4950.
12 Hu, Z., Lu, X., and Goa, J. (2001) *Adv. Mater.*, **13**, 1708–1712.
13 Debord, J.D. (2000) *J. Phys. Chem. B*, **104**, 6327.
14 Debord, J.D., Eustis, S., Debord, S.B., Lofye, M.T., and Lyon, L.A. (2002) *Adv. Mater.*, **14**, 658.
15 Fujii, S., Armes, S.P., Araki, T., and Ade, H. (2005) *J. Am. Chem. Soc.*, **127**, 16808–16809.
16 Murray, M.J. and Snowden, M.J. (1995) *Adv. Colloid Interface Sci.*, **73**, 73.
17 Morris, G.E., Vincent, B., and Snowden, M.J. (1997) *J. Colloid Interface Sci.*, **190**, 198–205.
18 Okano, T. (1998) *Biorelated Polymers and Gels*, Academic Press, San Diego.
19 Langer, R. and Tirrel, D.A. (2004) *Nature*, **428**, 487–492.
20 Anderson, D.G., Burdick, J.A., and Langer, R. (2004) *Science*, **305**, 1923–1924.
21 Pelton, R. (2000) *Adv. Colloid Interface Sci.*, **85**, 1–33.
22 Nayak, S. and Lyon, A.L. (2004) *Angew. Chem., Int. Ed. Engl.*, **116**, 6874–6877.
23 Berndt, J.S., Pederson, J.S., and Richtering, W. (2006) *Angew. Chem., Int. Ed. Engl.*, **118**, 1769–1773.
24 Rogach, L., Nagesha, D., Ostrander, J.W., Giersig, M., and Kotov, N.A. (2000) *Chem. Mater.*, **12**, 2676–2685.
25 Zhang, J., Xu, S., and Kumacheva, E. (2004) *J. Am. Chem. Soc.*, **126**, 7908–7914.
26 Pich, A., Karak, A., Lu, Y., Ghosh, A.K., and Adler, H.J.P. (2006) *Macromol. Rapid Commun.*, **27**, 344–350.
27 Shiga, K., Muramatsu, N., and Kondo, T. (1996) *J. Pharm. Pharmacol.*, **48**, 891–895.
28 Xiao, X.C., Chu, L.Y., Chem, W.M., Wang, S., and Xie, R. (2004) *Langmuir*, **20**, 5247–5253.
29 Kim, J.-W., Utada, A.S., Fernandez-Nieves, A., Hu, Z., and Weitz, A.D. (2007) *Angew. Chem., Int. Ed. Engl.*, **46**, 1819–1822.
30 Nagashima, S., Ando, S., Makino, K., Tsukamoto, T., and Ohshima, H. (1998) *J. Colloid Interface Sci.*, **197**, 377–382.
31 Nagashima, S., Ando, S., Tsukamoto, T., and Ohshima, H. (1998) *Colloids Surf. B*, **11**, 47–56.
32 Ogawa, K., Nakayama, A., and Kokufuta, E. (2003) *Langmuir*, **19**, 3178–3184.
33 Loos, W., Verbrugghe, S., Goethals, E.J., Du Prez, F.E., Bakeeva, I.V., and Zubov, V.P. (2003) *Macromol. Chem. Phys.*, **204**, 98–103.
34 Bronstein, L., Kostylev, M., Tsvetkova, I., Tomaszewski, J., Stein, B., Makhaeva, E.E., Khokhlov, A.R., and Okhapkin, A.R. (2005) *Langmuir*, **21**, 2652–2655.
35 Van Durme, K., Verbrugghe, S., Du Prez, F.E., and Van Mele, B. (2003) *Polymer*, **44**, 6807–6814.
36 Van Durme, K., Verbrugghe, S., Du Prez, F.E., and Van Mele, B. (2004) *Macromolecules*, **37**, 1054–1061.
37 Duracher, D., Elaissari, A., and Pichot, C. (1999) *J. Polym. Sci. A: Poly. Chem.*, **37**, 1823–1837.
38 Pelton, R. (2000) *Adv. Colloid Interface Sci.*, **85** (1), 1–33.
39 Kawaguchi, H., Yamada, Y., Kataoka, S., Morita, Y., and Ohtsuka, Y. (1991) *Polym. J.*, **23** (8), 955–962.
40 Meunier, F., Elaissari, A., and Pichot, C. (1995) *Polym. Adv. Tech.*, **6**, 489–496.
41 Wu, X., Pelton, R.H., Hamielec, A.E., Woods, D.R., and McPhee, W. (1994) *Colloid Polym. Sci.*, **272**, 467–477.
42 Kamijo, Y., Fujimoto, K., Kawaguchi, H., Yuguchi, Y., Urakawa, H., and Kajiwara, K. (1996) *Polym. J.*, **28**, 309–316.
43 Matyjaszewski, K. and Davis, T.P. (2002) *Handbook of Radical Polymerization*, Wiley Interscience, Hoboken.
44 Meunier, F. (1996) PhD-Thesis: Synthesis and characterization of hydrophilic polymeric support particle based on N-isopropylacrylamide. Study on the conjugation between particles/ODN and their use in medical diagnostic. University of Lyon-1, France.
45 Mahdavian, A.R. and Abdollahi, M. (2004) *Polymer*, **45**, 3233–3239.
46 Meunier, F., Pichot, C., and Elaissari, A. (2006) *Colloid Polym. Sci.*, **284**, 1287–1292.
47 Duracher, D., Elaissari, A., and Pichot, C. (1999) *Colloid Polym. Sci.*, **277**, 905–913.

48 Guillermo, A., Cohen-Addad, J.P., Bazil, J.P., Duracher, D., Elaïssari, A., and Pichot, C. (2000) *J. Polym. Sci. B: Polym. Phys.*, **38**, 889–898.
49 Duracher, D., Elaïssari, A., Mallet, F., and Pichot, C. (2000) *Macromol. Symp.*, **150**, 297–303.
50 Uguzdogan, E., Denkbas, E.B., and Tuncel, A. (2002) *Macromol. Biosci.*, **2**, 214–222.
51 çamli, S.T., Senel, S., and Tuncel, A. (2002) *Colloids Surf. A*, **207**, 127–137.
52 Elmas, B., Onur, M.A., Senel, S., and Tuncel, A. (2002) *Colloid Polym. Sci.*, **280**, 1137–1146.
53 Elmas, B., Onur, M.A., Senel, S., and Tuncel, A. (2004) *Colloids Surf. A*, **232**, 253–259.
54 Zhou, G., Elaissari, A., Delair, T., and Pichot, C. (1998) *Colloid Polym. Sci.*, **276**, 1131–1139.
55 Zhou, G., Veron, L., Elaissari, A., Delair, T., and Pichot, C. (2004) *Polym. Int.*, **53**, 603–608.
56 Lichti, G., Gilbert, R.G., and Napper, D.H. (1980) *J. Polym. Sci.*, **18**, 1297.
57 Ghielmi, A., Storti, G., Morbidelli, M., and Ray, W.H. (1998) *Macromolecules*, **31**, 7172.
58 Ghielmi, A., Storti, G., and Morbidelli, M. (2001) *Chem. Eng. Sci.*, **56**, 937.
59 Gilbert, R.G. (1995) *Emulsion Polymerization: A Mechanistic Approach*, Academic Press, Landon.
60 Tiobita, H. (1994) *Polymer*, **35**, 3023.
61 Arzamendi, G. and Asua, J.M. (1995) *Macromolecules*, **28**, 7479.
62 Ghielmi, A., Florentino, S., Storti, G., Mazzotti, M., and Morbidelli, M. (1997) *J. Polym. Sci. A: Polym. Chem.*, **35**, 827.
63 Butte, A., Ghielmi, A., Storti, G., and Morbidelli, M. (1999) *Macromol. Theory Simul.*, **8**, 498.
64 Sayer, C., Araujo, P.H.H., Arzamendi, G., Asua, J.M., Lima, E.L., and Pinto, J.C. (2001) *J. Polym. Sci. A: Polym. Chem.*, **39**, 3513.
65 Ledema, P.D., Grcev, S., and Hoefsloof, H.C. (2003) *J. Macromolecules*, **36**, 458.
66 Dias, R.C.S. and Costa, M.R.P.F.N. (2003) *Macromolecules*, **36**, 8853.
67 Tobita, H. (1999) *Colloids Surf. A*, **153**, 119.
68 Teymour, F. and Campbell, J.D. (1994) *Macromolecules*, **27**, 2460.
69 Florentino, S., Ghielmi, A., Storti, G., and Morbidelli, M. (1997) *Ind. Eng. Chem. Res.*, **36**, 1283.

3
New Functional Microgels from Microfluidics
Jin-Woong Kim and Liang-Yin Chu

3.1
Introduction

Stimuli-sensitive hydrogel microspheres, or microgels, are ubiquitous both in academia and in industry due to their potential applications in numerous fields [1–5], including controlled drug delivery [6–11], chemical separations [12, 13], sensors [14–16], catalysis [17], enzyme immobilization [18], and color-tunable crystals [19]. Poly(*N*-isopropylacrylamide) (PNIPAM) is one of the most studied stimuli-sensitive polymers since it exhibits a lower critical solution temperature (LCST); the amide groups of PNIPAM selectively associate with water molecules at low temperatures or repel them at high temperatures, which gives rise to a thermally controlled phase transition. The phase transition temperature is usually regulated by the hydrophobic isopropyl groups in the side chain [20]. Recently, numerous investigations have been made on the thermosensitive PNIPAM microgels [1–5, 7, 21–26].

Monodispersity is very important in order to improve the performance of stimuli-sensitive microgels in various applications. For example, a uniform particle size is important for drug delivery systems because the distribution of particles within the body and their interaction with biological cells are greatly affected by particle size [27]. In addition, if the microgels are monodisperse, the drug release kinetics can be better manipulated, thereby making it easier to design more sophisticated systems. To prepare monodisperse microgels with sizes in the micrometer range, emulsification is a necessary step before polymerization. Although membrane emulsification [28] and microchannel emulsification [29] have both been used to prepare thermosensitive microcapsules and microspheres, membrane emulsification cannot ensure high monodispersity because of the size distribution of membrane pores and the long-time stirring process [30]. Moreover, in both methods the emulsification and polymerization are carried out in two steps and in two separate devices, which is generally time consuming and results in low processability, thus hindering the widespread application of these synthesis methods. Therefore, it is necessary, and important, to develop new methods for preparation of monodisperse thermosensitive microgels within integrated devices.

Microgel Suspensions: Fundamentals and Applications
Edited by Alberto Fernandez-Nieves, Hans M. Wyss, Johan Mattsson, and David A. Weitz
Copyright © 2011 WILEY-VCH Verlag GmbH & Co. KGaA, Weinheim
ISBN: 978-3-527-32158-2

Recently, microfluidic methods have provided a promising approach to preparing monodisperse polymeric microspheres [31–47]. The microfluidic technique can precisely control the fluid flow and thus ensure high monodispersity of the prepared particles. Furthermore, both emulsification and polymerization can be carried out within the same device. However, in most previous studies polymerization is carried out inside the microfluidic devices and radical initiation is obtained by irradiation with UV light [31–47]. Unfortunately, the use of UV irradiation is not good for preparing homogeneous thermosensitive PNIPAM microgels since it creates local heating on the monomer droplets. This heating usually causes spinodal decomposition during polymerization, resulting in heterogeneities within or on the surface of the PNIPAM microgels.

In this chapter, we introduce a novel and straightforward way of fabricating monodisperse thermosensitive PNIPAM microgels by using microfluidics-based methods. This approach is truly advantageous in that it allows us to engineer monodisperse microgels that have controllable internal structures, thus enabling a widening of their applications [48–50].

3.2
Monodisperse Thermosensitive Microgels Fabricated in a PDMS Microfluidic Chip

An on-chip fabricating technique has been developed for preparation of highly monodisperse and homogeneous thermosensitive PNIPAM microgels, as shown in Figure 3.1. Instead of using UV irradiation, a redox reaction approach in the microfluidic chip is designed to initiate the polymerization of a NIPAM monomer. Since we can carry out polymerization below the LCST, which is typically 32 °C, the PNIPAM microgels obtained after polymerization are highly homogeneous. Another advantage of this approach is that all fabrication processes take place within one chip and without using any other supplementary instruments, such as a UV lamp; thus, this microfluidic reactor is more compact, making it more easily scalable.

The microfluidic chip is made of poly(dimethylsiloxane) (PDMS) by using a standard soft lithography method [51], which allows rapid replication of an integrated microchannel prototype. A flow-focusing geometry is introduced to generate monodisperse emulsion droplets, as shown in Figure 3.1. In using the emulsification-polymerization approach to prepare PNIPAM hydrogels, polymerization of NIPAM monomer is usually initiated (i) by adding accelerator to the monomer solution to start the redox reaction, (ii) by heating the monomer solution at a temperature above the LCST, or (iii) by irradiating with UV light. Generally, the redox reaction generates hydrogels with homogeneous internal microstructures, while the second and the third methods result in heterogeneous internal microstructures [52]. However, if the accelerator is directly added to the monomer solution, the polymerization will occur over a very short time, resulting in clogging of the microchannel by the polymerized PNIPAM hydrogels. To prevent this problem, the accelerator is injected into the oil phase downstream of the emulsion solution (Figure 3.1). The accelerator, N,N,N',N'-tetramethylethylenediamine (TEMED) is

3.2 Monodisperse Thermosensitive Microgels Fabricated in a PDMS Microfluidic Chip

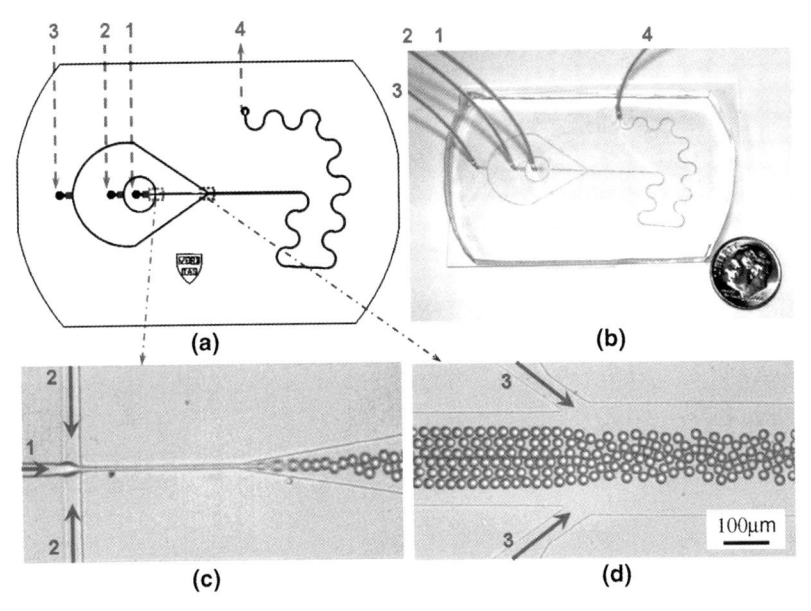

Figure 3.1 (a) Schematic illustration of the channel design in the microfluidic chip. (b) A photograph of the PDMS microfluidic system compared with a one-dime coin. The polyethylene tubing and microchannels are filled with a dye-labeled aqueous solution to increase the contrast of the image. Fluid 1 is an aqueous solution containing monomer, initiator, and cross-linker and is pumped through an inlet channel. Fluid 2 is a kerosene solution containing surfactant and is pumped through two flanking channels for the continuous phase. Fluid 3 is a kerosene solution containing surfactant and accelerator. (c) Optical microscope image of the drop formation in the emulsification step. (d) Optical microscope image showing the addition of accelerator solution downstream of emulsification.

readily soluble in water; thus, if it is added to the oil phase (Figure 3.1d), it diffuses into the water phase. When the accelerator meets with the initiator, ammonium persulfate (APS), inside the monomer emulsion droplets, the redox reaction is initiated, leading to polymerization of the NIPAM monomer. Thus, using this redox initiation system allows us to not only avoid the channel-clogging problem but also obtain homogeneous microstructures inside the microgels.

In the system shown in Figure 3.1, fluid 1 is an aqueous solution containing the monomer (NIPAM, 11.3% w/v), an initiator (APS, 1.13% w/v), and the cross-linker N,N'-methylenebisacryamide (BIS, 0.77% w/v). Fluid 2 is a kerosene solution containing the surfactant polyglycerol polyricinoleate (PGPR 90, 8% w/v) and fluid 3 is a kerosene solution containing both the surfactant (PGPR 90, 8% w/v) and an accelerator (TEMED, 10% v/v). The surfactant in fluids 2 and 3 is used to prevent drop coalescence. The solutions are supplied to the microfluidic device using polyethylene tubing attached to syringes operated by syringe pumps. A high-speed camera is used to record the drop formation processes.

An important feature of this microfluidic approach is that both the drop sizes and the size monodispersity can be precisely controlled for the monomer emulsion drops forming inside the channel; for this, we have designed a thin and long-throat channel (Figure 3.1c). In the experiments, monodisperse monomer droplets with sizes ranging from ~ 3 to $\sim 10\,\mu m$ by radius can be generated, as shown in Figure 3.2. When the viscosities of the fluids and the geometry of the device are given, the flow

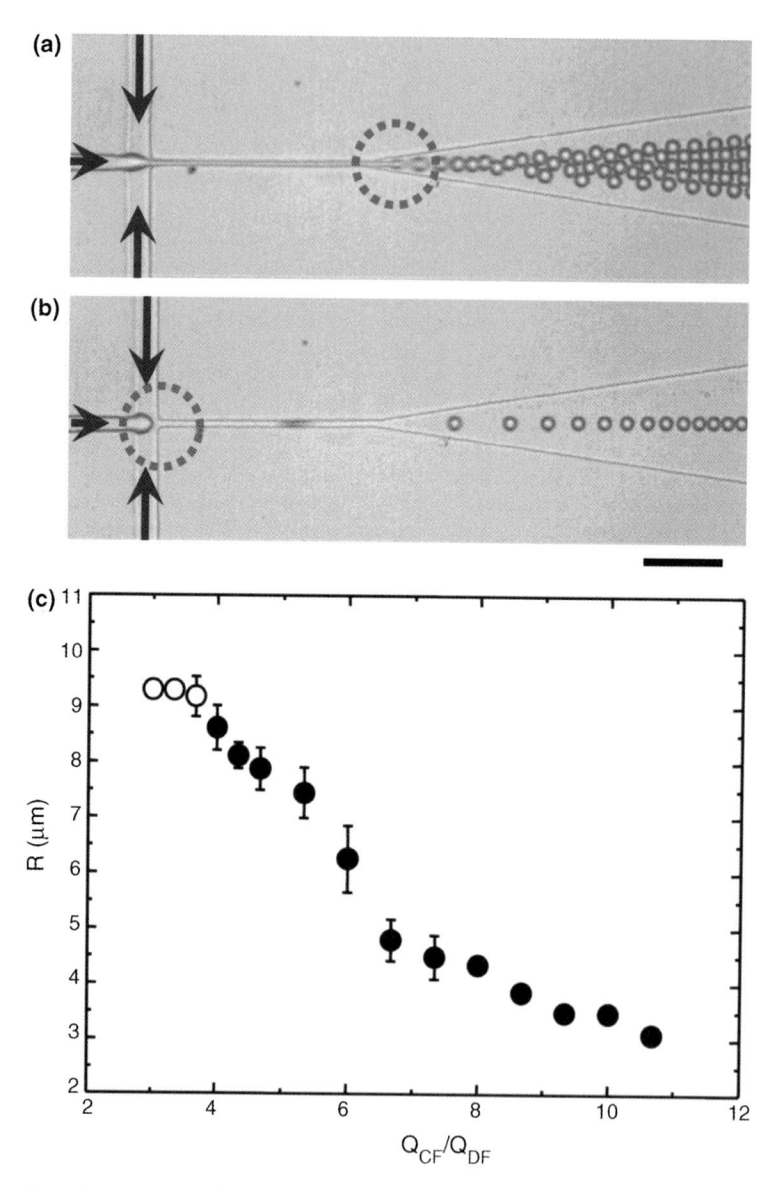

Figure 3.2 (a) Optical microscope image of the drop formation at a low flow rate ratio, Q_{CF}/Q_{DF}. (b) Optical microscope image of the drop formation at a high Q_{CF}/Q_{DF}. The scale bar is 25 μm. (c) Dependence of drop radii on Q_{CF}/Q_{DF}. Open circles correspond to the case of Figure 3.2a; and close circles correspond to the case of Figure 3.2b.

rate ratio (Q_{CF}/Q_{DF}), which is the flow rate of the continuous fluid (Q_{CF}) relative to the dispersed fluid (Q_{DF}), controls the resulting drop sizes [53, 54]. When Q_{CF}/Q_{DF} is less than ~4, the drop breakup occurs at the end of the throat channel (Figure 3.2a) and the resulting drop sizes are larger than the cross section of the throat channel by a factor of approximately two. In this case, the droplet size is still highly monodisperse, which is attributed to the fact that the solid wall of the throat channel makes the flow inside the channel stable. As Q_{CF}/Q_{DF} is increased, smaller drop sizes can be created. Moreover, when Q_{CF}/Q_{DF} is higher than ~6, the external flow creates a larger drag force, thus forming drops that are smaller than the dimension of the throat channel (Figure 3.2b). In our study, the drop size is controllable by adjusting Q_{CF}/Q_{DF} (Figure 3.2c). Owing to the presence of the surfactant, in this study, the coalescence of the drops is prevented as they flow downstream. Monomer drops, formed at either entrance or exit of the throat channel, are highly monodisperse with size standard deviations of less than ~2%.

After the polymerization, microgels are completely washed with a solvent, isopropanol, to remove oil from the microgel surface by repeated centrifugation. The washed microgel particles are subsequently dispersed in pure water. To test the thermosensitive volume-phase transition behavior, the PNIPAM microgels, dispersed in pure water, are put into a transparent sealed holder on top of a glass slide and placed on a microscope heating/cooling stage. The actual temperature inside the sample holder is measured by an infrared thermometer. A digital camera is used to record the thermo-responsivity. We have confirmed that the PNIPAM microgels prepared using this fabrication method have a homogeneous internal structure (Figure 3.3a) and excellent thermo-sensitivity; at 22 °C, the diameter of the PNIPAM microgel is approximately 2.7 times larger than that at 40 °C (Figure 3.3). The phase transition behaviors of the microgel particles in response to temperature changes are satisfactorily reversible and show sharp transitions at the LCST, as shown in Figure 3.3c.

We have demonstrated a microfluidic approach that is highly useful for simply and efficiently fabricating monodisperse thermosensitive microgels on a microfluidic chip. Moreover, we have shown that the approach can also be used to prepare other polymeric microspheres, as long as the polymerization can be initiated by a redox reaction. Furthermore, this method can also be used to prepare multifunctional microgels with small variations in size by incorporating functional substances into the monomer solution [50].

3.3
Monodisperse Thermosensitive Microgels Fabricated in a Capillary Microfluidic Device

Recently, a microcapillary microfluidic device has been developed with a flow-focusing geometry [55], which makes it both easily controllable and easily scalable. Using capillary-based microfluidic devices is advantageous since they have both nonpermeability and resistance to solvents. By using the same synthesis principle as mentioned above for the fabrication in the PDMS chip, monodisperse thermo-

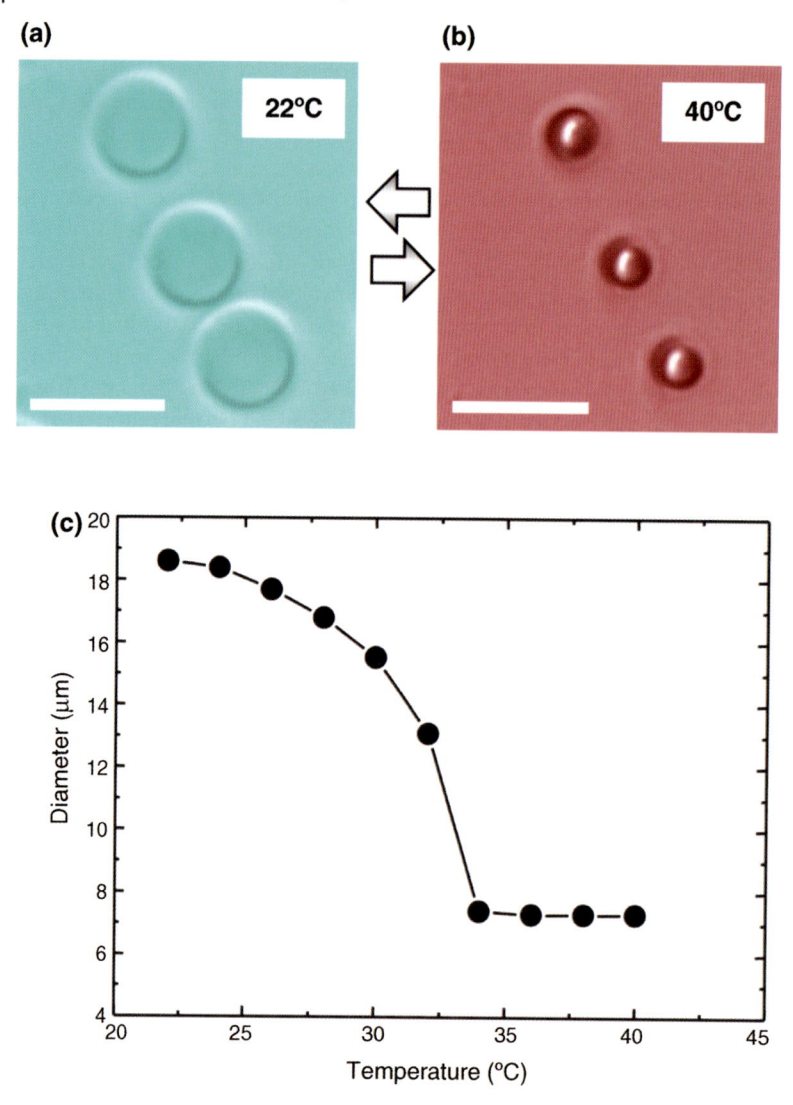

Figure 3.3 (a and b) Optical microscope images of the microgels prepared by the PDMS microfluidic chip in pure water at 22 °C and 40 °C, respectively. The scale bar is 25 μm. (c) Temperature dependence of the diameter of PNIPAM microgels in water.

responsive microgels have been fabricated in a capillary microfluidic device. The device consists of cylindrical glass capillary tubes nested within square glass capillary tubes, as shown in Figure 3.4. It features a coaxial, coflow geometry ensured by matching the outside diameters of the cylindrical tubes to the inner dimensions of the square tubes. An aqueous solution containing the monomer NIPAM, cross-linker BIS, and initiator APS (fluid A) is pumped into the left end of the left square tube. The continuous phase (fluid B), kerosene, containing a surfactant PGPR 90, is pumped

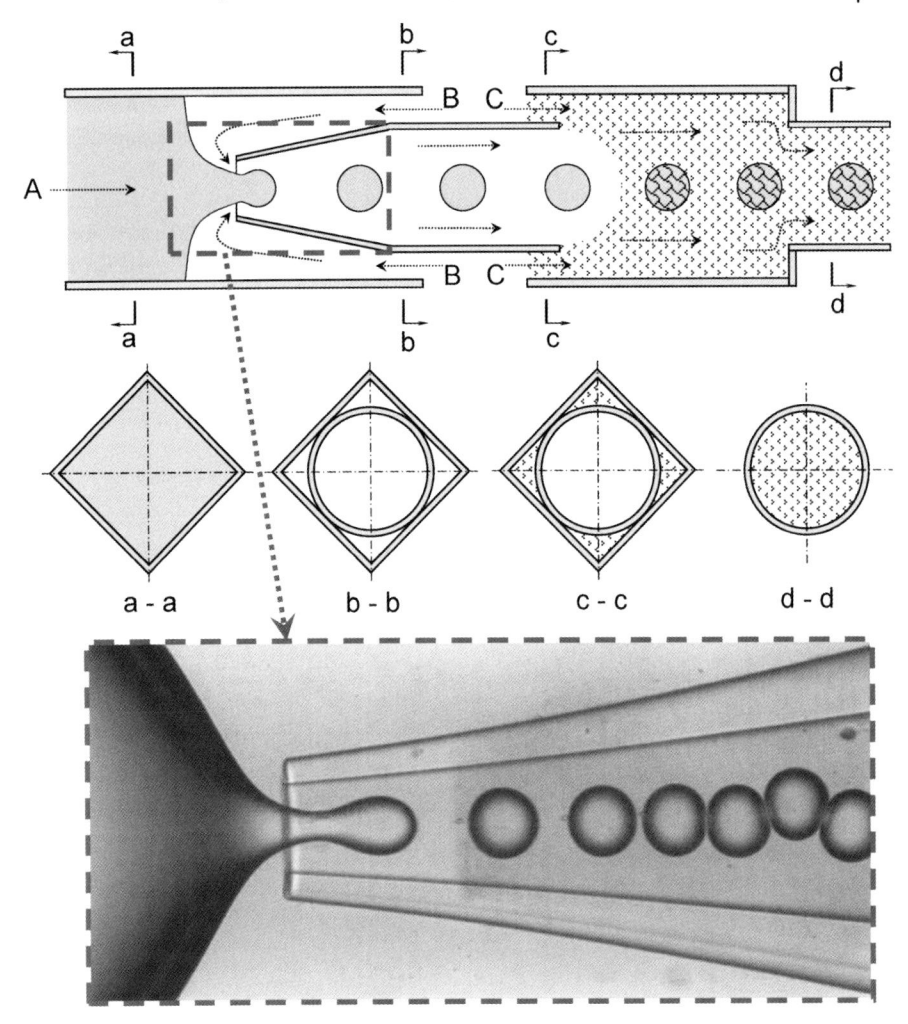

Figure 3.4 Schematic illustration of the capillary microfluidic device for fabricating monodisperse microgels. Fluid A is an aqueous solution containing the monomer, initiator, and cross-linker; fluid B is an oil phase containing a surfactant; and fluid C is a mixture of an oil phase and a reaction accelerator.

through the outer coaxial region between the left square tube and a tapered round microcapillary tube. The aqueous phase breaks into droplets at the entrance orifice of the tapered tube, forming monodisperse emulsion drops in the tube. The accelerator TEMED, dissolved in the continuous phase (fluid C), is pumped through the outer coaxial region between the right square tube and the right end of the tapered round tube. When the accelerator meets with the initiator, it starts the redox reaction that polymerizes the monomers, thus solidifying monomer drops into microgels.

Thermosensitive microgel particles have thus been fabricated using the same chemical recipe and characterization method as used for the previously described

60 | *3 New Functional Microgels from Microfluidics*

microgels prepared on a PDMS chip. We have found that the sizes of the PNIPAM microgels are tunable by changing the flow rates and the microgels have homogeneous structures (Figure 3.5a) and excellent thermosensitivity. At 20 °C, the diameter of the PNIPAM microgels is about 2.3 times larger than that at 40 °C (Figure 3.5). The

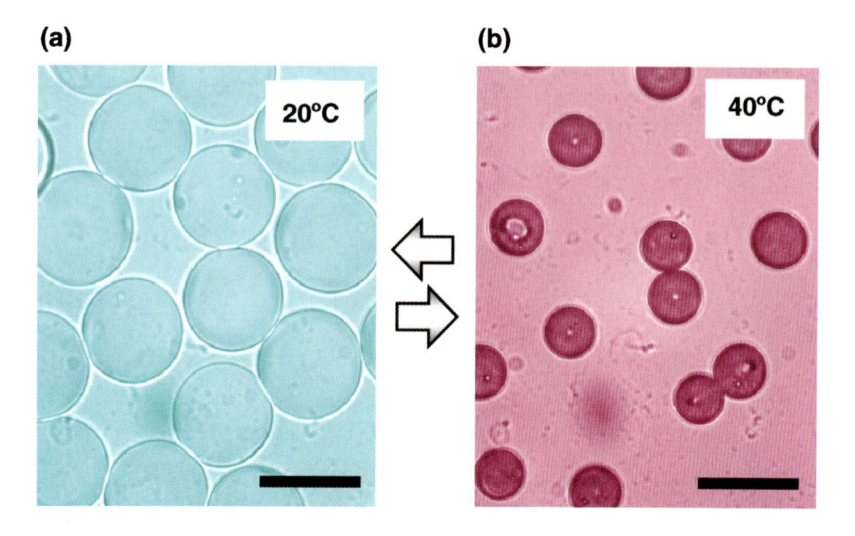

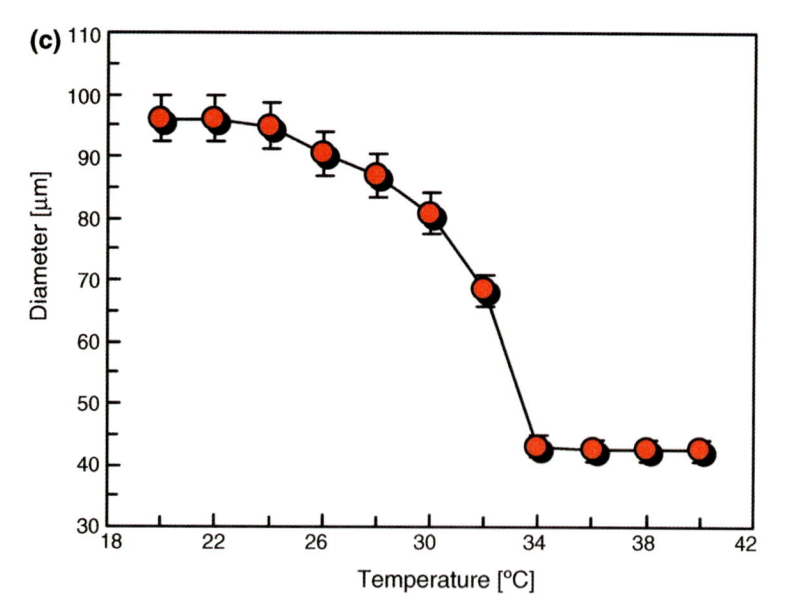

Figure 3.5 (a and b) Optical microscope images of the microgels prepared by the capillary microfluidic device in pure water at 20 and 40 °C, respectively. The scale bar is 100 μm. (c) Temperature dependence of the diameter of PNIPAM microgels in water.

3.3 Monodisperse Thermosensitive Microgels Fabricated in a Capillary Microfluidic Device

temperature-dependent diameter change of the microgels shows a sharp transition close to the LCST, as shown in Figure 3.5c, and is highly reversible.

To achieve a better control of the microgel size, we examine the physical mechanism of premicrogel drop formation. The premicrogel drops form within the collection tube at approximately one tube diameter downstream from the entrance [56–58]. When drops are formed close to the entrance of the tube, the drop size is controlled by the ratio of the flow rates of the combined inner and middle fluids (MFs) to the outer fluid (OF), which is given by

$$Q_{sum}/Q_{OF} = \pi R_{thread}^2 / \left(\pi R_{orifice}^2 - \pi R_{thread}^2\right) \quad (3.1)$$

where Q_{sum} is the sum of the inner and middle fluid flow rates, R_{thread} is the radius of the fluid thread that breaks into drops, and $R_{orifice}$ is the radius of the collection tube, where the drops are formed. This equation is valid for plug flow (Figure 3.6a), which is a reasonable assumption given the proximity of drop formation to the entrance of the collection tube. The experimentally measured diameters of different threads and the diameters of the corresponding drops that pinch from these threads are shown as the open and closed symbols, respectively, in Figure 3.6b. By solving for $R_{thread}/R_{orifice}$ in Equation 3.1 and plotting this ratio as a function of Q_{sum}/Q_{OF}, we can quantitatively predict the experimentally measured values, as shown by the dashed line. When a fluid thread breaks, the diameter of the resulting drop is proportional to the diameter of the thread. For a viscosity ratio of $\eta_{MF}/\eta_{OF} \approx 0.1$, the drop diameter is approximately twice the diameter of the thread [59]. By multiplying the predicted thread diameter by a factor of 2, we find very good agreement between the model and the data. These simple physical arguments highlight the versatility

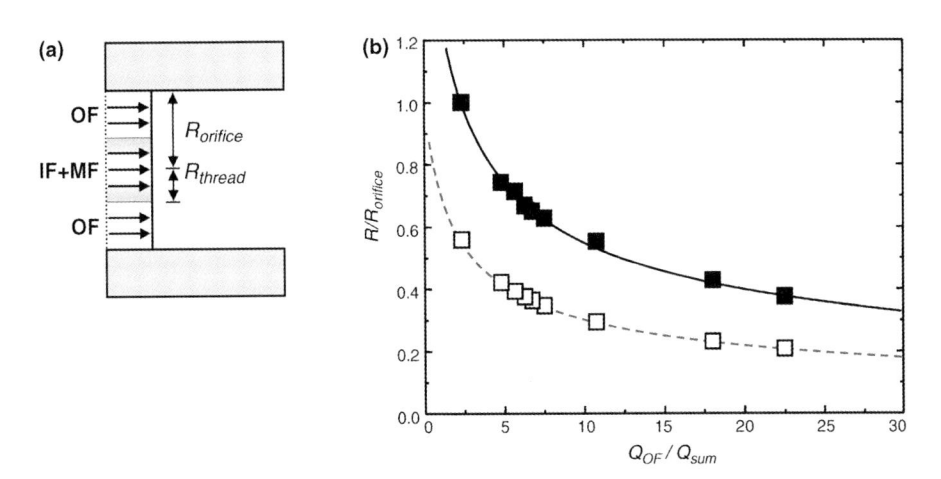

Figure 3.6 (a) The flat velocity profile of the flow as it enters the capillary tube. (b) Dependence of thread, R_{thread}, and drop, R_{drop}, radii on the scaled flow rate, Q_{OF}/Q_{sum}. The open symbols represent the R_{thread} for different premicrogel liquids: the premicrogel liquid only. The dashed line represents the predicted R_{thread}. R_{drop} are represented with solid symbols. The solid line represents the predicted R_{drop}.

of this method in generating microgels of different sizes without losing monodispersity of the suspension; typical standard deviations in size are less than 2.5%.

3.4
Monodisperse Thermosensitive Microgels with Tunable Volume-Phase Transition Kinetics

The kinetics of swelling and deswelling of stimuli-responsive polymeric hydrogels are typically governed by diffusion-limited transport of water in and out of the polymeric networks [60]. Hence, if a direct way to control this transport behavior of water molecules is developed, the volume-phase transition kinetics of the microgels can be precisely tuned. We have accomplished this by fabricating microgel particles with different types of voids [49]. A microgel internal structure containing voids offers less resistance to water transport compared to a voidless core. Thus, varying the size and number of voids of a microgel enables direct control over swelling and shrinking dynamics.

Microgels with internal voids can be prepared by using a two-step method [49]. First, monodisperse PNIPAM microgels containing different number/sizes of solid polystyrene microspheres are synthesized in a capillary microfluidic device, as shown in Figure 3.4. Subsequently, the embedded microspheres are chemically dissolved out of the gel phase, eventually forming voids inside the microgels. The experimental recipe is the same as that mentioned above. In this case, monodisperse polystyrene microspheres are dispersed in fluid A. Once the polymerization process is complete, the PNIPAM microgels are washed with isopropanol and subsequently immersed in xylene that is a good solvent for polystyrene, thereby leaving behind holes in the microgels. The microgels with internal voids are again rinsed with isopropanol and are finally dispersed in pure water.

Microgels containing different numbers of 25 μm diameter polystyrene beads and microgels with internal voids formed by dissolving the polystyrene beads in similar microgels are shown in Figure 3.7a. The number of voids in a microgel is determined by the number of encapsulated polystyrene beads, while their size is determined by how much they swell after dissolving the beads. The size of a void is larger than the beads and depends on the cross-linking density of the gel network. The precise circular periphery of the voids and unchanged size of the microgels also suggest that the chemical dissolution of the beads does not affect the structural integrity of the surrounding microgel structure. The shrinking and swelling of these microgels with internal voids are compared with the behavior of voidless microgels of the same size, as shown in Figure 3.7b and c. When the temperature is increased from 23 to 47 °C, the microgels with four voids shrink faster than all other samples with less number of voids, whereas the response times of voidless microgels are the slowest. The differences are more distinct during swelling. The inset in Figure 3.7c clearly reveals that the microgel with many voids requires shorter time for the inception of swelling. This suggests that the volume-phase transition kinetics of microgels can be tuned simply by varying the number of internal voids, which is achieved by employing microfluidic techniques [48].

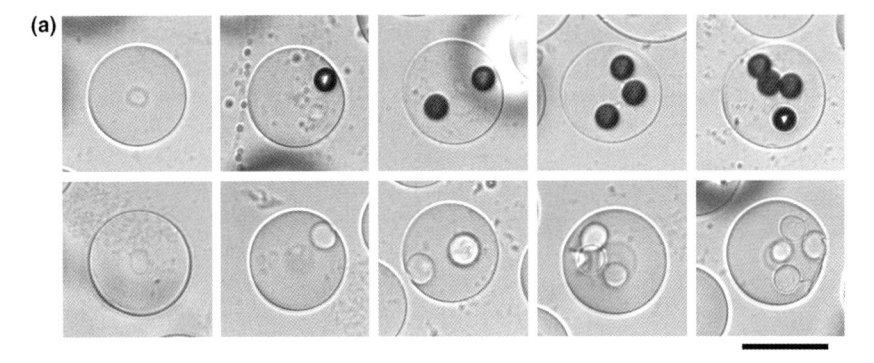

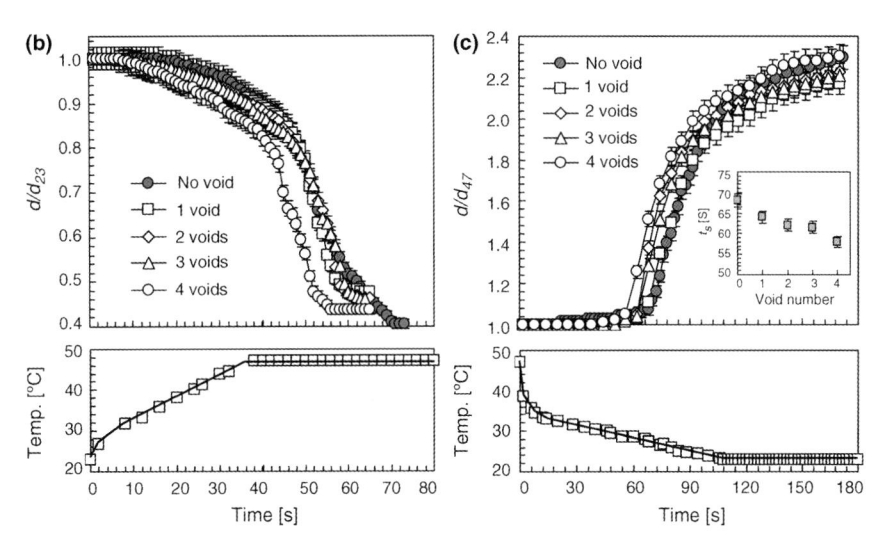

Figure 3.7 (a) Microgels with different number of embedded polystyrene beads (*top*), and microgels with internal voids formed by dissolving the embedded beads from such microgels (*bottom*). Scale bar = 100 μm. (b and c) Effect of the number of voids on the volume shrinking and swelling behavior of microgels. The average diameter of the voids is 25 μm. The samples were (b) heated from 23 to 47 °C and (c) cooled from 47 to 23 °C; d_{23} and d_{47} are the microgel diameters at 23 and 47 °C, respectively, and t_s is the time elapsed before the microgels begin to swell [49].

3.5
Monodisperse Thermosensitive Microgels with Core–Shell Structures Containing Functional Materials

With a capillary microfluidic device developed for generating double emulsions [55], spherical microgels are prepared in a single step, which allows us to freely incorporate functional materials into the polymeric network. The schematic illustration of

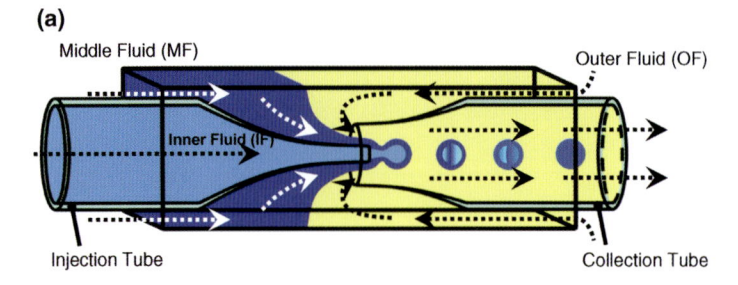

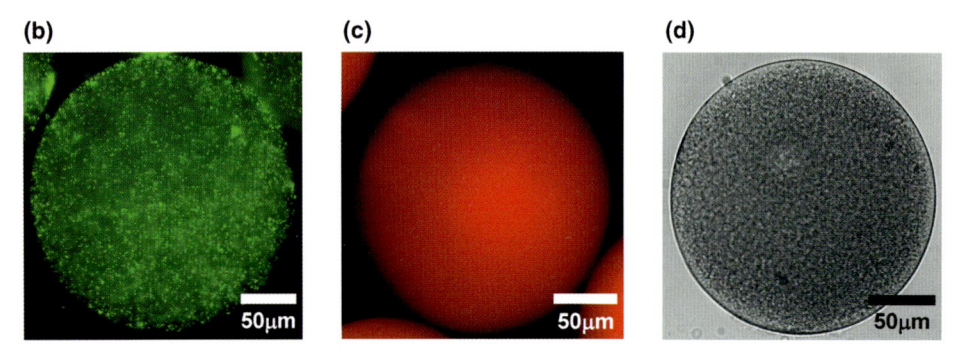

Figure 3.8 (a) Schematic geometry of a capillary microfluidic device for synthesizing functional microgels. The inner fluid is the aqueous initiator solution, the middle fluid is the monomer/accelerator aqueous solution, and the outer fluid is the continuous oil phase. (b) Fluorescence microscope image of a microgel containing fluorescent PS particles. (c) Fluorescence microscope image of a microgel containing quantum dots. (d) Bright field microscope image of a microgel containing magnetic nanoparticles.

the capillary microfluidic device is shown in Figure 3.8a, in which the two internal cylindrical tubes are served as injection and collection tubes and coaxially aligned inside a square tube. The alignment of the capillary tubes is the same as that illustrated in Figure 3.4. Silicon oil is used as the outer or continuous-phase liquid. The MF consists of an aqueous monomer solution (NIPAM, 15.5% w/v), a cross-linker (BIS, 1.5%w/v), a reaction accelerator (TEMED, 2 vol%), and two comonomers – 2-(methacryloyloxy) ethyl trimethyl ammonium chloride (METAC, 2 vol%) and allylamine (1 vol%). METAC is added to increase the phase transition temperature of PNIPAM, thereby facilitating a homogeneous polymerization at room temperature. Allylamine adds amine groups to the network, which can subsequently be labeled with dyes after the formation of the microgel particles. The inner fluid is an aqueous solution containing the initiator (APS, 3% w/v). Quantum dots, magnetic nanoparticles, and polymer microparticles are here used as examples of the materials that can be added to provide specific chemical, physical, or mechanical properties to the original microgels [50]. The microscopic images of microgels containing $d = 1$ μm fluorescent polystyrene particles, $d = 19$ nm quantum dots, and $d=10$ nm magnetic particles are shown in Figure 3.8b, c, and d, respectively [50]. We demonstrate that this microfluidic approach allows the engineering of microgels that encapsulate or immobilize colloidal materials within their network, thereby physically locking-in these particles and conferring unique properties on the original microgels. Unlike other methods that yield microgels with less sensitivity to stimuli, these complex microgels exhibit the same physical response to temperature changes as the original microgels, irrespective of the material incorporated into the gel matrix, which is due to the fact that the added materials are physically trapped within the gel network rather than being chemically linked through some specific interactions.

Using the same device, monodisperse microgels with oil drops have also been fabricated. This was achieved simply by changing the inner fluid (IF) from an aqueous initiator solution to an oil solution containing the accelerator, TEMED, and adding the initiator, APS, to the MF instead of to the IF, as shown in Figure 3.9a. The core–shell microgels result from the double emulsion structures. Using this approach allows us to generate a novel microgel shell that is suspended in water and has an aqueous core, as shown in Figure 3.9b. In this case, the initiator is located in the middle fluid while the accelerator is dissolved in the inner oil. Upon forming the double emulsion drops, the accelerator diffuses from the internal oil droplet into the surrounding aqueous monomer solution layer, initiating the polymerization. After collecting the microgel shell particles, we extract the oil by washing with an excess of isopropyl alcohol and finally transfer the particles into deionized water. Once these novel microgel shells are generated, we probe their thermosensitivity by measuring changes in the inner and outer radii as a function of temperature, as shown in Figure 3.9c. Both radii decrease with increasing temperature and show no signs of hysteresis after repeatedly cooling and heating the sample; however, while the core reaches a minimum volume at about 50 °C, the overall volume of the shell still decreases. This is more easily seen by plotting the volume of the core scaled by the overall particle volume as shown in Figure 3.9d; this ratio sharply increases at approximately 50 °C, indicating that above this temperature the shell itself shrinks while the core volume remains unchanged. We believe that the microgel shell becomes more hydrophobic and denser as the temperature increases, which reduces the permeability to water molecules and ultimately limits the degree to which the volume of the core can change.

3.6
Monodisperse Thermosensitive Microgels with Multiphase Complex Structures

Microgels with multiphase complex internal structures can encapsulate both water-soluble and oil-soluble substances, and can thus be important in a wide range of applications, such as stepwise drug release in pharmaceuticals and multiphase transitions of complex formulations in cosmetics. It is very difficult, nearly impossible, to fabricate such structured microgels using the available techniques. However, recent advances in microfluidic techniques have provided opportunities

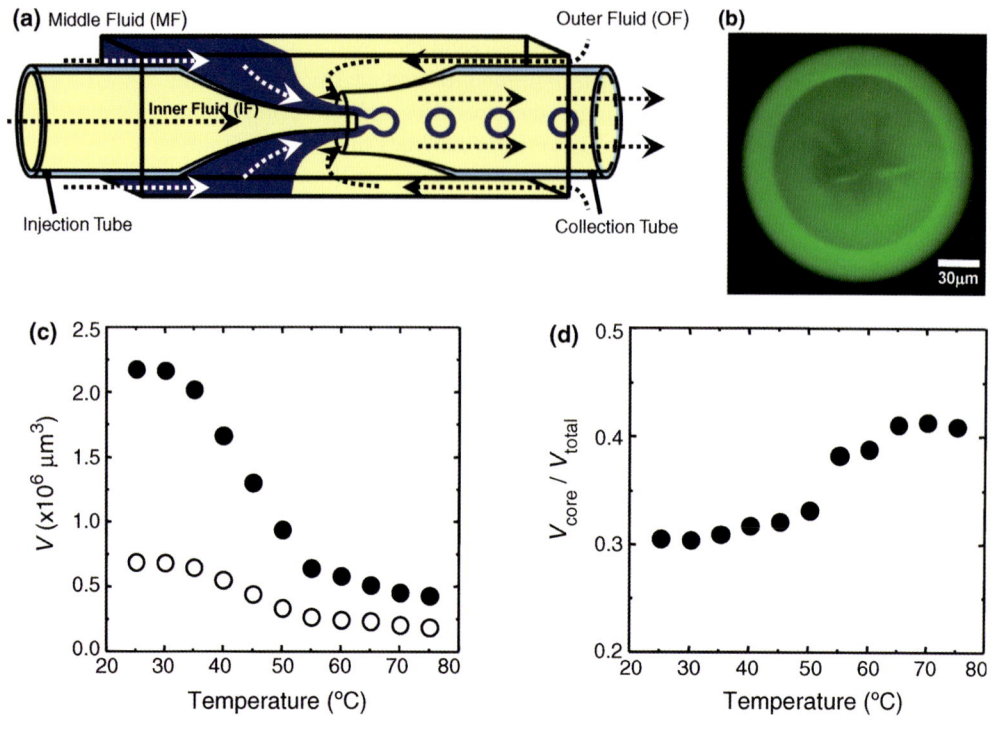

Figure 3.9 (a) Schematic geometry of a capillary microfluidic device for synthesizing microgel shells. IF is the oil solution containing accelerator, MF is the monomer/initiator aqueous solution, and OF is a continuous oil phase. (b) A fluorescence microscope image of an FITC-labeled microgel shell that is prepared from a premicrogel double emulsion that contained a single silicon oil droplet. (c) Volume of the overall core–shell PNIPAM microgel (●) and its internal void (○) as a function of temperature. (d) Volume of the internal void scaled by the volume of the whole microgel as a function of temperature.

to generate well-controlled multiple emulsions [48]. The high degree of control and scalability offered by this method makes it a flexible and promising route for engineering emulsions and microcapsules with multiphase structures. By using a capillary microfluidic device, shown in Figure 3.10a, well controllable triple emulsions have been generated, from which monodisperse thermosensitive microgels with multiphase complex structures have been synthesized. Uniform monodisperse triple emulsions are generated in three emulsification steps: Droplets of the innermost fluid are emulsified in the first stage of the device by coaxial flow of the middle fluid (I) and these emulsion drops are subsequently emulsified in the second stage through coaxial flow of the middle fluid (II). The double emulsion droplets are then emulsified in the third stage by coaxial flow of the outermost fluid, which is injected in the outer stream through the square capillary. In all the emulsification steps, droplets form immediately at the exit of the tapered

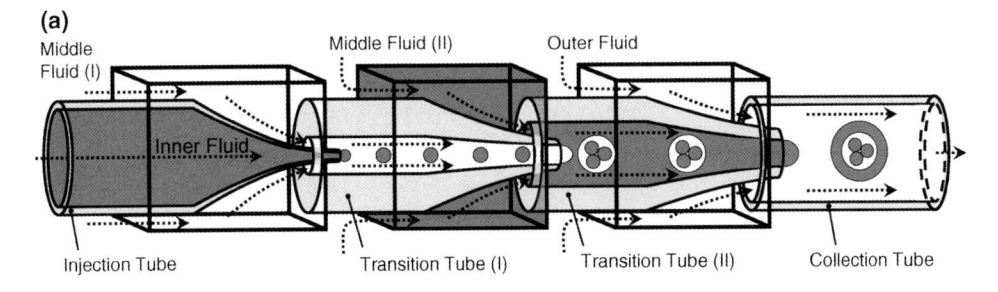

3.6 Monodisperse Thermosensitive Microgels with Multiphase Complex Structures | 67

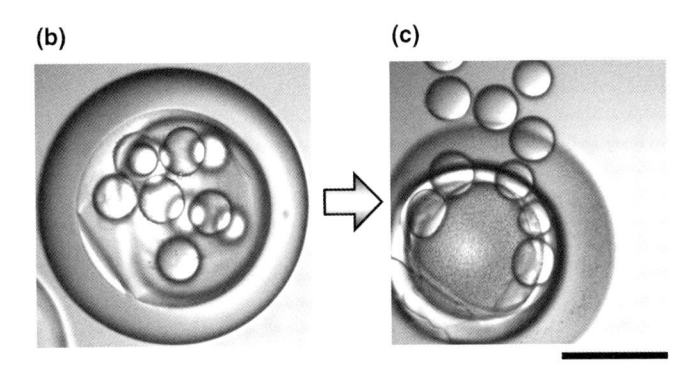

Figure 3.10 (a) Schematic diagram of the capillary microfluidic device for generating controllable monodisperse triple emulsions. (b) Optical micrograph of a microcapsule with a shell comprised of a thermosensitive hydrogel, containing aqueous droplets dispersed in oil. The capsule was generated from a triple emulsion. (c) Optical micrograph showing the forced expulsion of the oil and water droplets contained within the microcapsule when the temperature is increased from 25 to 50 °C. The extra layer, surrounding the microcapsule in (c), is water that is squeezed out from the hydrogel shell as it shrinks. The scale bar is 200 μm [48].

capillary; this dripping mechanism [61] ensures the formation of highly monodisperse drops.

To prepare a hydrogel microcapsule with a thermosensitive PNIPAM shell and multiphase internal content, a water-in-oil-in-water-in-oil (W/O/W/O) emulsion system has been used. Since each fluid layer is sandwiched between two immiscible fluids, a polymerization reaction can be performed in a specific layer. The outermost fluid is 0.1 mPa s PDMS oil containing 2 wt% Dow Corning 749 fluid as a surfactant. The outer middle fluid (II) is an aqueous solution containing 10 wt% glycerol, 2% (w/v) PVA, 11.3% (w/v) of the monomer NIPAM, 0.8% (w/v) of the comonomer sodium acrylate, 0.77% (w/v) of the cross-linker BIS, and 0.6% (w/v) of the initiator APS. The inner middle fluid (I) is 0.01 mPa s PDMS oil containing 5% wt Dow Corning 749 fluid and 8% (v/v) of the accelerator TEMED. The innermost fluid is an aqueous solution containing 10 wt% glycerol and 2%

(w/v) poly(vinyl alcohol) (PVA, 87–89% hydrolyzed). The accelerator in the middle fluid (I) diffuses into the outer aqueous shell and speeds up the polymerization of the hydrogel.

The fabricated microcapsule consisting of a shell of thermosensitive hydrogel that encapsulates an oil core containing several water droplets at 25 °C is shown in Figure 3.10b [48]. Upon heating from 25 to 50°, the thermosensitive hydrogel shell rapidly shrinks by expelling water. Although the shrinking increases the pressure within the hydrogel shell, the incompressibility of both the oil core and the water droplets causes the water droplets to remain stable until eventually a temperature-induced breakage of the hydrogel shell results in the spontaneous, pulsed release of the water droplets into the continuous oil phase, as shown in Figure 3.10c. This structure has a Trojan-horse-like behavior, protecting the innermost water droplets in the hydrogel shell until their temperature-induced release. This demonstrates the utility of this microfluidic technique to generate highly controlled microgel capsules with multiple internal volumes that remain separate from each other; it also highlights the potential of this microfluidic device to create highly engineered structures for controlled release of biologically active ingredients.

3.7
Conclusions

This chapter introduces some recent work focused on the fabrication of monodisperse thermosensitive microgels and the engineering of complex internal structures in these microgels, using microfluidics approaches. The unique utility of microfluidics-based methods in the area of synthesis of microgels is demonstrated. The microfluidic techniques provide simple and efficient methods not only for preparing monodisperse microgels but also for creating microgels with novel internal structures, for example, microgel capsules with controlled multiphase cores. The flexibility in tuning the size and monodispersity of microgels and the controllability over their internal structures, which are offered by the above-mentioned microfluidic method, allow promising new applications of microgels in the future.

References

1 Pelton, R. (2000) *Adv. Colloid Interface Sci.*, **85**, 1.
2 Jones, C.D. and Lyon, L.A. (2003) *Macromolecules*, **36**, 1988.
3 Gan, D.J. and Lyon, L.A. (2002) *Macromolecules*, **35**, 9634.
4 Gan, D.J. and Lyon, L.A. (2001) *J. Am. Chem. Soc.*, **123**, 8203.
5 Gan, D.J. and Lyon, L.A. (2001) *J. Am. Chem. Soc.*, **123**, 7511.
6 Kwon, I.C., Bae, Y.H., and Kim, S.W. (1991) *Nature*, **354**, 291.
7 Jeong, B., Bae, Y.H., Lee, D.S., and Kim, S.W. (1997) *Nature*, **388**, 860.
8 Ichikawa, H. and Fukumori, Y. (2000) *J. Control. Release*, **63**, 107.

9 Leobandung, W., Ichikawa, H., Fukumori, Y., and Peppas, N.A. (2003) *J. Appl. Polym. Sci.*, **87**, 1678.
10 Murthy, N., Thng, Y.X., Schuck, S., Xu, M.C., and Frechet, J.M.J. (2002) *J. Am. Chem. Soc.*, **124**, 12398.
11 Vihola, H., Laukkanen, A., Hirvonen, J., and Tenhu European, H. (2002) *J. Pharm. Sci.*, **16**, 69.
12 Kawaguchi, H. and Fujimoto, K. (1998) *Bioseparation*, **7**, 253.
13 Kondo, A., Kaneko, T., and Higashitani, K. (1994) *Biotechnol. Bioeng.*, **44**, 1.
14 Hu, Z.B., Chen, Y.Y., Wang, C.J., Zheng, Y.D., and Li, Y. (1998) *Nature*, **393**, 149.
15 Panchapakesan, B., DeVoe, D.L., Widmaier, M.R., Cavicchi, R., and Semancik, S. (2001) *Nanotechnology*, **12**, 336.
16 van der Linden, H., Herber, S., Olthuis, W., and Bergveld, P. (2002) *Sensor. Mater.*, **14**, 129.
17 Bergbreiter, D.E., Case, B.L., Liu, Y.S., and Caraway, J.W. (1998) *Macromolecules*, **31**, 6053.
18 Guiseppi-Elie, A., Sheppard, N.F., Brahim, S., and Narinesingh, D. (2001) *Biotechnol. Bioeng.*, **75**, 475.
19 Debord, J.D., Eustis, S., Debord, S.B., Lofye, M.T., and Lyon, L.A. (2002) *Adv. Mater.*, **14**, 658.
20 Ilmain, F., Tanaka, T., and Kokufuta, E. (1991) *Nature*, **349**, 400.
21 Matsuoka, H., Fujimoto, K., and Kawaguchi, H. (1999) *Polym. J.*, **31**, 1139.
22 Zhu, P.W. and Napper, D.H. (2000) *Langmuir*, **16**, 8543.
23 Gao, J. and Hu, Z.B. (2002) *Langmuir*, **18**, 1360.
24 Zha, L.S., Zhang, Y., Yang, W.L., and Fu, S.K. (2002) *Adv. Mater.*, **14**, 1090.
25 Xiao, X.-C., Chu, L.-Y., Chen, W.-M., Wang, S., and Xie, R. (2004) *Langmuir*, **20**, 5247.
26 Xiao, X.-C., Chu, L.-Y., Chen, W.-M., Wang, S., and Li, Y. (2003) *Adv. Funct. Mater.*, **13**, 847.
27 Shiga, K., Muramatsu, N., and Kondo, T. (1996) *J. Pharm. Pharmacol.*, **48**, 891.
28 Chu, L.-Y., Park, S.-H., Yamaguchi, T., and Nakao, S. (2002) *Langmuir*, **18**, 1856.
29 Ikkai, F., Iwamoto, S., Adachi, E., and Nakajima, M. (2005) *Colloid Polym. Sci.*, **283**, 1149.
30 Chu, L.-Y., Xie, R., Zhu, J.-H., Chen, W.-M., Yamaguchi, T., and Nakao, S. (2003) *J. Colloid Interface Sci.*, **265**, 187.
31 Whitesides, G.M. and Stroock, A.D. (2001) *Phys. Today*, **54**, 42.
32 Whitesides, G.M. (2006) *Nature*, **442**, 368.
33 Takeuchi, S., Garstecki, P., Weibel, D.B., and Whitesides, G.M. (2005) *Adv. Mater.*, **17**, 1067.
34 Xu, S.Q., Nie, Z.H., Seo, M., Lewis, P., Kumacheva, E., Stone, H.A., Garstecki, P., Weibel, D.B., Gitlin, I., and Whitesides, G.M. (2005) *Angew. Chem., Int. Ed.*, **44**, 724.
35 Zhang, H., Tumarkin, E., Peerani, R., Nie, Z., Sullan, R.M.A., Walker, G.C., and Kumacheva, E. (2006) *J. Am. Chem. Soc.*, **128**, 12205.
36 Nie, Z.H., Xu, S.Q., Seo, M., Lewis, P.C., and Kumacheva, E. (2005) *J. Am. Chem. Soc.*, **127**, 8058.
37 Abraham, S., Jeong, E.H., Arakawa, T., Shoji, S., Kim, K.C., Kim, I., and Go, J.S. (2006) *Lab. Chip*, **6**, 752.
38 De Geest, B.G., Urbanski, J.P., Thorsen, T., Demeester, J., and De Smedt, S.C. (2005) *Langmuir*, **21**, 10275.
39 Dendukuri, D., Tsoi, K., Hatton, T.A., and Doyle, P.S. (2005) *Langmuir*, **21**, 2113.
40 Huang, K.S., Lai, T.H., and Lin, Y.C. (2006) *Lab. Chip*, **6**, 954.
41 Kubo, A., Shinmori, H., and Takeuchi, T. (2006) *Chem. Lett. (Jpn)*, **35**, 588.
42 Liu, K., Ding, H.J., Liu, J., Chen, Y., and Zhao, X.Z. (2006) *Langmuir*, **22**, 9453.
43 Lewis, P.C., Graham, R.R., Nie, Z.H., Xu, S.Q., Seo, M., and Kumacheva, E. (2005) *Macromolecules*, **38**, 4536.
44 Nisisako, T., Torii, T., Takahashi, T., and Takizawa, Y. (2006) *Adv. Mater.*, **18**, 1152.
45 Shepherd, R.F., Conrad, J.C., Rhodes, S.K., Link, D.R., Marquez, M., Weitz, D.A., and Lewis, J.A. (2006) *Langmuir*, **22**, 8618.
46 Velev, O.D. and Bhatt, K.H. (2006) *Soft Matter*, **2**, 738.
47 Zourob, M., Mohr, S., Mayes, A.G., Macaskill, A., Perez-Moral, N., Fielden, P.R., and Goddard, N.J. (2006) *Lab. Chip*, **6**, 296.
48 Chu, L.-Y., Utada, A.S., Shah, R.K., Kim, J.-W., and Weitz, D.A. (2007) *Angew. Chem., Int. Ed.*, **46**, 8970.

49 Chu, L.-Y., Kim, J.-W., Shah, R.K., and Weitz, D.A. (2007) *Adv. Funct. Mater.*, **17**, 3499.
50 Kim, J.-W., Utada, A.S., Fernández-Nieves, A., Hu, Z.B., and Weitz, D.A. (2007) *Angew. Chem., Int. Ed.*, **46**, 1819.
51 Xia, Y.N. and Whitesides, G.M. (1998) *Angew. Chem., Int. Ed.*, **37**, 551.
52 Ju, X.J., Chu, L.Y., Zhu, X.L., Hu, L., Song, H., and Chen, W.M. (2006) *Smart Mater. Struct.*, **15**, 1767.
53 Garstecki, P., Gitlin, I., DiLuzio, W., Whitesides, G.M., Kumacheva, E., and Stone, H.A. (2004) *Appl. Phys. Lett.*, **85**, 2649.
54 Anna, S.L., Bontoux, N., and Stone, H.A. (2003) *Appl. Phys. Lett.*, **82**, 364.
55 Utada, A.S., Lorenceau, E., Link, D.R., Kaplan, P.D., Stone, H.A., and Weitz, D.A. (2005) *Science*, **308**, 537.
56 Ambravaneswaran, B., Subramani, H.J., Phillips, S.D., and Basaran, O.A. (2004) *Phys. Rev. Lett.*, **93**, 034501.
57 Clanet, C. and Lasheras, J.C. (1999) *J. Fluid Mech.*, **383**, 307.
58 Deen, W.M. (1998) *Analysis of Transport Phenomenon*, Oxford University Press, p. 265.
59 Tomotika, S. (1935) *Proc. Roy. Soc. Lond. A Mat.*, **150**, 322.
60 Matsuo, E.S. and Tanaka, T. (1988) *J. Chem. Phys.*, **89**, 1695.
61 Umbanhowar, P.B., Prasad, V., and Weitz, D.A. (2000) *Langmuir*, **16**, 347.

Part Two
Physical Properties of Microgel Particles

4
Swelling Thermodynamics of Microgel Particles

Benjamin Sierra-Martin, Juan Jose Lietor-Santos, Antonio Fernandez-Barbero, Toan T. Nguyen, and Alberto Fernandez-Nieves

4.1
Introduction

Microgels are cross-linked polymer networks of colloidal size immersed in a liquid. Perhaps, the most interesting feature exhibited by these systems is their volume phase transition: Microgels can reversibly swell and deswell in response to changes in external stimuli, which include temperature, pH, ionic strength, solvent composition, and presence of adsorbing and nonadsorbing polymers or surfactants. This response is thermodynamically identical to that of macroscopic gels, but remarkably, it occurs in much shorter timescales; an important fact for most industrial applications. It was Dusek and Patterson [1] who theoretically predicted, based on the work of Flory and Huggins with polymer solutions, the volume phase transition of macroscopic gels, and Tanaka *et al.* [2] who first observed it experimentally using poly(acrylamide) gels in acetone/water mixtures. Since these pioneering works, the volume phase transition of polymer gels has become a fundamental problem in polymer physics.

The volume phase transition of polymer gels results from the competition of attractive and repulsive interactions within the polymer network and between the polymer network and the solvent it is immersed in. There are usually four intermolecular forces involved in the swelling process: van der Waals, hydrophobic, hydrogen bonding, and electrostatic interactions [3, 4]. Van der Waals forces arise from interactions between the electric multipoles of the monomers forming the polymer network. Its strength is of the order of 10^{-2}–10^{-1} eV/atom [5] and depends on the solvent composition. Interestingly, the first volume transition observed in a polymer gel was induced via this interaction, by adding a nonpolar solvent to water in order to increase the strength of the van der Waals attraction between the polymer chains [2].

Hydrogen bonding is also a relevant interaction greatly affecting the swelling behavior of polymer gels. It is an attractive interaction, a van der Waals interaction between fixed dipoles. It has a strength of about 10^{-1} eV/atom and occurs when a hydrogen atom is close to an atom of high electronegativity, such as oxygen, nitrogen or fluor [5]. It is a directional interaction that requires an adequate orientation between the groups involved. For the case of polymer gels, due to the typical flexibility of polymer chains, this is usually not an impediment. In the presence of this interaction, microgels deswell with increasing temperature, as the hydrogen bonds are unable to resist thermal agitation and break.

In contrast to hydrogen bonding, hydrophobic interactions increase in strength with increasing temperature and drive particle deswelling. They occur between nonpolar parts of polymer chains in water or aqueous solvents and are related to the way these nonpolar groups, which are often referred to as the hydrophobic sites of the polymer, disrupt the preferred hydrogen bonding between water molecules [6]. It is a weak interaction, with typical strengths within the range 5.10^{-3}–5.10^{-2} eV/atom [5]; addition of organic solvents, such as alcohols, greatly modify its strength, as these molecules also affect the hydrogen bonding between water molecules [3, 5].

The electrostatic interaction is the relevant interaction when the polymer gel bears fixed charges along its backbone. The presence of this charge, either positive or negative, causes an imbalance between the ion distributions inside and outside the gel, thereby causing it to swell or deswell. In addition, there are direct Coulombic interactions between the charged groups and effects due to ionic and fixed charge correlations. However, the dominant contribution to the volume phase transition arises from the effect of the ions through the osmotic pressure they exert. In the presence of multivalent counterions, correlations might also be relevant, but no systematic investigations on how this effect affects microgel swelling are available at this stage.

Microgel particles are subjected to the same interactions as that of macroscopic gels. The single-particle behavior is thus controlled by the same physical mechanisms. However, microgels are always present in the form of particle suspensions and there are interactions between the particles, which can also contribute to their swelling behavior. Among most microgels, those based on poly-N-isopropylacrylamide (PNIPAM) have received most attention; perhaps because they are thermoresponsive and temperature changes are usually easy and convenient to achieve. Since Pelton and Chibante [7] published the synthesis of the first PNIPAM-based microgel, there has been considerable number of reviews related to this kind of systems [8–14]. PNIPAM microgels show a remarkable volume transition in response to temperature changes, as shown in Figure 4.1. Consistent with the coil-to-globule transition of PNIPAM polymers [15, 16], these microgels are characterized by a lower critical solution temperature (LCST) of approximately 32 °C [9, 10, 17]. Below this temperature, the microgel is swollen and contains a large amount of water. Above this temperature, the microgel is deswollen and has a much smaller water content. The dominant interaction for this volume transition is the hydrophobic interaction [3, 18, 19]. At low temperatures, the water molecules reorganize around the alkyl groups of the polymer, which are its hydrophobic sites, and hydrogen bond to the amide groups of PNIPAM [20, 21]. As the temperature is increased, the extent of water–water and water–monomer hydrogen bonding diminishes eventually inducing the collapse of the particle [22].

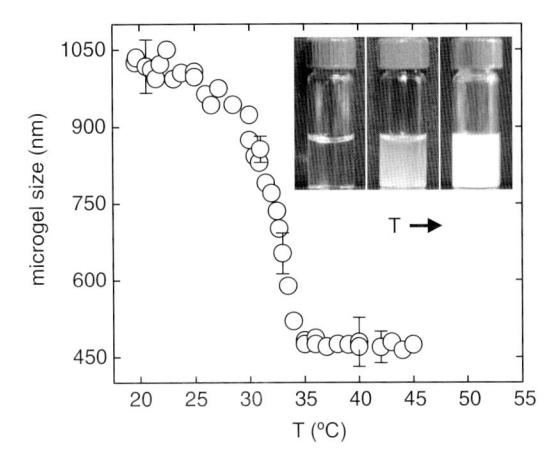

Figure 4.1 Deswelling of PNIPAM microgels as a function of temperature. The images show how the optical properties of the suspension change from transparent (left), when the particles are highly swollen, to highly scattering (right), when the particles are deswollen.

Interestingly, the volume change is accompanied by changes in other physical parameters. With deswelling, the refractive index of the particle also changes. As a result, the suspension transitions from a transparent state, when the refractive indexes of the particle and the solvent are equal, to a milky state resulting from the large amount of light scattered, when the refractive indexes of the particle and the solvent are dissimilar; this optical transition is shown by the image sequence in Figure 4.1 for a PNIPAM microgel. Additionally, the elastic properties of the microgel change as deswelling proceeds and there are also important internal changes happening, since the mesh size of the particle decreases with deswelling. These modifications are at the heart of why microgels are relevant systems in science and technology [11–13]; and they can all be traced back to the thermodynamic behavior of microgel particles.

For the case of ionic microgels, the fixed charge of the polymer network can be used to induce all these particle changes. As it turns out, the major role of this charge is to create a Donnan potential inside the particle that affects the ionic distributions. This, in turn, generates an osmotic pressure difference between the inside and the outside of the particle, causing either its swelling or its deswelling. Direct electrostatic interactions do not play a major role in this process, provided the distribution of fixed charge within the particle is reasonably homogenous. This was described by Barrat et al. [23] and will be discussed in this chapter, as these ionic and electrostatic contributions to microgel swelling are not usually covered in most treatises discussing the swelling behavior of polymer gels.

In this chapter, we review the fundamental thermodynamic aspects that control the swelling and deswelling of polymer networks and compare the expectations with existing experimental results for microgel particles. We hope this chapter will provide the reader with a profound knowledge of the physics behind the intrinsic nature of microgels, for many usual situations, which resembles that of their macroscopic counterparts. This insight is essential and provides the starting point for understanding the rich behavior that microgel suspensions exhibit, particularly at high particle volume fractions, as will be seen in subsequent chapters of this book.

4.2
Swelling Thermodynamics

4.2.1
Polymer/Solvent Mixing

From a thermodynamic point of view, a microgel particle can be treated as a binary mixture consisting of polymer chains in a solvent. For a particular microgel, the particle properties strongly depend on the mixing of the polymer and the solvent. If the solvent is a good solvent, the mixture is homogeneous; all components of the mixture are uniformly mixed. In contrast, if the solvent is a bad solvent, the mixture is heterogeneous and the two phases, the polymer and the solvent, remain unmixed. Whether the mixture is homogeneous or heterogeneous is determined by the internal energy of mixing, U_{mix}, which either promote or inhibit the mixing of both species, and by the entropy of mixing, S_{mix}, which always favors mixing.

The miscibility of polymer chains was first described by Flory using a lattice model [24]. Consider n_2 polymers, each made of x segments, immersed in a solvent with n_1 molecules. The system is a lattice with $n_0 = n_1 + xn_2$ sites, where each site is occupied by either a solvent molecule or a polymer segment. The configurational entropy of the polymer can be calculated using the Boltzmann law: $S_{mix} = k_b \ln \Omega_{mix}$, where Ω_{mix} is the number of configurations for the polymer solution and k_b is the Boltzmann constant. By subtracting the entropy of the solvent and the entropy of the pure polymer arranged in a crystalline state, which we take as the reference states, we can obtain the entropy change resulting from the mixing process [24, 25]:

$$\Delta S_{mix} = -k_b \left[n_1 \ln \phi_1 + n_2 \ln \phi_2 \right] \tag{4.1}$$

where $\phi_1 = n_1/n_0$ and $\phi_2 = xn_2/n_0$ are the volume fractions of the solvent and the polymer, respectively. Note that this expression also describes the change in entropy due to mixing of solvents if the volume fractions, ϕ_1 and ϕ_2, are substituted by the molar fractions of the solvents [26].

The interaction between species in polymer mixtures is short ranged, typically due to van der Waals forces or hydrogen bonding. Therefore, only the interaction between neighboring sites in the lattice needs to be considered. The energy of mixing is calculated as the energy change induced in the rearrangement: $\frac{1}{2}(s-s) + \frac{1}{2}(p-p) = (p-s)$; the mixing of the polymer and the solvent implies breaking polymer–polymer and solvent–solvent contacts to create polymer–solvent contacts. If the energy associated to these interactions are, respectively, e_{pp}, e_{ss}, and e_{ps}, the energy change associated with the mixing is $\Delta e = e_{ps} - (e_{pp} + e_{ss})/2$. The change in internal energy is then written as $\Delta U_{mix} = P_{12}\Delta e$, where P_{12} denotes the number of polymer–solvent contacts resulting from the probability of having a polymer chain in contact with a solvent molecule. After evaluating this probability, we get [24]:

$$\Delta U_{mix} = z n_1 \phi_2 \Delta e \qquad (4.2)$$

where z is the coordination number of a lattice site. This expression is usually written in terms of the Flory solvency parameter, $\chi_1 = z\Delta e/k_b T$, as:

$$\Delta U_{mix} = k_b T \chi_1 n_1 \phi_2 \qquad (4.3)$$

where $k_b T \chi_1$ can be interpreted as the energy change experienced by a solvent molecule inside a pure solvent solution that is placed inside a pure polymer solution.

In this calculation of ΔU_{mix}, we have neglected the entropic contribution associated with the mixing process. There is no *a priori* reason to ignore this contribution, which can be taken into account by assuming that Δe includes a pure energetic term and an entropic contribution. Considering this, $\Delta e = \Delta e_u - T\Delta e_s$ and $\chi_1 = \frac{z\Delta e}{k_b T} = \frac{z\Delta e_u}{k_b T} - \frac{z\Delta e_s}{k_b}$; in this way, χ_1 reflects the influence of the energy and entropy changes when both a polymer–polymer and a solvent–solvent contact are replaced by a polymer–solvent contact, consistent with our mixing process.

The entropic and energetic contributions to the polymer–solvent mixing can be recast in terms of the Helmholtz energy change, ΔF_{mix}, which is the natural thermodynamic potential if the volume and temperature of the system remain constant during the mixing process [27]:

$$\Delta F_{mix} = \Delta U_{mix} - T\Delta S_{mix} = k_b T \left[n_1 \ln \phi_1 + n_2 \ln \phi_2 + \chi_1 n_1 \phi_2 \right] \qquad (4.4)$$

Since the mixing is additionally achieved without changes in pressure, the change in the Helmholtz energy is equal to the change in the Gibbs free energy [24, 25, 28] and thus $\Delta F_{mix} = \Delta G_{mix}$.

In real experiments, the mixing of polymer chains with the solvent is usually achieved at constant pressure and also at constant temperature, since the system is usually in contact with a heat reservoir. In addition, since the volume change is, in most cases, small, the mixing process can be effectively treated as isochoric [25]. Consequently, Equation 4.4 provides a reasonable way to obtain ΔG_{mix}.

As a result of the mixing process, the chemical potential of the solvent is altered, resulting in an osmotic pressure that accounts for the extra pressure that would be required in order to equilibrate the solvent in the polymer/solvent mixture at a volume fraction ϕ_1 with a solution of pure solvent. This extra pressure was first noticed by Pfeffer in 1877 for sugar solutions. By connecting the solution and the water through a water-permeable membrane, Pfeffer noticed that the height of mercury in a manometer connected to the sugar solution rose a height h, which corresponded to an equivalent induced pressure of $\varrho g h$, where ϱ is the density of mercury and g is the gravitational constant near the Earth's surface; this identically corresponds to the osmotic pressure of the sugar solution [29].

By applying this result to our polymer solution, we conclude that in equilibrium:

$$\mu_1^{pure}(P, T) = \mu_1(P + \Pi, T) \tag{4.5}$$

where $\mu_1^{pure}(P, T)$ is the chemical potential of a solvent molecule in an only solvent state and $\mu_1(P + \Pi, T)$ is the chemical potential of a solvent molecule in the polymer solution. By using the excess chemical potential, $\Delta\mu_1$, we can write [28]:

$$\mu_1(P + \Pi, T) = \mu_1^{pure}(P + \Pi, T) + \Delta\mu_1 \tag{4.6}$$

Furthermore, since both the temperature and the volume of a solvent molecule, v_s, are constants, we can also write:

$$\mu_1^{pure}(P + \Pi, T) = \mu_1^{pure}(P, T) + \int_{P}^{P+\Pi} \frac{\partial \mu_1^{pure}}{\partial P'} dP' = \mu_1^{pure}(P, T) + v_s \Pi \tag{4.7}$$

where we have used the Gibbs–Duhem relation, $N d\mu = -S dT + V dP$, and $\partial \mu / \partial P = v_s$ [27]. By combining Equations 4.6 and 4.7, we obtain the usual expression for the osmotic pressure:

$$\Pi = -\frac{\Delta\mu_1}{v_s} \tag{4.8}$$

where the negative sign accounts for the fact that a decrease in the chemical potential induces an increase in the osmotic pressure.

Finally, by using the usual thermodynamic relation between the chemical potential and the Gibbs free energy, $\mu_1 = (\partial G / \partial n_1)_{T, P, n_{j \neq 1}}$, we can calculate the osmotic pressure that results from the polymer/solvent mixing [24, 25] by using Equation 4.4:

$$\Pi_{mix} = -\frac{1}{v_s}\left(\frac{\partial \Delta G_{mix}}{\partial n_1}\right)_{T, P, n_2} = -\frac{k_b T}{v_s}\left[\phi_2 + \ln(1-\phi_2) + \chi_1 \phi_2^2\right] \tag{4.9}$$

where we considered that $n_1 \gg n_2$, and therefore $n_2 \ln \phi_2 < n_1 \ln \phi_1 = n_1 \ln(1-\phi_2)$ [30]. [1)]

---

1) We note that scaling theory [30] yields different results for Π_{mix} in the semidilute regime, which is a relevant regime in the case of polymer gels. Instead of the $\sim \phi_2^2$ dependence of Equation 4.9, scaling theory predicts a $\sim \phi_2^{9/4}$ dependence due to additional correlations between the monomers. Depending on the ϕ_2 range spanned as the microgel swells, this correction can lead to differences with respect to Flory's prediction.

4.2.2
Rubber Elasticity

The presence of cross-link molecules randomly distributed within the microgel particle provides a permanent structure to the polymer network. As a result, the network is elastic and thus is able to recover after being deformed. If L is the stretch and f is the resultant restoring force, the infinitesimal Helmholtz energy change is $dF = SdT + PdV + fdL$ and $f = \left(\frac{\partial F}{\partial L}\right)_{T,V} = \left(\frac{\partial (U-TS)}{\partial L}\right)_{T,V} = \left(\frac{\partial U}{\partial L}\right)_{T,V} - T\left(\frac{\partial S}{\partial L}\right)_{T,V}$. The restoring force thus consists of an energetic and an entropic contribution. For a cross-linked polymer, the relevant contribution is entropic [25, 31]. This can be shown experimentally by measuring the restoring force as a function of temperature and determining the offset and the slope of the resulting curve; the results indicate that indeed $\left(\frac{\partial U}{\partial L}\right)_{T,V} \ll \left(\frac{\partial S}{\partial L}\right)_{T,V}$. In general, when this inequality is fulfilled, the polymer chains were referred to as ideal chains by Flory, since there is no relevant energy changes associated with the deformation process [24].

To account for the entropy associated with the elastic deformation of a polymer network, one usually relies on the affine approximation: The relative deformation of each polymer chain is the same as that of the whole polymer network. With this assumption in mind, consider a system of polymer chains with a random cross-link distribution, where a chain is defined as the polymer between two cross-link points. Let us call $\vec{r}_{c,i}$ the end-to-end displacement vector of the chain and further assume that the distribution of $\vec{r}_{c,i}$ is Gaussian, which inherently implies that the chains are flexible. [2)]

Consider now a homogeneous deformation of the chain that modifies its end-to-end distance from $\vec{r}_{c,i}\left(\frac{x_{c,i}}{\alpha_x}, \frac{y_{c,i}}{\alpha_y}, \frac{z_{c,i}}{\alpha_z}\right)$ to $\vec{r}_{c,i}(x_{c,i}, y_{c,i}, z_{c,i})$, where α_j is the deformation ratio in the j-direction. To calculate the entropy, we use the Boltzmann expression, $S_{el} = k_b \ln \Omega_{el}$, where Ω_{el} is the number of polymer and cross-link configurations consistent with the formation of a network with end-to-end distances $\vec{r}_{c,i}(x_{c,i}, y_{c,i}, z_{c,i})$. With all these ingredients, Flory and Rehner calculated the entropy change associated with this deformation [32, 33]:

$$\Delta S_{el} = S_{el}(\text{deformed}) - S_{el}(\text{undeformed}) = -\frac{k_b N_c}{2}\left[\alpha_x^2 + \alpha_y^2 + \alpha_z^2 - 3 - \ln(\alpha_x \alpha_y \alpha_z)\right]$$

(4.10)

with S_{el} (deformed) the entropy associated with the deformed network, S_{el} (undeformed) the entropy associated with the undeformed state, where $\alpha_x = \alpha_y = \alpha_z = 1$, and N_c is the number of chains. This equation constitutes the basis of rubber elasticity.

---

2) Flexible chains are modeled as ideal or freely jointed chains, which are characterized by a random walk configuration. As a result, the $\vec{r}_{c,i}$ distribution is not affected by any correlations between the chains.

From the entropy change, we get:

$$\Delta G_{el} = -T\Delta S_{el} = \frac{k_b T N_c}{2}\left[\alpha_x^2 + \alpha_y^2 + \alpha_z^2 - 3 - \ln(\alpha_x \alpha_y \alpha_z)\right] \quad (4.11)$$

where we have neglected the energetic contribution to the Gibbs free energy, which is strictly correct for ideal polymer chains. If, in addition, we consider the special case of homogeneous swelling, $\alpha_x = \alpha_y = \alpha_z = \alpha = (V/V_0)^{1/3}$, with V and V_0 the deformed and undeformed volumes, respectively, the resultant Gibbs free energy is:

$$\Delta G_{el} = \frac{3k_b T N_c}{2}\left[\alpha^2 - 1 - \ln \alpha\right] \quad (4.12)$$

From ΔG_{el}, we can obtain the osmotic pressure associated with the elastic deformation of the polymer network by simply taking the derivative with respect to the number of solvent molecules:

$$\Pi_{el} = -\frac{1}{v_s}\left(\frac{\partial \Delta G_{el}}{\partial n_1}\right)_{T,P,n_j \neq 1} = -\frac{1}{v_s}\left(\frac{\partial \Delta G_{el}}{\partial \alpha}\right)_{T,P,n_j \neq 1}\left(\frac{\partial \alpha}{\partial n_1}\right)_{T,P,n_j \neq 1}$$

$$= \frac{k_b T N_c}{V_0}\left[\frac{\phi_2}{2\phi_{2,0}} - \left(\frac{\phi_2}{\phi_{2,0}}\right)^{1/3}\right]pc \quad (4.13)$$

with $\phi_{2,0}$ the polymer volume fraction for a particle volume V_0. The undeformed state, $(\phi_{2,0}, V_0)$, corresponds to the state where the polymer chains forming the microgel have random-walk configurations [2]. Flory considered that this condition is satisfied when the polymer is cross-linked to form the permanent network. As a result, the state $(\phi_{2,0}, V_0)$ corresponds to the state of the system when it is synthesized. In the case of microgels, this would often correspond to the deswollen state of the particles, since most microgels are synthesized under *bad solvent* conditions.

In addition, for microgels the chains are usually composed of flexible polymers, as shown in Table 4.1, where we quote the persistence length, l_p, of some of the most widely used polymers in microgel particles. As a result, the swelling behavior of most microgels is reasonably well described by rubber elasticity [34]. Deviations from this behavior are expected for particles based on semiflexible or rigid polymers, which have persistence lengths on the order or much larger than the microgel chain size [35]; polymer networks based on biopolymers provide clear examples of this

Table 4.1 Persistence length of some of the polymers used in microgel preparations in the indicated solvents.

Polymer	Solvent	l_p (nm)
Poly(isopropylacrylamide) [45]	Water	0.35
Poly(acrylamide) [45]	Water	0.30
Poly(styrene) [46]	Toluene	0.37
Polyvinylpyridine [46]	THF	0.4

fact [10, 36, 37]. In addition, we expect that highly cross-linked microgel particles cannot be described using rubber elasticity. In fact, this has already been confirmed with stiff macroscopic gels [38]. Extensions of rubber elasticity to account for all these observed deviations include consideration of polymer entanglements [39, 40], of possible non-Gaussian chain statistics [38], of internal energy effects [38], and of possible internal nematic ordering [41, 42]. More recently, the effects of thermal fluctuations of the polymer network have been considered [43, 44], and shown to provide a large entropic contribution to the Gibbs free energy of the system. This contribution, however, becomes less important as the bulk modulus of the polymer gel decreases or equivalently when the swelling of the network is considerable.

4.2.3
Ionic Effects

For ionic microgels, there are additional contributions to the free energy arising from the screened repulsion between polymer chains and from the osmotic pressure exerted by the confined counterions, which are required to preserve electroneutrality. Both contributions naturally tend to swell the network, but their relevance in the swelling process is not comparable in most usual situations. We expand here on these aspects as they are not treated in detail in most polymer physics textbooks and reviews.

4.2.3.1 Ideal Gas Contribution
Let us first consider the case of a microgel particle of radius R in a solvent without any added salt, and let N be the number of monomers between cross-link points, r_c be the average end-to-end distance of the polymer chain, $c = N/r_c^3 = NN_c/R^3$ the monomer concentration inside the particle and f the fraction of charged monomers per chain. With these definitions, the total number of counterions inside the particle is $fcV = fNN_c$, where $V = R^3$ is the volume of the particle.

The entropy associated with the counterions confined inside the network can be calculated, in first approximation, as that of an ideal gas. In this case, there are no spatial correlations between the counterions and thus the probability of any one of them being found in a particular region of the available space is completely independent of the location of the other counterions. As a result, the total number of possibilities to distribute the fcV counterions in the system will simply be equal to the product of the number of possibilities for all the individual counterions. Provided that fcV and the energy per counterion, $U_{ideal} = 3/2\,k_b T$, are fixed, the number of possibilities is directly proportional to V and thus [47]:

$$\Omega \propto V^{fNN_c} \tag{4.14}$$

and the entropy:

$$S_{ionic} = k_b \ln \Omega = k_b f\, NN_c \ln(V/V^{cte}) \tag{4.15}$$

with $V^{cte} = [h^2/(2\pi m k_b T)]^{3/2} (fNN_c)/e^{5/2}$, where m is the mass of the counterion, and h is Planck's constant [48]. To obtain the osmotic pressure, we calculate the chemical potential of the solvent in a solution with counterions:

$$\mu_1 = -T\left(\frac{\partial S_{ionic}}{\partial n_1}\right)_{U,V,n_j \neq 1} = -k_b T f N N_c \frac{1}{V} v_s \qquad (4.16)$$

where we have used $V = v_s(n_1 + n_2 + fNN_c)$ and where we have assumed, for simplicity, that the volumes of a solvent molecule, counterion and polymer segment are all equal. Taking the chemical potential of a pure solvent solution as the reference state, $\mu_1^{pure} = 0$, we can write the osmotic pressure as:

$$\Pi_{ionic} = -\frac{\Delta \mu_1}{v_s} = k_b T f c \qquad (4.17)$$

where we have further used the definition of c. This is Van't Hoff's expression for the osmotic pressure of any ideal solution; it resembles that for the pressure of an ideal gas [48]. The osmotic pressure is $k_b T$ times the counterion concentration.

Let us now consider that a finite concentration of salt, n, is added to the solvent. In this case, a Donnan equilibrium is established between the inside and the outside of the microgel particle since the electroneutrality constraint on the gel plays the role of a semipermeable membrane keeping the fNN_c counterions from diffusing away. Let us call the salt concentration inside the microgel n_{in} and reserve the symbol n for the nominal salt concentration. In this situation, the concentration of counterions and coions inside the particle are $fc + n_{in}$ and n_{in}, respectively. To obtain the osmotic pressure difference between the inside and the outside of the microgel particle, we treat the ionic solutions as ideal gases, as we have done when treating the no salt case, and write this osmotic pressure difference in terms of the ionic concentrations inside and outside the particle:

$$\Delta \Pi = \Pi_{in} - \Pi_{out} = k_b T \sum_i (n_i - n_i^o) = k_b T \left[fc + 2n_{in} - 2n\right] \qquad (4.18)$$

where n_i and n_i^o are the concentration of the ith ionic specie inside and outside the particle, respectively, and where we have used that the amount of salt inside the particle is small compared to the nominal salt concentration, which implies that $n^o = \sum_i n_i^o \approx 2n$. The amount of salt inside the particle can be obtained as a function of the other two parameters, fc and n, by considering that in equilibrium, the chemical potential of the ions inside and outside the particle becomes equal [27]. We start by calculating the chemical potential of the ions as:

$$\mu_{ion} = \left(\frac{\partial G_{ionic}}{\partial (fNN_c)}\right)_{T,P,n} = -T\left(\frac{\partial S_{ionic}}{\partial (fNN_c)}\right)_{U,V,n} = k_b T \ln \frac{fNN_c}{V} + A \qquad (4.19)$$

by using Equation 4.15 and $A = k_b T \left\{\ln\left[\frac{h^2}{2\pi m k_b T}\right]^{3/2} - \frac{3}{2}\right\}$. By using this expression for μ_{ion} and further considering that in equilibrium there are $fc + n_{in}$ counterions and n_{in} coions inside the microgel and n counterions and n coions outside the microgel, we get the desired relation between n, n_{in}, and fc [47, 49]:

$$\mu_{\text{counterions outside}} + \mu_{\text{coions outside}} = \mu_{\text{counterions inside}} + \mu_{\text{coions inside}} \Leftrightarrow n^2 = n_{in}(n_{in} + fc) \tag{4.20}$$

which we can rewrite as:

$$n_{in} = \frac{1}{2}\left[-fc + \sqrt{(fc)^2 + 4n^2}\right] \tag{4.21}$$

where the negative solution is discarded, as it would correspond to an unphysical situation. Substituting this expression for n_{in} into Equation 4.18 results in an osmotic pressure difference between the inside and the outside of the microgel that can be written as:

$$\Delta\Pi = k_b T\left[\sqrt{(fc)^2 + 4n^2} - 2n\right] \tag{4.22}$$

If the amount of added salt is such that $fc \ll n$, this expression can be further reduced as [23]:

$$\Delta\Pi = k_b T \frac{f^2 c^2}{4n} \tag{4.23}$$

which is often used to describe the effect of salt on the swelling of ionic microgels. It is interesting to note that $\Delta\Pi$ is inversely proportional to the salt concentration. As n increases, the osmotic pressure difference between the inside and the outside of the microgel particle continuously decreases; as a result, with increasing n, ionic effects become less relevant and the microgel can eventually be treated as a nonionic particle.

4.2.3.2 Electrostatic Energy of a Homogeneously Charged Microgel

Up to this point, we have not considered direct electrostatic effects. The simplest way to calculate the electrostatic contribution to the free energy is to suppose that the counterions are uniformly distributed in an also uniform charge background, which is provided by the ionic polymer network. This is equivalent to treating the microgel particle as a uniform charged sphere with an effective charge that depends on the counterion distribution inside and outside the particle. Note that a nonzero effective charge will imply a violation of the electroneutrality condition inside the microgel, which we have consistently assumed so far; in this case, the whole suspension is electroneutral and the counterions are free to stay inside or outside the microgel particle. With these considerations, we can write the electrostatic energy as:

$$U_{el} = \int_0^Q \psi \, dq \tag{4.24}$$

where $\psi = \dfrac{Q_{\text{eff}}(r)}{4\pi\varepsilon_0\varepsilon_r r}$ is simply the electrostatic potential at a distance r from the particle center and $Q_{\text{eff}}(r) = q_e[Z_{\text{network}}(r) - Z_{in}(r)]$ is the effective charge of a sphere of radius r, with q_e the elementary charge, and $Z_{network}(r)$ and $Z_{in}(r)$ the number of network charges and number of counterions inside this sphere, respectively. Using the effective charge density, $\varrho_q = \frac{Q_{eff}(r)}{V(r)}$, with $V(r)$ the volume of a sphere of radius r, we can rewrite the integral over the charge in terms of an integral over the particle volume as:

$$U_{el} = \int_0^V \frac{\varrho_q(4\pi r^3/3)}{4\pi\varepsilon_0\varepsilon_r r} \varrho_q dV' = \int_0^R \frac{\varrho_q(4\pi r^3/3)}{4\pi\varepsilon_0\varepsilon_r r} \varrho_q 4\pi r^2 dr \qquad (4.25)$$

which can be easily performed to yield [50]:

$$U_{el} = \frac{3}{5} k_b T \frac{\lambda_b}{R} [Z_{network} - Z_{in}]^2 \qquad (4.26)$$

where $\lambda_b = \frac{q_e^2}{4\pi\varepsilon_0\varepsilon_r k_b T}$ is the Bjerrum length, with ε_0 and ε_r the vacuum permittivity and the relative dielectric constant of the solvent, respectively; physically, this length scale corresponds to the distance between two elementary charges at which the electrostatic interaction energy is equal to the thermal energy.

In most situations of interest, the large electrostatic attraction between the charges in the polymer network and the counterions dramatically reduces the number of counterions that actually leave the microgel [51]. As a result, the effective charge of a particle is usually several orders of magnitude smaller than the net charge of the polymer network [52]; Q_{eff} is thus essentially zero and this electrostatic contribution to the free energy is in most cases negligible.

4.2.3.3 Contribution from Counterion Correlations

Note that Equation 4.26 provides a purely mean-field contribution to the electrostatic energy. Additional contributions arise from local charge correlations between the counterions and from local density correlations within the polymer network, which induce nonuniformities in the distribution of counterions and nonuniformities in the distribution of fixed charges within the microgel particle [23]. Let us first consider the local correlations of the counterion density in a uniformly charged background [53], by following the ideas of Debye and Hückel for strong electrolytes [48, 52, 54]. The interaction energy associated with the correlations between counterions can be written in this case as the superposition of the electrostatic energy associated with the interactions between pairs of counterions:

$$U_{corr} = \frac{1}{2} \sum_i \sum_j \frac{q_i q_j}{4\pi\varepsilon_0\varepsilon_r |\vec{r}_i - \vec{r}_j|} = \frac{1}{2} \sum_j q_j \Psi_j(\vec{r}_j) \qquad (4.27)$$

where $\Psi_j(\vec{r}_j) = \sum_{i \neq j} \frac{q_i}{4\pi\varepsilon_0\varepsilon_r |\vec{r}_i - \vec{r}_j|}$ is the electrostatic potential acting on counterion j, located at $\vec{r}_j$, due to the presence of the other counterions, with q_i the charge of counterion i. Since the situation is analogous for each counterion and all of them have the same charge, q, we can further write U_{corr} in terms of the number of counterions per microgel particle, fcV:

$$U_{corr} = \frac{1}{2} fcV\, q\, \Psi \tag{4.28}$$

To determine Ψ, let us consider each counterion separately and realize that around each of them there is a nonuniformly charged and spherically symmetric ion cloud. Let $c_c(r)$ be the density distribution of counterions in this ion cloud. The potential ψ in the ion cloud is related to the charge density inside the cloud, $qc_c(r)$, by the Poisson–Boltzmann equation:

$$\nabla^2 \psi = -\frac{1}{\varepsilon_0 \varepsilon_r} q\, c_c(r) = -\frac{1}{\varepsilon_0 \varepsilon_r} qc \exp\left\{\frac{-q\psi}{k_b T}\right\} \tag{4.29}$$

For small interactions between counterions or at sufficiently low concentrations, $\frac{q\psi}{k_b T} \ll 1$ and the exponential can be expanded in Taylor series, resulting in the so-called Debye–Hückel equation:

$$\nabla^2 \psi = \varkappa^2 \psi \tag{4.30}$$

where $\varkappa = q\sqrt{\frac{fc}{\varepsilon_0 \varepsilon_r k_b T}} = \sqrt{4\pi \lambda_b fc}$ is the inverse Debye length. The spherically symmetric solution of this equation is given as:

$$\psi = \psi_0 \frac{e^{-\varkappa r}}{r} \tag{4.31}$$

Since in the immediate neighborhood of a counterion ($r \approx 0$), ψ must become the pure Coulomb potential of the considered counterion, $\psi = \frac{q}{4\pi \varepsilon_0 \varepsilon_r r}$, the constant ψ_0 must thus be equal to $\psi_0 = \frac{q}{4\pi \varepsilon_0 \varepsilon_r}$. Furthermore, we can expand the exponential in a Taylor series for small $\varkappa r$, to give:

$$\psi = \frac{q}{4\pi \varepsilon_0 \varepsilon_r r} - \frac{q}{4\pi \varepsilon_0 \varepsilon_r} \varkappa + \cdots \tag{4.32}$$

where the omitted terms vanish when $r = 0$. The first term is the Coulomb potential of the counterion itself. The second term is the electrostatic potential produced by all other counterions at the point occupied the considered counterion, and it is the quantity to be substituted in Equation 4.28, implying $\Psi = -\frac{q}{4\pi \varepsilon_0 \varepsilon_r} \varkappa$. As a result, the energy associated with counterion correlations, within the Debye–Hückel approximation becomes:

$$U_{corr} = -\frac{1}{2} fcV \frac{q^2}{4\pi \varepsilon_0 \varepsilon_r} \varkappa = -\frac{1}{8\pi} V k_b T \varkappa^3 \tag{4.33}$$

From the energy, by integrating the thermodynamic relation $U = -T^2\left[\frac{\partial(F/T)}{\partial T}\right]_{V,n_i}$ [27], we can obtain the Helmholtz energy as:

$$F = F_0 - \frac{1}{12\pi} V k_b T \varkappa^3 \tag{4.34}$$

where F_0 results from the integration and must equal the Helmholtz energy in the absence of counterion correlations, since for $T \to \infty$, $F = F_{\text{ideal}}$. From this expression, we can calculate the chemical potential of the solvent: $\mu_1 = \left(\frac{\partial F}{\partial n_1}\right)_{V,T,n_{j\neq 1}}$, with $V = v_s(n_1 + n_2 + fNN_c)$ and where we assume that macroscopically the volume does not change much even if some solvent molecules either enter or leave the microgel particle [25]. From μ_1, it is straightforward to calculate, by using Equation 4.8, the osmotic pressure as:

$$\Pi = \Pi_{\text{ionic}} + \Pi_{\text{corr}} = k_b T f c - k_b T \frac{\varkappa^3}{24\pi} \tag{4.35}$$

Note the minus sign, which accounts for the screening of the counterion interactions due to the presence of the remaining counterions; this screening tends to decrease the osmotic pressure due to the counterions. However, it is easy to check that the counterion-correlation contribution to the osmotic pressure is much smaller than the ideal gas contribution. For instance, for water as a solvent and for a microgel with a concentration of monovalent counterions of 10 mM, $\frac{fc}{\varkappa^3/(24\pi)} \sim 10$ and thus Π_{ionic} is about an order of magnitude larger than Π_{corr}.[3] As a result, the Debye–Hückel correction does not appreciably contribute to the counterion contribution to the osmotic pressure; this is the case even if higher order terms in $\varkappa$ are considered in Equation 4.35 [50].

4.2.3.4 Effect of a Slightly Inhomogeneous Fixed-Charge Distribution

The other nonuniformity in the charge distribution of a microgel results from correlations in the local monomer density within the particle; these are a direct consequence of the cross-linker, since around any cross-link point the local monomer density will be greater. Let $c_p(\vec{r})$ be the local monomer concentration and $f(c_p(\vec{r}) - c)$ the fixed charge nonuniformity, which is the local fixed charge density minus the average fixed charge concentration inside the particle, which due to electroneutrality, is equal to the counterion concentration inside the microgel particle. This fixed charge nonuniformity essentially reflects the spatial fluctuations of the fixed charge in the polymer network. The interaction between fixed charges located at positions $\vec{r}$ and $\vec{r}'$ within the particle is mediated through a mean-field potential, $v(\vec{r} - \vec{r}')$, which we write within the linear approximation of the Debye–Hückel theory [54] as:

---

[3] For $fc = 10$ mM, $\varkappa = 0.23$ nm^{-1} and $\varkappa r \approx 2.3$ for $r \approx 10$ nm. Since Debye–Hückel is valid for $\varkappa r \to 0$, this counterion concentration is reasonably high. For smaller fc, this ratio is even smaller.

$$v(r) = k_b T \frac{\lambda_b}{r} e^{-\varkappa r} \tag{4.36}$$

The total energy associated with this interaction can be calculated, to the lowest order in concentration fluctuations, by averaging over all possible statistical configurations of the chains [23]:

$$U_{\text{fluc}} = \frac{f^2}{2} \left\langle \int \int d\vec{r} d\vec{r}' \, [c_p(\vec{r}) - c] \, v(\vec{r} - \vec{r}') [c_p(\vec{r}') - c] \right\rangle \tag{4.37}$$

In order to solve this integral, we introduce the Fourier transform of the charge nonuniformity, $\delta c_p(\vec{r}) = c_p(\vec{r}) - c = 1/(2\pi)^3 \int d\vec{q} \, \delta c_p(\vec{q}) \, e^{i\vec{q} \cdot \vec{r}}$, and the Fourier transform of the interaction potential, $v(\vec{r} - \vec{r}') = k_b T \lambda_b / (2\pi)^3 \int \frac{4\pi}{\varkappa^2 + q^2} e^{i\vec{q} \cdot (\vec{r} - \vec{r}')} d\vec{q}$:

$$U_{\text{fluc}} = \frac{f^2}{2} k_b T \lambda_b \frac{4\pi}{(2\pi)^9} \left\langle \int \int \int \int \int d\vec{r} d\vec{r}' d\vec{q} d\vec{q}' d\vec{q}'' \, \delta c_p(\vec{q}) \, e^{i\vec{q} \cdot \vec{r}} \right.$$

$$\left. \delta c_p(\vec{q}') \, e^{i\vec{q}' \cdot \vec{r}'} \frac{1}{\varkappa^2 + q''^2} e^{i\vec{q}'' \cdot (\vec{r} - \vec{r}')} \right\rangle$$

$$= \frac{f^2}{2} k_b T \lambda_b \frac{4\pi}{(2\pi)^3} \left\langle \int \int \int d\vec{q} d\vec{q}' d\vec{q}'' \, \delta c_p(\vec{q}) \delta c_p(\vec{q}') \, \delta(\vec{q} + \vec{q}'') \delta(\vec{q}' - \vec{q}'') \frac{1}{\varkappa^2 + q''^2} \right\rangle$$

$$= \frac{f^2}{2} k_b T \lambda_b \frac{1}{(2\pi)^3} \left\langle \int d\vec{q}'' \, \delta c_p(\vec{q}'') \delta c_p(-\vec{q}'') \frac{1}{\varkappa^2 + q''^2} \right\rangle \tag{4.38}$$

where we have used the definition and properties of the Dirac delta. By further introducing the definition of the structure factor, $S(\vec{q}) = \frac{1}{NN_c} \left\langle \int \int d\vec{r} d\vec{r}' \, \delta c_p(\vec{r}) \times \delta c_p(\vec{r}') \, e^{i\vec{q} \cdot (\vec{r} - \vec{r}')} \right\rangle = \frac{1}{NN_c} \langle \delta c_p(\vec{q}) \delta c_p(-\vec{q}) \rangle$, we get:

$$U_{\text{fluc}} = NN_c k_b T \frac{f^2}{2} \lambda_b \frac{4\pi}{(2\pi)^3} \int d\vec{q} \, S(\vec{q}) \frac{1}{q^2 + \varkappa^2} \tag{4.39}$$

To evaluate this integral, we take into consideration that $S(q)$ is peaked at $q^* \sim 1/r_c$, reflecting the network structure of the polymer gel, which has a characteristic length scale of the order of the mesh size, r_c [55, 56]. In addition, $S(q^*)$ must be of order N, since the contributions from different chains having random orientations will add in an incoherent way [23]. As a result, we follow Joanny et al. and approximate the structure factor to $S(\vec{q}) \approx N \pi q^* \delta(\vec{q} - \vec{q}^*)$. With this [23], we get:

$$U_{\text{fluc}} \approx N^2 N_c k_b T \frac{f^2}{r_c} \lambda_b \frac{1}{1 + \varkappa^2 r_c^2} \tag{4.40}$$

This contribution is always smaller than the ideal gas contribution. To show this, let us consider the limiting cases of $\varkappa r_c \ll 1$ and $\varkappa r_c \gg 1$ in the absence of added salt. For

$\varkappa\, r_c \ll 1$, which is often referred to as the weak screening regime,

$$\frac{U_{\text{fluc}}}{U_{\text{ideal}}} \approx \frac{U_{\text{fluc}}}{\frac{3}{2}fNN_c k_b T} = \frac{2}{3} Nf \frac{\lambda_b}{r_c} = \frac{1}{6\pi}(\varkappa r_c)^2 \ll 1 \tag{4.41}$$

while for $\varkappa\, r_c \gg 1$, in the so-called strong screening regime,

$$\frac{U_{\text{fluc}}}{U_{\text{ideal}}} \approx \frac{1}{6\pi} < 1 \tag{4.42}$$

In both cases, U_{fluc} is smaller than U_{ideal}, emphasizing that the relevant contribution to the osmotic pressure arising from the charged character of the polymer network results from the ideal gas contribution of the ions, Equation 4.17.

For a certain salt concentration, n, in the weak screening regime,

$$\frac{U_{\text{fluc}}}{U_{\text{ideal},\,n}} \approx \frac{U_{\text{fluc}}}{k_b T \frac{f^2 c^2}{4n} N_c r_c^3} \sim (\varkappa r_c)^2 \ll 1 \tag{4.43}$$

and once more, the osmotic pressure is dominated by the ideal gas contribution of the ions, Equation 4.23. In contrast, in the strong screening regime,

$$\frac{U_{\text{fluc}}}{U_{\text{ideal},\,n}} \approx 1/\pi \approx 1 \tag{4.44}$$

which indicates that the contribution arising from fixed charge correlations and the ideal gas contribution of the ions are comparable in this case. However, in this situation other interactions can dominate over electrostatic interactions; the swelling behavior of the microgel is then dominated by these nonelectrostatic interactions.

4.2.4
Equilibrium: Equation of State

Thermodynamic equilibrium is attained when the net osmotic pressure, Π, in a microgel is zero:

$$\Pi = \Pi_{\text{mix}} + \Pi_{\text{el}} + \Pi_{\text{ionic}} = 0 \tag{4.45}$$

Explicitly:

$$\Pi = \frac{k_b T N_A}{v_s} \left\{ \frac{N_c v_s}{V_0 N_A} \left[\left(fN + \frac{1}{2}\right)\left(\frac{\phi_2}{\phi_{2,0}}\right) - \left(\frac{\phi_2}{\phi_{2,0}}\right)^{1/3} \right] - \phi_2 - \ln(1-\phi_2) - \chi \phi_2^2 \right\} = 0 \tag{4.46}$$

where we have used the ionic contribution to the osmotic pressure due to the counterions, Equation 4.17. This provides the equation of state for our microgel particle; it relates the particle size to the other relevant thermodynamic parameters. The particle size is determined by how well the polymer mixes with the solvent, by the

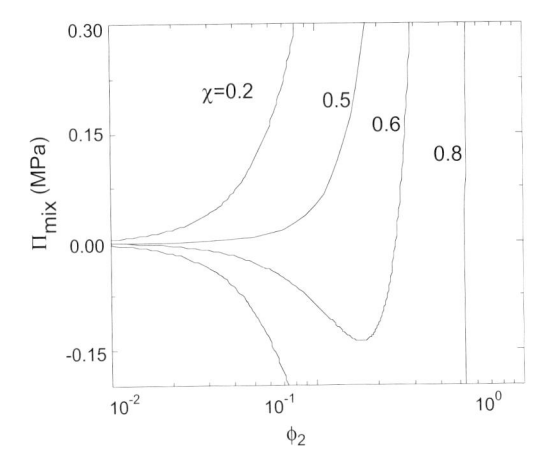

Figure 4.2 Mixing osmotic pressure, Π_{mix}, as a function of polymer volume fraction, ϕ_2, for different polymer solubilities, χ.

elasticity of the polymer network and by the ionic contribution to the total osmotic pressure. When the microgel is out of equilibrium, the polymer network either swells or deswells depending on the sign of the osmotic pressure; positive osmotic pressures tend to swell the microgel while negative osmotic pressures tend to deswell the microgel. In the end, it is the interplay between Π_{mix}, Π_{el}, and Π_{ionic} what determines the equilibrium size of the particle.

The mixing contribution to the total osmotic pressure exhibits a rich behavior depending on the value of the Flory solvency parameter, as can be seen in Figure 4.2. For $\chi < 0.5$, Π_{mix} is positive and monotonically increases with polymer volume fraction, thus contributing to the swelling of the microgel particle. In contrast, for $\chi \geq 0.5$, the energetic contribution to the mixing plays an important role and the monotonical behavior no longer holds. At low polymer volume fractions, Π_{mix} is negative contributing to particle deswelling. However, as ϕ_2 increases, Π_{mix} exhibits a minimum, and eventually becomes positive; at this point, the entropic contribution to mixing overwhelms the energetic contribution and swelling is favored. The balance between these two contributions greatly depends on χ, and thus on the solvent quality for the particular polymer.

The elastic contribution to the net osmotic pressure is in most cases negative for the case of microgel particles, as shown in Figure 4.3(a) for different cross-linking content. The network elasticity thus contributes to particle deswelling and its importance increases with the amount of cross-linker, as intuitively expected. In Figure 4.3(a), we have taken $\phi_{2,0} = 1$ and quote the cross-linker concentration, which is easily related to the approximate number of chains by using the amount of monomer added in the particle synthesis:

$$N_c = 2N_A X \tag{4.47}$$

where X denotes the number of moles of cross-linker. This assumes that all the added monomer reacts to form microgel particles and neglects the presence of knots and

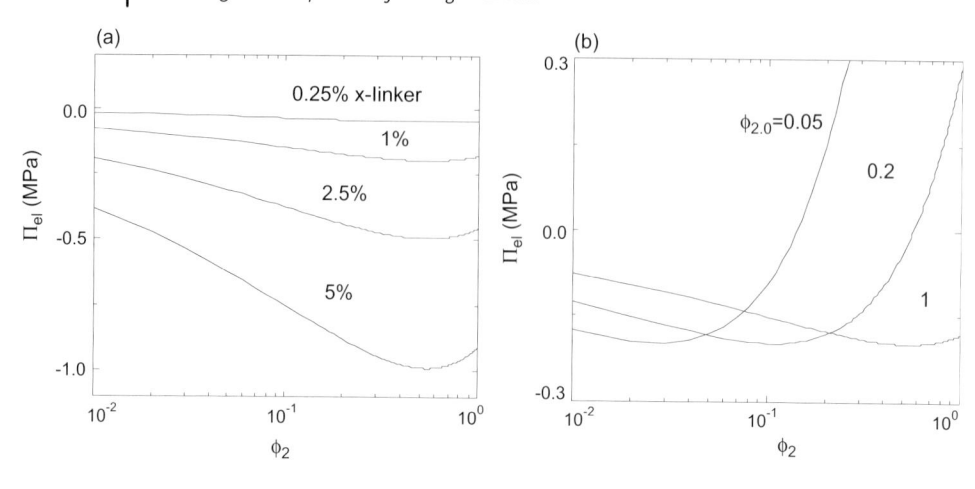

Figure 4.3 (a) Dependence of the elastic osmotic pressure, Π_{el}, with the polymer volume fraction, ϕ_2, for different cross-linking concentrations; $\phi_{2,0} = 1$ for all curves and N_c/V_0 (m^{-3}) = 2.2×10^{25} (0.25 wt%), 8.8×10^{25} (1 wt%), 2.2×10^{26} (2.5 wt%), and 4.4×10^{26} (5 wt%). (b) Influence of the polymer volume fraction in the undeformed state, $\phi_{2,0}$, on the Π_{el}–ϕ_2 curve; N_c/V_0 (m^{-3}) = 8.8×10^{25}.

other defects in the network topology. Despite these simplifications, Equation 4.47 gives a reasonable estimate for N_c.

The value of $\phi_{2,0}$ also plays a relevant role. In fact, careful inspection of the Π_{el} versus ϕ_2 curve reveals the presence of a minimum at $\phi_{2,\min} = 0.54$, irrespective of the number of chains. For higher polymer volume fractions, the magnitude of Π_{el} increases. The minimum in the curve is easily obtained by using Equation 4.13 and from the condition

$$\frac{d\Pi_{el}}{d\phi_2} = 0 \Rightarrow \phi_{2,\min} = (3/2)^{-3/2} \phi_{2,0} \tag{4.48}$$

The turnover, located at $\phi_{2,\min}$, is directly related to $\phi_{2,0}$. As a result, the minimum in the Π_{el} versus ϕ_2 curve shifts to smaller polymer volume fractions as $\phi_{2,0}$ decreases, as shown in Figure 4.3b. Interestingly, if $\phi_{2,0}$ is low enough, Π_{el} can turn positive for sufficiently high ϕ_2, as also shown in Figure 4.3b; this happens when

$$\frac{\phi_2/2\phi_{2,0}}{(\phi_2/\phi_{2,0})^{1/3}} = 1 \Rightarrow \phi_{\Pi_{el}=0} = 2^{3/2}\phi_{2,0} \tag{4.49}$$

and implies that for sufficiently deswollen particles, the network elasticity can, in principle, contribute to particle swelling. However, in the case of microgels, $\phi_{2,0}$ is the polymer volume fraction corresponding to the deswollen particle state, since this corresponds to the synthesis conditions in most cases. As a result, $\phi_{2,0}$ is relatively high and thus neither the turnover in Π_{el} nor its change of sign are typically possible to observe. In the case of macroscopic gels, since in many cases they are synthesized in conditions corresponding to low polymer volume fractions [57], this regime can be reached when the gels are sufficiently deswollen.

The total osmotic pressure that results from adding the mixing and elastic osmotic pressures, plus the ionic contribution due to the counterions, Equation 4.46, exhibits a rich behavior, essentially depending on the Flory χ parameter and the number of counterions, fN, provided N_c and ϕ_0 are fixed, which is the case for a particular microgel particle. For low fN and irrespective of χ, the shape of the total osmotic pressure curve resembles the shape of the mixing contribution: It decreases with increasing ϕ_2 up to the minimum in the curve and then increases for higher values of ϕ_2, as shown in Figure 4.4(a), which corresponds to $fN = 2$ and $\chi = 0.6$. The low ϕ_2 region is essentially controlled by the elastic contribution to Π, while at high ϕ_2, Π is essentially equal to Π_{mix}. Only the part of the Π–ϕ_2 curve with positive slope corresponds to thermodynamically stable phases, since otherwise the osmotic compressibility, $\varkappa_T = -\frac{1}{V}\left(\frac{\partial V}{\partial \Pi}\right)_T = \frac{1}{\phi_2}\left(\frac{\partial \phi_2}{\partial \Pi}\right)_T$, would be negative, corresponding to

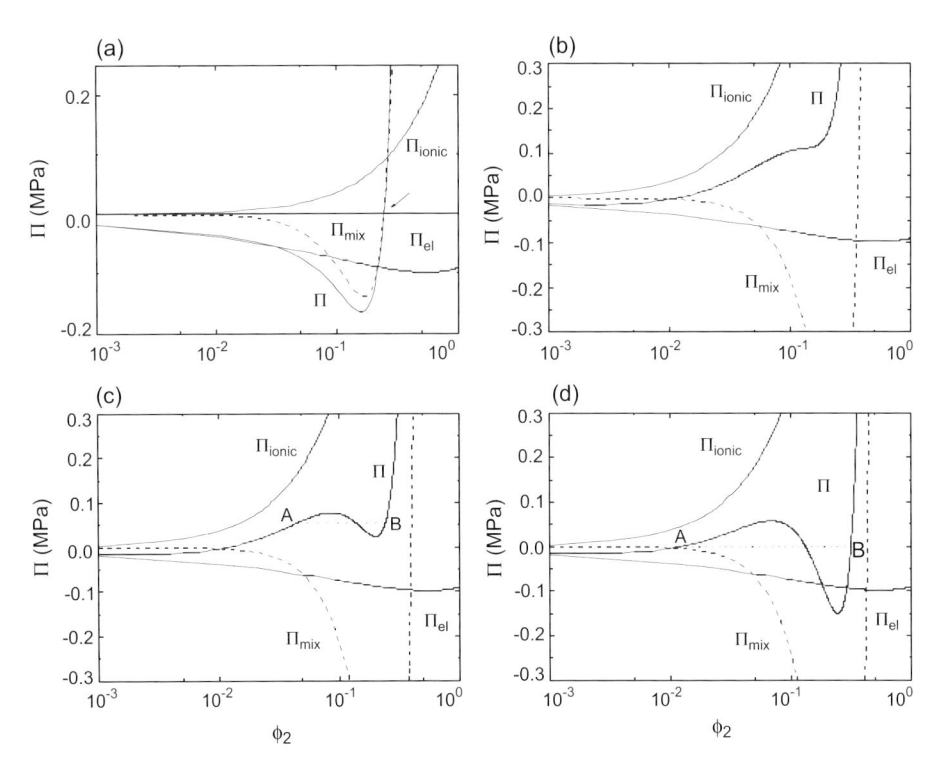

Figure 4.4 Total osmotic pressure, Π, versus polymer volume fraction, ϕ_2, for $\phi_{2,0} = 1$, a cross-linker concentration of 0.5% ($N_c/V_0 = 4.4 \times 10^{25}$ m^{-3}), and (a) $\chi = 0.6$, $fN = 2$, (b) $\chi = 0.669$, $fN = 20$, (c) $\chi = 0.69$, $fN = 20$, and (d) $\chi = 0.715$, $fN = 20$. We also show the individual contributions to Π.

unstable thermodynamic phases. In these situations, the system phase separates into two phases for which $\Pi = 0$, corresponding to a pure fluid ($\phi_2 = 0$) and a microgel with a certain polymer volume fraction, as indicated by the arrow in Figure 4.4a; this corresponds to the only equilibrium state of the system. If, within the thermodynamically stable region, $\Pi > 0$, the microgel swells and ϕ_2 decreases until $\Pi = 0$. Equivalently, if within this region also, $\Pi < 0$, the microgel deswells and ϕ_2 increases until equilibrium is reached and $\Pi = 0$ again. The ionic contribution to the total osmotic pressure in cases such as the one that we have just discussed only slightly affects the overall magnitude of Π, without significantly affecting its overall dependence with ϕ_2. As a result, we can affirm that the driving mechanism for the swelling or deswelling of neutral microgels is mainly incorporated into the χ parameter.

In contrast, for charged microgels, the values of fN and thus of Π_{ionic}, can qualitatively affect the influence of χ on the overall Π-ϕ_2 behavior, as shown in Figure 4.4b–d, which correspond to $fN = 20$ and χ values equal to 0.669, 0.69, and 0.715, respectively. For this value of fN, if $\chi < 0.669$, the Π–ϕ_2 behavior is similar to that obtained for low fN, except that the minimum of the curve becomes located at very low values of ϕ_2. Interestingly, for $\chi = 0.669$ there is a critical point, where $\frac{\partial \Pi}{\partial \phi_2} = 0 = \frac{\partial^2 \Pi}{\partial \phi_2^2}$, as shown in Figure 4.4b, and for $\chi > 0.669$ the Π–ϕ_2 curve displays van der Waals loops, as shown in Figure 4.4c and d, indicating occurrence of a phase transition. In these cases, the thermodynamically stable regions are obtained by using the Maxwell construction. The end points of the Maxwell line represent volume fractions along the coexistence line corresponding to swollen and deswollen phases. The volume fractions between these points and the maximum and the minimum of the van der Waals loops are metastable states, while the volume fractions between the extremes of the loops correspond to unstable states inside the spinodal region, where no metastability is allowed. For $\chi = 0.69$, the phase transition occurs for positive Π and the microgel phase separates into two different phases, as shown in Figure 4.4c; in this case, the two microgel phases A and B will not be in equilibrium with the surrounding liquid. However, while phase A will be able to swell by decreasing ϕ_2 until $\Pi = 0$, phase B will remain out of equilibrium with the ambient solvent. In contrast, for $\chi = 0.715$, the Maxwell line lies in the $\phi_2 = 0$ axis, as shown in Figure 4.4d, and the two coexisting microgel phases, A and B, will be in equilibrium with the surrounding solvent. For higher χ, the phase transition occurs at negative osmotic pressure and, as for the positive Π case, the resultant microgel phases will not be in equilibrium with the surrounding liquid. In this case, however, both phases evolve towards equilibrium to yield a dense microgel phase corresponding to $\Pi = 0$ in coexistence with a pure liquid phase, where both Π and ϕ_2 are equal to zero.

The overall behavior of the osmotic pressure with polymer volume fraction is reminiscent of the pressure dependence on volume for atomic fluids, with the Flory parameter playing the role of temperature. To emphasize this analogy, let us consider the virial expansion of Equation 4.46, which we can derive by Taylor-expanding the logarithm for small ϕ_2:

4.2 Swelling Thermodynamics

$$\Pi = \frac{k_b T N_A}{v_s} \left\{ \frac{N_c}{V_0} \frac{v_s}{N_A} \left[\left(fN + \frac{1}{2} \right) \left(\frac{\phi_2}{\phi_{2,0}} \right) - \left(\frac{\phi_2}{\phi_{2,0}} \right)^{1/3} \right] \right.$$

$$\left. + \left(\frac{1}{2} - \chi \right) \phi_2^2 + \frac{\phi_2^3}{3} + \frac{\phi_2^4}{4} + \cdots \right\} \tag{4.50}$$

By further introducing a reduced polymer volume fraction, $\varrho = \phi_2/\phi_{2,0}$, and defining $A = \frac{k_b + 2\Delta S}{2k_b}$, $\Theta = \frac{2\Delta H}{k_b + 2\Delta S}$, and $\chi = \frac{1}{2} - A\left(1 - \frac{\Theta}{T}\right)$, we can rewrite this equation as:

$$\Pi = \frac{k_b T N_A}{v_s} \left\{ -\left(\frac{v_s N_c}{V_0 N_A} \varrho^{1/3} \right) + \left(\frac{v_s N_c}{V_0 N_A} \right) \left(fN + \frac{1}{2} \right) \varrho + A\phi_0^2 \left(1 - \frac{\Theta}{T} \right) \varrho^2 \right.$$

$$\left. + \frac{\phi_{2,0}^3}{3} \varrho^3 + \cdots \right\} \tag{4.51}$$

which is analogous to the virial expansion for a van der Waals fluid,

$$P_r = T_r \{ \varrho_r + \varrho_r^2 [1 - (1/T_r)] + \varrho_r^3 + \varrho_r^4 + \cdots \} \tag{4.52}$$

where P_r, T_r, and ϱ_r are the reduced pressure, temperature, and density of the fluid, respectively. In this analogy, swollen and deswollen microgel states would correspond to the gas and liquid phases of a fluid, as schematically shown in Figure 4.5. Furthermore, the Boyle temperature of the fluid corresponds to the so-called Θ temperature of the microgel. For $T = T_r$ or for $T = \Theta$, the second virial coefficient vanishes and binary interactions among constituents are negligible. At this temperature, either system thus behaves ideally, provided ϱ_r or ϱ is sufficiently small.

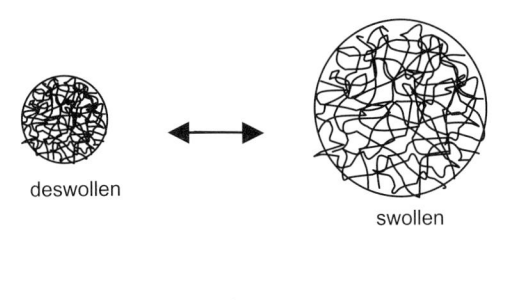

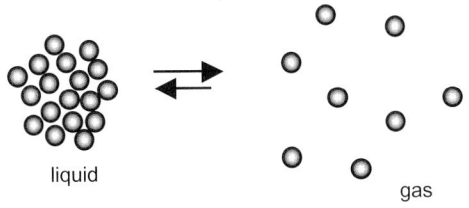

Figure 4.5 Analogy between microgels and atomic fluids. The deswollen and swollen states of a microgel particle correspond to the liquid and gas states of a fluid of van der Waals.

Despite the similarities, there is a key difference between Equations 4.51 and 4.52; for microgels, there is a negative $\varrho^{1/3}$ term that has no analogue in the van der Waals fluid case. This contribution to the osmotic pressure results from the network elasticity. For a microgel in a good solvent, the balance between all positive contributions to Π, and the $\varrho^{1/3}$ contribution fixes the equilibrium particle size. In contrast, a gas cannot withhold its own volume and always expands to occupy all available space. For a microgel, the network elasticity restricts this expansion, restricting the maximum particle size that can be achieved.

4.3
Theory Versus Experiment

The swelling behavior of microgels has been widely investigated and there is a rich phenomenology that results when the relevant experimental parameters, including particle properties, are changed. In this section, we review some of the most relevant experimental results and describe them based on the thermodynamic theory presented above. These results all pertain to measurements of the particle size, which is usually determined by dynamic light scattering (DLS) in the case of microgels [58, 59].[4]

4.3.1
Role of Flory Solubility Parameter

The typical size dependence of PNIPAM microgels with temperature essentially results from changes in the polymer network solubility. At low temperatures, the particle is swollen and highly mixed with the surrounding solvent, as shown in Figure 4.1. As the temperature increases, this mixing degrades and the particle shrinks to reach a terminal deswollen size for $T > 35\,°C$, as also shown in Figure 4.1. Since solubility is the driving mechanism for this transition, we use the equation of state, Equation 4.46, and solve for χ taking into account that $\Pi_{ionic} = 0$, since PNIPAM is a neutral microgel. We obtain:

$$\chi = \frac{1}{\phi_2^2}\left\{\frac{N_c}{V_0}\frac{v_s}{N_A}\left[\frac{\phi_2}{2\phi_{2,0}} - \left(\frac{\phi_2}{\phi_{2,0}}\right)^{1/3}\right] - \phi_2 - \ln(1-\phi_2)\right\} \quad (4.53)$$

To compute the T-dependence of the Flory parameter, we need to relate the particle size, which is what is usually measured, to the polymer volume fraction, which is the relevant quantity in the previous equation. This is easy to do if the swelling is homogeneous and isotropic. In this case,

$$\frac{\phi_2}{\phi_{2,ref}} = \frac{V_{ref}}{V} = \left(\frac{d_{ref}}{d}\right)^3 \quad (4.54)$$

---

4) Pelton et al. [59] were the first to perform light scattering experiments with PNIPAM microgels.

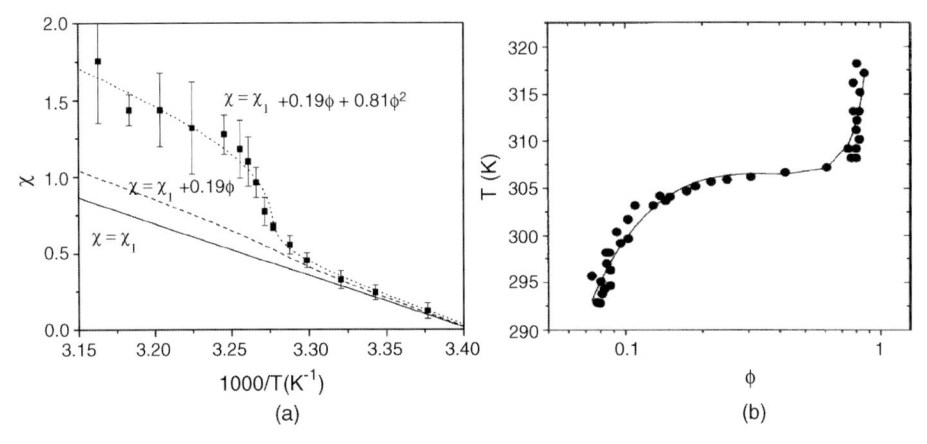

Figure 4.6 (a) Flory parameter versus $1/T$ for a PNIPAM microgel. The curves correspond to $\chi = \chi_1$ (solid), $\chi = \chi_1 + \chi_2\phi_2$ (dashed) and $\chi = \chi_1 + \chi_2\phi_2 + \chi_3\phi_2^2$ (dotted). The second order approximation provides the best fit of the experimentally obtained χ-values. (Reprinted with permission from Ref. [60]. Copyright (2007) by the American Physical Society.) (b) Temperature-induced volume transition of a different PNIPAM microgel. The solid line is the theoretical prediction including the first-order concentration dependence of χ, with $\chi_2 = 0.9$. (Reprinted with permission from Ref. [68]. Copyright (2002) by the American Physical Society.)

with $\phi_{2,\text{ref}}$ the polymer volume fraction of the microgel particle in a reference state where the particle volume and diameter are V_{ref} and d_{ref}, respectively. In the case of microgels, this reference state is often taken as the deswollen state, which further corresponds to the conditions of particle synthesis; as a result, for microgels, $\phi_{2,\text{ref}}$ is often taken as $\phi_{2,0}$. Using Equations 4.53 and 4.54, it is straightforward to calculate χ from the experimentally measured sizes. An example of this is provided in Figure 4.6a, where the Flory parameter is plotted as a function of $1/T$ for a PNIPAM microgel [60]. At low temperatures, the particles are swollen and χ is linear with $1/T$, as expected from its definition; the slope and intercept characterizing this linear trend correspond to $\Delta H/k_b$ and $\Delta S/k_b$, respectively. In this temperature region, $\chi < 1/2$, as expected for a polymer in a good solvent. However, for sufficiently high temperatures or low $1/T$, this linearity no longer holds and χ exhibits more pronounced temperature dependence. This feature, which is often found for most PNIPAM microgels and gels, is attributed to a concentration dependence of χ [61]. Consistent with this, solvent activity measurements with a number of different systems reveal a nonlinear dependence of χ with polymer volume fraction [62–64]. In addition, the phase transitions in neutral gels can only be explained if a $\chi = \chi(\phi_2)$ dependence is adopted, which is taken as a further support for the concentration dependence of χ [65–67].

In the original theory, χ is related to the change in free energy when pairs of polymer and solvent contacts are replaced by mixed polymer–solvent pairs. However, Erman and Flory suggested the inadequacy of treating exchange interactions merely in terms of contact pairs and asserted that interactions of higher order must be considered [61]. They accounted for these many-body interactions by proposing a power series expansion of χ as a function of ϕ_2:

$$\chi(T,\phi) = \chi_1(T) + \chi_2\phi_2 + \chi_3\phi_2^2 + \cdots \tag{4.55}$$

where $\chi_1(T)$ is the original Flory parameter that is given in Equation 4.9, and $\chi_2, \chi_3, \ldots$ are temperature independent coefficients. Since their proposal, it is customary to use this $\chi = \chi(\phi_2)$ relation with the original Flory theory to interpret the swelling of neutral polymer gels and microgels. The $\chi = \chi(T)$ data shown in Figure 4.6a is well described with Equation 4.53 if the series expansion is taken up to second order, with $\chi_2 = 0.19$ and $\chi_3 = 0.81$. For large particle sizes, $\phi_2 \to 0$ and $\chi \approx \chi_1$. However, for higher values of ϕ_2 and thus for larger T, this is no longer true and the full $\chi = \chi(\phi_2)$ dependence is required in order to justify the swelling behavior.

An alternative, though equivalent, approach to interpret the swelling behavior of typical PNIPAM microgels, is to introduce the power series expansion for χ into the equation of state, Equation 4.46, and perform a least square fit of the experimental T–ϕ_2 data, as shown in Figure 4.6b for another PNIPAM microgel [68]. In this case, the fit is performed for different values of $\phi_{2,0}$[5] leaving as free parameters the constants A, Θ, N_c, and χ_i, with $i \geq 2$. In this example, the best fit is obtained by only retaining the linear χ dependence with ϕ_2; the result is $\chi_2 = 0.9 \pm 0.1$. The values of additional parameters are $A = -49$ and $\Theta = 308$ K, which result in values for the entropy and enthalpy changes when a solvent–solvent contact is replaced by a polymer–solvent contact of $\Delta H = -(3 \pm 1) \times 10^{-20}$ J and $\Delta S = -(1.1 \pm 0.8) \times 10^{-23}$ J/K. These values are negative, indicating the LCST character of PNIPAM microgels: Increasing temperature lowers the Gibbs free energy causing the observed particle deswelling. In addition, the magnitude of ΔH and ΔS are representative of what is usually measured for this kind of microgel particles [68–70]. Similarly, the resultant value for the Θ-temperature always roughly coincides with the temperature where swelling begins, as shown in Figure 4.6b.

In contrast to the agreement in the values of some of the parameters, the values of χ_i, with $i \geq 2$, are more widespread, perhaps reminiscent of the lack of a true physical meaning for these high order coefficients. This is already clear from the two examples we have discussed above, and becomes clearer after looking at the published literature on PNIPAM gels. For instance, Hirotsu has reported differences in the phase transition behavior of two PNIPAM gels with identical composition but synthesized using different reaction initiators [71, 72]. Remarkably, one system exhibited a phase transition at a certain temperature while the other showed a progressive and continuous deswelling. It is known that in order to describe the phase transition of a neutral gel a strong concentration dependence of χ is required. For the gel exhibiting the phase transition, Hirotsu found an experimental value of $\chi_2 = 0.6$. The origin of this difference was attributed to the presence of imperfections induced on the network during the synthesis process, a proposal that has not been investigated since then. The particularities of the $\chi = \chi(\phi_2)$ relation thus depends on

---

5) The reference state is usually unkown; thus, different values of $\phi_{2,0}$ are tested in order to obtain the best fit.

details of the synthesis that are typically not included in any theory. It is the complexity of the many-body interactions that has prevented a microscopic understanding of these effects, which continues to be an open question within all polymer science. The $\chi = \chi(\phi_2)$ relation given in Equation 4.55 must be taken as an empirical relation that aims to account for the various many-body molecular interactions in the system but that lacks a clear physical meaning. However, the use of this relation in the context of Flory theory correctly describes the swelling behavior of neutral gels, providing a unifying framework to understand the broad range of behaviors exhibited by these fascinating materials.

4.3.2
Influence of Cross-Linking Density

The extent of swelling is mainly controlled by the elasticity of the polymer network, which is directly related to the cross-linker concentration. The influence of the cross-linking content has been extensively studied. McPhee et al. [73] were the first to report the influence of the cross-linker methylenbisacrylamide (BA) on the swelling of PNIPAM microgels and since then, many others have published similar results on PNIPAM-based microgels [74–80]. As intuitively expected, the swelling capability of a microgel decreases with increasing cross-linker concentration and the volume transition broadens, as shown in Figure 4.7, where we plot the particle size normalized by the deswollen size versus temperature [81]. Increasing the cross-linker content of the particle induces a more gradual size change. The transition temperature, however, is nearly independent of the cross-linker concentration, indicating no significant change in Π_{mix}. Similar results have been obtained with macroscopic gels [82] and ionic microgels [83–85].

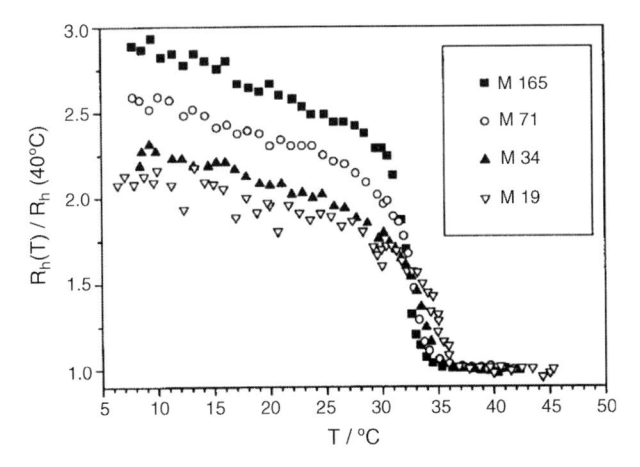

Figure 4.7 Microgel radius, normalized with the high-temperature radius, as a function of temperature for different cross-linker content. The particles are made of PNIPAM cross-linked with methylenbisacrylamide (BA). The name of the sample indicates the molar ratio of NIPAM to BA. (Reprinted with permission from Ref. [81]. Copyright (2000) by Springer.)

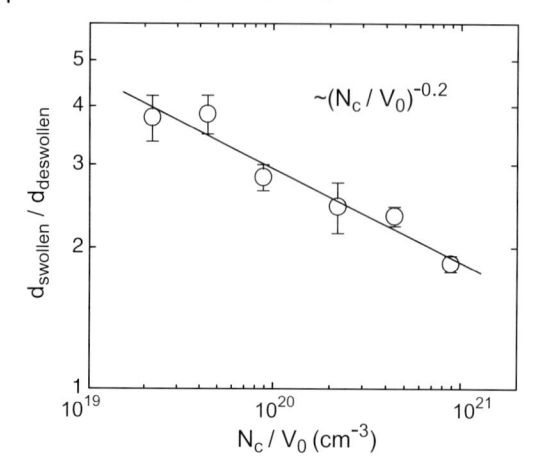

Figure 4.8 Low temperature size, where the particles are swollen, normalized with the high temperature size, where the particles are deswollen, as a function of the density of chains, N_c/V_0. The particles are all based on PNIPAM, cross-linked with BA. The line reflects the theoretically expected scaling, $d \sim (N_c/V_0)^{-0.2}$. (Reprinted with permission from Ref. [13]. Copyright (2009) by Elsevier.)

To quantify the lower T behavior of the particle size with the cross-linker concentration, we plot the size versus the density of chains, N_c/V_0, in Figure 4.8, where N_c is obtained from the cross-linker concentration using Equation 4.47. The microgel deswells with increasing N_c/V_0 following a power law behavior within the N_c/V_0 range explored [13]: $d \sim (N_c/V_0)^{-0.2}$. To account for this result, we consider the relevant contributions to the total osmotic pressure; we thus balance Π_{el} and Π_{mix}, taking into account that $\phi_2 \ll 1$ for the experimental situation of Figure 4.8, which corresponds to swollen particles. As a result, $(\phi_2/\phi_{2,0}) \ll (\phi_2/\phi_{2,0})^{1/3}$ in Equation 4.13 and $\ln(1-\phi_2) \approx -\phi_2 - \phi_2^2/2$ in Π_{mix}. Finally,

$$-\frac{N_c v_s}{V_0}\left(\frac{\phi_2}{\phi_{2,0}}\right)^{1/3} \approx \left(\frac{1}{2}-\chi\right)\phi_2^2 \tag{4.56}$$

which provides the relation between the particle size and the number of chains. By using Equation 4.54, this balance predicts that $d \sim (N_c/V_0)^{-0.2}$, consistent with the experimental result and confirming the interplay between mixing and elasticity to determine the equilibrium size of neutral microgels.

4.3.3
Effect of Charge Density

In the case of ionic microgels, Π_{ionic} is an important contribution to the total osmotic pressure. As a result, the particular value of χ can dramatically affect the phase behavior; the particle can swell, when χ is below the critical χ, or it can experience a phase transition, when χ is above the critical χ. In equilibrium, $\Pi = 0$.

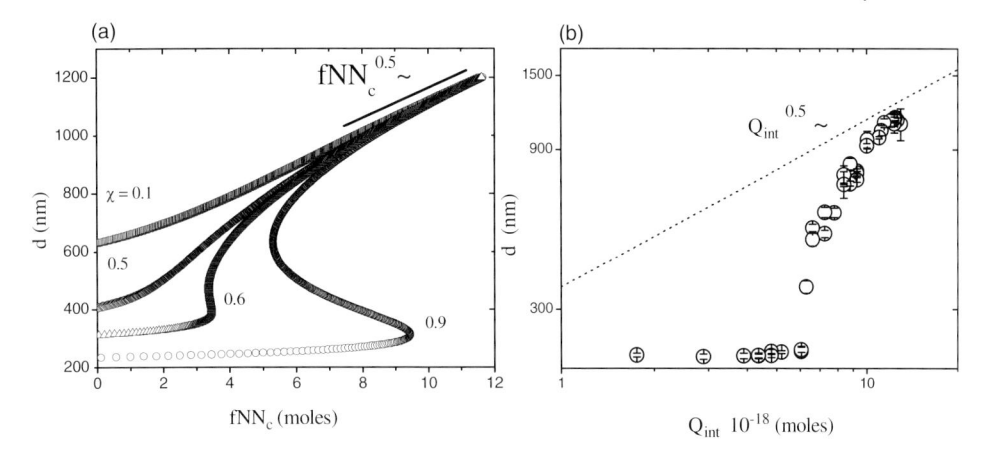

Figure 4.9 Influence of the fixed charge density on the swelling behavior of ionic microgels. (a) Theoretical expectations for the particle size as a function of the number of counterions inside the particle, for different values of χ. While at low fNN_c, the particle size greatly depends on χ, for large fNN_c, the particle size is independent of χ. In this large fNN_c limit, $d \sim [fNN_c]^{0.5}$. The curves are obtained using $d_0 = 200$ nm; $\phi_{2,0} = 1$ and a cross-linker concentration of 0.25 wt%. (b) Experimental relation between particle size and fixed charge, Q_{int}, for polyvinylpyridine microgels. We note that in the experimental situation, $Q_{int} \approx fNN_c$, since $n \ll fc$. The data is consistent with the theoretical expectations at high fNN_c. (Reprinted with permission from Ref. [86]. Copyright (2000) by the American Chemical Society.)

This equation provides an explicit relation between the microgel size and the number of counterions per particle, as shown in Figure 4.9a. As expected, the value of χ determines whether the system continuously changes size or experiences a phase transition. For low fNN_c values, there is a strong dependence of the particle size with χ, indicating that swelling is controlled, in this region, by the mixing and elastic contributions to Π, as the case of neutral microgels. In addition, for low χ values, the size increases monotonically with fNN_c. In contrast, for larger χ values, the microgel size is not very much dependent on fNN_c, for low fNN_c; this results from the poor mixing between the polymer and the solvent, which renders the increase in Π_{ionic} with fNN_c insignificant in this region. In addition, the size exhibits a nonmonotonic dependence with fNN_c, indicative of a phase transition. Interestingly, for high fNN_c, the microgel size is no longer affected by χ and thus it is insensitive to the polymer–solvent mixing. In this case, the microgel size results from the balance between the ionic contribution to Π, which tends to swell the particle, and the elastic contribution to Π, which prevents swelling. The balance of these two contributions yields the scaling $d \sim (fNN_c)^{1/2}$, as shown by the line in Figure 4.9a for large fNN_c. In fact, by considering the dominant term in Π_{el} and balancing it with Π_{ionic} we obtain

$$\Pi_{ionic} \approx \Pi_{el} \Leftrightarrow \frac{fNN_c k_b T}{V} \approx \frac{k_b T N_c}{V_0}\left(\frac{\phi_2}{\phi_{2,0}}\right)^{1/3} \quad (4.57)$$

which results in the previous scaling relation after using the size-ϕ_2 relation provided by Equation 4.54. Interestingly, this scaling relation has been experimentally confirmed with ionic microgels based on polyvinylpyridine [86] and for PNIPAM-co-Acrylic acid microgels [87]. In these cases, the solution pH provides an external trigger to induce changes in the particle size. For microgels based on vinylpyiridine (VP), which is a weak base with a pK_a of ~ 5 [88] that ionizes at low pHs, changing the pH, changes the charge of the polymer network, Q_{int}, which in turn changes the number of counterions inside the microgel particle, since $Q_{int} = fNN_c$ due to electroneutrality. For low Q_{int}, the particle size does not appreciably increase with Q_{int}, consistent with the theoretical expectations. In this situation, the size is essentially determined by the mixing of the polymer with the solvent since $\Pi_{el} \approx 0$, as $\phi_2 \approx \phi_{2,0}$ for deswollen microgels. For larger Q_{int}, the microgel size increases due to Π_{ionic} and eventually for sufficiently large Q_{int}, d grows with Q_{int} in a way that is consistent with the expected scaling behavior, as show in Figure 4.9b.

4.3.4
Salt Effects

The influence of any added salt depends on its relation to the counterion concentration. If $n < fNN_c/V$, the added salt is not significant and Π_{ionic} is essentially given by the ideal gas contribution due to the counterions, as shown in Equation 4.17. In contrast, if $n > fNN_c/V$, this is no longer true and Π_{ionic} is given by Equation 4.23. Consistent with this, the size of polyvinylpyridine microgel particles at pH = 3.4, when most network charges are ionized, exhibits two different regimes when measured as a function of salt concentration. For low salt concentration, the size is insensitive to the value of n, as shown in Figure 4.10; this corresponds to the $n < fNN_c/V$ limit, where Π_{ionic} is given by Equation 4.17 and is thus independent of n. For larger salt concentrations, the particle deswells with increasing n, as also shown in Figure 4.10. In this region, it is experimentally found that $d \sim n^{-1/5}$. To account for this scaling, we balance the relevant contributions to Π; we thus balance the leading term in the elastic osmotic pressure and the ionic contribution given by Equation 4.23, which is correct provided that $n > fNN_c/V$:

$$\frac{k_b T \phi_2^2 (fN)^2}{n} \approx \frac{N_c k_b T}{V_0} \left(\frac{\phi_2}{\phi_{2,0}}\right)^{1/3} \tag{4.58}$$

This balance provides the experimentally found relation between d and n, provided Equation 4.54 is used to relate d to ϕ_2. The microgel size is thus obtained by a balance between the network elasticity and the ionic contribution to the total osmotic pressure, which must account for the uneven salt distribution inside and outside the microgel particle. Interestingly, when the polymer network is partially ionized, for pH = 4.2 in the experiments shown in Figure 4.10, a maximum in the d–n curve is observed. Above the maximum, the microgel size decreases following the $d \sim n^{-1/5}$ scaling relation. In contrast, for salt

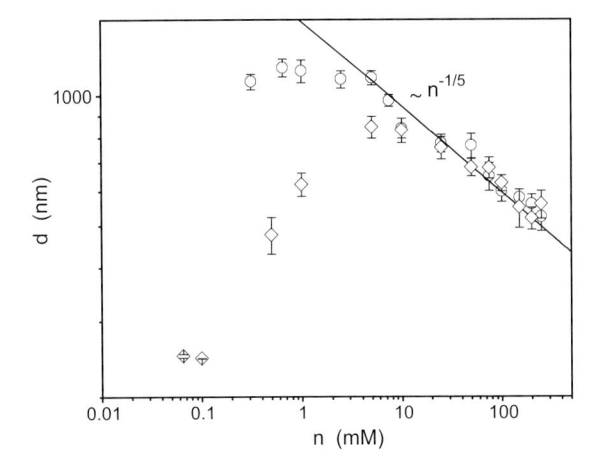

Figure 4.10 Effect of salt concentration on the particle size of ionized microgel particles at different charge states (○) pH = 3.4 (totally charged) and (◇) pH = 4.2 (partially charged). The decreasing part of the d–n curves is consistent with the expected scaling law based on the dominant contributions to the total osmotic pressure. (Reprinted with permission from Ref. [90]. Copyright (2001) by the American Institute of Physics.)

concentrations below the maximum, the microgel size increases with n. Remarkably, this happens in the salt concentration range where the particle size does not change with n, when the polymer network is fully ionized, as shown in Figure 4.10. For a partially ionized network, charge regulation is active, since there are possible groups in the polymer backbone, which are susceptible of becoming ionized [87, 89]. The addition of salt causes ionic redistributions between the inside and the outside of the microgel particles and these redistributions involve H^+ ions, which participate in the charge regulation mechanism that is responsible for charging of the VP groups. As a result, the addition of salt causes the ionization of some previously uncharged groups in the network, which additionally drive more ions into the microgel particle causing the observed swelling [90]. This behavior persists until $n > fNN_c/V$, where the salt concentration causes deswelling of the microgel particle.

Overall, the swelling of ionic microgels is controlled by the osmotic pressure of the ions. The ionic redistribution inside and outside the microgel determines the ionic osmotic pressure, which can either swell or deswell the microgel, depending on whether the network is or is not fully ionized, and whether the salt concentration is larger or not than the counterion concentration inside the particle. The theory of Flory for polymer gels with the consideration of these ionic effects is able to capture the essential physics behind the experimental observations. However, there are situations where this has not been the case. Recently, Pelton and Hoare [89] have found that microgels with different charged group distributions show dramatically different swelling behaviors as a function of the network ionization degree. In addition, Lyon et al. [91] observed sharp differences in the swelling kinetics of PNIPAM microgels copolymerized with acrylic acid depending on the distribution of the ionizable groups. This is closely in line with the results of Pelton et al., despite the fact that they are kinetic and not equilibrium results. The theory of Flory is unable to describe the observed effects, as it does not account for any inhomogeneities within the gel microstructure. It is, however, well known from light [92, 93] and neutron [68, 94–96] scattering that most microgels are inhomogeneous to a certain extent, since the cross-linker typically reacts faster than all other monomers [74, 97]. Nevertheless, the assumption of homogeneous swelling seems to correctly capture the experimental results in many cases. However, when the charge distribution is not homogeneous, this assumption dramatically fails since in this case the electrostatic contribution to the free energy is no longer negligible compared to the ideal gas contribution of the ions. In fact, by considering the particular distribution of fixed charge within the polymer network, the electrostatic contribution to Π can be derived and included in the equation of state. Following this approach, Pelton and Hoare were able to describe their swelling data [89].

We note that within these new theoretical developments, as with the theory of Flory, the various contributions to the total osmotic pressure are treated independent from each other. However, there are possible couplings that one could easily imagine. For instance, ionization of the polymer network could easily be thought of as also changing the polymer solubility. These couplings are not accounted for in any theory and it has been suggested recently that they could play a relevant role in the resultant swelling behavior [87].

4.3.5
Effect of Added Polymer

When a polymer is added to a microgel suspension, the equilibrium condition must account for the resulting external osmotic pressure, Π_{ext}. As a result, the equation of state becomes

$$\Pi = \Pi_{mix} + \Pi_{el} + \Pi_{ionic} + \Pi_{ext} = 0 \qquad (4.59)$$

Kiefer et al. were the first to look at the effect of polymer addition on the swelling of microgels [98]. Later, Saunders and Vincent [99–101] performed a detailed study of the osmotic deswelling of PNIPAM and polystyrene microgels in the presence of poly(ethyleneglycol), PEG, of high molecular weight. They found that the particles deswelled with increasing polymer concentration and explained their results by considering that the added free polymer exerted an external osmotic pressure which contributed to particle deswelling.

For the case of ionic microgels, the presence of an external osmotic pressure can also cause deswelling [102]. However, there is a threshold Π_{ext} that is required before this can be achieved, as shown in Figure 4.11, where we plot the size of VP microgels as a function of the external osmotic pressure. In fact, deswelling can only begin when Π_{ext} is comparable to the microgel bulk modulus, K, which is the inverse of the osmotic compressibility [25]; only when $\Pi_{ext} > K$, deswelling can

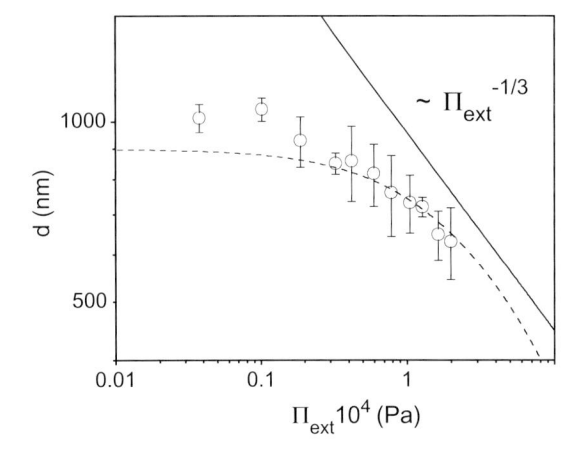

Figure 4.11 Size dependence on external osmotic pressure for polyvinypyridine microgels. The external osmotic pressure is induced using dextrans as stressing polymers. The dash line corresponds to the theoretical prediction vertically rescaled to emphasize it qualitatively captures the overall d–Π_{ext} behavior. The limiting behavior $d \sim \Pi_{ext}^{-1/3}$, expected at high osmotic pressures, is also plotted. (Reprinted with permission form Ref. [102]. Copyright (2003) by the American Institute of Physics.)

occur. In this region, Π_{ext} competes with Π_{ionic}, which is the relevant contribution to the total osmotic pressure opposing the shrinkage of the particle. From this balance, we get

$$\Pi_{ionic} \approx \Pi_{ext} \Rightarrow d \sim \Pi_{ext}^{-1/3} \qquad (4.60)$$

by using Equation 4.59 and considering that Π_{ionic} is given by the ideal gas contribution. Consistent with this, the experimental data shows that this limiting behavior is indeed approached at sufficiently high Π_{ext}.

Qualitatively different behavior can be obtained if the polymer can penetrate the microgel particle [103]. For instance, PNIPAM microgels copolymerized with acrylic acid in the presence of polyethyleneglycol (PEO) display a maximum size as a function of PEO concentration, as shown in Figure 4.12a; the microgel swells with increasing PEO concentration and after reaching a maximum size, the microgel deswells with any further increase in PEO concentration. Interestingly, this only happens for PEO of sufficiently low molecular weight. When the PEO molecular weight is larger than a certain threshold, the maximum is lost and the microgel simply deswells with increasing PEO concentration, as also shown in Figure 4.12a. The overall behavior results from the interaction between the polymer and the microgel particle, which is related to hydrogen bonding of the PEO to the acrylic acid [104]. This is the driving force that unbalances the concentration of PEO inside and outside the particle, resulting in a net osmotic pressure that contributes to particle swelling. The PEO continues to enter the particle until it no longer physically fits inside it; for PEO concentrations above this point, deswelling is observed as a result of an increased Π_{ext}. Despite this attraction-driven hydrogen bonding, if the

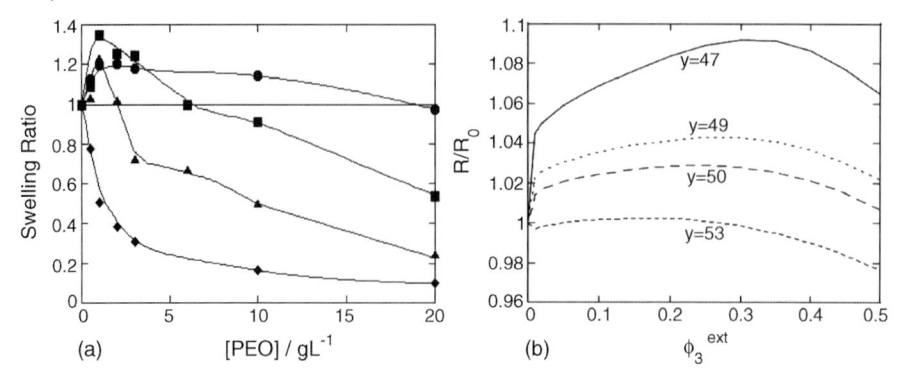

Figure 4.12 Influence of polymer concentration on the swelling of microgel particles. (a) Experimentally determined swelling ratio, defined as the particle size normalized with the particle size for zero polymer concentration, as a function of polymer concentration. The particles are based on PNIPAM-AAc and the polymer is PEO at pH = 3. The different symbols reflect different PEO molecular weights, M_w (g/mol): (●) 2000, (■) 200 00, (τ) 100 000, (♦) 300 000. (Reprinted with permission from Ref. [104]. Copyright (2005) by the American Chemical Society.) (b) Theoretical predictions for the swelling ratio, R/R_0, as a function of polymer concentration, ϕ_3^{ext}, for different polymer molecular weights, which are directly related to y; larger y, implies larger M_w. (Reprinted with permission from Ref. [103]. Copyright (2006) by the American Chemical Society.).

PEO has a sufficiently large molecular weight, it cannot penetrate the microgel. In this case, the polymer remains outside the particle, only contributing to Π_{ext} and causing the observed deswelling.

To account for these results, new contributions to the total osmotic pressure are required. There is mixing of the PEO with the polymer network and the solvent. In particular, depending on the Flory parameter associated with the PEO–microgel interaction, it may be or it may not be favorable for the PEO to enter the microgel particle. When this is favorable from a mixing point of view, there is always an entropic penalty, since the PEO chain would have to stretch in order to fit inside the polymer network. By incorporating these contributions into the equation of state [103], the maximum of the particle size versus PEO concentration can be qualitatively reproduced as shown in Figure 4.12b. Additionally, the height of the maximum gradually decreases with increasing molecular weight of the polymer as shown in Figure 4.12b and consistent with the experimental observations.

4.3.6
Cononsolvency: Swelling in Solvent Mixtures

The so-called cononsolvency is related to the ability of solvents within a solvent mixture to mix with a certain polymer or polymer network. The resulting competition can result in a poorer solubility of the polymer in the solvent mixture than in the individual solvents [105–107]; this depends on the relative affinity between the solvents themselves and between the solvents and the polymer. Cononsolvency was

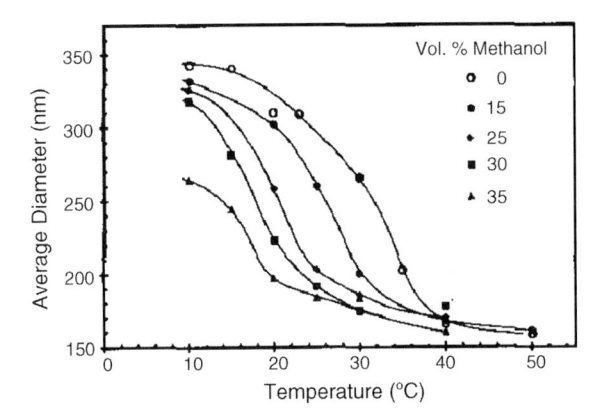

Figure 4.13 Size change of PNIPAM microgels as a function of temperature, for different methanol concentrations. By increasing the alcohol concentration of the water/methanol mixtures, the solvent becomes progressively poorer and the particle size decreases. (Reprinted with permission from Ref. [73]. Copyright (1993) by Elsevier.)

first reported by Winnik et al. [108] for the case of PNIPAM polymer solutions and by McPhee et al. [73] for the case of PNIPAM microgels in alcohol–water mixtures. In these mixtures, the transition temperature shifts to lower temperatures and the microgels deswell as the alcohol concentration increases, as shown in Figure 4.13. Similar results were also published by Mielke et al. for PNIPAM in methanol/water, ethanol/water, and propanol/water mixtures [109].

Interestingly, if the volume fraction of the alcohol is increased above ~ 0.4, a reentrant swelling is observed, as shown in Figure 4.14a for PNIPAM [110, 111] and poly(methylmethacrylate-co-methacrylic acid) microgels in isopropanol/water mixtures [101]. As the volume fraction of isopropanol increases, the particle size initially decreases, reaches a minimum value and then increases with any further increase in the alcohol volume fraction. In addition, increasing temperature decreases the particle size, as expected for any thermosensitive microgel with a LCST. For the case of macroscopic gels, this reentrant volume transition was first observed by Tanaka et al. in mixtures of dimethyl sulfoxide and water [112].

These experimental observations can be qualitatively interpreted on the basis of Flory theory by introducing additional contributions to Π_{mix} due to solvent–solvent and solvent–polymer interactions [113]. A way to do this is to redefine the Flory solvency parameter in terms of these interactions and the solvent composition:

$$\chi = x_1\chi_{1p} + x_2\chi_{2p} - x_1x_2\chi_{12} \qquad (4.61)$$

where x_1 and x_2 are the molar fractions of each component in the mixture and χ_{1p}, χ_{2p}, and χ_{12} are the Flory parameters associated with solvent 1–polymer, solvent 2–polymer, and solvent 1–solvent 2 interactions, respectively. The minus sign in this equation arises from the decrease in the polymer–solvent solubility due to miscibility

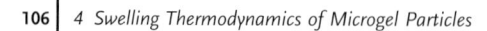

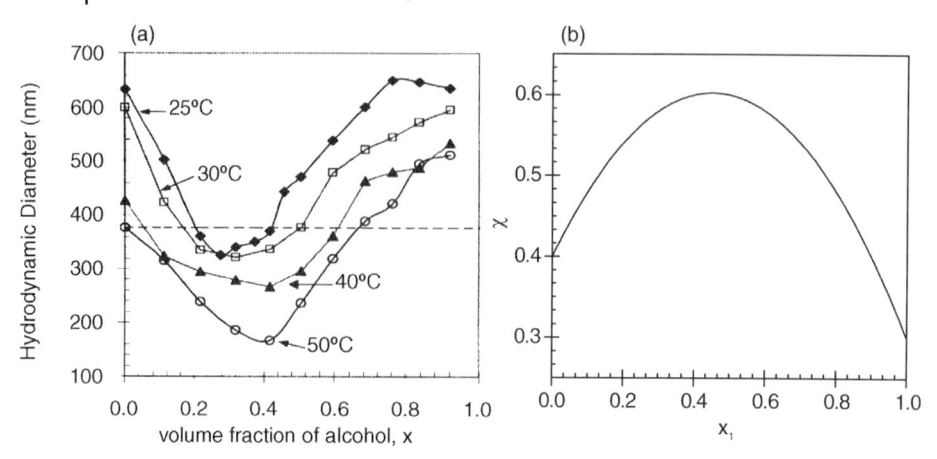

Figure 4.14 (a) Microgel particle size versus alcohol volume fraction, for different temperatures. At each temperature the swelling degree of the particles changes, as indicated by the size change for zero alcohol content. The experimental system is based on PNIPAM microgels dispersed in propanol/water mixtures. The dashed line corresponds to the particle size measured in pure water at 50 °C. (b) Prediction for χ as a function of the molar fraction of alcohol, x_1, when $\chi_{1p}=0.3$, $\chi_{2p}=0.4$, and $\chi_{12}=-1$. The nonmonotonous behavior of χ qualitatively describes the experimental observations, with the minimum in the particle size corresponding to the maximum in χ. (Reprinted with permission from Ref. [110]. Copyright (1998) by Springer.)

of the two solvents. For $\chi_{1p}=0.3$, $\chi_{2p}=0.4$, and $\chi_{12}=-1$, χ exhibits a maximum as a function of x_1, as shown in Figure 4.14b. The initial increase in χ decreases the polymer solubility in the solvent mixture causing particle deswelling. In contrast, above the maximum, χ decreases, the mixing in the solvent mixture increases and the particle swells. Note that the maximum only appears for negative values of χ_{12},[6] which reflects the required attractive interaction among water and alcohol at intermediate concentrations.

There are more sophisticated theories to explain these cononsolvency effects. For example, the lattice fluid hydrogen bound theory (LFHB) [114–116] accounts for specific hydrogen bonding between the solvents and the solvent and polymer; this theory has been successfully applied to the swelling of PNIPAM macrogels [117, 118], and should in principle also be applicable to microgel systems.

4.3.7
Surfactant Effects

The effect of surfactant on microgel swelling has been intensively studied [119–122], perhaps due to its industrial relevance in processes and products based on microgel/surfactant mixtures. For instance, such systems are useful to control the rheological

---

6) Note that the second derivative, $\partial^2\chi/\partial x_1^2 = 2\chi_{12}$, is only negative in case that $\chi_{12}<0$.

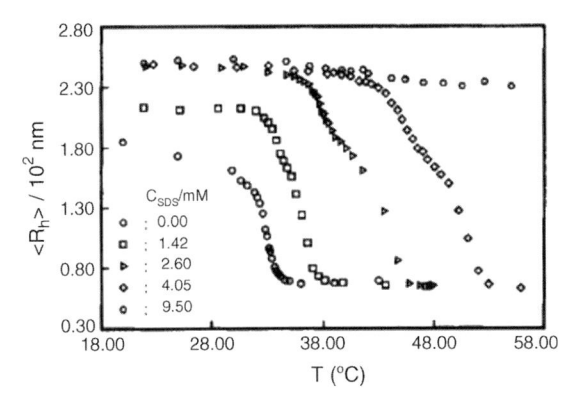

Figure 4.15 Effect of the surfactant sodium dodecyl sulfate concentration on the size of PNIPAM microgels. (Reprinted with permission from Ref. [127]. Copyright (1996) by John Wiley & Sons, Inc.).

properties of coatings [123, 124] or toothpaste [125] and play a crucial role in oil recovery processes [126]. For the case of PNIPAM, the presence of sodium dodecyl sulfate (SDS) increases the magnitude of the volume change at the transition, also causing, at sufficiently high SDS concentrations, a two-step deswelling process, as shown in Figure 4.15 [127]. Additionally, the presence of SDS increases the transition temperature, as also shown in Figure 4.15. These results are easy to understand by considering that SDS interacts with PNIPAM hydrophobically and that additionally it is a charged molecule [127]. The hydrophobic interaction provides the driving force that unbalances the SDS concentration inside and outside the microgel particle, and its charge causes the redistribution of ions inside and outside the microgel particle, giving rise to an ionic contribution to the total osmotic pressure. It is this Π_{ionic} that induces the observed swelling with SDS concentration. The two-step deswelling with temperature can then be rationalized by the competing effects of Π_{mix}, which tends to deswell the microgel, and Π_{ionic}, which tends to swell the microgel. As T increases, the solubility decreases and the particle deswells, as evidenced from the first deswelling step. However, after this reduction in size, the particle volume has dramatically decreased while the charge inside the particle has essentially remained constant; the charge density and thus the counterion concentration inside the microgel particle have risen considerably causing a slowdown in the deswelling process, as observed experimentally. At even higher temperatures, the mixing dominates again over this Π_{ionic} causing the second deswelling step. Similar results have also been obtained with macroscopic PNIPAM gels and explained using a modified ionization degree for the gel that depends on the SDS organization as single molecules, micelles, and whether these are free or bound to the polymer network [128–130]. The influence of surfactants has been also studied in PNIPAM microgels copolymerized with different monomers [131–133], microgels based on vinylcaprolactam [134, 135], and PNIPAM polymers [136–140].

4.4
Additional Aspects

4.4.1
Elastic Moduli

The interest of microgel particles is not just limited to their size tunability, since changes in the particle size also imply changes in the physical properties of the particles. For example, the refractive index of a microgel decreases with swelling. Interestingly, the elasticity of the polymer network also changes with swelling. Two elastic constants are required to describe the elasticity of isotropic materials [141]. For polymer gels, it is convenient to choose the bulk modulus, which is related to the compressibility of the particle, and the shear modulus [65]. Both moduli strongly depend on the polymer volume fraction. By using Equation 4.46, the osmotic modulus is defined as

$$K = -V\left(\frac{\partial \Pi}{\partial V}\right)_T = \phi_2 \left(\frac{\partial \Pi}{\partial \phi_2}\right)_T = \frac{k_b T}{v_s}\left\{\frac{N_c}{V_0}v_s\left[\left(fN+\frac{1}{2}\right)\left(\frac{\phi_2}{\phi_{2,0}}\right) - \frac{1}{3}\left(\frac{\phi_2}{\phi_{2,0}}\right)^{1/3}\right] + \frac{\phi_2^2}{1-\phi_2} - 2\chi\phi_2^2\right\} \quad (4.62)$$

where it depends on the number of chains in the microgel particle, on the number of counterions and on the Flory solubility parameter. In contrast, the shear modulus, μ, is essentially determined by the number of chains per unit volume, which provides connectivity to the polymer network. As a result [25], we get

$$\mu \approx \frac{N_c k_b T}{V} \sim N_c k_b T \phi_2 \quad (4.63)$$

Remarkably, while the shear modulus increases monotonically with ϕ_2, as shown in Figure 4.16, the bulk modulus can exhibit a marked minimum at a certain polymer volume fraction, depending on the values of χ and fNN_c, as also shown in Figure 4.16. This suggests the rich variety of elastic behaviors expected for this class of materials. The remarkable bulk softening of K has been experimentally observed with PNIPAM gels [65] and microgels [142] and could open the way to tune in surprising ways the elastic response of these materials using external triggers.

4.4.2
Brief Remarks on Swelling Kinetics

Perhaps the key difference between microgels and their macroscopic counterparts is the swelling kinetics, which is much faster for microgels than it is for macroscopic gels [85, 143, 144]. This process involves transport of solvent and relaxation of the polymer chains in the network, which must relax to accommodate the gain or lost of solvent by the particle. Based on these considerations, Tanaka et al. derived a diffusion

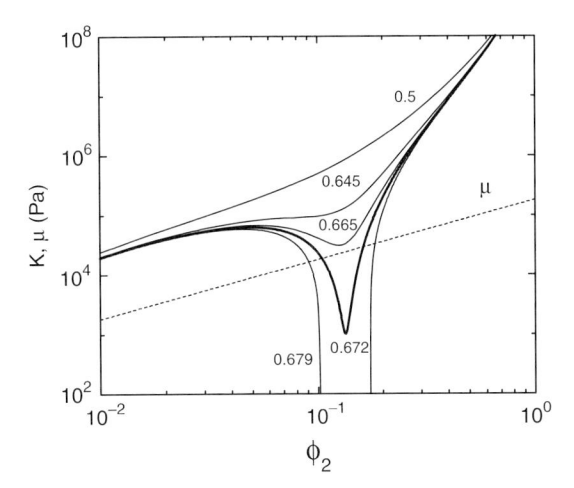

Figure 4.16 Theoretical expectations for the elastic moduli of microgel particles as a function of polymer volume fraction. The dashed line shows the monotonical behavior expected for the shear modulus, μ, whereas the solid lines represent the behavior expected for the bulk modulus, K, depending on the value of χ. For these calculations, we assume no volume fraction dependence for χ and consider $fN = 20$ and $\phi_{2,0} = 1$. For $\chi = 0.679$ the system experiences a phase transition and K becomes negative, which lacks physical meaning; this is the case for a certain range of polymer volume fractions.

model for the polymer in the gel; the key ingredient in this model is the diffusion coefficient of the network, $D = (K + 4\mu/3)/\gamma$, which is determined by its elastic moduli, K and μ, and by the friction coefficient associated with the drag experienced by the network, γ [145, 146]. The kinetics of swelling depends on D and not on the diffusion of the solvent, as the polymer relaxation is always slower than solvent diffusion. For a spherically symmetric swelling process, the diffusion equation can be solved to obtain the time dependence of the particle radius [145]:

$$R(t) = R_f - \left(\frac{6}{\pi^2}\right) \Delta R \sum_{n=1}^{\infty} \frac{\exp(-n^2 t/\tau)}{n^2} \quad (4.64)$$

where R_f is the final equilibrium radius, ΔR is the total radius change for the whole swelling process, and τ is the characteristic swelling time. Note that although this equation is expressed as an infinite series, it rapidly converges, due to its strong dependence on n. As a result, in many practical cases, only the first term suffices to roughly describe the change in particle radius. The characteristic swelling time associated with this change is given by

$$\tau = \frac{R_f^2}{\pi^2 D} \quad (4.65)$$

and thus depends on the square of the particle radius; this explains why microgels are able to change size much faster than macroscopic gels.

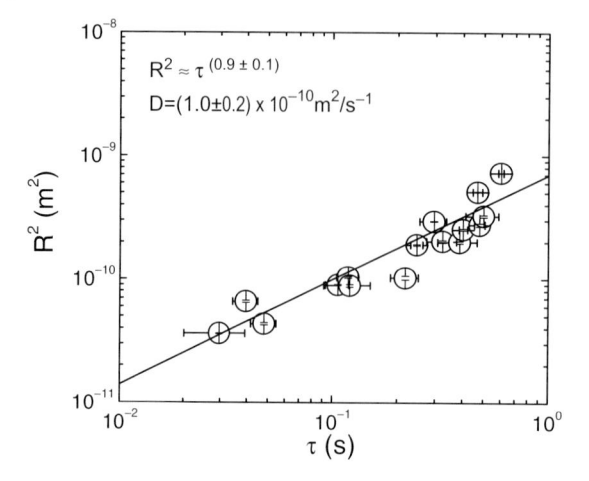

Figure 4.17 Square of the equilibrium particle size for PNIPAM minigels versus the characteristic swelling time. The scaling, $R^2 \sim \tau^{0.9 + 0.1}$, is consistent with the kinetic model of Tanaka for the swelling of polymer gels [145]. (Reprinted with permission from Ref. [143]. Copyright (2006) by the American Chemical Society.)

This simple model for the swelling kinetics correctly captures the experimental observations, as shown in Figure 4.17, where we plot the size squared of PNIPAM minigels as a function of the characteristic swelling time, τ [143]. There is a linear relationship between the two, consistent with Equation 4.65. Similar results were first found with larger polymer gels [147–151]. In addition, there are other measurements of the swelling kinetics reported for a variety of polymer chemistries [104, 152–154]. The values of D found in the literature range between that of the solvent, $D = 2.3 \times 10^{-9}$ m^2/s at 25 °C for water, and values which are 2–3 orders of magnitude smaller, depending on the swollen or deswollen character of the polymer network and on the amount and distribution of the cross-linker.

4.5
Summary

In this chapter, we have provided a description of the thermodynamics required to describe the swelling of microgels, although the presented formalism can very well apply to other polymer networks. In addition to the mixing between the polymer and the solvent, as described by Flory, and the elasticity of the cross-linked network, which is treated within the frame of rubber elasticity, we have extensively considered electrostatic effects within a mean-field approximation. In the case of microgels with a reasonably homogeneous distribution of both cross-linkers and fixed charges, the relevant electrostatic effects arise from the ideal gas contribution of the ions; correlations between ions and between fixed charges in the polymer network only provide negligible contributions compared to these ideal gas contributions. A wealth of experimental results can be described based on the interplay between these three relevant contributions to the free energy. This suggests that the lack of a homogeneous distribution of cross-linker within the microgel particle can be neglected, at least to describe the swelling behavior. Equivalently, small inhomogeneities in the fixed charge distribution, at the level treated here, do not contribute significantly to the osmotic pressure. It is possible that strong correlations due to multivalent ions, either in solution or distributed along the polymer backbone, might have significant impact, but this still constitutes to be an unexplored domain. The range of applications and fundamental studies that take advantage of the responsiveness of microgel particles is continuously increasing. Since much of the macroscopic behavior exhibited by suspensions based on microgels depends on the single-particle properties, it is important to understand microgels as polymer-network entities.

Symbols

c	average monomer concentration inside a particle
c_c	counterion concentration around a counterion
c_p	local monomer concentration
d	particle diameter
d_{ref}	particle diameter in a reference state
D	diffusion coefficient of the polymer network
e_{pp}	energy associated to a polymer–polymer interaction
e_{ps}	energy associated to a polymer–solvent interaction
e_{ss}	energy associated to a solvent–solvent interaction
f	fraction of charged monomers per chain
F	Helmholtz energy
F_{mix}	Helmholtz energy of mixing
G	Gibbs energy
G_{el}	Gibbs energy due to network elasticity
G_{ionic}	Gibbs energy due to ionic contributions
G_{mix}	Gibbs energy of mixing
h	Planck's constant
k_b	Boltzmann constant
K	osmotic modulus
l_p	persistence length
L	stretch of a chain
n	salt concentration
n_{in}	salt concentration inside a particle
n^o	salt concentration outside a particle.
n_1	number of solvent molecules
n_2	number of polymer molecules
N	number of monomers per chain
N_c	number of chains

P	pressure
P_{12}	number of polymer–solvent contacts
q_e	elementary charge
Q_{eff}	effective charge
Q_{int}	fixed charge inside a polymer network
r_c	end-to-end distance of a polymer chain
R	particle radius
S	entropy
S_{el}	elastic entropy
S_{ionic}	entropy due to ionic contributions
S_{mix}	entropy of mixing
$S(q)$	structure factor
T	temperature
U	internal energy
U_{corr}	interaction energy associated to correlation between counterions.
U_{el}	electrostatic internal energy
U_{fluc}	energy due to nonuniformities in the fixed charge distribution inside a particle
U_{ideal}	internal energy of an ideal gas of counterions
$U_{ideal,n}$	internal energy of an ideal gas of counterions with salt concentration n
U_{mix}	internal energy of mixing
v	mean-field potential acting between fixed charges
v_s	volume of a lattice site (solvent molecule)
V	volume of a particle
V_0	volume of a particle in the synthesis conditions
V_{ref}	volume of a particle in a reference state
x_i	molar fraction of the ith component
X	number of moles of cross-linker molecules
z	coordination number of a lattice site
Z_{ion}	number of counterions inside the particle
$Z_{network}$	number of network charges inside the particle
α_i	deformation ratio in the ith direction
γ	drag coefficient experience by the network
Δe	energy change associated to mixing
Δe_u	energetic term of Δe
Δe_s	entropic term of Δe
ΔH	change in enthalpy
ε	relative dielectric permittivity
ε_0	vacuum dielectric permittivity
Θ	theta temperature
$\varkappa$	inverse Debye length
λ_b	Bjerrum length
μ	shear modulus
μ_{ion}	chemical potential of an ion
μ_1	chemical potential of a solvent molecule

μ_1^{pure}	chemical potential of a solvent molecule in a pure solvent solution.
Π	osmotic pressure
Π_{corr}	osmotic pressure due to correlations between counterions
Π_{el}	osmotic pressure due to elasticity
Π_{ext}	osmotic pressure externally induced by a polymer solution
Π_{ionic}	osmotic pressure due to ionic contributions
Π_{mix}	osmotic pressure of mixing
ϱ_q	charge density
τ	characteristic swelling time
ϕ_1	volume fraction of the solvent
ϕ_2	volume fraction of polymer
$\phi_{2,0}$	volume fraction of polymer in the synthesis conditions
$\phi_{2,ref}$	volume fraction of polymer in a reference state
ϕ_3^{ext}	volume fraction of external added polymer
χ_1	Flory solvency parameter
χ_{ip}	Flory parameter associated to the interaction between the ith component of a mixture and the polymer
χ_{ij}	Flory parameter associated to the interaction between the i and the j components of a mixture
ψ	electrostatic potential
Ψ_j	mean-field electrostatic potential acting on counterion j.
Ω_{el}	number of polymer and cross-linker configurations compatible with the formation of a network.
Ω_{mix}	number of configurations for a polymer solution.

Acknowledgments

We are grateful to Y. Levin, X.J. Xing, J. Mattson, and H.M. Wyss for their useful comments and suggestions. We thank FQM-03116 and DPI2008-06624-C03-03.

References

1 Dusek, K. and Patterson, D. (1968) *J. Polym. Sci. B Polym. Phys.*, **6**, 1209.
2 Tanaka, T. (1978) *Phys. Rev. Lett.*, **40**, 820.
3 Shibayama, M. and Tanaka, T. (1993) Volume phase transition and related phenomena in polymer gels, in *Responsive Gels, Volume Transitions I, Advances in Polymer Science*, vol. 109 (ed. K. Dusek), Springer Verlag.
4 Kokufuta, E. (2001) Transitions in polyelectrolyte gels, in *Physical Chemistry of Polyelectrolytes* (ed. T. Radeva), Taylor & Francis Ltd.
5 Israelachvili, J.N. (1992) *Intermolecular and Surface Forces*, Academic Press.
6 Chandler, D. (2005) *Nature*, **437**, 640.
7 Pelton, R.H. and Chibante, P. (1986) *Colloid Surface*, **20**, 247.
8 Funke, W., Okay, O., and Joos-Muller, B. (1998) *Microencapsulation–Microgels–Iniferters*, vol. 33, Springer-Verlag Berlin, Berlin, p. 139.
9 Saunders, B.R. and Vincent, B. (1999) *Adv. Colloid Interface Sci.*, **80**, 1.

10 Pelton, R. (2000) *Adv. Colloid Interface Sci.*, **85**, 1.
11 Nayak, S. and Lyon, L.A. (2005) *Angew. Chem. Int. Ed.*, **44**, 7686.
12 Das, M., Zhang, H., and Kumacheva, E. (2006) *Ann. Rev. Mater. Res.*, **36**, 117.
13 Fernandez-Barbero, A. et al. (2009) *Adv. Colloid Interface Sci.*, **147–48**, 88.
14 Galaev, I. and Mattiasson, B. (2007) *Smart Polymers: Applications in Biotechnology and Biomedicine*, Taylor & Francis, Inc.
15 Tiktopulo, E.I. et al. (1994) *Macromolecules*, **27**, 2879.
16 Wang, X.H., Qiu, X.P., and Wu, C. (1998) *Macromolecules*, **31**, 2972.
17 Sierra-Martin, B. et al. (2006) *Langmuir*, **22**, 3586.
18 Otake, K. et al. (1990) *Macromolecules*, **23**, 283.
19 Cho, E.C., Lee, J., and Cho, K. (2003) *Macromolecules*, **36**, 9929.
20 Woodward, N.C. et al. (2003) *Langmuir*, **19**, 3202.
21 Hirashima, Y. and Suzuki, A. (2004) *J. Phys. Soc. Jpn.*, **73**, 404.
22 Schild, H.G. (1992) *Prog. Polym. Sci.*, **17**, 163.
23 Barrat, J.L., Joanny, J.F., and Pincus, P. (1992) *J. Phys. II*, **2**, 1531.
24 Flory, P.J. (1953) *Principles of Polymer Chemistry*, Cornell University Press, London.
25 Rubinstein, M. and Colby, R.H. (2003) *Polymer Physics*, Oxford University Press, USA.
26 Nicholson, J.W. (2006) *The Chemistry of Polymers*, RSC Publishing.
27 Zemansky, M.W. and Dittman, R.H. (1996) *Heat and Thermodynamics*, McGraw-Hill.
28 Teraoka, I. (2002) *Polymer Solutions: An Introduction to Physical Properties*, John Wiley & Sons, Inc.
29 Whetham, W.C.D. (1924) *Matter and Change*, Cambridge.
30 de Gennes, P.-G. (1979) *Scaling Concepts in Polymer Physics*, Cornell University Press
31 Rodriguez, F. (1996) *Principles of Polymer Systems*, Taylor & Francis.
32 Flory, P.J. and Rehner, J. (1943) *J. Chem. Phys.*, **11**, 512.

33 Flory, P.J. and Rehner, J. (1943) *J. Chem. Phys.*, **11**, 521.
34 Flory, P.J. and Erman, B. (1982) *Macromolecules*, **15**, 800.
35 Sperling, L.H. (2005) *Introduction to Physical Polymer Science*, Wiley-Interscience.
36 Mackintosh, F.C., Kas, J., and Janmey, P.A. (1995) *Phys. Rev. Lett.*, **75**, 4425.
37 Head, D.A., Levine, A.J., and MacKintosh, F.C. (2003) *Phys. Rev. Lett.*, **91**, 108102.
38 Treloar, L.R.G. (1975) *The Physics of Rubber Elasticity*, Clarendon Press, Oxford.
39 Edwards, S.F. and Vilgis, T. (1986) *Polymer*, **27**, 483.
40 Kavassalis, T.A. and Noolandi, J. (1987) *Phys. Rev. Lett.*, **59**, 2674.
41 Warner, M. and Terentjev, E.M. (1996) *Prog. Polym. Sci.*, **21**, 853.
42 Xing, X.J. et al. (2008) *Phys. Rev. E*, **77**, 51802.
43 Xing, X.J., Goldbart, P.M., and Radzihovsky, L. (2007) *Phys. Rev. Lett.*, **98**, 75502.
44 Hansen, R., Skov, A.L., and Hassager, O. (2008) *Phys. Rev. E*, **77**, 11802.
45 Zhang, W.K. et al. (2000) *J. Phys. Chem. B*, **104**, 10258.
46 Bemis, J.E., Akhremitchev, B.B., and Walker, G.C. (1999) *Langmuir*, **15**, 2799.
47 Hill, T.L. (1987) *An Introduction to Statistical Thermodynamics*, Dover Publications.
48 McQuarrie, D.A. (2000) *Statistical Mechanics*, University Science Books.
49 Groot, R.D. (1991) *J. Chem. Phys.*, **94**, 5083.
50 Levin, Y. et al. (2002) *Phys. Rev. E*, **65**, 36143.
51 Borrega, R. et al. (1999) *Europhys. Lett.*, **47**, 729.
52 Levin, Y. (2002) *Rep. Prog. Phys.*, **65**, 1577.
53 Landau, L.D. and Lifshitz, E.M. (1984) *Statistical Physics: Course of Theoretical Physics*, vol. 5, Butterworth-Heinemann.
54 Debye, P. and Hückel, E. (1923) *Physikalische Zeitschrift*, **24**, 185.
55 Shibayama, M., Tanaka, T., and Han, C.C. (1992) *J. Chem. Phys.*, **97**, 6829.
56 Shibayama, M., Tanaka, T., and Han, C.C. (1992) *J. Chem. Phys.*, **97**, 6842.

57 Dusek, K. (1993) *Responsive Gels: Volume Transitions I, Advances in Polymer Science*, vol. 109 Springer Verlag.
58 Berne, B.J. and Pecora, R. (2000) *Dynamic Light scattering: With Applications to Chemistry, Biology and Physics*, Dover Publications, Inc.
59 Pelton R.H. et al. (1989) *Langmuir*, **5**, 816.
60 Lopez-Leon, T. and Fernandez-Nieves, A. (2007) *Phys. Rev. E*, **75**, 011801.
61 Erman, B. and Flory, P.J. (1986) *Macromolecules*, **19**, 2342.
62 Baulin, V.A. and Halperin, A. (2002) *Macromolecules*, **35**, 6432.
63 Kamide, K., Matsuda, S., and Saito, M. (1985) *Polym. J.*, **17**, 1013.
64 Baulin, V.A. and Halperin, A. (2003) *Macromol. Theory Simul.*, **12**, 549.
65 Hirotsu, S. (1991) *J. Chem. Phys.*, **94**, 3949.
66 Hirotsu, S. (1993) *Adv. Polym. Sci.*, **110**, 1.
67 Be, E. and Flory F P.J. (1968). *Trans. Faraday Soc*, **64** 2035.
68 Fernandez-Barbero, A. et al. (2002) *Phys. Rev. E*, **66**, 011801.
69 Pinkrah, V.T. et al. (2004) *Langmuir*, **20**, 8531.
70 Lietor-Santos, J.J. et al. (2009) *Macromolecules*, **42**, 6225.
71 Hirotsu, S., Hirokawa, Y., and Tanaka, T. (1987) *J. Chem. Phys.*, **87**, 1392.
72 Hirotsu, S. (1988) *J. Chem. Phys.*, **88**, 427.
73 McPhee, W., Tam, K.C., and Pelton, R. (1993) *J. Colloid Interface Sci.*, **156**, 24.
74 Wu, X. et al. (1994) *Colloid Polym. Sci.*, **272**, 467.
75 Kratz, K. and Eimer, W. (1998) *Ber. Bunsen Ges. Phys. Chem.*, **102**, 848.
76 Oh, K.S. et al. (1998) *Macromolecules*, **31**, 7328.
77 Hellweg, T. et al. (2000) *Colloid Polym. Sci.*, **278**, 972.
78 Varga, I. et al. (2001) *J. Phys. Chem. B*, **105**, 9071.
79 Sierra-Martin, B. et al. (2005) *Macromolecules*, **38**, 10782.
80 Omari, A. et al. (2006) *J. Colloid Interface Sci.*, **302**, 537.
81 Senff, H. and Richtering, W. (2000) *Colloid Polym. Sci.*, **278**, 830.
82 Inomata, H. et al. (1995) *Polymer*, **36**, 875.
83 Cloitre, M. et al. (2003) *C. R. Phys.*, **4**, 221.
84 Morris, G.E., Vincent, B., and Snowden, M.J. (1997) *J. Colloid Interface Sci.*, **190**, 198.
85 Loxley, A. and Vincent, B. (1997) *Colloid Polym. Sci.*, **275**, 1108.
86 Fernandez-Nieves, A. et al. (2000) *Macromolecules*, **33**, 2114.
87 Capriles-Gonzalez, D. et al. (2008) *J. Phys. Chem. B*, **112**, 12195.
88 Perrin, D.D. (1965) *Dissociation Constants of Organic Bases in Aqueous Solution*, Butterworths, London.
89 Hoare, T. and Pelton, R. (2007) *J. Phys. Chem. B*, **111**, 11895.
90 Fernandez-Nieves, A., Fernandez-Barbero, A., and de las Nieves, F.J. (2001) *J. Chem. Phys.*, **115**, 7644.
91 Jones, C.D. and Lyon, L.A. (2003) *Macromolecules*, **36**, 1988.
92 Fernandez-Nieves, A., de las Nieves, F.J., and Fernandez-Barbero, A. (2004) *J. Chem. Phys.*, **120**, 374.
93 Reufer, M. et al. (2009) *Eur. Phys. J. E*, **28**, 165.
94 Saunders, B.R. (2004) *Langmuir*, **20**, 3925.
95 Stieger, M. et al. (2004) *J. Chem. Phys.*, **120**, 6197.
96 Lopez-Cabarcos, E. et al. (2004) *Phys. Chem. Chem. Phys.*, **6**, 1396.
97 Guillermo, A. et al. (2000) *J. Polym. Sci. B Polym. Phys.*, **38**, 889.
98 Kiefer, J. et al. (1993) *Colloid Polym. Sci.*, **271**, 253.
99 Saunders, B.R. and Vincent, B. (1996) *J. Chem. Soc. Faraday Trans.*, **92**, 3385.
100 Saunders, B.R. and Vincent, B. (1997) *Colloid Polym. Sci.*, **275**, 9.
101 Saunders, B.R., Crowther, H.M., and Vincent, B. (1997) *Macromolecules*, **30**, 482.
102 Fernandez-Nieves, A. et al. (2003) *J. Chem. Phys.*, **119**, 10383.
103 Routh, A.F. et al. (2006) *J. Phys. Chem. B*, **110**, 12721.
104 Bradley, M., Ramos, J., and Vincent, B. (2005) *Langmuir*, **21**, 1209.
105 Schild, H.G., Muthukumar, M., and Tirrell, D.A. (1991) *Macromolecules*, **24**, 948.
106 Winnik, F.M. et al. (1992) *Macromolecules*, **25**, 6007.
107 Tanaka, F., Koga, T., and Winnik, F.M. (2008) *Phys. Rev. Lett.*, **101**, 28302.

108 Winnik, F.M., Ringsdorf, H., and Venzmer, J. (1990) *Macromolecules*, **23**, 2415.
109 Mielke, M. and Zimehl, R. (1998) *Ber. Bunsen Ges. Phys. Chem.*, **102**, 1698.
110 Crowther, H.M. and Vincent, B. (1998) *Colloid Polym. Sci.*, **276**, 46.
111 Saunders, B.R. et al. (1999) *Colloids Surf. A Physicochem. Eng. Asp.*, **149**, 57.
112 Katayama, S., Hirokawa, Y., and Tanaka, T. (1984) *Macromolecules*, **17**, 2641.
113 Amiya, T. et al. (1987) *J. Chem. Phys.*, **86**, 2375.
114 Panayiotou, C. and Sanchez, I.C. (1991) *J. Phys. Chem.*, **95**, 10090.
115 Lele, A.K., Devotta, I., and Mashelkar, R.A. (1997) *J. Chem. Phys.*, **106**, 4768.
116 Lele, A.K. et al. (1995) *Chem. Eng. Sci.*, **50**, 3535.
117 Lele, A.K. et al. (1997) *Macromolecules*, **30**, 157.
118 Lele, A.K. et al. (1997) *J. Chem. Phys.*, **107**, 2142.
119 Schild, H.G. and Tirrell, D.A. (1991) *Langmuir*, **7**, 665.
120 Gao, Y.B. et al. (1997) *J. Macromol. Sci. Phys.*, **B36**, 417.
121 Andersson, M. and Maunu, S.L. (2006) *J. Polym. Sci. B Polym. Phys.*, **44**, 3305.
122 Bradley, M., Vincent, B., and Burnett, G. (2007) *Langmuir*, **23**, 9237.
123 Wright, H.J., Leonard, D.P., and Etzell, R.A. (1981) US Patent 4290932.
124 Ishii, K. (1999) *Colloids Surf. A*, **153**, 591.
125 Aizawa, T., Nakamura, H., and Yamaguchi, T. (1994) US Patent 5338815.
126 Snowden, M.J., Vincent, B., and Morgan, J.C. (1993) Patent GB 226 2117A.
127 Wu, C. and Zhou, S.Q. (1996) *J. Polym. Sci. B Polym. Phys.*, **34**, 1597.
128 Kokufuta, E. et al. (1993) *Macromolecules*, **26**, 1053.
129 Saito, S., Konno, M., and Inomata, H. (1993) *Adv. Polym. Sci.*, **109**, 207.
130 Kokufuta, E. et al. (1995) *Macromolecules*, **28**, 1704.
131 Huang, J. and Wu, X.Y. (1999) *J. Polym. Sci. A Polym. Chem.*, **37**, 2667.
132 Bradley, M. and Vincent, B. (2008) *Langmuir*, **24**, 2421.
133 Matsukata, M. et al. (1998) *Colloid Polym. Sci.*, **276**, 11.
134 Peng, S.F. and Wu, C. (2001) *Macromolecules*, **34**, 568.
135 Gao, Y.B., Au-Yeung, S.C.F., and Wu, C. (1999) *Macromolecules*, **32**, 3674.
136 Goddard, E.D. (1986) *Colloid Surface*, **19**, 255.
137 Ricka, J. et al. (1990) *Phys. Rev. Lett.*, **65**, 657.
138 Meewes, M. et al. (1991) *Macromolecules*, **24**, 5811.
139 Zhu, P.W. and Napper, D.H. (1996) *Langmuir*, **12**, 5992.
140 Walter, R. et al. (1996) *Macromolecules*, **29**, 4019.
141 Crandall, S. and Lardner, T. (1999) *An Introduction to the Mechanics of Solids*, McGraw-Hill.
142 Hashmi, S.M. and Dufresne, E.R. (2009) *Soft Matter*, **5**, 3682.
143 Suarez, I.J., Fernandez-Nieves, A., and Marquez, M. (2006) *J. Phys. Chem. B*, **110**, 25729.
144 Yoshida, R. et al. (1995) *Nature*, **374**, 240.
145 Tanaka, T. and Fillmore, D.J. (1979) *J. Chem. Phys.*, **70**, 1214.
146 Peters, A. and Candau, S.J. (1988) *Macromolecules*, **21**, 2278.
147 Peters, A. and Candau, S.J. (1986) *Macromolecules*, **19**, 1952.
148 Siegel, R.A. (1993) *Adv. Polym. Sci.*, **109**, 233.
149 Li, Y. and Tanaka, T. (1990) *J. Chem. Phys.*, **92**, 1365.
150 Wang, C.J., Li, Y., and Hu, Z.B. (1997) *Macromolecules*, **30**, 4727.
151 Takahashi, K., Takigawa, T., and Masuda, T. (2004) *J. Chem. Phys.*, **120**, 2972.
152 Dupin, D. et al. (2007) *Langmuir*, **23**, 4035.
153 Yin, J. et al. (2008) *Langmuir*, **24**, 9334.
154 Yan, H. et al. (2005) *Angew. Chem. Int. Ed.*, **44**, 1951.

5
Determination of Microgel Structure by Small-Angle Neutron Scattering

Walter Richtering, Ingo Berndt, and Jan Skov Pedersen

5.1
Introduction

A microgel particle is a chemically cross-linked polymer, which in contrast to a latex sphere is swollen by a good solvent. A cross-linker is incorporated in the microgel network, which on the one hand can give rise to a complex and inhomogeneous internal structure. On the other hand, cross-linking provides the topological integrity and thus microgels differ from colloidal supramolecular aggregates [1]. Microgel dispersions gained great interest as promising candidates for various applications as, for example, in the printing and pharmaceutical industries as well as for drug delivery, separation, catalysis, and microoptics. Microgels are also interesting model systems in the research area of soft condensed matter, where they allow the investigation of the structure and dynamics of concentrated colloidal suspensions. Since the particle interaction forces can be controlled by the properties of the particle, microgels are ideal systems to study the relation between the interaction potential, the phase behavior, and the flow properties.

Many aqueous polymer systems get even more complex as the solubility often changes upon variation of the temperature. The most widely studied water-swellable, temperature-sensitive microgel system is based on poly-*N*-isopropylacrylamide (PNIPAM) first prepared by Pelton *et al.* [2]. Upon heating, above the volume phase transition temperature (VPTT) the size of PNIPAM microgel particles decreases sharply and higher concentrated aqueous PNIPAM solutions display macroscopic phase separation. The macroscopic volume phase transition has been thoroughly studied employing various methods including static and dynamic light scattering, differential scanning calorimetry, viscometry, and rheology [3, 4].

Often, swollen microgel particles exhibit a structure different from that of a homogenous sphere. Scattering methods including small-angle neutron scattering (SANS) and small-angle X-ray scattering (SAXS) as well as static light scattering (SLS) and dynamic light scattering (DLS) are well suited to investigate in detail the structure of PNIPAM microgels.

Dynamic light scattering is probably the most frequently used technique to determine the size of microgel particles. DLS detects temporal fluctuations of the scattered light intensity and the collective diffusion coefficient is obtained from the correlation function. The Stokes–Einstein equation connects the diffusion coefficient at infinite dilution, D_0, with the hydrodynamic radius R_h; in other words R_h is the radius of a sphere with the same diffusion coefficient as the species that are probed in the DLS experiment. Microgels are strongly swollen by the solvent and usually the nondraining limit is reached, that is, the solvent is immobilized inside the microgel. Consequently, DLS is not sensitive to the internal structure of the microgels and the hydrodynamic radius that is obtained from DLS is further influenced by the polymer chains pendant at the particle surface.

5.2
Form Factor of Microgels

From static scattering experiments, a form factor can be obtained, which describes the structure of microgel particles accounting for both the inhomogeneous internal network structure and the overall particle shape. Depending on the particle size, one needs to employ light, neutron, or X-ray scattering or a combination of them. Basically, these methods provide similar information; however, the experimental range of momentum transfer q and the contrast are different. Thus, a suitable choice has to be made depending on microgel properties. Often SANS method is most suitable and therefore, in this chapter, we will focus on SANS. Deuterated solvents or mixtures of deuterated and hydrogenated solvents are used in SANS studies in order to enhance or vary the contrast. However, one should keep in mind that phase boundaries are often slightly different between H- and D-containing solvents due to the changes in the hydrogen bonding to water. General information on scattering methods can be found, for example, in Ref. [5].

SANS experiments yield information about an intensity distribution $I(q)$ in reciprocal space as a function of q where $q = (4\pi/\lambda) \sin(\theta/2)$ denotes the magnitude of the scattering vector (momentum transfer) with wavelength λ and scattering angle θ. In order to obtain the corresponding real space structure considerable effort must be invested in the SANS-data analysis [6]. In this chapter, we discuss a direct modeling approach in which a direct real space model for the structure of the microgel is suggested and the corresponding scattering intensity is calculated. The structural parameters are then optimized by least square fitting and the methods thus allows determining the radial density distribution of polymer segments inside the microgel particle from the experimental scattering curve as schematically illustrated in Figure 5.1.

A convenient way to express $I(q)$ is to introduce the differential scattering cross section $d\sigma(q)/d\Omega$ since it is independent of transmission and form of the sample. For a suspension of monodisperse particles with spherical symmetry, the differential scattering cross section is given by

$$\frac{d\sigma(q)}{d\Omega} = n\Delta\varrho^2 V_{poly}^2 P(q) S(q) \tag{5.1}$$

5.2 Form Factor of Microgels

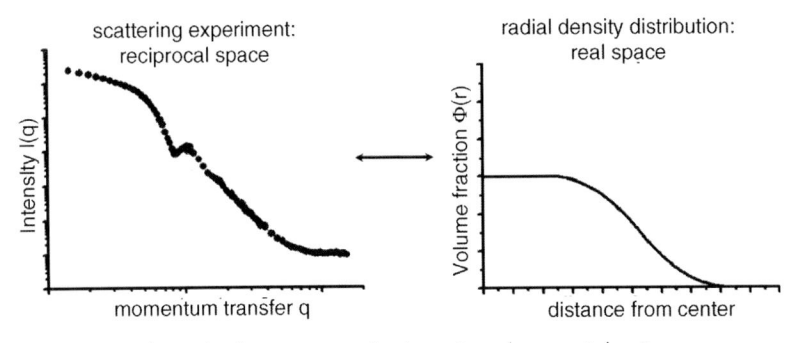

Figure 5.1 Relationship between scattering intensity and segment density.

where n denotes the particle number density of the microgels, $\Delta\varrho$ describes the difference in scattering length density between the polymer and the solvent, and V_{poly} accounts for the volume of polymer in a particle as obtained from the apparent partial specific volume or density of the polymer. The apparent specific volume/density ascribes all volume/density changes in the solution to the solute (polymer network), although changes in the solvent volume/density in the neighborhood of the solute contribute to the measured solution volume/density. The (normalized) form factor $P(q)$ describes the structure of a single particle and the structure factor $S(q)$ accounts for the interference of scattering from different particles. The difference in scattering length density $\Delta\varrho = \varrho_{polymer} - \varrho_{solvent}$, which corresponds to the scattering contrast, can be calculated for every microgel suspension accounting for the particle composition (e.g., fraction of cross-linking molecules) and the apparent specific density at a given temperature. The scattering length density of the particle or the solvent is given by

$$\varrho = \frac{\varrho_d}{M_w} N_A \sum_{i=1}^{m} b_i \tag{5.2}$$

where ϱ_d the apparent specific density, M_w is the molecular weight, N_A is Avogadro's number, and b_i is the bound coherent scattering length of a nucleus. The particles are swollen and contain solvent. The volume of a particle is $V(R) = (4/3)\pi R^3$ and therefore the volume fraction of polymer in the particle is $\phi_{poly} = V_{poly}/V$. The number density of particles can be calculated as $n = c/(V_{poly}\varrho_{poly})$, where c is the mass concentration of polymer in the sample and $V_{poly}\varrho_{poly}$ is the mass of polymer in one particle. For c given as a weight fraction, one has to use $n = c/(V_{poly}\varrho_{poly})/(c/\varrho_{poly} + [1-c]/\varrho_{D2O})$, where ϱ_{D2O} is the density of D_2O.

Scattering methods such as SANS, SAXS, DLS, and SLS have been applied to temperature-sensitive microgels in several studies [7–11]. However, scattering intensity profiles are often investigated in the high-q regime only. In this case, the applied model equations are those of the limiting behavior and do not describe the structure of the entire particle.

For a homogeneous spherical particle with the radius R, the distribution of scattering material is uniform throughout the sphere. Therefore, the radial scattering length density distribution is described by a box function and the form factor is given by

$$P_{\text{hom}}(q) = \left(\frac{3[\sin(qR) - qR\cos(qR)]}{(qR)^3}\right)^2 \tag{5.3}$$

For microgel particles, a higher degree of cross-linking density is expected inside the particle than in the periphery, when the cross-linker is consumed faster than the monomer during polymerization [12]. This leads to a fuzziness of the particle surface that has to be included in the model [13]. This can be done by convoluting the radial box profile with a Gaussian. In reciprocal space the convolution is just a product yielding the form factor $P_{\text{inho}}(q)$ that describes scattering from a monodisperse sphere with an interface that gradually decreases at the sphere surface with a shape similar to that of an error function. The form factor becomes

$$P_{\text{inho}}(q) = \left[\frac{3[\sin(qR) - qR\cos(qR)]}{(qR)^3} \exp\left(-\frac{(\sigma_{\text{surf}} q)^2}{2}\right)\right]^2 \tag{5.4}$$

where σ_{surf} denotes the width of the smeared particle surface. A similar approach, in terms of error functions, was applied by Pedersen et al. [14] to describe the scattering of block copolymer micelles.

Figure 5.2 schematically depicts the radial profile corresponding to $P_{\text{inho}}(q)$. The core of the microgel exhibits a higher degree of cross-linking density and is characterized by the radial box profile up to a radius of about $R_{\text{box}} = R - 2\sigma_{\text{surf}}$. The subsequent decrease in cross-linking density is described by σ_{surf} and the profile has decreased to half the core density at R. At $R_{\text{SANS}} = R + 2\sigma_{\text{surf}}$, the profile approaches zero and thus, the overall size of the particle obtained by SANS is approximately given by R_{SANS}. A small number of chains reaching outside the particle will contribute only to the hydrodynamics of the particle and therefore the size obtained by SANS is expected to be slightly smaller than the hydrodynamic radius R_{h} determined by DLS. A slightly different model was suggested by Richtering et al., who used a parabola-based shape, that has the advantage that the profile decays to zero at finite distance [15].

To consider size polydispersity of the particles one can assume the number distribution with respect to the particle radius R to be a Gaussian function

$$D(R, \langle R \rangle, \sigma_{\text{poly}}) = \frac{1}{\sqrt{2\pi\sigma_{\text{poly}}^2 \langle R \rangle^2}} \exp\left(-\frac{(R - \langle R \rangle)^2}{2\sigma_{\text{poly}}^2 \langle R \rangle^2}\right) \tag{5.5}$$

with $\langle R \rangle$ describing the average particle radius and σ_{poly} denoting the relative particle size polydispersity.

At infinite dilution colloidal systems reveal no position–position correlations. The influence of the structure factor $S(q)$ on the scattering distribution can be disregarded and $S(q) = 1$ can be assumed. However, at higher concentrations the interference of scattering from different particles cannot be neglected and $S(q)$ needs to be included in a model expression for $I(q)$.

For monodisperse, spherical particles that interact with a spherically symmetric hard-sphere interaction potential $S(q)$ can be obtained from the liquid state theory employing, for example, the Percus–Yevick approximation for the closure relation. Neglecting the influence of polydispersity on the structure factor, an expression for

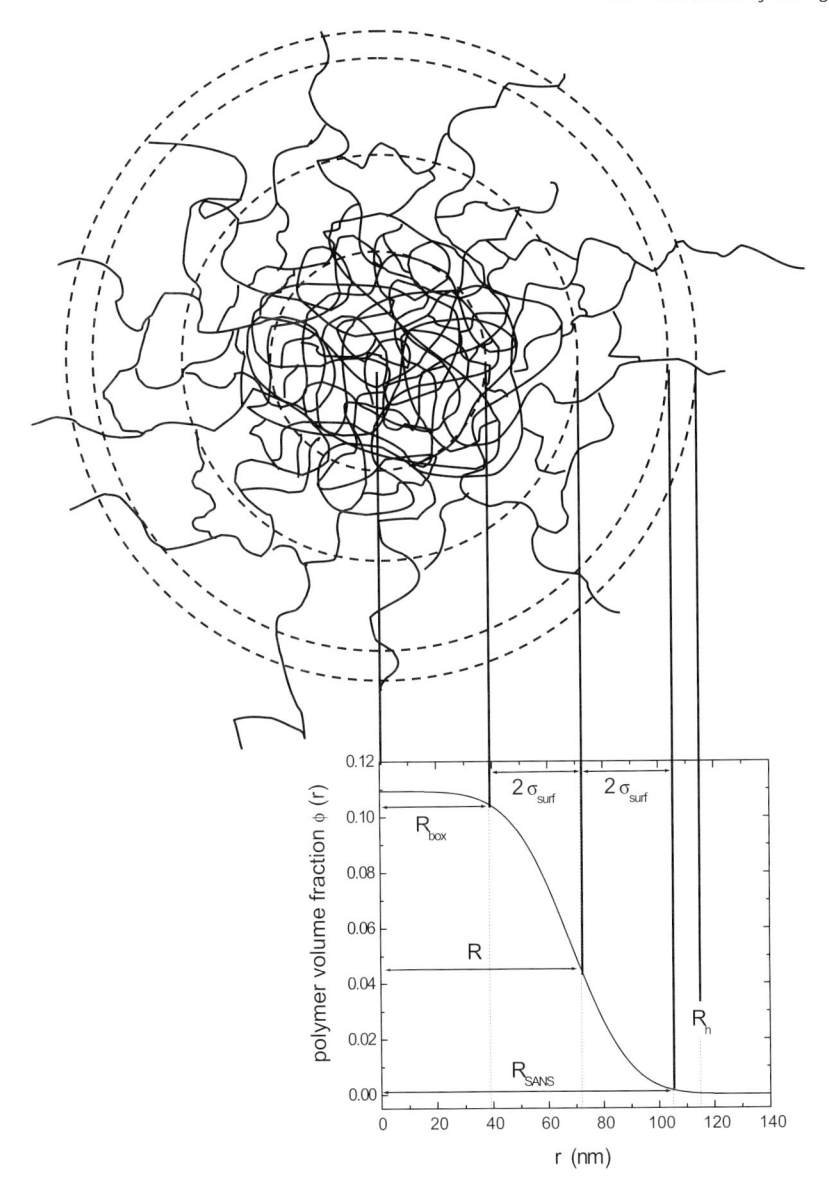

Figure 5.2 The structure of PNIPAM microgels. A highly cross-linked core is characterized by a radial box profile up to $R_{box} = R - 2\sigma_{surf}$. The cross-linking density decreases with increasing distance to the core and is described by σ_{surf}. At R, the profile has decreased to half the core density. The overall size obtained by SANS is approximately given by $R_{SANS} = R + 2\sigma_{surf}$, where the profile approaches zero. R_{SANS} is often slightly smaller than the hydrodynamic radius R_h obtained by DLS. Reused with permission from Stieger et al. [13]. Copyright 2004, American Institute of Physics.

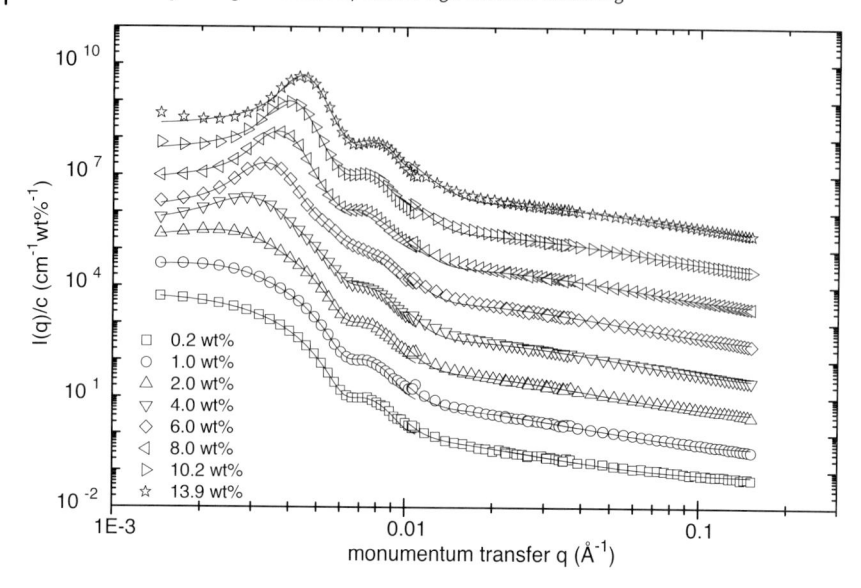

Figure 5.3 Intensity distribution normalized on concentration, $I(q)/c$, versus momentum transfer q for various concentrations of a PNIPAM microgel at 25 °C. For clarity, the data sets 1.0 to 13.9 wt% were vertically shifted by one order of magnitude. The lines represent fits according to a model that takes the particle interaction into account [16].

the differential scattering cross section

$$\frac{d\sigma(q)}{d\Omega} = n\Delta\varrho^2 S(q,\langle R\rangle) \int_0^\infty D(R,\langle R\rangle,\sigma_{poly}) V_{poly}(R)^2 P_{inho}(q,R) dR \quad (5.6)$$

is obtained, where the structure factor effects are treated in a simple decoupling approximation.

The influence of particle interaction is illustrated in Figure 5.3. The scattering intensity versus q for different concentrations of a PNIPAM microgel in aqueous solution is plotted. A structure factor peak is observed at low q, which becomes more pronounced the higher the concentration. Obviously, particle interaction contribute significantly to the scattering intensity and $S(q)$ needs to be considered for quantitative analysis. Details of such an analysis are beyond the scope of this contribution but are described in detail by Stieger et al. [16].

One should note that the usual procedure of measuring the form factor at low concentration and calculating the structure factor at higher concentration by simple division by this form factor is not valid for soft microgels, since the particles can deform and interpenetrate at high concentration. We wish to note that temperature-sensitive microgels are interesting model systems to investigate particle interaction as they provide two routes to increase the effective concentration: (i) by increasing the mass concentration, as shown in Figure 5.3 and (ii) by lowering the temperature. In the latter case, the microgels swell upon cooling leading to an increasing effective concentration while the particle number density stays constant. The static and dynamic structure factor of concentrated suspensions of PNIPAM microgels was discussed in detail by Stieger et al. [16] and Eckert and Richtering [17].

In order to describe the scattering contributions arising from polymer-related fluctuations of the microgel network, a Lorentzian function can be used [7]

$$I_{\text{fluct}}(q) = \frac{I_{\text{fluct}}(0)}{1+\xi^2 q^2} \qquad (5.7)$$

where $I_{\text{fluct}}(0)$ denotes the contribution of the network fluctuations to the intensity at $q=0$ and ξ is the correlation length of the fluctuations, which can be considered to be related to the blob or mesh size [18]. However, due to the overlap of the contribution with the scattering from the microgel particle itself and the presence of the incoherent background at high q it is often rather difficult to determine ξ with high accuracy from SANS data. Finally, a constant background $\text{const}_{\text{back}}$ needs to be added to account for residual incoherent scattering from the microgel, which is difficult to estimate and subtract.

Experimental SANS data are always smeared by the instrument since a distribution of radiation with scattering vectors q around the nominal scattering vector $\langle q \rangle$ contributes. The distribution is due to the finite collimation of the beam, the wavelength spread of the incoming neutrons and the finite spatial resolution of the detector. It was shown that all these three contributions can be approximated separately by Gaussians [19]. The combined resolution function $R(\langle q \rangle, q)$ describes the distribution of scattering vectors q contributing to the scattering at the setting $\langle q \rangle$ as

$$R(\langle q \rangle, q) = \frac{q}{\sigma^2_{\text{smear}}} \exp\left[-\frac{1}{2}\left(q^2 + \frac{\langle q \rangle^2}{\sigma^2_{\text{smear}}}\right)\right] I_0\left(\frac{\langle q \rangle q}{\sigma^2_{\text{smear}}}\right) \qquad (5.8)$$

where σ_{smear} denotes the width of the instrumental smearing, which is different for the different instrumental settings, and I_0 is a modified first kind and zeroth order Bessel function. The contribution to σ_{smear} from wavelength smearing is included as well as the contributions from collimation and detector resolution. The latter can be determined by fitting the attenuated direct beam profile with a Gaussian. To account for instrumental smearing the resolution function is included in the model as

$$I^{\text{mod}}(\langle q \rangle) = \int_0^\infty R(\langle q \rangle, q) \frac{d\sigma(q)}{d\Omega} dq \qquad (5.9)$$

Finally, after incorporating all contributions the exhaustive model expression for the intensity distribution is found:

$$I^{\text{mod}}(\langle q \rangle) = n\Delta\varrho^2 \int_0^\infty R(\langle q \rangle, q)$$
$$\times \left[S(q, \langle R \rangle) \int_0^\infty (D(R, \langle R \rangle, \sigma_{\text{poly}}) V_{\text{poly}}(R)^2 P_{\text{inho}}(q, R)) dR + I_{\text{fluct}}(q) \right] dq + \text{const}_{\text{back}}$$
$$(5.10)$$

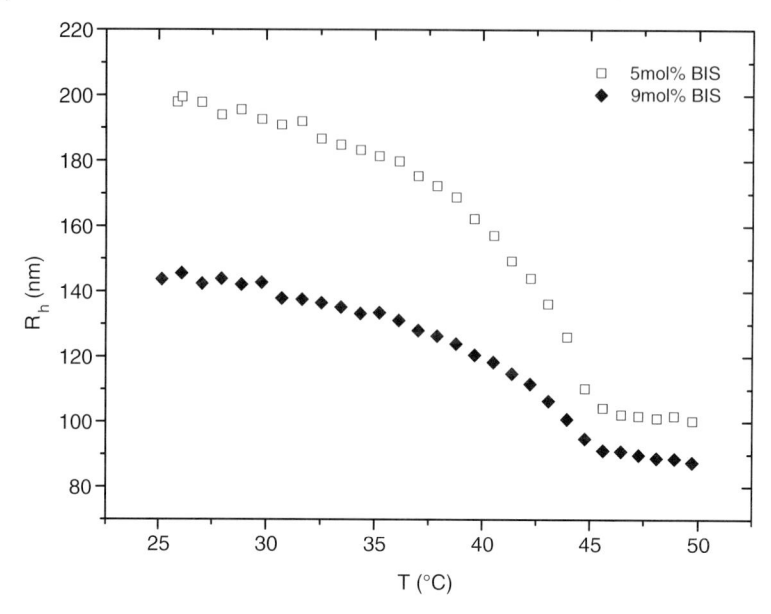

Figure 5.4 Hydrodynamic radii versus temperature of PNIPMAM-microgels in D_2O. Two samples with different cross-linker content of 5 and 9% BIS, respectively, are shown.

After the model is fitted to the experimental data, the amplitude of the form factor can be calculated and a numerical Fourier transformation leads to the radial density profile of the particle providing detailed information about the structure of the particle.

We will illustrate the possibilities that are provided by SANS for the investigation of microgel structure by using as an example aqueous, temperature-sensitive poly-N-isopropylmethyacrylamide (PNIPMAM) microgel solutions [20]. PNIPMAM microgels are similar to PNIPAM, however, the volume phase transition temperature is at a higher temperature as can be seen from the temperature-dependent hydrodynamic radius as determined from dynamic light scattering in D_2O solutions shown in Figure 5.4.

For PNIPMAM microgels the hydrodynamic radii R_h decreased slightly with increasing temperature but when the VPTT is approached a strong volume transition occurred. As anticipated, the microgel with a higher degree of cross-linking density has a smaller swelling ratio.

SANS experiments were performed at the instrument D11 of the Institute Laue-Langevin (ILL) in Grenoble, France, typically at a microgel concentration of 0.2 wt%. The neutron wavelength was $\lambda = 6$ or 12 Å with a spread of $\Delta\lambda/\lambda = 9\%$. The data were collected on a two-dimensional multidetector (64×64 elements of $1 \times 1\,cm^2$) and corrected for background and empty cell scattering. Sample–detector distances of 36.7, 10.5, and 2.5 m were employed. The incoherent scattering of H_2O was used for absolute calibration according to standard procedures and software available at the ILL ($GRAS_{ans}P$ V. 3.25). H_2O calibration measurements were performed at sample–detector distances of 10.5 and 2.5 m. The H_2O scattering obtained at 10.5 m was

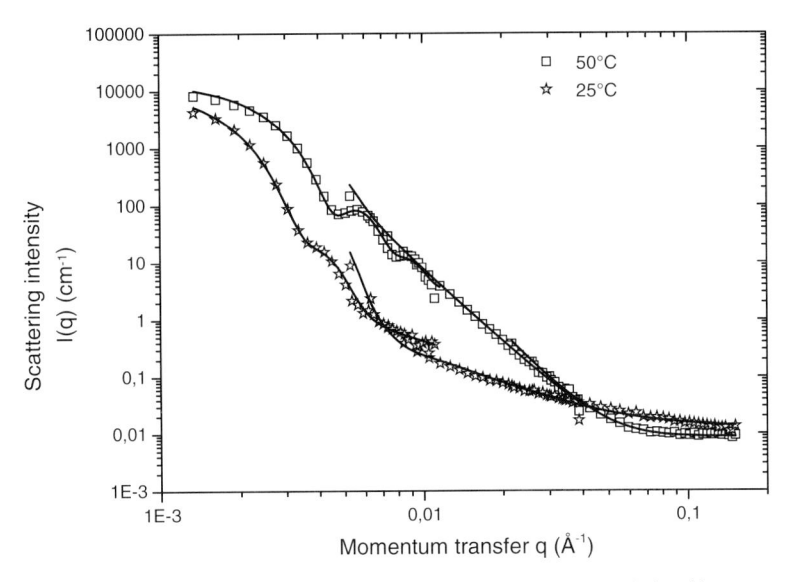

Figure 5.5 SANS data from a PNIPMAM microgel with 5 mol% cross-linker dilute D_2O suspension.

used to calibrate the sample data measured at 36.7 m since calibration by H_2O cannot realistically be performed with high accuracy for long sample–detector distances. However, the changes in the incident neutron flux at the sample position can be calculated from the sample–detector distance and the collimation length. Further processing was done by radial averaging to obtain a one-dimensional data set. All experiments were carried out at full contrast using D_2O as the solvent.

Several characteristic features of the scattering curves, shown in Figure 5.5 are obvious and already provide important qualitative information on the microgel properties. Particle form factor minima up to second order demonstrate the narrow size distribution of the microgel particles. The shift of the form factor minima to higher q values with increasing temperature is directly related to the particle shrinkage. The angular dependence of the scattering intensity at low q is connected with the radius of gyration. The slope decreases with increasing temperature also indicating the change in size. At 50 °C, Porod scattering ($I \propto q^{-4}$) is observed in the intermediate q range indicating scattering from a sharp interface when the particles are in the collapsed state.

Some data points deviate from the others in the q range around $q = 0.006$ Å^{-1} where measurements with different sample–detector distances overlap. This is due to the different instrumental resolution and as mentioned above, the experimental resolution can be taken into account (see Equation 5.8) when the data are fitted with the model. Fits are shown as full lines in Figure 5.5 and in the overlap regions of two sample–detector distances two lines are shown. These fits correspond to the model discussed above (see Equation 5.10) and as the resolution is taken into account, two lines are obtained in the overlap region. They agree well with the experimental data.

The quantitative analysis of the scattering curves provides more detailed information on the structure of microgel particles and the density profiles are shown in Figure 5.6. The plots contain results from fits involving two slightly different models: (i) Dashed lines represent data from the Gaussian interface model which was discussed above and includes the form factor give in Equation 5.4. (ii) Full lines represent data from a parabola-based interface model from the paper by Berndt et al. [15]. For this profile, half a parabola is used for describing the profile from the half value of the profile at R to the maximum extent of the profile to $R + s_{surf}$. A corresponding half inverted parabola is subtracted from the profile from $R - s_{surf}$ to R to yield a profile that is antisymmetric around the half value at R leading to

$$\begin{aligned} \varrho(r) &= 1 & r &\leq (R-s_{surf}) \\ \varrho(r) &= 1 - \frac{1}{2}\frac{[(r-R)+s_{surf}]^2}{s_{surf}^2} & (R-s_{surf}) &< r \leq R \\ \varrho(r) &= \frac{1}{2}\frac{[(R-r)+s_{surf}]^2}{s_{surf}^2} & R &< r \leq (R+s_{surf}) \\ \varrho(r) &= 0 & r &> (R+s_{surf}) \end{aligned} \quad (5.11)$$

In this model, the density profile thus decays to zero at finite distance. Obviously, the two models lead to very similar results.

Above the transition temperature, the density profile resembles a box profile of a homogenous sphere with a sharp particle surface. Interestingly, the volume fraction of polymer segments is still well below unity, and obviously the particles still contain a lot of solvent even in the collapsed state. Remarkably, the water content in the collapse state is largely independent of the cross-linker content.

When the temperature is reduced, the particles swell and the density profile changes significantly; it is no longer a box function but the profile decays smoothly at the surface. This reflects that the cross-link density is inhomogeneous throughout the particle, which leads to a fuzzy surface. Particle size as well as segment density depend on the cross-linker content; the more cross-linker was used during synthesis, the smaller the swelling, thus a smaller size and a higher segment density is found. The inhomogeneous cross-linker distribution is due to different reaction kinetics of cross-linker and monomer, respectively. With higher cross-linker content, the particles can swell less; consequently, the decrease in segment density in the particle interior with decreasing temperature (i.e., increasing solvent quality) is less pronounced for the more cross-linked microgels.

In order to investigate the importance of the details of the model, we have compared two different approaches for describing the radial density profile in Figure 5.6 and Table 5.1. One of the models is the solid sphere profile smeared by a Gaussian as described above (Equation 5.4). The other one is the profile based on a parabola as described in Ref. [15] and given by Equation 5.11. For this profile, the SANS-derived radius R_{SANS} that can be compared to the hydrodynamic radius is $R_{SANS} = R + s_{surf}$. Please note also that the width of the smeared surface is defined differently in the two models and one has to compare s_{surf} with $2 \times \sigma_{surf}$. The two

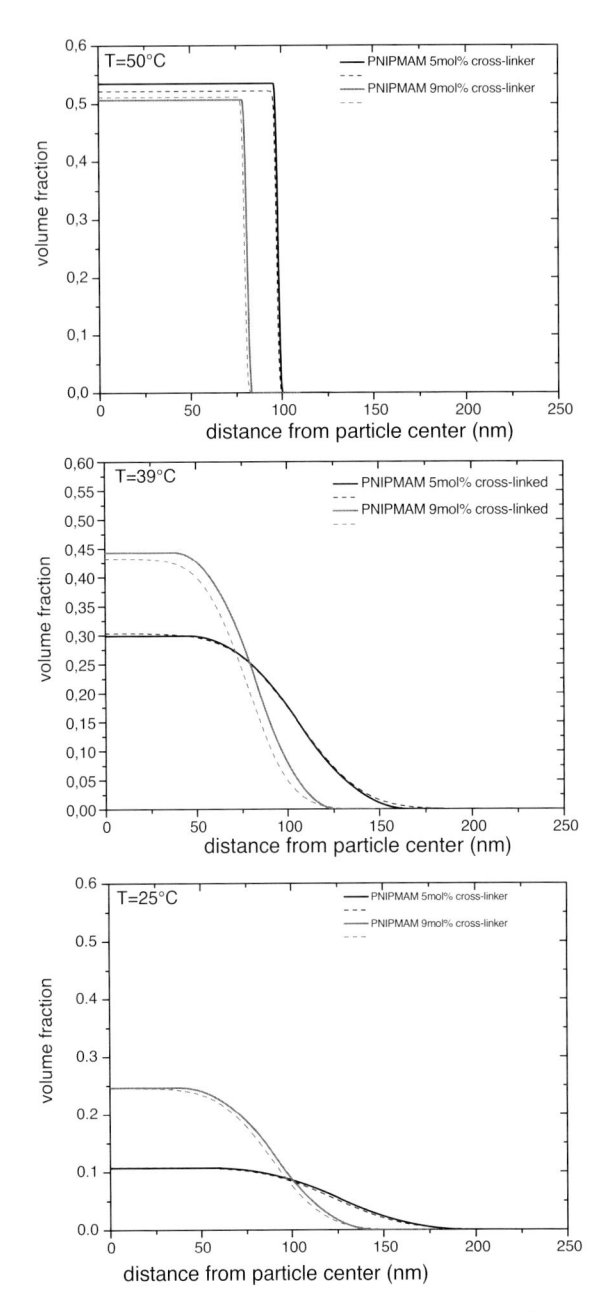

Figure 5.6 Density profiles for PNIPMAM microgels at different temperatures. Two samples with different cross-linker content are shown. Dashed lines represent data from the Gaussian interface model; full lines represent data from the parabola-based interface model.

5 Determination of Microgel Structure by Small-Angle Neutron Scattering

Table 5.1 Parameters obtained from modeling the SANS curves of PNIPMAM microgels at different temperatures.

Sample	T (°C)	Parabola-based interface					Gaussian interface					R_h (nm)
		R (nm)	s_{surf} (nm)	Φ_{vol} (%)	σ_{rel} (%)	R_{SANS} (nm)	R (nm)	σ_{surf} (nm)	Φ_{vol} (%)	σ_{rel} (%)	R_{SANS} (nm)	
5 mol%	50	98.1	2.2	53.4	10.4	100	97.3	0.9	52.2	10.7	99	100
	39	105.7	60.2	29.9	12.4	166	111.9	26.7	30.4	13.0	165	168
	25	126.3	70.1	10.7	12.3	196	129.9	29.6	10.8	12.5	189	199
9 mol%	50	80.9	2.5	50.6	12.9	83	79.5	1.0	51.2	13.9	82	88
	39	82.3	44.2	44.3	11.4	126	81.6	18.3	43.3	14.9	118	123
	25	91.7	53.3	24.7	13.7	145	94.2	22.3	24.6	13.8	139	145

models provide very similar results demonstrating that both approaches are well suited to describe the internal structure of the microgel particles.

At low temperature, the overall particle size obtained by SANS (R_{SANS}) is smaller than the hydrodynamic radius R_h obtained by DLS. This is due to few long dangling polymer chains that contribute significantly to the hydrodynamic radius but only little to the scattering intensity and thus are difficult to detect by SANS. This is also the reason that the ϱ parameter, which is the ratio of radius of gyration and hydrodynamic radius $\varrho = \frac{R_g}{R_h}$, is much smaller for microgel as compared to the value for hard spheres. Senff and Richtering reported on the temperature dependence of $\varrho = \frac{R_g}{R_h}$ in the case of PNIPAM microgels [21]. At low temperatures, $\varrho \approx 0.6$ was observed but approached the hard sphere limit of $\sqrt{3/5}$ above the transition temperature (see Figure 5.7).

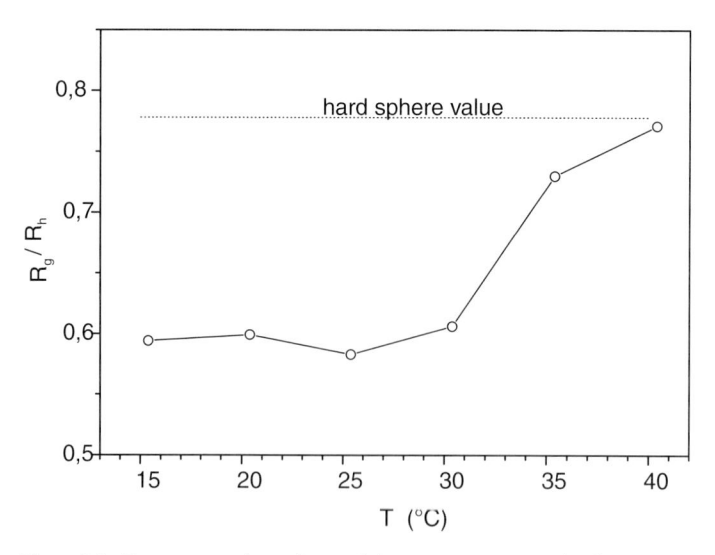

Figure 5.7 Temperature dependence of the parameter $\varrho = R_g/R_h$ for a PNIPAM microgel in H_2O [21].

As the form factor analysis provides detailed insight into the particle structure, one can use this information in order to adapt the polymerization conditions, as for example, batch, semi-batch synthesis or controlled monomer feed, in order to tailor particle morphology and thus microgel properties [22]. In addition one can employ SANS data to investigate the influence of chemical reactivity on the microstructure by comparing for example, different cross-linkers, initiators or monomers.

5.3
Core–Shell Particles

More complex microgel particles can be obtained when different comonomers are employed during polymerization or when seed particles are used for the preparation of core–shell microgels, which have spatially separated regions with different polymers. In the latter case, the swelling of core and shell, respectively, can be different due to a different chemical composition leading to microgels that are multisensitive [23–25].

Again SANS is a powerful tool to determine the structure of such core–shell microgels. Data evaluation follows the same routes as outlined above; however, the form factor model needs to take the core–shell morphology into account. Richtering et al. developed a suitable model, which is schematically shown in Figure 5.8 [15]. For details, see Equations 1–3 in Ref. [15]. The core–shell model includes additional parameters that take the special morphology of such particles into account. W_{core} and W_{shell} describe the widths of central core and shell boxes. The interpenetration layer of core and shell is characterized by $2s_{in}$ and s_{out} denotes the half width of the outer surface. The dotted line in Figure 5.8 represents the total of core and shell.

In Figure 5.9, the density profiles of a PNIPAM-core–PNIPMAM-shell microgel as obtained from SANS experiments in D_2O for three different temperatures are shown. With such doubly temperature-sensitive microgels, varying the temperature is like

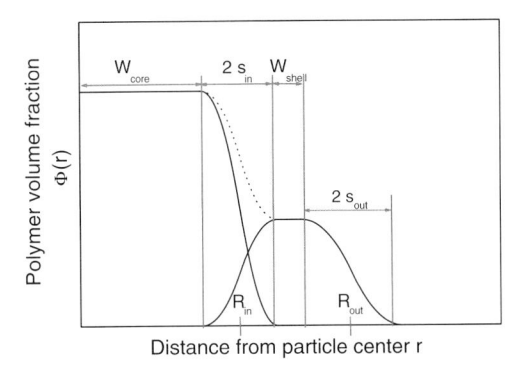

Figure 5.8 Schematic illustration of the internal structure of core–shell microgels. Redrawn from Ref. [15] with permission from the American Chemical Society.

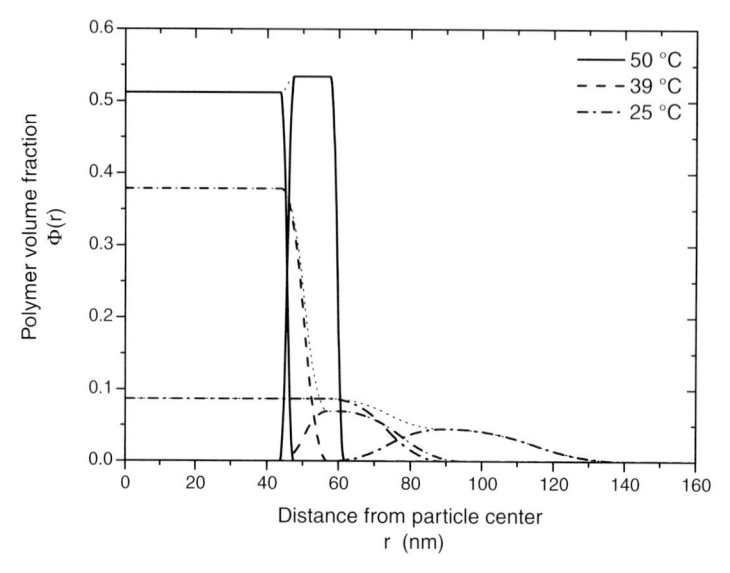

Figure 5.9 Radial density profiles of a PNIPAM-core–PNIPMAM-shell microgel in D$_2$O at different temperatures. Redrawn from Ref. [15] with permission from the American Chemical Society.

doing contrast variation. There is little contrast between core and shell at low and high temperatures, but at the intermediate temperature the contrast is high allowing for a good identification of the two components.

The structure of the core–shell particles when both core and shell are in the collapsed state can be well described by two-box profiles with a narrow core–shell interface and a sharp outer surface. The width of the core–shell interfaces can be compared with the surface thickness of the collapsed pure "naked" core. The width of the outer surface can be compared with that of pure PNIPMAM microgels that are prepared under similar conditions.

Most interesting is the structure of the core–shell microgel at intermediate temperature that is in between the VPTTs of the two components. In comparison to the structure obtained at 50 °C, where core and shell are collapsed, one observes a large difference in the local volume fractions of core and shell. Although the temperature is well above the core LCST, the size of the core is slightly increased due to the force developed in the swollen shell, which expands the core. The 25 °C profile shows low polymer volume fractions in the core and shell and smoothly decays from the particle center in a similar manner observed for pure PNIPAM or PNIPMAM microgels well below their VPTT. Remarkably is the fact that the dimensions of the core are significantly reduced as compared to the naked core, that is, one has a shell-restricted swelling of the core due to the fully swollen shell network.

Thus, the density profiles of core–shell microgels obtained at different swelling conditions provide detailed information on the internal particle structure and the mutual influence of core and shell swelling, respectively. This information can be compared, for example, with data obtained from other techniques such as fluorescence [26, 27], calorimetry [28], spectroscopy [29], or electron microscopy [30].

The core–shell form factor model as been developed further in order take a nonsymmetric interface between core and shell, respectively, into account [31]. In addition, we wish to mention that random copolymer microgels containing different temperature-sensitive moieties can also reveal peculiar behaviors [32] and for example, an internally phase-separated structure [33]. Such special examples are, however, far beyond the scope of this article.

5.4
Summary

The structure of microgels can be investigated in detail by SANS in dilute suspension. A direct modeling expression for the scattering intensity distribution describes very well the experimental data over an extensive q range. The overall particle shape as well as the internal structure of the microgel network can be deduced from SANS. Data analysis provides information on the distribution of segment density inside the particles. With thermosensitive microgels structural changes induced by temperature, cross-linking density and particle size are revealed by the density profiles. The polymer volume fraction inside the particle increased dramatically with increasing temperature accompanied by a decrease of the smearing of the particle surface. Above the VPTT, the particle surface sharpens significantly and a radial profile with an almost box shape was observed.

SANS from core–shell microgels composed of the temperature-sensitive polymers poly-N-isopropylacrylamide and poly-N-isopropylmethacrylamide, that is, with different transition temperatures in core and shell, respectively, reveals characteristic differences as a function of temperature. Data analysis provides detailed information on the mutual influence on swelling of core and shell, respectively.

References

1 Cohen Stuart, M.A. (2008) *Colloid Polym. Sci.*, **286**, 855.
2 Pelton, R.H. and Chibante, P. (1986) *Colloids Surf. A*, **20**, 247.
3 Schild, H.G. (1992) *Prog. Polym. Sci.*, **17**, 163.
4 Saunders, B.R. and Vincent, B. (1999) *Adv. Colloid Interface Sci.*, **80**, 1.
5 Lindner, P. and Zemb, Th. (eds) (2002) *Neutrons, X-Rays and Light: Scattering Methods Applied to Soft Condensed Matter*, North-Holland.
6 Pedersen, J.S. (1997) *Adv. Colloid Interface Sci.*, **70**, 171.
7 Mears, S.J., Deng, Y., Cosgrove, T., and Pelton, R. (1997) *Langmuir*, **13**, 1901–1906.
8 Crowther, H.M., Saunders, B.R., Mears, S.J., Cosgrove, T., Vincent, B., King, S.M., and Yu, G.-E. (1999) *Colloids Surf. A*, **152**, 327–333.
9 Kratz, K., Hellweg, T., and Eimer, W. (2001) *Polymer*, **42**, 6631–6639.
10 Fernández-Babero, A., Fernández-Nieves, A., Grillo, I., and López-Cabarcos, E. (2002) *Phys. Rev. E*, **66**, 051803–051813.
11 Saunders, B.R. (2004) *Langmuir*, **20**, 3925–3932.

12 Wu, X., Pelton, R.H., Hamjelec, A.E., Woods, D.R., and McPhee, W. (1994) *Colloid Polym. Sci.*, **272**, 467–477.
13 Stieger, M., Richtering, W., Pedersen, J.S., and Lindner, P. (2004) *J. Chem. Phys.*, **120**, 6197–6206.
14 Pedersen, J.S., Svaneborg, C., Almdal, K., Hamley, I.W., and Young, R.N. (2003) *Macromolecules*, **36**, 416–433.
15 Berndt, I., Pedersen, J.S., and Richtering, W. (2005) *J. Am. Chem. Soc.*, **127**, 9372.
16 Stieger, M., Richtering, W., Pedersen, J.S., and Lindner, P. (2004) *Langmuir*, **20**, 7283.
17 Eckert, T. and Richtering, W. (2008) *J. Chem. Phys.*, **129**, 124902 (1–6).
18 Shibayama, M. (1998) *Macromol. Chem. Phys.*, **199**, 1.
19 Pedersen, J.S., Posselt, D., and Mortensen, K. (1990) *J. Appl. Crystallogr.*, **23**, 321.
20 Berndt, I. (2005) Doctoral thesis, Christian-Albrechts University, Kiel.
21 Senff, H. and Richtering, W. (2000) *Colloid Polym. Sci.*, **278**, 830.
22 Meyer, S. and Richtering, W. (2005) *Macromolecules*, **38**, 1517–1519.
23 Jones, C.D. and Lyon, L.A. (2000) *Macromolecules*, **33**, 8301–8306.
24 Berndt, I. and Richtering, W. (2003) *Macromolecules*, **36**, 8780–8785.
25 Blackburn, W.H. and Lyon, L.A. (2008) *Colloid Polym. Sci.*, **286**, 563–569.
26 Gan, D. and Lyon, L.A. (2001) *J. Am. Chem. Soc.*, **123**, 8203.
27 Müller, C.B., Loman, A., Richtering, W., and Enderlein, J. (2008) *J. Phys. Chem. B*, **112** (28), 8236–8240.
28 Berndt, I., Popescu, C., Wortmann, F.-J., and Richtering, W. (2006) *Angew. Chem.*, **118**, 1099–1102; Berndt, I., Popescu, C., Wortmann, F.-J., and Richtering, W. (2006) *Angew. Chem. Int. Ed.*, **45**, 1081–1085.
29 Keerl, M., Smirnovas, V., Winter, R., and Richtering, W. (2008) *Angew. Chem*, **120**, 344–347; Keerl, M., Smirnovas, V., Winter, R., and Richtering, W. (2008) *Angew. Chem. Int. Ed.*, **47**, 338–341.
30 Crassous, J.J., Wittemann, A., Siebenbürger, M., Schrinner, M., Drechsler, M., and Ballauff, M. (2008) *Colloid Polym. Sci.*, **286**, 805.
31 Berndt, I., Pedersen, J.S., and Richtering, W. (2006) *Angew. Chem.*, **118**, 1769; Berndt, I., Pedersen, J.S., and Richtering, W. (2006) *Angew. Chem. Int. Ed.*, **45**, 1737.
32 Keerl, M. and Richtering, W. (2007) *Colloid Polym. Sci.*, **285** (4), 471–474.
33 Keerl, M., Pedersen, J.S., and Richtering, W. (2009) *J. Am. Chem. Soc.*, **131**, 3093–3097. DOI: 10.1021/ja807367p.

6
Interactions and Colloid Stability of Microgel Particles
Brian Vincent and Brian Saunders

6.1
Theoretical Background

6.1.1
Introduction

The pairwise interactions between soft microgel particles may be repulsive or attractive. Net attractive interactions may lead to particle aggregation. However, in the case of microgel particles, the net attraction is usually weak, so that any resulting aggregation is generally reversible. This, in principle, leads to an equilibrium state for the dispersion, a colloidal phase coexistence, which also occurs for weakly attractive *hard* particles [1]. Such colloidal phase coexistence is similar to that for molecules, and colloidal analogues of the gas, liquid, and solid (crystalline or amorphous) states exist.

One may deconvolute the total pairwise interaction free energy, $G(r)$, where r is the center–center separation between any two microgel particles (diameter, d) into various contributions; these are, in this chapter, for simplicity, assumed to be additive. The primary, *long-range* interactions (for $r > d$ or $h > 0$, where $h = r - d$) are the van der Waals attraction, G_A, and if the particles contain surface and/or bulk charged groups an electrostatic repulsion, G_E.

In addition, since microgel particles are soft, they may interpenetrate when they touch and it is thus possible for r to take values less than d (corresponding to $h < 0$). One way of thinking about this is to consider such an interaction as a "steric" interaction, G_S, similar in nature to that for hard particles carrying adsorbed or grafted polymer chains. Here, $G_S(h)$ is the steric interaction associated with any interpenetration of the microgel particles ($h < 0$). For *hard* spheres, $G_S(h)$ is a step (delta) function, that is, $G_S \rightarrow \infty$ for $h < 0$, but for soft microgel particles the magnitude of $G_S(h)$ will depend on the polymer segment density distribution near the periphery of the particles and the Flory polymer–solvent interaction parameter χ. For the steric interaction to be repulsive, that is, for G_S to be positive, then χ must be <0.5. If χ is adjusted to be >0.5 so that the interactions are attractive and thus G_S is

Microgel Suspensions: Fundamentals and Applications
Edited by Alberto Fernandez-Nieves, Hans M. Wyss, Johan Mattsson, and David A. Weitz
Copyright © 2011 WILEY-VCH Verlag GmbH & Co. KGaA, Weinheim
ISBN: 978-3-527-32158-2

negative, for example, by a suitable change in the temperature or the solvency of the medium, the microgel particles may show incipient weak aggregation.

Other interactions between microgel particles may be induced if polymers are added to the dispersion. Adding polymers to a dispersion of microgel particles invariably results in changes in the particle size (which may directly lead to changes in G_A and G_E). Whether one observes a (net) expansion or contraction in the size of the microgel particles depends on two considerations: (i) whether the diameter of the added polymer molecules (taken, for example, as $2R_g$, where R_g is the radius of gyration of the polymer chains in free solution) is larger than or less than the average mesh size, L, of the microgel polymer network and (ii) the interplay of the three Flory pair-interaction parameters: χ_{21}, χ_{31}, and χ_{32}, which refer to the [microgel polymer]/[solvent], the [added polymer]/[solvent], and the [added polymer]/[microgel polymer] interactions, respectively.

Let us first consider the case where $2R_g > L$. In this case, no deep penetration of the added polymer molecules into the interior of the microgel particles occurs and the microgel particles will contract in size. The reason for this is that the osmotic pressure outside the microgel particles in free solution is increased. This, in turn, leads to an increase in the osmotic pressure also within the microgel particles, by exclusion of some solvent, in order to maintain equilibrium. Depending on whether χ_{32} is negative or positive, two different situations can arise with regard to the introduced interparticle interactions. If χ_{32} is negative, the polymer chains in solution can "adsorb" onto the outer periphery of the microgel particles. At low adsorbed amounts (less than full coverage), interparticle polymer bridging interactions, G_{br}, might thus, in principle, be observed. However, this effect has not been observed for microgel particles to date. On the other hand, the effect has been observed for sterically stabilized particles by Cawdery and Vincent [2], who investigated the stability of polystyrene particles carrying terminally grafted poly(ethylene oxide) (PEO)] chains, to which poly(acrylic acid) (PAAc) was added. At *low* pH values (<4), a strong H-bonding interaction occurs between the ether oxygen atoms of the added PEO chains and the $-OH$ groups of the carboxylic acid moieties of the chains attached to the particles (i.e., χ_{32} is negative), leading to bridging flocculation of the PS-g-PEO particles by the PAAc chains, at low concentrations of added PAAc.

For concentrations of added polymer above that for full coverage of the microgel particles, it is, in principle, possible for depletion interactions, G_{dep} to occur, but *only if* the adsorbed and free chains are not in free exchange equilibrium. The latter is similar to the situation with polymers adsorbed onto the surface of *hard* particles; here, depletion interactions cannot occur if the adsorbed and free polymer molecules are in free exchange. Depletion interactions could, in principle, be observed, however, if the added polymer was irreversibly adsorbed onto the microgel particles.

A depletion interaction between microgel particles is much more likely to be observed if $2R_g > L$ and χ_{32} is *positive*, which is a situation similar to that of nonadsorbing polymers added to a suspension of hard particles. It is interesting to note that Cawdery and Vincent [2] observed depletion flocculation in the systems described in the previous paragraph, at *high* pH values, when the H-bonding interaction disappears through ionization of the $-OH$ moieties (i.e., χ_{32} becomes positive). More examples of depletion-induced flocculation for microgel particles, where $2R_g > L$ and χ_{32} is positive, will be discussed in Section 6.2.2.

The situation where $2R_g < L$ is more complex. In this case, deep penetration of the added polymer chains into the interior of the microgel particles may occur. Again, one must distinguish the cases where χ_{32} is either negative or positive. One example of the former situation has been described by Bradley et al. [3]. These authors showed that for microgel particles based on copolymers of N-isopropylacrylamide (PNIPAM) and acrylic acid (AAc), added PEO chains below a critical molecular weight (MW) (which depends on the cross-link density) entered the microgel particles for concentrations up to some maximum absorption limit, beyond which the concentration of free PEO chains started to build up in solution. The entry of PEO chains into the particles is, in this case, driven by the strong H-bonding interaction between the ether oxygen atoms of the PEO chains and the H-atoms of the $-COOH$ groups of the AAc groups (at low pH values) and, to a lesser extent, those of the $-NH$ groups of the NIPAM. For these systems, the size of the microgel particles passes through a maximum with increasing PEO concentration. The microgel particle size initially increases beyond the expanded equilibrium state of the microgel particles in pure water at ambient temperature, since the intruding PEO chains "force open" the microgel network. However, beyond a concentration corresponding to maximum absorption of the PEO chains, the size of the microgel particles decreases again. This is because, as discussed earlier, the subsequent increase in the osmotic pressure in the bulk solution has to be matched by a corresponding increase in the osmotic pressure within the microgel particles, which occurs by partial exclusion of solvent.

In the case when $2R_g < L$ and χ_{32} is negative, bridging interactions are unlikely to be induced, but depletion interactions could in principle occur if the microgel particles are not in full exchange equilibrium with the particles in solution (i.e., at concentrations beyond the absorption maximum). A depletion interaction may occur between microgel particles if the polymer molecules that have penetrated into the interiors of the microgel particles are "trapped" within the microgel polymer network and the concentration of free polymer in solution increases beyond some minimum value. Relative timescales become important here: the timescale of "trapping" needs to be longer than the timescale of the microgel particle collisions.

If $2R_g < L$, such that deep penetration of the microgel particles by added polymer molecules occurs, but χ_{32} is *positive*, the situation is more complex to analyze. With regard to the effect on the size of the microgel particles, it is probably simpler to regard the added polymer, in effect, as a second "solvent" species, and then one has to analyze the combined effect of all three χ parameters (see above) on the osmotic pressure difference between the inside and the outside of the microgel particles. Crowther and Vincent [4] showed that the response of aqueous dispersions of PNIPAM particles to the addition of short-chain alkanol molecules is complex: the particle size initially decreased, then increased again.

With regard to the particle interactions, when χ_{32} is positive, bridging is ruled out and depletion is much more unlikely since the absorbed and free polymer chains are now much more likely to be in exchange equilibrium (i.e., compared to the case where χ_{32} is *negative* and there is strong association between the added polymer and the microgel polymer).

Having outlined the various possible interactions between microgel particles, which may occur, we now consider them individually. Note that no theoretical analyses have been carried out to date to estimate G_S or G_{br} for microgel particles, referred to earlier. A major problem is that such interactions tend to be highly system specific and, in general, occur under nonequilibrium conditions.

6.1.2
Van der Waals Interactions

The simplest analytical equation for the long-range van der Waals interaction, $G_A(h)$, between two identical spherical particles of radius, a, as a function of surface–surface separation, h, (valid for $h \ll a$) is that given by Hamaker [5]:

$$G_A = - \frac{\left(A_{part}^{1/2} - A_{med}^{1/2}\right)^2 a}{12h} \tag{6.1}$$

Here, A_{part} and A_{med} are the Hamaker constants of the particles and the medium, respectively. The microgel particles consist of a *mixture* of polymer and solvent molecules at a given polymer volume fraction (ϕ_p).

For *homogeneous* mixtures of two components (1 + 2), Vincent [6] has introduced the following equation for the Hamaker constant (A),

$$A = \left[\phi_2 A_2^{1/2} + (1-\phi_2) A_1^{1/2}\right]^2 \tag{6.2}$$

where A_1 and A_2 are the Hamaker constants of components 1 and 2, respectively, and ϕ_2 is the volume fraction of component 2.

Hence, if the Hamaker constant of the solvent, A_S, and that of the polymer, A_P, are known and assuming that there is a uniform distribution of polymer segments throughout each microgel particle, using Equations 6.1 and 6.2, one obtains

$$G_A = - \frac{\left(A_P^{1/2} - A_S^{1/2}\right)^2 \phi_p^2 a}{12h} \tag{6.3}$$

where ϕ_p is the volume fraction of polymer within the microgel particles. $\phi_p = q^{-1}$, where q is the volume swelling ratio of the particles, that is,

$$q = \left(\frac{d}{d_0}\right)^3 \tag{6.4}$$

where d is the particle diameter in some given state and d_0 is the particle diameter in the *dry* state. d may be equated to the hydrodynamic diameter, as determined, for example, using dynamic light scattering. Routh and Zimmerman [7] have shown that the hydrodynamic value of d is a reasonable approximation to the actual *physical* diameter of the microgel particles.

It is not straightforward to obtain d_0, since the hydrodynamic particle size at conditions of maximum deswelling *in dispersion*, cannot be directly used because some solvent will inevitably still be present within the microgel particles. This problem has been discussed further in the context of PNIPAM-based microgel particles in water by Rasmusson *et al.* [8]. It is better to dry the particles completely under vacuum and use electron microscopy to evaluate the particle diameter. However, care must be taken and it is better to dry the particles from the deswollen state than from the swollen state in dispersion, since drying of swollen particles on an EM grid may lead to some distortion of their shape through partial spreading on the surface. This is illustrated in Figure 6.1 that is taken from an earlier SEM study by the authors [9] on PNIPAM microgel particles. It is found that at 60 °C (deswollen state) the microgel particles retain their spherical shape, but at 25 °C (swollen state), the microgel particles spread somewhat. The extent of spreading will depend on the cross-link density of the particles.

One interesting feature, which is relevant for microgel particles, but not for hard-sphere dispersions, is that one is able to tune $G_A(h)$ by varying q, which for PNIPAM particles, for example, can be achieved by varying the temperature. Near room temperature, PNIPAM particles are generally highly swollen and ϕ_p is very small ($\phi_p \ll 1$). The Hamaker constant of the particles thus approaches that of the medium ($A_{part} \sim A_{med}$) and the interparticle van der Waals attraction becomes negligible, see Equation 6.1.

6.1.3
Electrostatic Interactions

In principle, one could consider using the standard equations in the literature for the electrostatic interaction, $G_E(h)$, between hard sphere particles [10]. All such equations require knowledge of the Stern potential of the particles [10] (for which the experimentally more accessible zeta potential is usually substituted) and the bulk ionic strength of the medium. However, microgel particles are more complex and less well defined than hard sphere type particles. They may contain charged groups within the interior of the particle, but these would be effectively neutralized by counterions. It is the charged groups in the *surface region*, including charged groups arising from the initiator used in the synthesis, which need to be considered. Their distribution is unlikely to be well defined. This makes $G_E(h)$ difficult to calculate, because of problems firstly in establishing the relevant, required electrostatic potential term to ascribe to the particles and secondly in determining the true ionic strength of the bulk solution. If counterions are able to penetrate into the microgel particles there will be an equilibrium Donnan partition [11] established between the interior and the exterior of the microgel particles. The latter effect will be more significant the greater the microgel particle number concentration. To determine the electrostatic potential is difficult because one needs to know the electrostatic potential distribution from the *center* of the microgel particle into the bulk solution. Knowing the potential distribution from the "*surface*" which is often ill defined for microgel particles is not sufficient, in contrast to the case for hard particles. Moreover, a directly

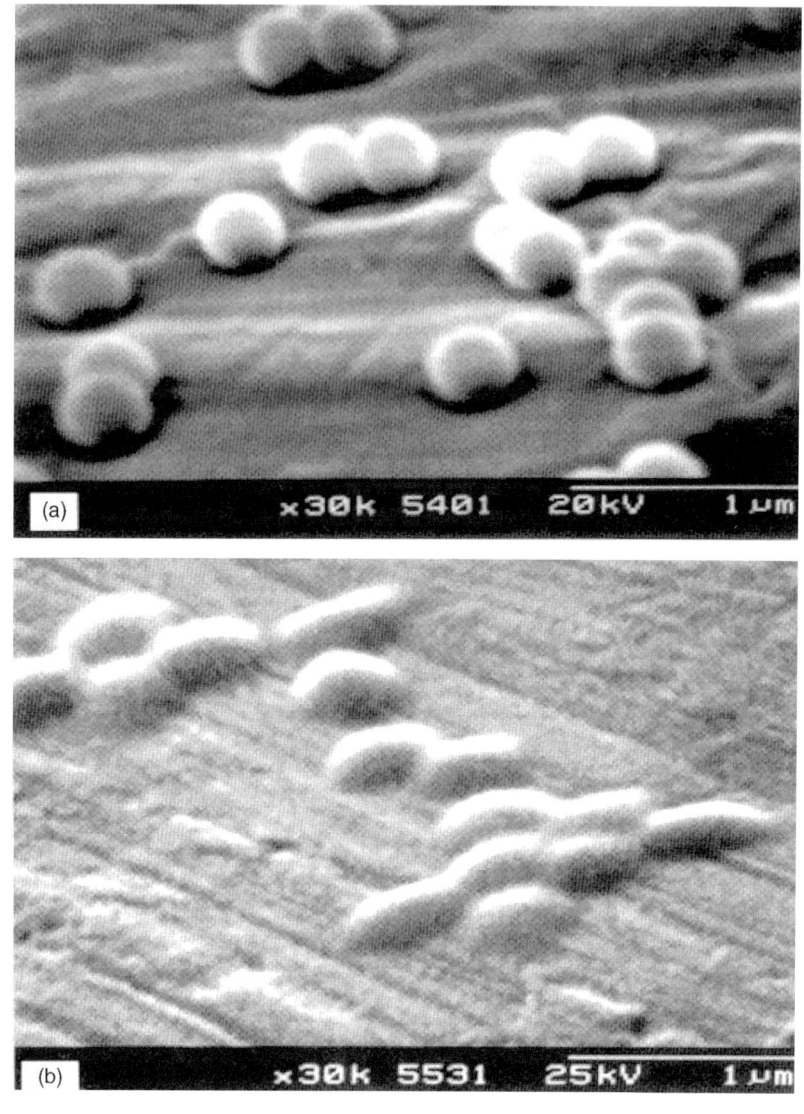

Figure 6.1 SEM pictures of PNIPAM microgel particles deposited on a grid at (a) 60 °C (deswollen state), and (b) 25 °C (swollen state), and then dried under vacuum [9].

related problem exists in applying the concept of a zeta potential to microgel particles, since the location of the plane of shear, which is the plane at which the zeta potential should be defined for hard particles, is not obvious; both solvent molecules and ions may diffuse within the interior of microgel particles. The electrophoresis of microgel particles in applied electric fields has been considered theoretically by Oshima [11], Fernández-Nieves et al. [12], and others. We will not discuss this problem further in this chapter, as the issues involved are still not fully resolved. It suffices to say that one should be wary of simply applying the classical equations relating the zeta potential to the electrophoretic mobility, such as those of O'Brien and White [13], for microgel particles since, in general, these have been derived for charged hard sphere particles.

The only paper to date that discusses the electrostatic interaction between charged microgel particles (together with the interaction between star-branched polyelectrolytes) is that by Denton [14], who applies a second-order perturbation theory. Interestingly he finds that counterions do not penetrate charged microgel particles as readily as they do for star-branched polyelectrolytes.

6.1.4
Depletion Interactions

Expressions for the depletion interaction, G_{dep}, between two *hard* spheres (radius a), as a function of their separation, h, in the presence of nonadsorbing polymer chains in solution have been provided in many studies. The simplest form [15–17] is given in Equation 6.5:

$$G_{dep} = 2\pi a \Pi (2\delta - h)^2 \quad [\text{for } h \ll a \text{ and } \delta \ll a] \tag{6.5}$$

where Π is the osmotic pressure of the exterior polymer solution and δ is the depletion layer thickness. If the particles carry a layer of chemically grafted (or irreversibly physically adsorbed) polymer chains, then a depletion interaction may occur, but one has to take into account the fact that the free chains may interpenetrate somewhat with the grafted or adsorbed chains. This situation has been analyzed by Vincent et al. [18, 19] and has been termed "soft depletion." The basic concept is that the effective depletion layer thickness is reduced by an amount, p, which corresponds to the distance of penetration of the free chains into the anchored chains. Various analytical expressions for p were derived for G_{dep} [18, 19]. Reasonable agreement was found with corresponding numerical calculations for G_{dep} based on the self-consistent mean-field theory of Scheutjens and Fleer [20].

For microgel particles a similar situation occurs, at least for the cases where $2R_g > L$ and χ_{23} is positive. No *deep* interpenetration of the microgel particles occurs, but there is still, for entropic reasons, some possible penetration of the *periphery* of the microgel particles by the added chains even if the added and microgel polymers do not associate. Thus, in general, the depletion interaction between microgel particles will be weaker than that between hard particles of the same size and at similar polymer concentrations. A modified form of Equation 6.5, as modified by Vincent et al. [18, 19] and incorporating the parameter, p, may be used in this case. If $2R_g < L$, so that deep penetration of the microgel by the added polymer may occur and χ_{23} is again positive, a similar approach could be taken to calculate G_{dep}. This approach is, however, only valid if one can assume that the absorbed polymer is "trapped" and not in free exchange with the polymer in solution, as previously discussed. As discussed in Section 6.1.1, this assumption is quite dubious when χ_{23} is positive. Moreover, for the case where χ_{23} is *negative*, no simple analytical approaches are known for calculating G_{dep} for microgels.

6.1.5
Criteria for Dispersion Stability

The classical approach that has evolved for describing the stability/aggregation behavior of (dilute) *hard sphere* colloidal dispersions is based on the original ideas of Derjaguin and Landau [21], and of Verwey and Overbeek [22], the "DLVO" theory. In this theoretical approach, one considers the magnitudes of potential energy barriers (maxima) and potential energy minima in the total pairwise interaction $G(h)$, that is the sum of all the separate interactions described in the previous sections. The reader is referred to any of the major textbooks on colloid science for a full description of these concepts [23]. An important question is how far these ideas can be applied to *soft* microgel particles? Generally, the answer is that there should be no intrinsic problem of applying the same ideas to soft microgels, provided one remembers the difficulties in calculating the various component interactions, as described in the previous sections.

For fully *swollen* microgel particles, $G_A(h)$ is, in most cases, negligible. Hence, in the absence of added polymers that could introduce attractive bridging or depletion interactions, aggregation cannot be induced, as for hard sphere particles, simply by reducing $G_E(h)$ by either decreasing the total charge of each microgel particle, for example, by a pH change, or by adding sufficient inert electrolyte.

Even though there is no driving force to induce aggregation for highly swollen microgels a $G_A(h)$ contribution, of controlled magnitude, can be "tuned in" by increasing ϕ_p; for PNIPAM particles, this can be achieved by increasing the temperature. This theme was investigated by Rasmusson *et al.* [8], and their work will be described in Section 6.2.1. It is sufficient to state here that this increase in attraction can be described in terms of the depth of the free energy minimum, G_{min}, introduced into $G(h)$, where

$$G(h) = G_A(h) + G_E(h) \quad \text{(for } h > 0\text{)} \tag{6.6}$$

We note that a similar situation arises if one introduces any additional attraction term, for example, $G_{dep}(h)$. The value of G_{min} required to achieve flocculation has been discussed in many studies. For example, Napper [24] suggested that weak flocculation would occur if G_{min} were somewhat greater than the thermal energy ($\sim 3/2$ kT) available to the particles. Bevan and Scales [25] and Scales *et al.* [26] suggested that reversible flocculation occurs when G_{min} is less than 5 kT and irreversible flocculation occurs when G_{min} is greater than 10 kT. Vincent *et al.* [27], however, have claimed that for cases of weak, *reversible* flocculation such arguments are spurious and a thermodynamic approach should be used instead, in line with the analogy between weak particle flocculation and molecular condensation processes. They postulated that the free energy of flocculation of particles ΔG_{floc} may be split into two contributions, as shown in Equation 6.7.

$$\Delta G_{floc} = \Delta G_i + \Delta G_{hs} \tag{6.7}$$

ΔG_{hs} ($= -T\Delta S_{hs}$) is the entropic contribution associated with the flocculation of hard spheres in the absence of interparticle interactions. ΔG_{hs} per particle is positive and therefore opposes the flocculation process. ΔS_{hs} depends on the particle volume fraction, ϕ_{part}; its magnitude decreases with increasing ϕ_{part}. Feigin et al. [28] have calculated ΔG_{hs} based on a simple lattice model. They estimated that, at 20 °C, ΔG_{hs} decreases from 12.2 to 3.0 kT, over the ϕ_{part} range from 1×10^{-5} to 1×10^{-1}.

ΔG_i is the interaction free energy term associated with the (nonhard sphere) interactions between the particles. It is a function of the floc structure and the depth of the relevant free energy minimum, G_{min}, in the interaction free energy–particle separation curve for two interacting particles. ΔG_i may be related to G_{min} at low ϕ_{part} through $\Delta G_i = zG_{min}$, where z is the average coordination number of a particle in a floc. For low values of ϕ_{part}, it is reasonable to assume that ΔG_i is independent of ϕ_{part}. For strongly aggregating (coagulating) dispersions, ΔG_i (negative) $\gg \Delta G_{hs}$ (positive) in magnitude, and the aggregation process is spontaneous and irreversible. However, for weakly interacting particles ($G_{min} < \sim 10$ kT), ΔG_i and ΔG_{hs} are of the same order of magnitude. If $\Delta G_i < \Delta G_{hs}$, the dispersion is thermodynamically stable, since ΔG_{floc} is positive, see Equation 6.7 On the other hand, if $\Delta G_i > \Delta G_{hs}$ then weak reversible flocculation occurs, leading to colloidal phase separation. Such colloidal phase separation has been widely reported for dispersions of weakly flocculating particles, since the first systematic study by Long et al. in 1973 [29]. Since that time, the theory relating the equilibrium phase behavior of weakly interacting particles to the pair potential has also been developed extensively. A review of this work is outside the scope of this chapter, but the reader is referred, for example, to the article by Gögelein and Tuinier [30] for a recent discussion of this topic, particularly in the context of weak aggregation and phase separation between charged particles with added nonadsorbing polymer.

6.2
Experimental Studies

6.2.1
Temperature- and Electrolyte-Induced Aggregation

We initially consider PNIPAM "homopolymer" microgels. Note that the term "homopolymer" is not strictly correct because all microgel particles contain a cross-linking comonomer. However, the concentration of the cross-linking comonomer is usually much lower than that of the other monomers present. PNIPAM dispersions were first reported by Pelton and Chibante [31], who studied the critical flocculation temperature (CFT) as a function of $CaCl_2$ concentration. Snowden and Vincent [32] made similar studies as a function of NaCl concentration. These authors showed that the CFT decreased with increasing electrolyte concentration. However, it was Daly and Saunders [33] who conducted the first comprehensive study of the effect of added electrolyte on the particle swelling and stability of PNIPAM particles.

Figure 6.2 shows the variation of the hydrodynamic diameter of PNIPAM microgel particles, as a function of temperature, in the presence of various concentrations of added NaCl. The addition of NaCl causes a substantial decrease in the diameter, in

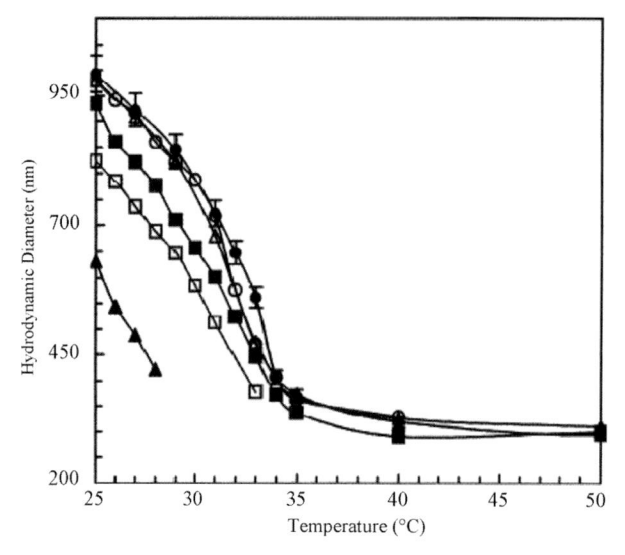

Figure 6.2 Variation of the hydrodynamic diameter, as a function of temperature, for PNIPAM microgel particle dispersions, at different NaCl concentrations: 0 (●), 10^{-4} (○), 10^{-3} (△), 10^{-2} (■), 10^{-1}(□), and 0.5 M (▲) [33].

particular at concentrations in excess of $0.1\,\text{mol}/\text{dm}^3$. This is due to the poorer solvency conditions established for the PNIPAM chains (i.e., χ is increased). Na^+ and Cl^- ions (omitted the comment about high charge density) promote the formation of extensive ion hydration spheres due to their relatively strong ion–dipole interactions with water molecules. Hence, at high NaCl concentrations a proportion of the water molecules, otherwise available for hydrating the PNIPAM chains, are no longer available.

A second important consequence of salt addition is the suppression of electrostatic interactions between neighboring PNIPAM particles. This results in temperature-induced flocculation. Flocculation may be followed using the wavelength exponent method [29]. The wavelength exponent (n) is the magnitude of the gradient obtained from a plot of log(optical density) versus log(wavelength) for light passing through a given dispersion. This parameter is very sensitive to particle aggregation [29].

Figure 6.3, taken from the work by Daly and Saunders [33], shows the variation of n with temperature for PNIPAM microgel dispersions. The onset of aggregation is apparent from an abrupt decrease in the value of n at a given temperature. This demonstrates that flocculation does occur at a critical temperature, namely, the CFT. The CFT moves to lower temperatures with increasing NaCl concentration, as observed earlier by Snowden and Vincent [32]. Zha et al. [34] have similarly used the wavelength exponent of the turbidity to investigate the critical flocculation [NaCl] concentration for a variety of PNIPAM microgels containing different initiator-derived charge groups, both anionic and cationic, at ambient temperature.

Daly and Saunders [33] also compared the dependence of the CFT values of the PNIPAM particle dispersions on NaCl concentration with the corresponding depen-

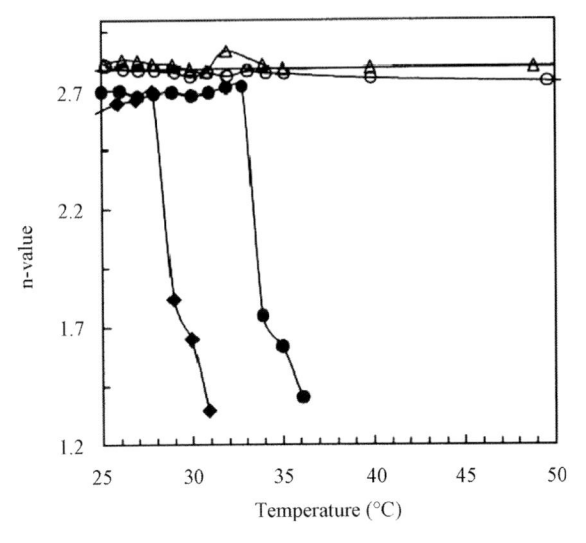

Figure 6.3 Wavelength exponent (*n*), as a function of temperature, for PNIPAM particle dispersions, at different NaCl concentrations: 10^{-2} (△), 10^{-1} (●), 0.5 M (◆), the absence of electrolyte (○) [33].

dence of the LCST values for homopolymer PNIPAM solutions in water [35]; these results are shown in Figure 6.4.

These data indicate that flocculation of the microgel particles occurs at about the same temperature as the LCST for PNIPAM (except perhaps for the case of pure water). Napper [24] has suggested that the CFT value for sterically stabilized particles

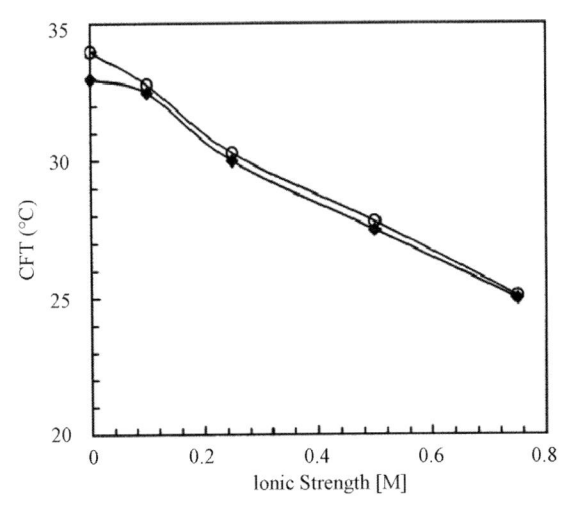

Figure 6.4 Variation of CFT for PNIPAM microgel particle dispersions with ionic strength (NaCl concentration). The CFT data from this chapter are shown (○), as well as LCST values obtained for homopolymer PNIPAM from Ref. 53 (◆) [33].

with terminally grafted polymer chains should correlate with the *theta*-temperature of the corresponding linear polymer chains in solution under corresponding conditions (the theta-temperature is the extrapolated value of the LCST in the limit of infinite MW). In the Daly and Saunders experiments, the PNIPAM particles may be flocculating under better-than-theta (i.e., $\chi < 0.5$) solution conditions due to influence of the longer range van der Waals attraction between the particles (cf. incipient flocculation discussed in Section 6.1.1).

Daly and Saunders [33] have also investigated the nature of the added electrolyte on the CFT of PNIPAM dispersions. There is a long history [36] of the so-called "Hofmeister" or lyotropic series regarding different ions and their effects on particle aggregation. Figure 6.5 shows the effect of anion type on the hydrodynamic diameter of PNIPAM microgel particles, measured at 25 °C.

The anions have the following order in terms of their ability to "salt-out" hydrophilic colloids [37]: $Cit^{3-} > Cl^- > SCN^-$. The data shown in Figure 6.5 are consistent with this order.

Figure 6.6 shows the CFT values for these PNIPAM dispersions in the presence of 0.5 mol/dm^3 concentration of different electrolytes containing a range of anions. The order of the CFTs also generally follows the lyotropic series, that is, the CFT values increase in the order: $Cit^{3-} < Cl^- < Br^- < I^- < SCN^-$. The more strongly structure-breaking anions (e.g., Cit^{3-}) more effectively compete for water. The effect of different cations on microgel particle size and CFT values, at equivalent concentrations, has also been investigated [33]. However, the results did not follow any particular trend. This is commonly observed for cations and may be a consequence of factors other than direct ion hydration by water molecules.

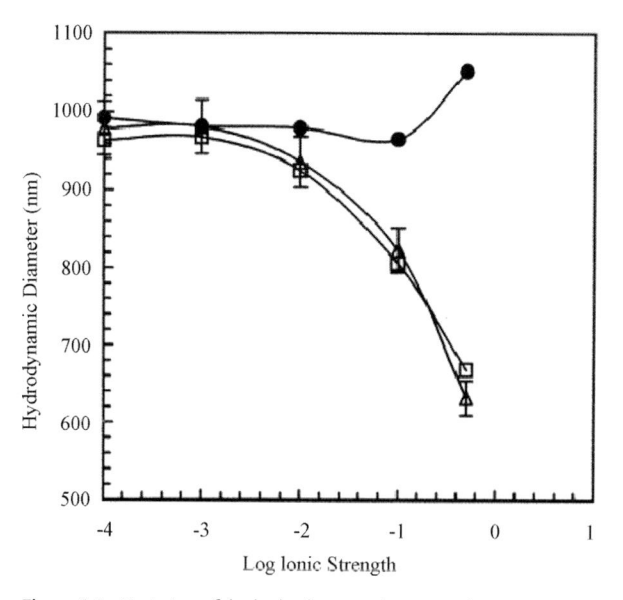

Figure 6.5 Variation of the hydrodynamic diameter of PNIPAM microgel particles, as a function of ionic strength, for NaCl (△), NaSCN (●), and Na$_3$Cit (□), at 25 °C [33].

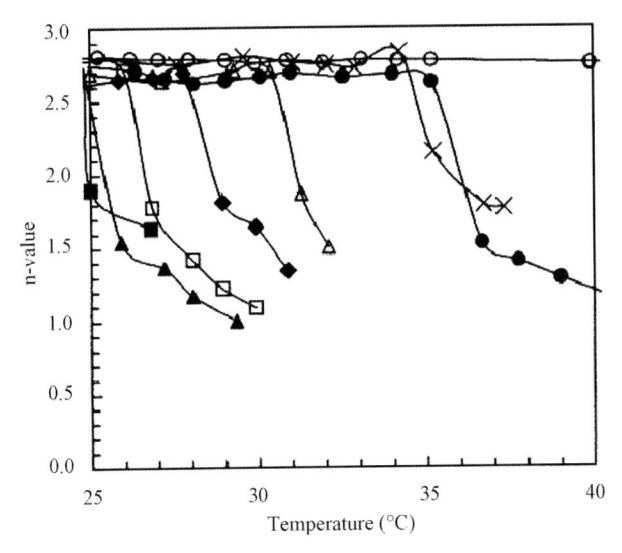

Figure 6.6 Wavelength exponent (n) as a function of temperature for PNIPAM particle dispersions in the presence of NaCl (◆), NaSCN (●), Na$_3$Cit (□), NaI (×), NaBr (△), Na$_2$CO$_3$ (▲), NaCH$_3$COO$^-$ (■), and no electrolyte (○). The ionic strengths were all 0.5 M [33].

Duracher et al. [37] investigated the flocculation behavior of poly(N-isopropyl-methacrylamide) (PNIPMAM) microgel particles. Linear PNIPMAM has a LCST of ~44 °C (cf. the corresponding value for PNIPAM is ~32 °C). They showed, like Daly and Saunders [33] that the there is a correlation between the CFT of the microgel particles and the LCST of the homopolymer in solution, as a function of added salt concentration. They also showed that the CFT of the microgel particles depends on their degree of internal cross-linking.

Cross-linked poly(2-vinylpyridine) (P2VP) microgel particles swell at pH values below ~4 due to protonation of the N atoms in the pyridine rings; they also have a (positive) surface charge arising from the initiator used [38]. Fernández-Nieves et al. [39, 40] have investigated the aggregation of dispersions of P2VP microgel particles as a function of NaCl concentration at pH 9 (i.e., in the *unswollen* state). They showed, from small-angle static light scattering studies, that the fractal dimension (d_f) of the aggregates decreased, from 2.13 at 0.25 mol/dm^3 NaCl to 1.82 at 3.0 mol/dm^3 NaCl. The explanation offered was that the depth of the free energy minimum, G_{min}, in the pair interaction potential was increasing with increasing salt concentration. At high salt concentrations, the value of d_f reached the limiting value generally found for irreversible diffusion-controlled aggregation (d_f ~1.7–1.8). Indeed, these authors [39, 40] calculated that the Hamaker constant of the P2VP microgel particles changed from 6.0×10^{-20} J at 0.25 mol/dm^3 NaCl to 6.6×10^{-20} J at 3 mol/dm^3, and the value of G_{min} increased from 2.5 to 12.5 kT over the same NaCl concentration range. Results on these systems by the same group [39] from *dynamic* light scattering studies were less easy to explain, however, especially at very high salt concentrations (>2 mol/dm^3 NaCl).

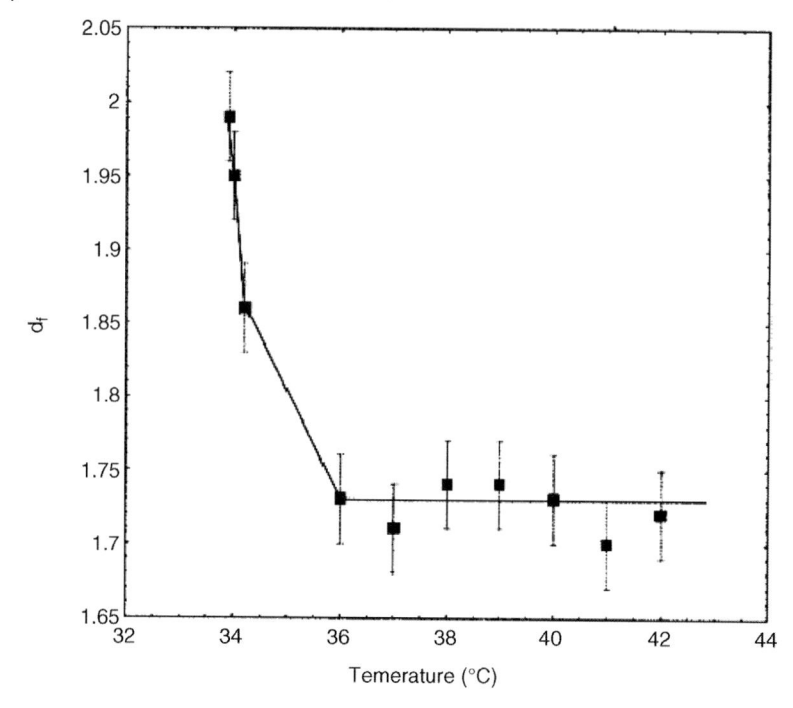

Figure 6.7 Fractal dimension of PNIPAM flocs, in 1 M NaCl, as a function of temperature [41].

Routh and Vincent [41] determined the fractal dimensions of PNIPAM aggregates as a function of temperature, at a fixed NaCl concentration (1 mol/dm^3) and pH 9, from static small-angle light scattering studies. Their results are shown in Figure 6.7.

The PNIPAM particles contained a small amount of acrylic acid as comonomer. However, although these microgel particles would therefore be strongly charged at pH 9, the electrostatic repulsion between the particles would be negligible at 1 mol/dm^3 salt concentration. The corresponding CFT value was 29 °C. It is clear from Figure 6.7 that as the temperature increases the value of d_f decreases, reaching the diffusion-controlled limiting value around 36 °C. Again, the most likely explanation is that, as the particles contract for increasing temperatures, the van der Waals attraction between the particles becomes stronger, corresponding to a deeper G_{min}.

Rasmusson et al. [8] have also studied the aggregation of PNIPAM microgel particles as a function of temperature and NaCl concentration. Figure 6.8 shows a similar plot to that of Daly and Saunders ([33]) (Figure 6.2), but here the hydrodynamic size is plotted as a function of NaCl concentration at different temperatures. This figure very clearly illustrates the strong decrease in solvency of water for PNIPAM with increasing NaCl concentration, particularly at 25 °C.

The CFT of the PNIPAM microgel dispersions, as a function of NaCl concentration, is shown in Figure 6.9 (cf. Figure 6.4 from Daly and Saunders [33]). There are three distinct regions. Below ~10 mM NaCl, no flocculation occurred at any temperature (over the range studied). At these very low NaCl concentrations, the

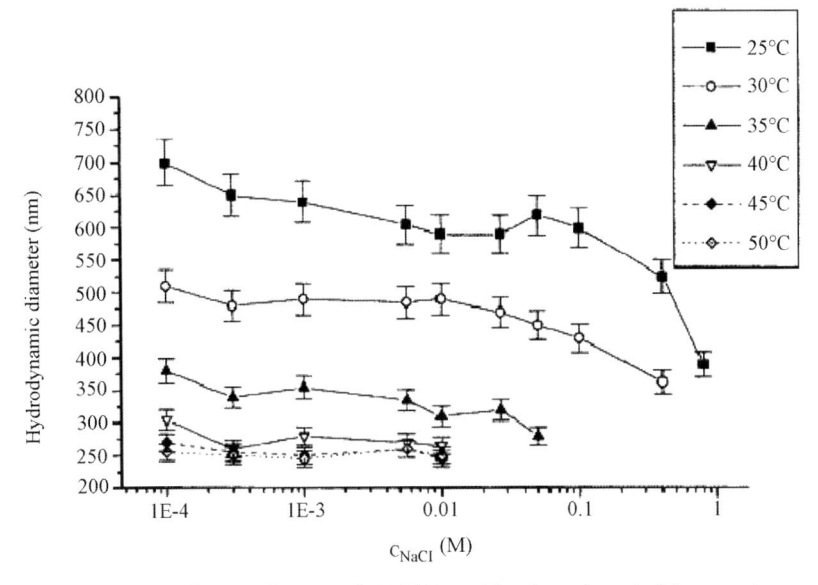

Figure 6.8 Hydrodynamic diameter of PNIPAM particles, from dynamic light scattering measurements, as a function of (log) NaCl concentration, at different temperatures [8].

electrostatic repulsion is sufficiently strong to prevent any aggregation. At NaCl concentrations greater than ~100 mM, the electrostatic repulsion is eliminated and the decrease in CFT values simply reflects the decrease in solvency of water for PNIPAM at these high salt concentrations and the corresponding increase in Hamaker constant of the particles, referred to earlier. In between 10 and 100 mol/dm^3 NaCl concentration, there is a trade-off between the electrostatic and van der Waals interactions that determines the CFT. Rasmusson et al. [8] analyzed

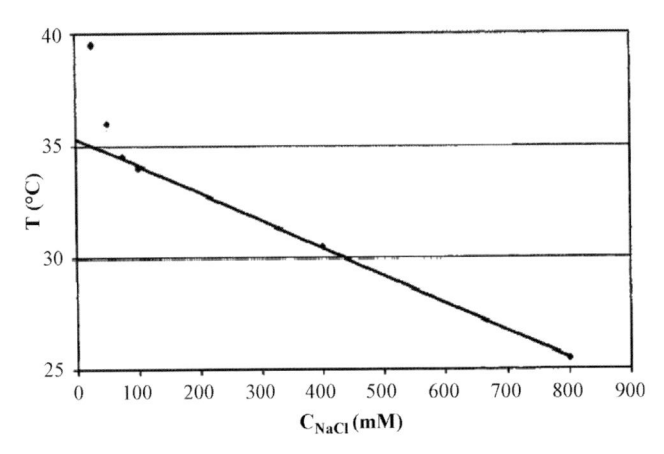

Figure 6.9 Critical flocculation temperature of PNIPAM particle dispersions as a function of NaCl concentration [8].

their data in terms of the theory presented in Section 6.1.5. They showed that, at the used particle concentration the critical value of G_{min} for the onset of flocculation (i.e., where $\Delta G_{floc} = 0$) was $\sim$8 kT, in line with the calculations reported above by Fernández-Nieves et al. [39, 40].

Wu et al. [42] have studied, both experimentally and theoretically, the equilibrium phase-coexistence behavior of PNIPAM particles in water as a function of temperature. They have modeled the fluid–solid coexistence boundary, using a first-order perturbation model for the fluid and a cell model for the solid (crystal). These authors [42] predict and observe that a solid phase may appear (i.e., freezing) at both low and high temperatures. The low temperature case is the classical case, as observed for hard spheres at high ϕ_p, which is entropically driven. The high temperature case, however, is associated with the flocculation of the particles when they deswell and is energetically driven. There are several other examples in the literature of crystalline phases forming in microgel dispersions [43–45].

There are a number of investigations of the effects of electrolyte and temperature on the stability of microgel particles containing NIPAM plus other comonomers. For example, Zhu and Napper [46] studied the CFT behavior of microgels based on P(NIPAM/PEO) copolymers, and showed that the CFT value depended strongly on the composition of the particles. Benee et al. [47] investigated P(NIPAM/VL) microgel particles (VL is vinyl laurate), using a range of techniques. It was found that the critical aggregation concentrations of NaCl for those microgel particles were lower than for PNIPAM microgel particles. This indicates that the Hamaker constant is greater for these copolymer microgel particles, presumably reflecting the greater hydrophobicity of VL compared to NIPAM.

Garciá-Salinas et al. [48] investigated the effects of added electrolyte and temperature on the aggregation of P(NIPAM/AMPS) microgel particles (AMPS is 2-acrylamido 2-methylpropane sulfonic acid). AMPS is a more hydrophilic monomer than NIPAM. They monitored the aggregation process continuously using dynamic light scattering and studied the rate of aggregation, as a function of temperature. They found good agreement between experimental and theoretical rate constants, based on model core–shell particle interactions. They also showed that the temperature-induced aggregation of the microgel particles, in the presence of 0.5 M NaCl, was not always reversible. when the temperature was reduced below the CFT value, in particular if the system was left in the aggregated state for any length of time. They suggested that this observed irreversibility is due to entanglements between polymer chains that belong to particles in contact within an aggregate. This concept had been proposed previously by Zhu and Napper [46] for PNIPAM chains grafted to the surface of polystyrene particles after they have flocculated.

Kratz et al. [49] investigated P(NIPAM/AAc) (AAc is acrylic acid) microgel particles. Interestingly, they reported a two-stage swelling transition for the microgel particles containing the highest AAc contents, that is, those prepared using 12.5 wt% AAc. A low temperature swelling transition at around 30 °C was found at low pH values, which is expected for this type of system. The higher temperature swelling transition at about 50 °C is intriguing. It may be due to a second-stage collapse associated with the microgel periphery. This idea was suggested at about the same time by Daly and

Saunders [50]. The P(NIPAM/AAc) dispersions of Kratz et al. exhibited reversible flocculation when heated in the presence of 0.2 M NaCl. They found that the CFT increased with increasing AAc incorporated into their microgels. This reflects the increased electrostatic repulsion between particles with higher AAc incorporation.

Zha et al. [51] investigated P(NIPAM/DMAEMA) (DMAEMA is dimethylaminoethylmethacrylate) microgel particles. Less than 2.5 mol% DMAEMA was incorporated into the microgel particles. They found similar trends to Saunders and Daly [33]. Structure-making anions led to greater particle deswelling and lower CFT values than structure-breaking anions. The CFT values for the P(NIPAM/DMEAEMA) microgels were similar in magnitude to those for PNIPAM microgels reported by Daly and Saunders [33], reflecting the low level of DMAEMA incorporated into the microgel particles.

Peng and Wu [52] studied the aggregation of copolymer microgel particles based on vinylcaprolactam and acrylic acid, induced by the addition of Ca^{2+} ions. The time evolution of the weight–average molecular weight and the hydrodynamic radius of the microgel particles were studied using static and dynamic light scattering, respectively. The fractal dimensions of the aggregates changed with temperature. Subsequently, Cheng et al. [53] studied the kinetics of aggregation of similar copolymer microgel particles, also based on vinylcaprolactam and acrylic acid. They showed that thermally induced aggregation, in the presence of Ca^{2+} ions, is reversible.

An interesting microgel system would be one that shows the *reverse* electrolyte effect, meaning that it swells rather than deswells upon addition of electrolyte. Polyampholyte based microgel particles, which contain an equal number of moles of a strong cationic monomer and a strong anionic monomer would be expected to show this effect. Such a system has been investigated by Neyret and Vincent [54]. They synthesized microgel particles containing equal numbers of moles of AMPS and MADQUAT (2-(methacryloyloxy)ethyltrimethylammonium chloride) using an inverse microemulsion polymerization route. These particles were strongly aggregated in water, but redispersed on adding sufficient electrolyte. This led to swelling of the microgel particles, as counterions entered the microgel particles to reduce the mutual electrostatic attraction between the positive and the negative segments. Hence, a reduction in the particle Hamaker constant occurred, leading to weaker interparticle attraction.

The internal structure of PNIPAM microgel particles has recently been studied using small-angle neutron scattering (SANS) [55]; the reader is directed to that publication for more information about the technique and its usefulness for studying the structure of microgel particles. Here, we simply present a brief discussion [55] of the use of SANS to obtain qualitative insights about the nature of the flocculated particles.

Figure 6.10 shows the effect of added NaCl concentration on the neutron scattering profiles for PNIPAM dispersions in D_2O, at 25 °C. Data are plotted as the product of the scattered intensity $I(q)$ and the square of the scattering vector, q, versus q. Results are shown for two NaCl concentrations: 0.60 mol/dm^3, where the dispersion is stable,

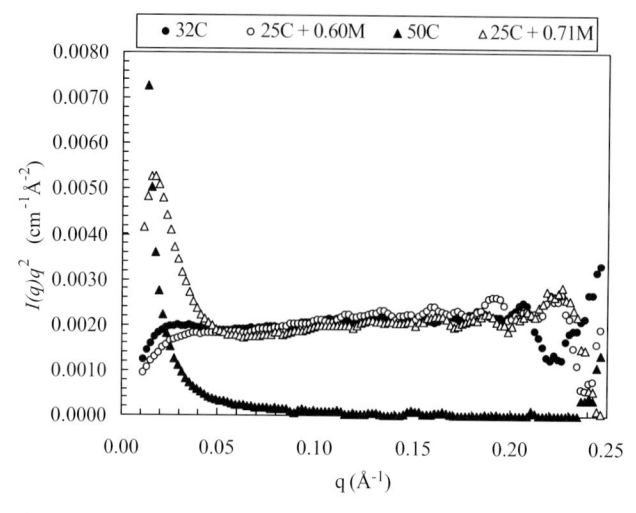

Figure 6.10 Kratky plots for PNIPAM microgel dispersions in D_2O, measured in the presence of NaCl at 25 °C, and in the absence of NaCl at 32 and 50 °C [55].

and 0.71 mol/dm^3, where the dispersions had been observed to flocculate at 25 °C. Corresponding data for the microgel particles dispersed in pure D_2O, measured at 32 and 50 °C, are also shown for comparison. The profiles at 32 and 50 °C, in the absence of NaCl, correspond to the partially expanded and collapsed states, respectively. At high q values, the gradient is ~ zero at 32 °C, indicating $I(q) \sim q^{-2}$, corresponding to Lorentzian scattering, which in turn suggests that the PNIPAM chains within the microgel particles are in a solution-like environment. On the other hand, at 50 °C, $I(q) \sim q^{-4}$, over the whole q range, which corresponds to Porod scattering that is expected for harder particles. At 25 °C, in the presence of NaCl at both 0.60 and 0.71 mol/dm^3, both SANS profiles show strong scattering at high q, again indicative of solution-like polymer chains within the microgel particles. This result was found despite the fact that the dispersion containing 0.71 mol/dm^3 NaCl was aggregated, which was also evidenced by the high scattering at low q. These data show that aggregation occurs for PNIPAM microgel particles while the particles are only partially collapsed. This implies that, in the presence of sufficient salt at 25 °C, the interparticle van der Waals interactions become strong enough to induce flocculation of the microgel particles. These results are consistent with the conclusions from the work of Routh and Vincent [41] and Rasmusson et al. [8] discussed above, where it was shown that microgel particles retain a considerable volume fraction of water under conditions when flocculation is induced, either by increasing the temperature or by increasing the electrolyte concentration.

Another interesting result from SANS experiments on PNIPAM particle dispersions is that the segment volume fraction is not uniform across these microgel particles and that ϕ_p, therefore, must be considered as an *average* value [55, 56]. As mentioned in Section 6.1.2, this has implications for the calculation of the van der Waals attraction between the particles.

6.2.2
Depletion-Induced Aggregation

The first reported systematic study of the induced flocculation of microgel particles by adding nonadsorbing polymer is that of Clarke and Vincent [57]. These authors prepared a series of cross-linked polystyrene (PS) particles in water, containing different percentages of divinylbenzene as the cross-linking monomer. These particles were then redispersed in ethylbenzene. In this solvent, the average value of ϕ_p for the swollen microgel particles varied from 0.10 to 0.63, with increasing cross-linker concentration. The effect of adding homopolymer PS was then investigated. It was found that weak reversible flocculation occurred above a critical free polymer concentration, which was strongly dependent on the microgel particle concentration; this critical polymer concentration decreased with increasing particle concentration. These results are consistent with the theory of weak flocculation discussed in Section 6.1.5 in terms of the interplay of ΔG_i (determined by G_{min}) and ΔG_{hs} (determined by the particle concentration). In this particular case, G_{min} is controlled by the concentration of free PS, through the depletion interaction, rather than the van der Waals interaction that remains negligible in these systems. Eckert and Bartsch [58] have recently revisited this system; they investigated the depletion flocculation of *concentrated* cross-linked polystyrene particles in a good solvent upon addition of linear polystyrene chains. Interestingly, they found that in a certain range of particle volume fraction, where the microgel particles formed a glass, small additions of the homopolymer could lead to melting of the glass. The result is a reentrant glass transition observed at high homopolymer concentrations. Similar observations of reentrant glass formation have been observed for hard sphere dispersions by Pham *et al.* [59]. It is significant that Bartsch's work with polystyrene microgel particles [60, 61], relating the glass transition in such dispersions to mode-coupling theory, illustrates the importance of microgel dispersions as model systems.

Depletion flocculation of PNIPAM microgel particles in aqueous media was first investigated by Snowden and Vincent [32, 62, 63]. These authors showed that such microgel particles could not be flocculated at 25 °C by adding sodium poly(styrene sulfonate) (NaPSS) of MW 50 000, at least up to concentrations of NaPSS of 0.8 wt%, but did so at 40 °C, and that the minimum amount of NaPSS required to do so decreased if NaCl was also present.

Fernández-Nieves *et al.* [64] have shown, using the wavelength exponent turbidity method described earlier, that P2VP microgel particles deswell and eventually reversibly flocculate upon the addition of dextran (MW 70 000) to the dispersion Rasmusson *et al.* [8] have extended the studies of Snowden and Vincent [32, 62, 63]. They investigated the minimum concentration of NaPSS (MW 100 000) required to induce flocculation of PNIPAM microgel particles as a function of temperature in the absence of added NaCl. Their results are shown in Figure 6.11.

Over the relatively small temperature range studied (23–33 °C) the particles do shrink (by about a factor of 2), but this is still not sufficient for the van der Waals

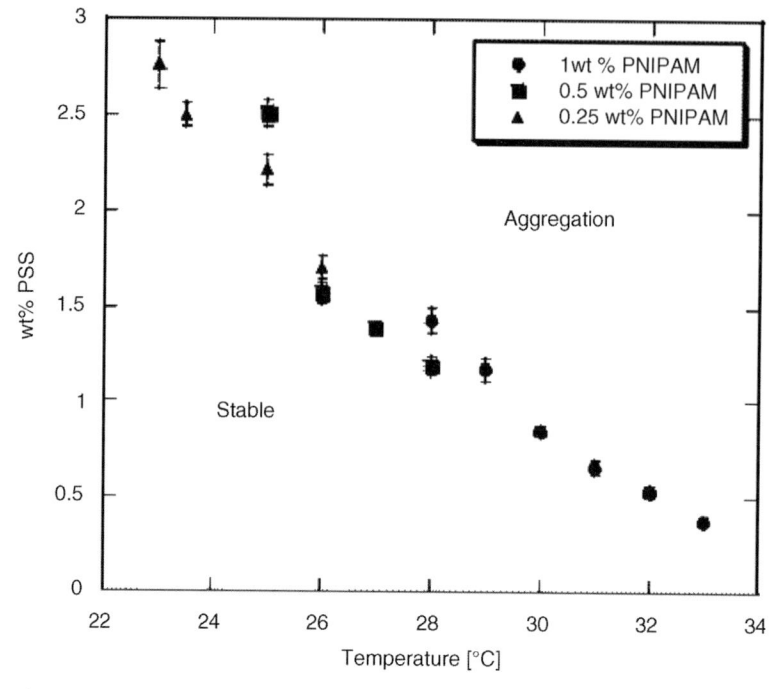

Figure 6.11 Minimum wt% of added PSS (molecular weight, 100 000 g/mol^1) required to induce flocculation of the PNIPAM particles, as a function of temperature, for three PNIPAM concentrations [8].

attraction to play a dominant role. Thus, a depletion attraction must be the more significant interaction occurring here. The question then arises as to why *less* NaPSS is required to induce flocculation at higher temperatures? As the particles become smaller with increasing temperature the depletion interaction should be smaller at a given NaPSS concentration, so that *more* NaPSS would seemingly be required to reach a condition where G_{min} is large enough for ΔG_{floc} to be zero (the boundary condition for flocculation). One possible explanation has to do with the variation in "softness" of the periphery of the PNIPAM particles with temperature. As mentioned in Section 6.1.4, Vincent *et al.* [18, 19, 65] have discussed the depletion interaction for "soft" particles, in particular for particles carrying terminally grafted polymer chains. They demonstrated that for a given core size, the weaker the depletion interaction, the softer the grafted layer is. This is due to (partial) interpenetration of the free polymer chains into the periphery of the grafted layer. A similar situation could arise for swollen microgel particles. Any interpenetration of the free chains into the soft particles reduces the effective depletion layer thickness of the free chains. It might be that with increasing temperature the microgel particles become harder, which means that it is more difficult for free NaPSS chains to penetrate the periphery of the microgel particles. Hence, the depletion interaction becomes stronger and less NaPSS is required to reach the $\Delta G_{floc} = 0$ condition.

6.2.3
Heteroaggregation

Heteroaggregation is the aggregation of particles of type A with those of type B in a mixed dispersion containing both A and B particles. Of course, homoaggregation between A and A, or between B and B, particles may also still occur. Mixtures of anionic (PNIPAM) and cationic (PNIPAM-*co*-PVP) microgel particles have been studied by Hall *et al.* [66]. Their main finding was that mixtures of the two sets of particles is stabilized against heteroaggregation at room temperature when suspended in a 10^{-1} M NaCl solution, which considerably reduces the interparticle electrostatic attraction. Homoaggregation does not occur, since there is no effective van der Waals attraction if the particles are sufficiently swollen. However, if the temperature is sufficiently increased both sets of particles shrink in size leading to stronger van der Waals forces and both homo- and heteroaggregation occur.

Fernández-Barbero *et al.* [67, 68] have studied the heteroaggregation of (equal numbers of) hard, negatively charged PS particles and soft, positively charged P2VP microgel particles, as a function of ionic strength, at room temperature. They used static light scattering to obtain the heteroaggregate fractal dimensions and dynamic light scattering to follow the kinetics. At low electrolyte concentrations, the heteroaggregates were observed to be highly branched ($d_f \sim 1.6$) because both repulsive and attractive forces are present in the structures. The kinetics was essentially diffusion controlled. However, at very high electrolyte concentrations ($\sim 1\,\mathrm{mol/dm^3}$), very compact clusters were obtained ($d_f \sim 2.3$), reflecting the weakness of the attractive forces under these conditions. This behavior is similar to the homoaggregation of microgel particles in this respect, as discussed earlier.

Snoswell *et al.* have also recently [69] studied the heteroaggregation of cationic P2VP microgel particles and anionic PS particles. The resulting heteroaggregated particles were concentrated by vacuum filtration, freeze-dried, and characterized by mercury porosimetry and electron microscopy. The growth of the structures formed by heteroaggregation at a constant KCl concentration of 0.01 mM was "arrested" at various time intervals by the addition of anionic silica nanoparticles. The average pore volume fraction increased from 0.58 to 0.63 as the aggregation time prior to "arrest" was increased from 15 to 120 s. In a second set of experiments, the aggregation time prior to arrest was maintained at 120 s, while the KCl concentration was varied between 0.01 and 10 mM. The pore volume fraction of the aggregates decreased from 0.63 to 0.54 as the electrolyte concentration increased, in accord with the changing Debye length, that is, the decreased thickness of the diffuse counterion layer around the charged particles. The inclusion of soft deformable microgels resulted in aggregates with higher mechanical strength and porosity than those obtained with heteroaggregates of anionic and cationic hard latex particles. Furthermore, incorporation of swellable microgels within a porous structure offers a potential for creating novel structures, suitable for controlled release applications.

Interestingly, but somewhat surprisingly, Armes *et al.* [70] have been able to form composites of anionic PNIPAM microgel particles (diameter ~ 800 nm) and much smaller anionic silica nanoparticles (diameter ~ 25 nm), by carrying out a conventional surfactant-free emulsion polymerization of NIPAM, using ammonium persulfate as the initiator, in the presence of the silica particles. From the SEM pictures shown in their paper [70], it appears that the silica particles form an even, close-packed coating around each microgel particle, but it is not at all clear why the silica particles become attached to the microgel network, since both sets of particles are negatively charged.

6.2.4
Probing Interactions between Microgel Particles

The two main techniques that have been used to probe the interactions between microgel particles are rheology and scattering techniques. The extraction of pair potentials from either type of measurement poses significant challenges, except perhaps in the limit of rather dilute dispersions, where Huggins coefficients from viscosity measurements and second virial coefficients from scattering experiments may be used to give information on the particle interactions. For example, in the work of Wu *et al.* [42], on the equilibrium phase-coexistence behavior of PNIPAM particles in water as a function of temperature, the authors also measured the second virial coefficient, B, of these systems as a function of temperature. On deswelling, B becomes negative in the region of the LCST for PNIPAM.

Section 4.3 deals specifically with the rheological aspects of microgel dispersions, but in this chapter we focus on using rheological measurements as an *indirect* tool for probing interactions between microgel particles. Of course, it would be useful to also have *direct* measurements of the interaction between microgel particles using for instance an AFM method, but to our knowledge no such measurements have yet been carried out. However, the interaction between an immobilized layer of P(NIPAM-coacrylic acid) microgel particles and a "colloidal probe" silica particle attached to the cantilever of an AFM has been investigated by Woodward *et al.* [71]. They found that the interaction was repulsive for all the pH and ionic strength values studied. They also found that both the magnitude and the range of the measured forces were much greater than DLVO theory predictions, which the authors attribute to the steric-like repulsion of the chains comprising the outer microgel surface.

Before embarking on a discussion of some the experimental data on concentrated dispersions of microgel particles, it should be noted that different definitions may be used to express microgel particle "concentration": (i) the *particle number concentration*, n_{part}, is the number of microgel particles per unit volume of dispersion; (ii) the *particle volume fraction*, ϕ_{part}, is defined as n_{part} times the volume, v_{part}, of each particle ($v_{part} = \pi d^3/6$, where d is the measured microgel particle diameter in a given state); (iii) the (average) *polymer* volume fraction, ϕ_{pol}, which is defined as the volume of dry microgel (i.e., the mass divided by the appropriate polymer mass density) per unit volume of dispersion. ϕ_{pol} can be obtained by drying a given volume of microgel dispersion and determining the weight change.

Note that ϕ_{part} and ϕ_{pol} are related through Equation 6.8,

$$\phi_{part} = \phi_{pol} \left(\frac{d}{d_0}\right)^3 \qquad (6.8)$$

$$\text{or} \quad \phi_{part} = \phi_{pol} q$$

where q is the volume swelling ratio of the particles, see Equation 6.4. For a given microgel dispersion, n_{part} and ϕ_{pol} are fixed but ϕ_{part} varies, as the value of d (and hence also q) varies, according to the local thermodynamic conditions. For dispersions of monodisperse, noninteracting, *hard sphere* particles, maximum close-packing occurs at $\phi_{part} = 0.74$, although the ordered crystalline phase occurs beyond ϕ_{part} values of 0.55, with phase coexistence between the crystalline and the disordered phases occurring over the range of ϕ_{part} values between 0.50 and 0.55. The transition to a glass state normally occurs at $\phi_{part} \sim 0.58$; beyond this value, the particles effectively "jam" and crystals cannot form. If the particles carry a charge, then jamming will occur at a significantly lower value of ϕ_{part} due to long-range electrostatic repulsions between the particles. This state is to be contrasted with a gel, which can form at lower values of ϕ_{part} as a result of interparticle *attractions* leading to network formation of aggregated particles. Under these conditions, at a sufficiently high value of ϕ_{part}, a percolated structure can form, which in turn may result in the formation of a macroscopic gel. Both the (repulsive) glasses and the (attractive) gels exhibit significant elastic properties.

The situation is very similar for dispersions of microgel particles. Gelling transitions, where the dispersion changes from being predominantly viscous to predominantly elastic, have been observed and discussed in many studies and some examples will be given in this chapter. One difference between microgel dispersions and dispersions of hard particles, however, is that changes in ϕ_{part} may be induced not only by increasing n_{part} but also by changing d. Changes in d can be induced by changing the local thermodynamic conditions such as temperature and solvency, and for charged microgels and also by changing the pH or ionic strength. Both glasses and gels may form.

Glasses can be formed from microgels dispersed in a good solvent, by simply increasing n_{part} beyond some critical value. This process is normally reversible upon dilution. Because microgel particles are "soft" and may be compressed (but cannot overlap significantly because of the cross-linking) higher values of n_{part} and ϕ_{part} may be reached than for conventional hard particles. It should be noted, however, that, although d is essentially independent of n_{part} at low values, d may decrease as the critical packing concentration is reached (as observed, for example, for PNIPAM particles by SANS measurements [72]), meaning that ϕ_{part} decreases correspondingly in this concentration range.

As mentioned at the beginning of this section, trying to obtain information regarding the particle interactions from rheological measurements on concentrated dispersions is a difficult exercise. One example here for microgels is the work of Berli et al. [73, 74], who studied the repulsive interactions between PNIPAM microgel

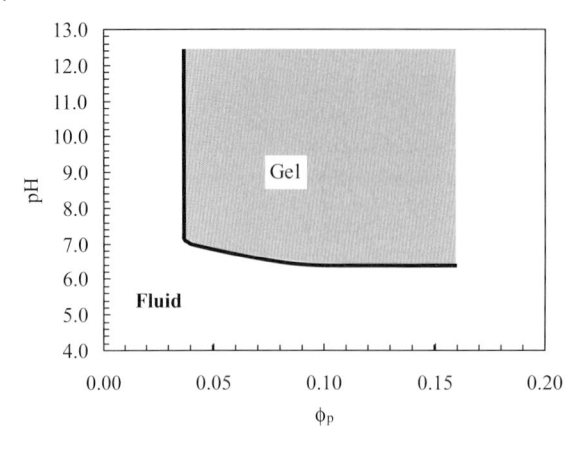

Figure 6.12 The pH-polymer volume phase "map," indicating the fluid and "gel" (strictly "glass" as used in this paper) regions, for microgel particles consisting of copolymers of methyl methacrylate (64.0 wt%) and methacrylic acid (35.6 wt%), cross-linked with ethylene glycol dimethacrylate (0.4 wt%) [75].

particles at room temperature. They compared various model potentials in attempting to fit the dynamic modulus, viscosity, and yield stress data they had obtained. The reader is referred to the original papers for details.

With regard to systems where a glass may be generated, *in situ*, by increasing d at a fixed value of n_{part}, Lally et al. [75] studied the rheology of microgel particles consisting of copolymers of methyl methacrylate (64.0 wt%) and methacrylic acid (35.6 wt%), cross-linked with ethylene glycol dimethacrylate (0.4 wt%). For these microgel particles, d increases with increasing pH, as the degree of dissociation of the carboxylic acid groups increases.

Figure 6.12 shows the "gel" (strictly "glass" to be consistent with the terminology used in this article) and fluid regions, in pH–ϕ_{pol} space, for these microgel particle dispersions. At sufficiently low values of ϕ_{pol}, a glass does not form on increasing the pH as the value of ϕ_{part} remains below the jamming limit even when the microgel particles are fully pH-swollen. However, beyond $\phi_{pol} = 0.035$ glasses form in the region of pH 6–7, as d increases and ϕ_{part} reaches the jamming limit.

Dynamic rheological experiments can nondestructively probe interparticle interactions that exist within microgel dispersions. The key parameters that are measured are the storage (G') and elastic (G'') moduli. The ratio of the two moduli (tan $\delta = G''/G'$) gives a qualitative measure of the gel strength.

Figure 6.13 shows G' and tanδ values, as a function of oscillation frequency, ω, at several pH values for a dispersion of the microgel particles described in the previous paragraph (at a fixed ϕ_{pol} value of 0.10) [75]. Clearly, a marked transition occurs over the narrow range of pH values between 6.2 and 6.4; this corresponds to the jamming transition shown in Figure 6.12. At pH 6.2 the G' values are two orders of magnitude lower than at pH 6.4. At pH 6.2, the tanδ values are greater than 1, indicative of a system exhibiting primarily viscous behavior, whereas at pH 6.4 the values of tanδ are

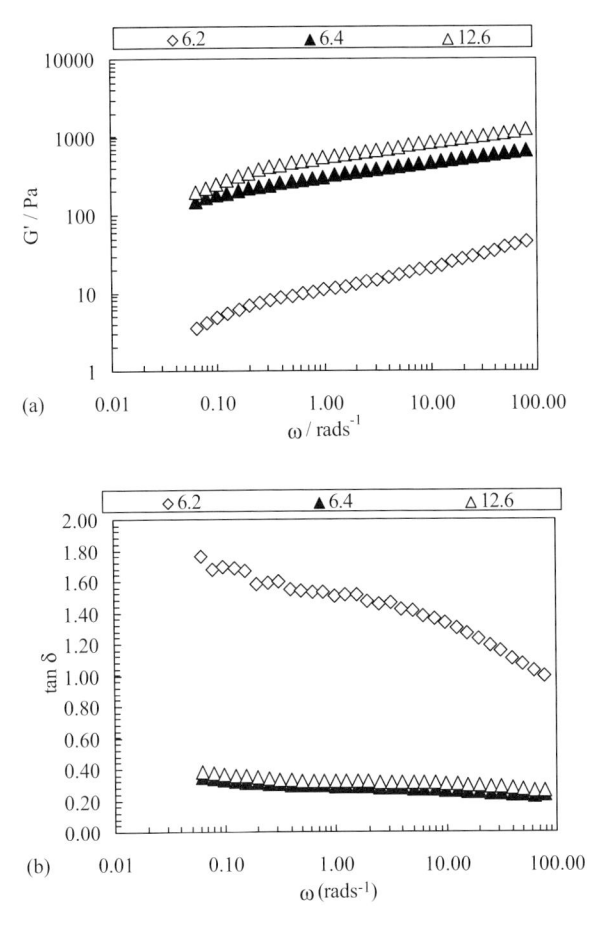

Figure 6.13 Variation of (a) elastic modulus and (b) tanδ with oscillation frequency and pH for microgel particles consisting of copolymers of methyl methacrylate (64.0 wt%) and methacrylic acid (35.6 wt%), cross-linked with ethylene glycol dimethacrylate (0.4 wt%). The polymer volume fraction was 0.10. pH values are indicated in the boxes [75].

lower than 1, indicative of a system exhibiting primarily elastic behavior. In the glass region G' was found to increase very strongly, at a given pH value, with increasing ϕ_{pol} for these microgel dispersions [75]. This is thought to be due to a combination of an increase in the (steric) confinement of the close-packed microgel particles and an increase in the ionic component of the osmotic pressure inside the charged microgel particles.

Tan *et al.* [76] have demonstrated a good correlation between the structure factor at zero scattering angle, $S(0)$ (i.e., the osmotic compressibility), as obtained from light turbidity measurements, and the rheological measurements, for microgel suspensions with repulsive interparticle interactions. $S(0)$ is a measure of the strength of the interparticle repulsion.

Aqueous PNIPAM microgel dispersions are the classic examples of a system that can form *attractive* gels at sufficiently high ϕ_{pol} values [72, 77–80], through a change in solvency; this is typically achieved by increasing the temperature beyond the LCST. Perhaps the most interesting among such of these studies, with regard to understanding the attractive forces operating between the microgel particles, is the one by Howe *et al.* [80]. These authors investigated the stress dependence of the viscosity of PNIPAM microgel particles which carried strong anionic (persulfate) charged groups, originating from the initiator used in their dispersion polymerization. At a ϕ_{pol} value of 0.04 the low stress (0.01 Pa) viscosity increased slowly as the temperature decreased from 50 to $\sim$35 °C, reflecting the increase in d and therefore also ϕ_{part}. Upon further decrease of temperature, the viscosity increased rapidly to a very high (immeasurable) value at the LCST of PNIPAM, at $\sim$32 °C, which is indicative of a sol-to-gel transition. In the sol state at high temperatures there is no indication of particle aggregation even though the interparticle attractive forces increase with increasing temperature. The reason for this behavior is that the electrostatic repulsion is too high, unless sufficient electrolyte is added to reduce the repulsive forces, as discussed in Section 6.2.1.

At low temperatures, the interparticle van der Waals attraction is very weak and the system is effectively a *glass*, formed by particle jamming. More subtle effects are observed when ϕ_{pol} is reduced to 0.01. Now jamming cannot occur on cooling and one might expect the viscosity to simply increase steadily with decreasing temperature, reflecting the increase in d. This is not the case, however, as Figure 6.14 illustrates.

The viscosity of the PNIPAM dispersions is plotted as a function of temperature for various applied stresses, both for heating and cooling, in Figure 6.14a and b, respectively. For both heating and cooling, at very low applied stresses, an anomalous viscosity "peak" is superimposed on the otherwise steady change in viscosity with temperature; the effect is much more pronounced on heating than on cooling. The peak occurs over a narrow temperature range in the vicinity of the LCST for PNIPAM and disappears when the stress is increased above $\sim$0.04 Pa. Thus, if it is due to interparticle attractive forces these must be very weak indeed.

Why is there a maximum in the viscosity? The explanation suggested by Howe *et al.* [80] is based on the structure of the PNIPAM particles. They give supporting evidence from the literature which indicates that PNIPAM particles prepared by batch dispersion polymerization are essentially core–shell structures with a highly cross-linked core (containing most of the cross-linking monomer) and with largely uncross-linked PNIPAM chains (effectively polymer "tails") forming the shell. Moreover, they suggest that much of the persulfate charge is located in or near the core of the particles and not at the outer periphery. This has to do with the greater propagation rate for bisacrylamide compared to NIPAM itself. At low temperatures, the PNIPAM particles are swollen and the shell tails are extended, which means that the van der Waals interaction is very weak and the particles are sterically stabilized with regard to aggregation. As the temperature increases close to the LCST, the tails begin to collapse and the solvent becomes poorer, leading to weak incipient

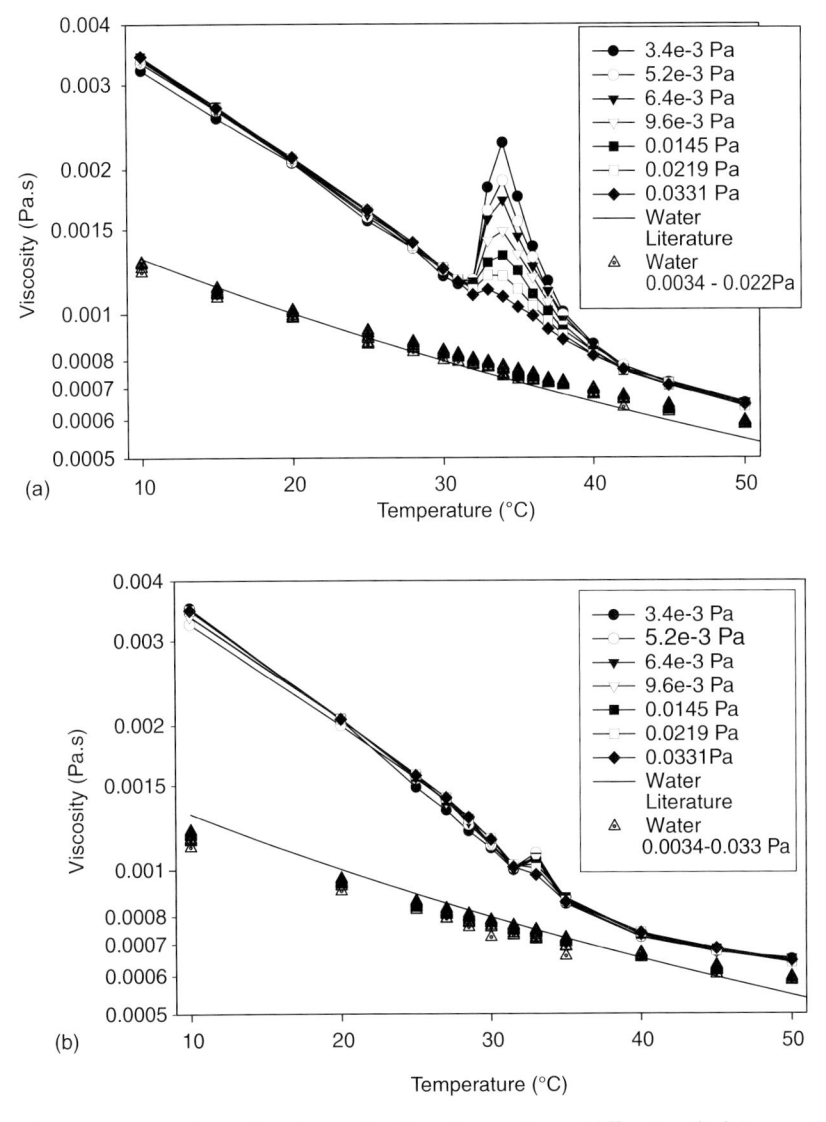

Figure 6.14 Viscosity as a function as a function of temperature, at different applied stresses (see inset), for PNIPAM microgel dispersion of polymer volume fraction 0.01, on (a) heating and (b) cooling [80].

flocculation; this is the origin of the rise in viscosity. As the temperature increases further, the van der Waals attraction becomes stronger and the solvency for the PNIPAM chains becomes worse. The reason suggested for the deaggregation of the dispersion on further heating is related to the charge on the particles; as the shell collapses, the persulfate groups within the core come closer to the periphery and they become electrostatically stabilized.

Another question, comparing Figure 6.14a and b, is why the peak in viscosity is more pronounced on heating than on cooling? Clearly, this is a nonequilibrium phenomenon and the reason probably has to do with differences in the rates of conformational relaxation of the PNIPAM tails in the shell on heating and cooling, respectively.

Some aspects of the use of SANS as a tool for probing aggregated PNIPAM microgel particle structures have been discussed in Section 6.2.1. Steiger et al. [72] have used both SANS and rheology to investigate interactions between PNIPAM particles as a function of temperature. From SANS measurements, it is shown that, as expected, the particle form factor changes with temperature as the size changes. From the structure factor, it appears that the interparticle pair potential does not change significantly between 25 and 32 °C, and that even 1 °C below the LCST the microgel particles behave effectively as hard spheres, when an equivalent hard-sphere particle size and volume fraction are used. The microgel particles resemble hard sphere behavior up to effective volume fractions of 0.35. Buscall [81] had earlier demonstrated the validity of using a hard sphere approach for scaling the viscosity of soft sphere dispersions, and Eckert and Bartsch [58] had previously come to a similar conclusion concerning the hard sphere-like behavior of cross-linked polystyrene particles in a good solvent. At higher particle volume fractions, strong deviations from the behavior of hard spheres were obtained by Steiger et al. [72]. Only at temperatures well above the LCST does the interaction potential become strongly attractive. This is in agreement with the findings reported in Section 6.2.1 that the van der Waals forces between PNIPAM particles are very weak until they have shrunk considerably.

We have seen that scattering and rheological measurements can give subtle insight into the interparticle interactions in microgel systems. It would, of course, be desirable to back up such experiments with *direct* measurements of the interparticle forces between microgel particles. This will be a challenge for the AFM experts in the future.

Acknowledgments

BV and BS would like to thank all their former students and coworkers, who have worked with them in this field over the years, and all the companies and government agencies who have funded this work. We would also like to thank the reviewers for helpful comments and suggestions for additional references.

References

1 Edwards, J., Everett, D.H., O'Sullivan, T., Pangalou, I., and Vincent, B. (1984) *J. Chem. Soc. Faraday Trans.*, **80**, 2599.
2 Cawdery, N. and Vincent, B. (1995) in *Colloidal Polymer Particles* (eds J.W. Goodwin and R. Buscall), Aademic Press, p. 245.
3 Bradley, M., Ramos, J., and Vincent, B. (2005) *Langmuir*, **21**, 1209.
4 Crowther, H.M. and Vincent, B. (1998) *Colloid Polym. Sci.*, **276**, 46.
5 Hamaker, H.C. (1937) *Physica*, **4**, 1058.
6 Vincent, B. (1973) *J. Colloid Interface Sci.*, **42**, 270.

7 Routh, A.F. and Zimmerman, W.B. (2003) *J. Colloid. Interface Sci.*, **261**, 547.
8 Rasmusson, M., Routh, A., and Vincent, B. (2004) *Langmuir*, **20**, 3536.
9 Saunders, B.R. and Vincent, B. (1997) *Colloid Polym. Sci.*, **275**, 9.
10 Shaw, D.J. (1993) *Introduction to Colloid and Surface Chemistry*, 4th edn, Butterworth-Heinemann, Oxford, p. 212.
11 Oshima, H. (1995) *Adv. Colloid Interface Sci.*, **62**, 189.
12 Fernández-Nieves, A., Fernández-Barbero, A., de las Nieves, F.J., and Vincent, B. (2000) *J. Phys. Condens. Matter.*, **12**, 3605.
13 O'Brien, R.W. and White, L.R. (1978) *J. Chem. Soc. Faraday Trans. II*, **74**, 1607.
14 Denton, A.R. (2003) *Phys. Rev. E*, **67**, 011804.
15 Fleer, G.J., Scheutjens, J.M.H.M., and Vincent, B. (1984) *Am. Chem. Soc. Symp. Ser.*, **240**, 245.
16 Sperry, P.R. (1981) *J. Colloid Interface Sci.*, **87**, 375.
17 de Hek, H. and Vrij, A. (1981) *J. Colloid Interface Sci.*, **84**, 409.
18 Vincent, B., Edwards, J., Emmett, S., and Jones, A. (1986) *J. Colloid interface Sci.*, **17**, 261.
19 Jones, A. and Vincent, B. (1989) *Colloids Surf.*, **42**, 113.
20 Scheutjens, J.M.H.M. and Fleer, G.J. (1979) *J. Phys. Chem.*, **83**, 1619; Scheutjens, J.M.H.M. and Fleer, G.J. (1980) *J. Phys. Chem.*, **84**, 178.
21 Derjaguin, B.V. and Landau, L.D. (1944) *Acta Physiochim.*, **14**, 1073.
22 Verwey, E.J.W. and Overbeek, J.Th. (1948) *Theory of the Stability of Lyophobic Colloids*, Elsevier, Netherlands.
23 Shaw, D.J. (1993) *Introduction to Colloid and Surface Chemistry*, 4th edn, Butterworth-Heinemann, Oxford, p. 219.
24 Napper, D.H. (1983) *Polymeric Stabilization of Colloidal Dispersions*, Academic Press, London.
25 Bevan, M. and Scales, P. (2002) *Langmuir*, **18**, 1474.
26 Scales, P., Grieser, F., Furlong, D., and Healy, T.W. (1986) *Colloids Surfaces*, **21**, 55.
27 Vincent, B., Luckham, P.F., and Waite, F.A. (1980) *J. Colloid Interface Sci.*, **73**, 508.
28 Feigin, R., Dodd, J., and Napper, D.H. (1981) *Colloid Polym. Sci.*, **259**, 1027.
29 Long, J.A., Osmond, D.W.J., and Vincent, B. (1973) *J. Colloid Interface Sci.*, **42**, 545.
30 Gögelein, C. and Tuinier, R. (2008) *Eur. Phys. J. E*, **27**, 171.
31 Pelton, R.H. and Chibante, P. (1986) *Colloids Surf. A*, **20**, 247.
32 Snowden, M.J. and Vincent, B. (1992) *J. Chem. Soc. Chem. Commun.*, **16**, 1103.
33 Daly, E. and Saunders, B.R. (2000) *Langmuir*, **16**, 5546.
34 Zha, L., Li, L., and Bao, L. (2007) *J. App. Polym. Sci.*, **103**, 3893.
35 Park, T.G. and Hoffman, A.S. (1993) *Macromolecules*, **26**, 5045.
36 Shaw, D.J. (1993) *Introduction to Colloid and Surface Chemistry*, 4th edn, Butterworth-Heinemann, Oxford, p. 235.
37 Duracher, D., Elaïssar, A., and Pichot, C. (1999) *Colloid Polym. Sci.*, **277**, 905.
38 Loxley, A. and Vincent, B. (1997) *Colloid Polym. Sci.*, **275**, 1108.
39 Fernández-Nieves, A., Fernández-Barbero, A., Vincent, B., and de las Nieves, F.J. (2001) *Langmuir*, **17**, 1841.
40 Fernández-Nieves, A., van Duijneveldt, J.S., Fernández-Barbero, A., Vincent, B., and de las Nieves, F.J. (2001) *Phys. Rev. E*, **64**, 1.
41 Routh, A. and Vincent, B. (2002) *Langmuir*, **18**, 5366.
42 Wu, J., Huang, G., and Hu, Z. (2003) *Macromolecules*, **36**, 440.
43 Senff, H. and Richtering, W. (1999) *J. Chem. Phys*, **111**, 1705.
44 Helleweg, T., Dewhurst, C.D., Brückner, E., Kratz, K., and Eimer, W. (2000) *Colloid Polym. Sci.*, **278**, 972.
45 Alsayed, A.M., Islam, M.F., Zhang, J., Collings, P.J., and Yodh, A.G. (2005) *Science*, **309**, 1207.
46 Zhu, P.W. and Napper, D.H. (2004) *J. Colloid Interface Sci.*, **268**, 380.
47 Benee, L.S., Snowden, M.J., and Chowdhry, B.Z. (2002) *Langmuir*, **18**, 6025.
48 Gargiá-Salinas, M.J., Romaro-Cano, M.S., and de las Nieves, F.J. (2002) *J. Colloid Interface Sci.*, **248**, 54.
49 Kratz, K., Hellweg, T., and Eimer, W. (2000) *Coloid Surf. A*, **170**, 137.

50 Daly, E. and Saunders, B.R. (2000) *Phys. Chem. Chem. Phys.*, **2**, 3187.
51 Zha, L., Hu, J., Wang, Z., Fu, S., and Luo, M. (2002) *Colloid Polym. Sci.*, **280**, 1116.
52 Peng, S. and Wu, C. (2001) *Macromolecules*, **34**, 6795.
53 Cheng, H., Wu, C., and Winnik, M.A. (2004) *Macromolecules*, **37**, 5127.
54 Neyret, S. and Vincent, B. (1997) *Polymer*, **38**, 6129.
55 Saunders, B.R. (2004) *Langmuir*, **20**, 3925.
56 Crowther, H.M., Saunders, B.R., Meares, S.J., Cosgrove, T., Vincent, B., King, S.M., and Yu, G.E. (1999) *Colloids Surf.*, **152**, 327.
57 Clarke, J. and Vincent, B. (1981) *J. Chem. Soc. Faraday Trans.*, **77**, 1831.
58 Eckert, T. and Bartsch, E. (2003) *Faraday Disc.*, **123**, 51.
59 Pham, X. et al. (2002) *Science*, **296**, 104.
60 Bartsch, E., Antionetti, M., Scupp, W., and Sillescu, H. (1992) *J. Chem. Phys.*, **97**, 3950.
61 Bartsch, E., Frenz, V., Baschnagel, J., Schärtl, W., and Sillescu, H. (1997) *J. Chem. Phys.*, **106**, 3743.
62 Snowden, M.J. and Vincent, B. (1993) Colloid–Polymer Interactions, in *ACS Symposium Series*, vol. 532, (eds P. Dubin and P. Tong), American Chemical Society, Washington, D.C., p. 153.
63 Snowden, M.J., Marston, N., and Vincent, B. (1994) *Colloids Polym. Sci.*, **272**, 1273.
64 Fernández-Nieves, A., Fernández-Barbero, A., Vincent, B., and de las Nieves, F.J. (2000) *Prog. Colloid Polym. Sci.*, **115**, 134.
65 Milling, A., Vincent, B., Emmett, S., and Jones, A. (1991) *Colloids Surf.*, **57**, 185.
66 Hall, R., Radford, J., Pinkrah, V.T., Chowdhry, B.Z., and Snowden, M.J. (2004) *Colloids Surf. A*, **233**, 25.
67 Fernández-Barbero, A., Loxley, A., and Vincent, B. (2000) *Prog. Colloid Polym. Sci.*, **115**, 84.
68 Fernández-Barbero, A. and Vincent, B. (2000) *Phys. Rev. E*, **63**, 1.
69 Snoswell, D.R.E., Rogers, T.J., Howe, A.M., and Vincent, B. (2005) *Langmuir*, **21**, 11439.
70 Percy, C., Barthet, J., Lobb, M., Khan, S., Lascelles, A., Vamvakaki, M., and Armes, S.P. (2000) *Langmuir*, **16**, 6913.
71 Woodward, N.C., Snowden, M.J., Chowdhry, B.Z., Jenkins, P., and Larson, I.P. (2002) *Langmuir*, **18**, 2089.
72 Stieger, M., Pederson, J.S., Lidner, P., and Richtering, W. (2004) *Langmuir*, **20**, 7283.
73 Berli, C.L.A. and Quemada, D. (2000) *Langmuir*, **16**, 7968.
74 Berli, C.L.A. and Quemada, D. (2000) *Langmuir*, **16**, 10509.
75 Lally, S., Mackenzie, P., Le Maitre, C.L., Freemont, A.J., and Saunders, B.R. (2007) *J. Colloid Interface Sci.*, **316**, 367.
76 Tan, B.H., Tam, K.C., Lam, Y.C., and Tan, C.B. (2005) *Langmuir*, **21**, 4283.
77 St. John, A.N., Breedveld, V., and Lyon, L.A. (2007) *J. Phys. Chem. B*, **111**, 7796.
78 Kiminta, D.M.O., Luckham, P.F., and Lenon, S. (1995) *Polymer*, **36**, 4827.
79 Senff, H. and Richtering, W. (2000) *Colloid Polym. Sci.*, **278**, 830.
80 Howe, A.M., Desrousseaux, S., Lunel, L.S., Tavacoli, J., Yow, H.N., and Routh, A.F. (2009) *Adv. Colloid Interface Sci.*, **124**, 147.
81 Buscall, R. (1994) *Colloids Surf. A*, **83**, 33.

Part Three
Phase Behavior and Dynamics of Microgel Suspensions

7
Structure and Thermodynamics of Ionic Microgels
Christos N. Likos

7.1
Introduction

Cross-linking polymer chains leads to the formation of polymeric gels, a versatile system with flexible characteristics and at the same time one of the oldest subjects of investigation in polymer science [1]. The term "gel" is used for cross-linked polymers in a solvent (as opposed to rubber); thus, gels are primarily two-component, polymer/solvent, systems. In the absence of ionizable groups along the polymer's backbones, one obtains neutral gels, whereas charged polymer networks result from the dissociation of ionizable groups when the material is dissolved in polar, aqueous solvents. The literature on the topic of uncharged networks is very rich and for a review the reader may consult Ref. [2]. Closely related to these are networks formed by polyelectrolyte (PE) chains. Early work on such systems, which are also referred to as *polyelectrolyte gels*, focused on the swelling behavior in the presence of salt [3, 4]. Active interest in the swelling of these gels remains to date due to their ability to absorb large amount of water and act as superabsorbers or drug delivery systems. Theoretical work on the swelling has been summarized in the review article of Khokhlov *et al.* [5]. With the advent of modern computers, simulation of ionic gels has also experienced a rapid growth. As a representative, we mention here the recent work of Schneider and Linse toward understanding the conformations of charged gels [6, 7], including the role played by short-range attractions [8].

The same type of networks arises also on a smaller scale, whereupon one obtains mesoscopically sized particles, synthesized by cross-linking of polymers, that are known as *microgels* [9, 10]. The most common polymer of which microgels are synthesized is poly(N-isopropylacrylamide) (PNIPAM), although other polymers such as polyacrylic acid [11] or polystyrene [12] can also be used [9], and the preparation of novel, starch-based microgels has recently been reported as well [13]. Similar to their macroscopic counterparts, microgels can be distinguished into neutral and ionic. Depending on the monomer concentration in their volume, they are classified as *uniform* or *core–shell* microgels.

Microgel Suspensions: Fundamentals and Applications
Edited by Alberto Fernandez-Nieves, Hans M. Wyss, Johan Mattsson, and David A. Weitz
Copyright © 2011 WILEY-VCH Verlag GmbH & Co. KGaA, Weinheim
ISBN: 978-3-527-32158-2

The mechanical, structural, and thermodynamic properties of such microgels are manifold and versatile; concomitantly, a large body of literature has grown, in which key properties of such mesoscopic particles have been examined, mainly from the experimental point of view. In view of the fact that it is unfeasible to present a full review and do justice to all aspects of recent work on microgels, we focus below on some key properties and applications that highlight the microgels' role as building blocks of materials with tailored properties.

Much in similarity to their macroscopic counterparts, microgels can swell in a good solvent and this property makes them promising drug delivering agents [11, 14], once they have been designed to swell in the vicinity of target sites [9]. In recent years, quite a bit of work has been done on the internal conformations and swelling properties of microgels, and in particular the dependence of the latter on parameters such as temperature, pH, solvent quality, net charge, and salt concentration [15–21]. In this way, microgels arise as tunable soft colloids that interpolate between hard spheres [20] and soft particles and allow tuning of their rheological behavior by changing any of the parameters mentioned above. In this context, the extensive work done by Richtering and coworkers on these systems [21–25] seems particularly relevant. A great deal of current technological interest in microgels focuses primarily on their usage in surface-coating applications, as demonstrated, for example, in the work of Lyon and collaborators [26–28] and Dong *et al.* [29] on ionic hydrogels. Another active area of research pertains to the fabrication of smart, responsive interfaces that react to external stimuli [30, 31]. Other applications of complex formation between soft particles and hard interfaces are related to the possibility of manipulation of the near fields of laser light [32, 33] or the control of charge transfer processes [34, 35].

Concomitant with their nature as colloidal systems, microgel solutions show structural behavior akin to hard sphere solutions. Hellweg *et al.* [36] have identified colloidal crystal formation of microgels, similar to that occurring in hard sphere systems. Gröhn and Antonietti have performed static light scattering experiments, finding scattering intensities typical of dense, liquid-like colloidal systems [37]. Formation of structured clusters from soft microgels has been reported by Fernández-Nieves *et al.* [38], whereas Fernández-Barbero and Vincent have focused on the complexation between charged microgels and oppositely charged colloids [39].

Theoretical developments aiming at the understanding of the collective, equilibrium behavior of microgel solutions are rarer since one is faced with a complex system and a description at the microscopic level is way too complex. To overcome this difficulty, a suitable approach is to coarse grain the particles, leading to the introduction of an *effective interaction potential* between suitably chosen coordinates that characterize the macromolecular aggregates as a whole [40]. Proposals for the effective potential acting between dense, hard sphere-like microgels have been put forward by Wu *et al.* [41, 42] and by Berli and Quemada [43], who introduced empirical dependencies of potential parameters on temperature in order to describe specific features in the phase behavior of thermosensitive microgels. Ionic microgels, on the other hand, are dominated by electrostatic interactions and by the presence of counterions and require a treatment that differs from that of hard spheres and hard-sphere-like interactions. For the case of loosely cross-linked, ionic microgels, an effective potential has been derived by Denton [44] within the formalism of linear response theory, which allows an (approximate) tracing out of the counterion degrees of freedom. The contents of this chapter are based on the aforementioned effective interaction, which allows us to make concrete predictions on the structural and phase behavior of ionic microgels using standard tools from the statistical mechanics of classical fluids.

The rest of this chapter is organized as follows: In Section 7.2, we present and discuss the effective interaction potential, on which are based all further investigations. The properties of the fluid phase are discussed in Section 7.3, whereas the evolutionary algorithm employed for finding the optimal crystal phases, as well as the properties of the solid phases, is discussed in Section 7.4. The resulting phase diagram is presented and discussed in Section 7.5. Finally, in Section 7.6 we summarize and draw our conclusions.

7.2
Effective Interparticle Potentials

Microgels display various types of interactions, whose importance depends on the chemistry, ambient conditions, and degree of cross-linking. The dispersion van der Waals interactions are ubiquitous and should be relevant for highly cross-linked microgels with a high inner monomer density, especially in poor solvent conditions. However, in this chapter we limit ourselves to the opposite case of weakly cross-linked, charged, and swollen microgels, for which the van der Waals forces will be suppressed by the presence of large amounts of solvent inside the particles. Elastic deformation forces can be modeled by means of the Hertz potential $V_H(r)$ between two overlapping spheres of radius a at center-to-center separation r, which reads as

$$V_H(r) = \frac{\sqrt{2a}}{5D}(2a-r)^{5/2}, \quad r \leq 2a \qquad (7.1)$$
$$= 0 \quad \text{otherwise}$$

In Equation 7.1 above, D is the quantity

$$D = \frac{3}{2}\left(\frac{1-v^2}{Y}\right) \qquad (7.2)$$

where v is Poisson's ratio and Y is Young's modulus of the microgel material, related to each other via

$$Y = 3K(1-2v) \qquad (7.3)$$

where K is the bulk modulus. Once more, loosely cross-linked, soft microgels have a low value of K and the elastic deformation contribution can be ignored compared to the electrostatic and counterion entropy contributions presented below. Naturally, deformation contributions can always be added, in the same fashion as monomer steric contributions, whose role is briefly discussed at the end of this chapter. Since elastic contributions are repulsive and have exactly the same range as the steric ones, we expect that their effects will be similar to the latter. Hence, we focus in what follows to the stronger and more involved effective interactions that arise by a canonical trace of the counterion degrees of freedom in ionic microgels.

We consider N_m spherical microgels of radius a (diameter $d = 2a$) enclosed in a macroscopic volume V that also contains a total of N_c counterions. The valences of microgels and counterions, respectively, are denoted as Z and z and the total electroneutrality condition is fulfilled:

$$ZN_m = zN_c \tag{7.4}$$

In dealing with highly complex systems, such as a solution of microgels, which includes counterions and (possibly) salt ions, as well as solvent (water), certain simplifying assumptions are necessary to reduce the problem complexity. In the first place, the aqueous solvent is not considered explicitly, that is, in a molecular fashion. It is rather treated as a dielectric continuum, modeled by its macroscopic dielectric constant $\varepsilon = 78$. This gives rise to screened Coulomb interactions and to a Bjerrum length λ_B:

$$\lambda_B \equiv \frac{e^2}{\varepsilon k_B T} = 7.14 \text{ Å} \tag{7.5}$$

at room temperature $T = 300$ K, where e denotes the elementary charge and k_B stands for the Boltzmann constant.

Furthermore, it is desirable to effectively eliminate from the Hamiltonian of the system any explicit reference to the microions and derive an *effective Hamiltonian* in which only the momenta and coordinates of the microgels show up, interacting by means of effective potentials that take into account the microions' degrees of freedom that have been traced out. Depending on the specific problem at hand, various techniques exist to derive effective Hamiltonians, all based on a precise, statistical–mechanical definition of the latter [40]. For the case of microgels, a very elegant and efficient approach has been developed by Denton [44]. Below we reproduce the basic steps and suggest the reader to consult the original publication for details.

The starting point is the full, two-component Hamiltonian that invokes the collective coordinates and momenta $\{\mathbf{P}, \mathbf{R}\}$ of the macroions (microgels) and $\{\mathbf{p}, \mathbf{r}\}$ of the microions (counterions) and reads as

$$\mathcal{H} = \mathcal{H}_{mm}(\{\mathbf{P}, \mathbf{R}\}) + \mathcal{H}_{cc}(\{\mathbf{p}, \mathbf{r}\}) + \mathcal{H}_{mc}(\{\mathbf{P}, \mathbf{R}\}; \{\mathbf{p}, \mathbf{r}\}) \tag{7.6}$$

Here, $\mathcal{H}_{mm}$, $\mathcal{H}_{cc}$, and $\mathcal{H}_{mc}$ denote the terms in the Hamiltonian that describe the microgel, the counterion, and the cross interactions, respectively. Explicitly, these terms are of the form

$$\mathcal{H}_{\alpha\beta} = \delta_{\alpha\beta} K_\alpha + V_{\alpha\beta} \qquad \alpha, \beta = m, c \tag{7.7}$$

where K_α stands for the kinetic energy and $V_{\alpha\beta}$ denotes the interactions. The latter can be expressed as

$$V_{mm}(\{\mathbf{R}\}) = \sum_{i<j}^{N_m} \phi_{mm}(|\mathbf{R}_i - \mathbf{R}_j|) \tag{7.8}$$

$$V_{cc}(\{\mathbf{r}\}) = \sum_{i<j}^{N_c} \phi_{cc}(|\mathbf{r}_i - \mathbf{r}_j|) \tag{7.9}$$

$$V_{mc}(\{\mathbf{R}\}, \{\mathbf{r}\}) = \sum_{i=1}^{N_m} \sum_{j=1}^{N_c} \phi_{mc}(|\mathbf{R}_i - \mathbf{r}_j|) \tag{7.10}$$

where $\phi_{\alpha\beta}$ are the microscopic (bare) interaction potentials between the degrees of freedom, whose positions are denoted by $\mathbf{R}_i$ and $\mathbf{r}_j$. Their functional form depends on the shape of the objects involved. The counterions, being modeled as point particles, interact by means of the Coulomb potential:

$$\phi_{cc}(r) = \frac{z^2 e^2}{\varepsilon r} \tag{7.11}$$

The microgels are modeled as *penetrable, uniformly charged spheres* of diameter d. The latter assumption is a very reasonable one for microgels, at least as long as the degree of cross-linking does not vary along the microgels' radius. The former is justifiable for loosely cross-linked microgels, which allow considerable interpenetration. Under these conditions, the remaining interactions take the form:

$$\phi_{mm}(r) = \frac{Z^2 e^2}{\varepsilon r}, \quad (r > d) \tag{7.12}$$

$$\phi_{mm}(r) = \frac{2Z^2 e^2}{\varepsilon d} \left[\frac{6}{5} - 2\left(\frac{r}{d}\right)^2 + \frac{3}{2}\left(\frac{r}{d}\right)^3 - \frac{1}{5}\left(\frac{r}{d}\right)^5 \right], \quad (r \leq d) \tag{7.13}$$

$$\phi_{mc}(r) = -\frac{Zze^2}{\varepsilon r}, \quad (r > a) \tag{7.14}$$

$$\phi_{mc}(r) = -\frac{Zze^2}{2\varepsilon a}\left(3 - \frac{r^2}{a^2}\right), \quad (r \leq a) \tag{7.15}$$

The partition function $\mathcal{Z}$ of the system is the canonical double trace over all microgel and counterion degrees of freedom; by tracing out, however, only the counterions for any *fixed* configuration of the microgels, one defines the effective Hamiltonian $\mathcal{H}_{\text{eff}}$, which depends on microgel degrees of freedom, as

$$\mathcal{Z} = \langle\langle \exp(-\beta\mathcal{H})\rangle_c\rangle_m \equiv \langle \exp(-\beta\mathcal{H}_{\text{eff}})\rangle_m \tag{7.16}$$

where $\beta = (k_B T)^{-1}$ and $\mathcal{H}_{\text{eff}} = \mathcal{H}_{\text{mm}} + \mathcal{F}_{\text{cc}}$, the latter quantity being the free energy of the counterions in the presence of the microgels, that is,

$$\mathcal{F}_{\text{cc}} = -k_B T \ln \langle \exp[-\beta(\mathcal{H}_{\text{mc}} + \mathcal{H}_{\text{cc}})] \rangle_c \tag{7.17}$$

As it stands, the partial Hamiltonian $\mathcal{H}_{\text{mc}} + \mathcal{H}_{\text{cc}}$ in Equation 7.17 is ill defined because it contains counterion–microgel and counterion–counterion Coulombic interactions without the repulsive, long-range bare microgel–microgel ones. It thus violates charge neutrality and the calculation of the free energy $\mathcal{F}_{\text{cc}}$ becomes pathological. To deal with this technical difficulty, the self-energy of a uniform, neutralizing background, $E_b = N_c n_c \tilde{\phi}_{\text{cc}}(0)/2$ is added to $\mathcal{H}_{\text{mc}}$ and subtracted from $\mathcal{H}_{\text{cc}}$, where $\tilde{\phi}_{\text{cc}}(k)$ is the Fourier transform of $\phi_{\text{cc}}(r)$ and $n_c = N_c/V$ is the counterion density. In this way, the counterion system in the presence of the frozen microgels can be treated as a perturbation around the one-component plasma (OCP) [45–47] and the sought-for quantity $\mathcal{F}_{\text{cc}}$ can be calculated via an expansion around this reference state. In fact, $\mathcal{F}_{\text{cc}}$ takes the form

$$\mathcal{F}_{\text{cc}} = \mathcal{F}_{\text{OCP}} + \int_0^1 d\lambda \, \langle \mathcal{H}_{\text{mc}} + E_b \rangle_\lambda \tag{7.18}$$

where λ is a charging parameter that scales charge Z of the micorgels, that is, in Equation 7.18 above the mapping $\phi_{\text{mc}}(r) \to \lambda \phi_{\text{mc}}(r)$ is understood, whereas the expectation value $\langle \cdots \rangle$ is carried over the counterions degrees of freedom.

Additional progress is now made by employing linear response theory [44] that allows to calculate the counterion response to the microgel density in the form

$$\tilde{\varrho}_c(k) = \chi(k) \tilde{\phi}_{\text{mc}}(k) \tilde{\varrho}_m(k) \tag{7.19}$$

where the tilde always denotes a Fourier transform, $\tilde{\varrho}_c(k)$ and $\tilde{\varrho}_m(k)$ are the density operators of the counterions and the microgels in inverse space, respectively, and $\chi(k)$ is the linear response function of the OCP. The latter is evaluated in the framework of the random phase approximation, yielding the expression

$$\chi(k) = -\frac{\beta n_c k^2}{\varkappa^2 + k^2} \tag{7.20}$$

where

$$\varkappa = \sqrt{4\pi(n_c + 2n_s)z^2 \lambda_B} \tag{7.21}$$

is the inverse Debye screening length, with the salt ion concentration n_s (if any salt is present in the solution).

The overall result of this procedure is twofold. On the one hand, the calculation of the expectation value in the integrand of Equation 7.18 generates an *induced interaction* $\phi_{\text{ind}}(r)$ acting between the microgels, which is caused by the presence of the counterions. This induced interaction can be added to the direct microgel potential $\phi_{\text{mm}}(r)$, yielding the effective interaction potential $\Phi_{\text{eff}}(r) = \phi_{\text{mm}}(r) + \phi_{\text{ind}}(r)$, to be presented below. On the other hand, a structure-independent, volume term E_0 is generated, consisting of $\mathcal{F}_{\text{OCP}}$ and additional terms from the integration. Accordingly, the microgel effective Hamiltonian (see Equation 7.16) takes the form

$$\mathcal{H}_{\text{eff}} = \sum_{i=1}^{N_m} \frac{P_i^2}{2M} + \sum_{i<j} \Phi_{\text{eff}}(|\mathbf{R}_i - \mathbf{R}_j|) + E_0 \qquad (7.22)$$

with the mass M of the microgels, which plays no role in the subsequent investigations of structure and thermodynamics, since gravity can be ignored.

The effective potential $\Phi_{\text{eff}}(r)$ is described by two different mathematical expressions, one valid for separations $r \leq d$ (overlaps) and one for $r > d$. For overlapping particles, it has the form

$$\Phi_{\text{eff}}(r) = \frac{2Z^2 e^2}{\varepsilon d}\left[\frac{6}{5} - 2\left(\frac{r}{d}\right)^2 + \frac{3}{2}\left(\frac{r}{d}\right)^3 - \frac{1}{5}\left(\frac{r}{d}\right)^5\right] - \frac{72 Z^2 e^2}{\varepsilon \varkappa^4 d^4 r}\phi_{\text{ind}}(r), \qquad (r \leq d) \qquad (7.23)$$

where $\phi_{\text{ind}}(r)$ is given by the expression

$$\phi_{\text{ind}}(r) = \left(1 - e^{-\varkappa r} + \frac{1}{2}\varkappa^2 r^2 + \frac{1}{24}\varkappa^4 r^4\right)\left(1 - \frac{4}{\varkappa^2 d^2}\right) + \frac{4}{\varkappa d}e^{-\varkappa d}\sinh(\varkappa r)$$

$$+ \left[e^{-\varkappa d}\sinh(\varkappa r) + \varkappa^2 dr + \frac{1}{6}\varkappa^4(d^3 r + r^3 d)\right]\left(1 + \frac{4}{\varkappa^2 d^2}\right)$$

$$- \frac{4r}{d}\left(1 + \frac{1}{2}\varkappa^2 d^2 + \frac{1}{30}\varkappa^4 d^4\right) - \frac{8r^3}{3d^3}\left(\frac{\varkappa^2 d^2}{4} + \frac{\varkappa^4 d^4}{12}\right) - \frac{1}{180}\frac{\varkappa^4}{d^2}r^6 \qquad (7.24)$$

On the other hand, for nonoverlapping distances the effective interaction crosses over to a screened electrostatic (Yukawa) potential of the form

$$\Phi_{\text{eff}}(r) = \frac{144 Z^2 e^2}{\varepsilon \varkappa^4 d^4}\left[\cosh(\varkappa d/2) - \frac{2\sinh(\varkappa d/2)}{\varkappa d}\right]^2 \frac{e^{-\varkappa r}}{r}; \qquad (r > d) \qquad (7.25)$$

The volume term E_0 mentioned above reads as

$$E_0 = Z N_m k_B T [\ln(Z\varrho \Lambda^3) - 1]$$

$$- N_m \frac{6Z^2 e^2}{\varepsilon d}\left\{\frac{1}{5} - \frac{2}{\varkappa^2 d^2} + \frac{6}{\varkappa^3 d^3}\left[1 - \frac{4}{\varkappa^2 d^2} + \left(1 + \frac{4}{\varkappa d} + \frac{4}{\varkappa^2 d^2}\right)e^{-\varkappa d}\right]\right\}$$

$$- Z N_m \frac{k_B T}{2} \qquad (7.26)$$

where Λ is the thermal de Broglie wavelength and $\varrho = N_m/V$ is the number density of the microgels. The volume term E_0 for a few selected values of the charge Z is shown in Figure 7.1. Though E_0 has no influence on the correlation functions of

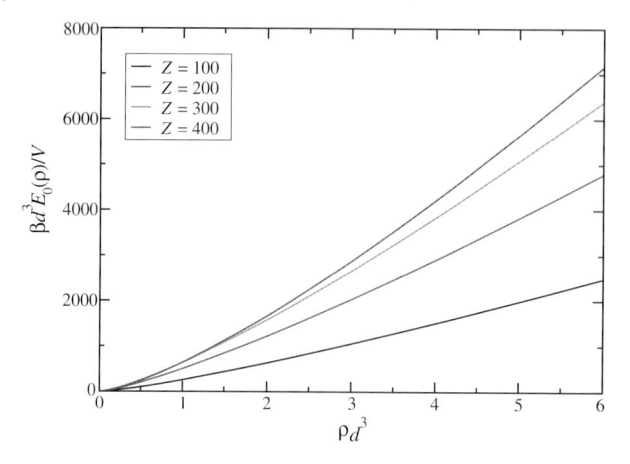

Figure 7.1 The volume term $\beta d^3 E_0(\varrho)/V$ of microgel solutions, Equation 7.26, as a function of the density ϱ and for different Z-values. Adapted with permission from Ref. [51], copyright 2005 American Institute of Physics.

macroions, it forms an integral part of the thermodynamics of the system, as is clear from Equation 7.22. Such extensive terms are quite common when microscopic degrees of freedom are traced out [40] and they appear, for example, also both for metals as a result of integrating out the electrons [48] and for other classical charged systems, such as polyelectrolyte stars [49] and charge-stabilized hard colloids [40, 50]. It is very important to point out, however, that their specific form depends both on the system under consideration and on the approximations involved in deriving the effective Hamiltonian. A crucial difference between micorgels and charge-stabilized hard colloids is that in the former case counterions can penetrate the colloidal microgel, whereas in the latter case they cannot. For a critical discussion on the influence that this property has on the form of the volume terms, the reader may consult the appendix of Ref. [51].

Let us now return to a concise discussion on the effective interaction $\Phi_{\text{eff}}(r)$ described by Equations 7.23–7.25 above. Through its dependence on the inverse Debye screening length $\varkappa$, Equation 7.21, the effective potential $\Phi_{\text{eff}}(r)$ acquires an explicit density (and temperature) dependence; the microgel density $\varrho = N_m/V$ is coupled to that of the counterions, n_c, via the electroneutrality condition, Equation 7.4. In Figure 7.2, we show some selected plots of the effective potential $\Phi_{\text{eff}}(r)$, where it can be seen that this quantity is ultrasoft and bounded. Moreover, both the range and the strength of $\Phi_{\text{eff}}(r)$ shrink with increasing concentration ϱ of the microgels. The reason lies in the concomitant increase of $\varkappa$, which renders the Yukawa part of the potential for $r > d$ shorter ranged and also decreases its prefactor. Associated with the latter is the decrease in the *effective charge* Z_{eff} of the microgels, which can be calculated as the sum of the bare charge Z and that of the *adsorbed* counterions that carry opposite charge. As Denton has shown [44], this quantity is given by the expression

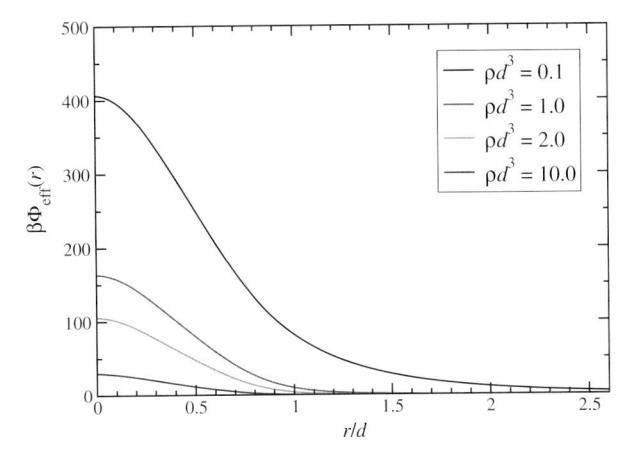

Figure 7.2 Plot of the effective interaction potential $\Phi_{\text{eff}}(r)$, in $k_B T$ units, Equations 7.23–7.25, for a fixed charge $Z = 300$ and various values of the density. Here, the microgel diameter is $d = 100$ nm. Adapted with permission from Ref. [51], copyright 2005 American Institute of Physics.

$$Z_{\text{eff}} = Z \frac{6}{\varkappa d}\left(1 + \frac{2}{\varkappa d}\right) e^{-\varkappa d/2}\left[\cosh(\varkappa d/2) - \frac{2\sinh(\varkappa d/2)}{\varkappa d}\right] \qquad (7.27)$$

In Figure 7.3, we show the dependence of the ratio Z_{eff}/Z on the quantity $\varkappa d$, the latter corresponding via Equation 7.21 to a variation of the density. It can be seen that for low densities, Z_{eff} essentially coincides with Z, since the counterions remain free outside the microgels, on entropic grounds. As the density grows, however, the attractive interaction $\phi_{\text{mc}}(r)$ between microgels and counterions drives an enhanced

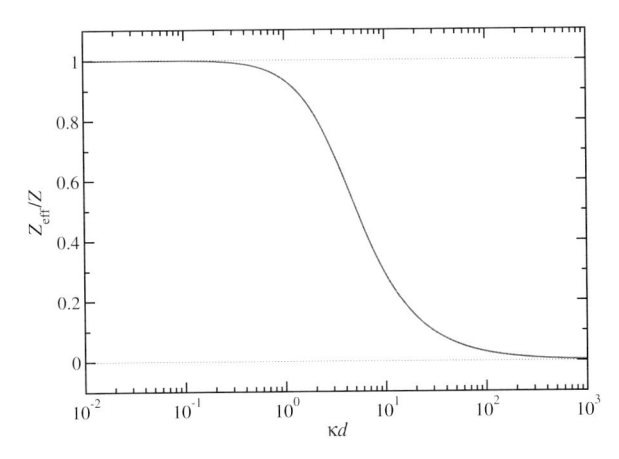

Figure 7.3 Ratio of the effective to nominal charge of the microgels, Equation 7.27, as a function of $\varkappa d$, corresponding to an increase in the density via Equation 7.21.

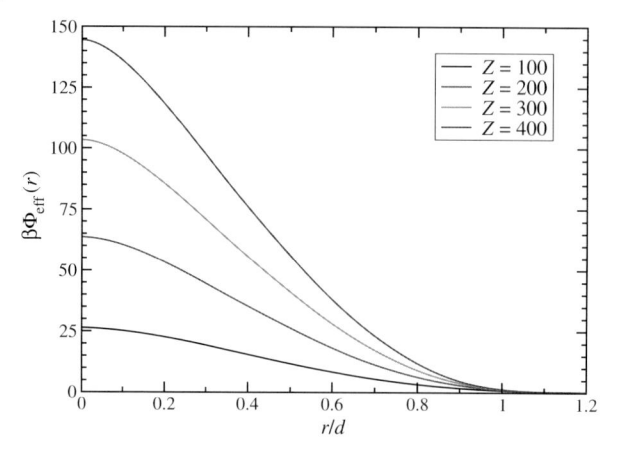

Figure 7.4 Dependence of the effective potential $\Phi_{\text{eff}}(r)$, in $k_B T$ units, on the charge Z for fixed density $\rho d^3 = 2.0$ and microgel diameter $d = 100$ nm. Adapted with permission from Ref. [51], copyright 2005 American Institute of Physics.

absorption of the latter inside the former and results into a rapid decrease in their net charge. This phenomenon, which has also been independently established for star-branched polyelectrolytes [52, 53], is absent in the case of hard charged colloids that are impenetrable to counterions. Here, the "renormalized charge" Z_{eff} can be explicitly calculated within the framework of linear response theory (see Equation 7.27). It is this reduced charge that leads to a reduction in the effective interaction $\Phi_{\text{eff}}(r)$ seen in Figure 7.2. On the other hand, as can be seen in Figure 7.4, an increase in the charge number Z on the microgels renders the potential more steeply repulsive. This is expected and physical since an increase in Z leads to enhanced repulsions between the microgels.

7.3
The Fluid Phase of Ionic Microgels

Under conditions of sufficiently low densities, microgel particles in solution are fluid, whereas at high concentrations they self-organize into periodic crystals, as will be shortly demonstrated. The investigation on phase transitions between the two states requires knowledge of the associated Helmholtz free energies. For the fluid state, a detailed statistical–mechanical theory of its thermodynamics has to start from the determination of the structural correlations present in the system [40, 54]; these then also offer a route to thermodynamics. It is sufficient for this purpose to focus on *pair* structural correlations, which are encoded in real space in the total correlation function $h(r)$ that measures the correlations in the density fluctuations located at two points separated by a distance r. A closely related quantity is the radial distribution function $g(r) \equiv h(r) + 1$, which is proportional to the conditional probability of finding a particle at a distance r from another one, which is "clamped down" at the origin. Finally, a third quantity, which also allows direct contact with scattering experiments, is the static structure factor $S(k)$, connected to the Fourier transform $\tilde{h}(k)$ of $h(r)$ via

$$S(k) = 1 + \varrho \tilde{h}(k) \tag{7.28}$$

Apart from being proportional to the experimentally measurable coherent scattering intensity $I(k)$ at scattering wave vector k, the structure factor also has a deep significance as a *response function* of the fluid. Indeed, $S(k)$ is a measure of the propensity of the fluid to sustain spontaneous density fluctuations of wavenumber k. As such, the amount of structure encoded in $S(k)$ gives information on the proximity of a possible transition to a spontaneously modulated state, for example, a crystal. Here, the value $S(k) = 1$ attained for a noninteracting system provides a baseline since ideal gases show no tendency to order under any wavelength.

It is clear from Equations 7.23–7.25 that the effective interaction at hand is density dependent. As it has been repeatedly pointed out in the literature [55–58], special care should be taken when dealing with such effective potentials. Their crucial difference with the usual, microscopic pair potentials is that they are derived through a coarse-graining procedure of one sort or the other. Thereby, the *context* in which the coarse graining has been carried out has always to be kept in mind. The usual, thermodynamic integration routes that lead to the free energy for microscopic interactions need not deliver consistent results when one works with effective potentials [55]. In particular, when an effective potential with *explicit* density and temperature dependence, $v(r; \varrho, T)$, is used, it is not a priori clear whether in taking density and/or temperature derivatives to calculate, for example, the pressure or energy from the correlation functions, these derivatives should also explicitly act in the potential parameters. Accordingly, the usual expressions for pressure, compressibility, and internal energy, as we know them from simple liquids [54], need not carry over to the case of effective, state-dependent potentials.

In the case at hand, the procedure involved in the derivation of the effective interaction is an approximate tracing out of the counterion degrees of freedom, which has been carried out with the goal of leaving the total free energy of the system unchanged. In other words, the sum of the interaction free energy and the volume terms should be the same as the original free energy of the system. We will shortly utilize this property in a particular procedure for calculating the free energy of the fluid. Since the compressibility and virial routes to the free energy [54] do not yield identical results [55], neither the "fluctuation compressibility" χ_{fl} given by the $k \to 0$-limit of the static structure factor $S(k)$ nor the "virial compressibility" χ_{vir}, obtained by differentiation of the virial pressure with respect to density, represents the true compressibility of the system. Notwithstanding this difficulty, we can make a "mental copy" of our system into a reference system that interacts with a state-*independent* interaction potential $v_0(r) = v(r; \varrho, T)$ for any given conditions of density and temperature. This system will, evidently, have exactly the same correlation functions $g(r)$ and $S(k)$ as the real one. Moreover, its thermodynamics will be free of ambiguities since its interaction potential is density-independent. We can thus set out to calculate as accurately as possible the correlation functions $g_0(r) = g(r)$ and

$S_0(k) = S(k)$ by imposing, for example, consistency of the fluctuation and virial compressibilities of the *reference* system, that is, by imposing $\chi_{fl}^0 = \chi_{vir}^0$. This is achieved by closing the Ornstein–Zernike relation [54] with a thermodynamically consistent closure, such as the Rogers–Young (RY) expression [59]. We reiterate, of course, that the compressibilities, energies, and pressures of the reference and the real system are *not* equal to each other. Another choice to derive correlation functions is by employing the simpler, (and thermodynamically inconsistent) hypernetted chain (HNC) closure to the Ornstein–Zernike relation. A comparison between the results of RY and HNC in obtaining the phase diagrams will be shown in Section 7.5.

The radial distribution function shows unusual dependence on the density ϱ; some representative results are shown in Figure 7.5. The amount of correlations encoded in $g(r)$ and expressed through the height of the correlation peaks *decreases* with increasing density, a feature unknown for diverging and density-independent potentials but most certainly present for bounded density-independent interactions, such as the Gaussian core model [60]. Moreover, the unusual behavior is seen that upon increasing the density the height of the peak corresponding to the first coordination shell *decreases*. Physically, we can understand this by noticing that the effective interaction potential $v_{eff}(r)$ softens (becomes less repulsive) upon increasing the density. In this sense, this correlation loss observed in $g(r)$ is akin to similar phenomenon seen for the related system of solutions of star-shaped polyelectrolytes [49, 61]. It is, however, *absent* in neutral star polymer solutions, whose effective interactions are density independent and do not soften with increasing concentration [62]. Above the overlap density, $g(r)$ develops a substructure inside diameter d; the physical reasons behind this behavior have been discussed in detail in Refs [49, 62]. All three physical systems, star polymers, polyelectrolyte stars, and ionic microgels, belong thus to the same class of soft matter systems that can be termed

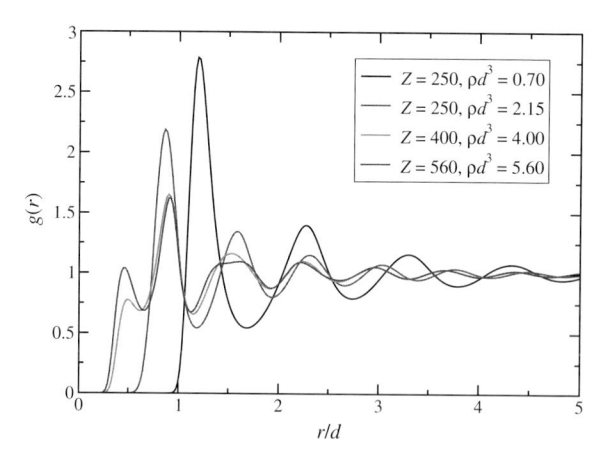

Figure 7.5 Microgel radial distribution functions $g(r)$ for various values of Z and ϱ, shown in the legend. The (Z, ϱ) values are chosen to lie close to the phase boundaries between the fluid and the incipient crystal structures, as shown in Figures 7.11 and 7.12. Adapted with permission from Ref. [51], copyright 2005 American Institute of Physics.

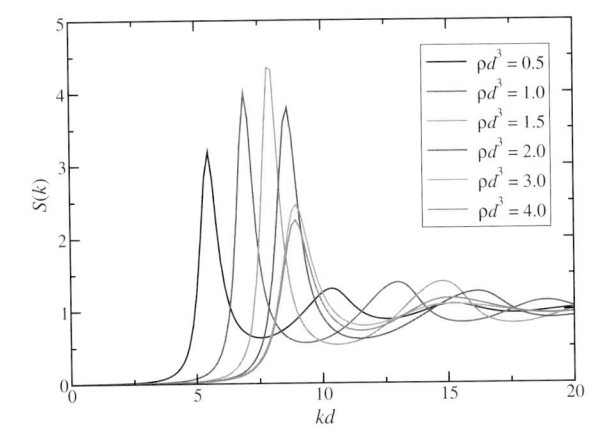

Figure 7.6 Microgel structure factors $S(k)$ for fixed $Z = 250$, varying the density ϱ. Here, $d = 100$ nm. Adapted with permission from Ref. [51], copyright 2005 American Institute of Physics.

ultrasoft [63]. Dendrimers, which interact by means of a Gaussian effective potential between their centers of mass, are also a member of this family [64–66].

The anomaly in $g(r)$ is also reflected in its reciprocal space counterpart, the structure factor $S(k)$, as obtained by the RY-closure, shown in Figure 7.6. The structure factor shows the fingerprint of the ultrasoft interaction, namely, the growth of the height of the main peak of $S(k)$ up to, roughly, the overlap density $\varrho_\star$ of the microgels, that is then followed by an anomalous behavior: on the one hand, the height of the main peak starts *decreasing* as the density grows and, on the other hand, its position changes very weakly with density above $\varrho_\star$. The same has been seen for solutions of star polymers [62] and polyelectrolyte stars [49, 61]. Since $S(k)$ is a measure of the tendency of the fluid to develop and enhance density fluctuations (modes) of wavelength $\lambda = 2\pi/k$, the nonmonotonic behavior of the main peak height with density points to re-entrant melting behavior. This prediction is also confirmed by explicit free energy calculations, to be presented in Section 7.5.

Finally, we address the issue of the appropriate determination of the free energy of the fluid, which we touched upon in the discussion of the peculiarities of density-dependent potentials. The availability of structural data, in particular the function $g(r)$, offers a way to the thermodynamics of the system via the λ-integration route [58], which corresponds to a gradual turning on of the pair interactions from a noninteracting system ($\lambda = 0$) to a fully interacting one ($\lambda = 1$). To be concrete, let us consider the Hamiltonian $\hat{\mathcal{H}} = \mathcal{H} - E_0$, with $\mathcal{H}$ and E_0 being defined in Equation 7.22. It can be shown [67] that the excess free energy density $\hat{f}_{\text{ex}}(\varrho)$ associated with $\hat{\mathcal{H}}$ can be evaluated by

$$\beta \hat{f}_{\text{ex}}(\varrho) = \frac{1}{2} \varrho^2 \int d^3 r \beta \Phi_{\text{eff}}(r; \varrho) \int_0^1 d\lambda g^{(\lambda)}(r; \varrho) \qquad (7.29)$$

where $g^{(\lambda)}(r; \varrho)$ is the radial distribution function corresponding to a fluid interacting by means of the "scaled" potential $\Phi_{\text{eff}}^{(\lambda)}(r; \varrho) = \lambda \Phi_{\text{eff}}(r; \varrho)$. The total free energy density $f(\varrho)$ is then obtained by adding to $\hat{f}_{ex}(\varrho)$ the ideal and volume-term contributions,[1] that is,

$$\beta f(\varrho) = \varrho \left[\ln(\varrho\Lambda^3) - 1\right] + \beta \hat{f}_{ex}(\varrho) + \frac{\beta E_0}{V} \tag{7.30}$$

The sum of the first two terms on the right-hand side of Equation 7.30 is shown in Figure 7.7a, whereas the total free energy in Figure 7.7b, for typical values of the charge parameter Z. Note that the free energy density is free of concave parts, that is, there is no spontaneous liquid–gas phase separation in the system, in agreement with the fact that the structure factors $S(k)$ of the microgel particles show no divergence at the limit $k \to 0$. The liquid free energy calculated in the fashion described above has been employed to draw the phase diagrams to be presented in Section 7.5, in conjunction with the free energies of the solid phases, which are the subject of the following section.

7.4
Genetic Algorithms for the Crystal Structures

Although steeply diverging interactions lead to crystals whose lattice structure is determined mainly by packing considerations, for soft interactions things are much less trivial. A variety of unusual, "exotic" structures can be stabilized even by simple, spherically symmetric ultrasoft pair potentials [68–70]. In this sense, simple guessing of the possible crystal structures is of no help for ultrasoft interactions, as almost certainly some competitive ones will be missed because they are difficult to guess. One needs an unbiased search strategy within the parameter space to find those crystal structures that minimize, for example, the $T = 0$ free energy (i.e., the internal energy U of the system), which then can also be employed as candidates for a $T \neq 0$ calculation of the free energies. A very efficient tool for that purpose is offered by genetic algorithms (GA) [51, 71, 72], whose logic and implementation are briefly described below.

GAs can be considered as optimization strategies that use features of evolutionary processes as key elements; their purpose is to find optimal solutions for a given problem [73]. Originally, they were developed by Holland and coworkers in the context of engineering science [74] and have been applied in many fields [73], such as economics, immunology, biology, or computer sciences. The pioneering work of Deaven and Ho [75] played a pivotal role in introducing GAs into physics with the specific purpose of geometry optimization for given interactions. Recently, these

---

1) For charged systems, the sum of the excess free energy $\hat{f}_{ex}(\varrho)$ and the volume term E_0/V can also be calculated from the $k \to 0$-limit of the structure factor $S(k)$. This is due to the long-range character of the Coulomb interaction and the electroneutrality condition that give rise to certain sum rules for charged systems. Direct comparison between the sum-rule route and the λ-integration route to the free energy has been shown to yield good agreement between the two in the case of polyelectrolyte star solutions. See the appendix of Ref. [49] for details.

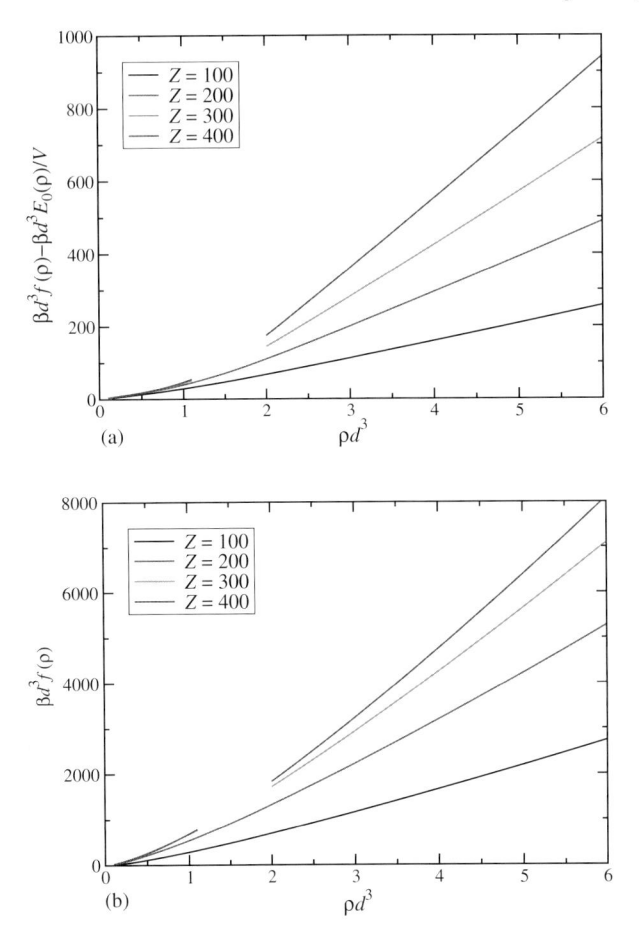

Figure 7.7 (a) The sum of excess plus ideal free energy density of the microgel fluid, drawn against the density ρd^3, excluding the volume term E_0. (b) The same including the volume term. In both panels, the density gap lies in the region of stability of crystals. Adapted with permission from Ref. [51], copyright 2005 American Institute of Physics.

tools have been extended to the determination of crystal structures for *infinite* systems as well [51, 76, 77].

The structural unit of a GA is an individual $\mathcal{I}$ that, in turn, is built up by a fixed number of "genes." In problems of 3d-crystallization, this individual encodes both the three lattice vectors of the fundamental unit cell and the vectors of the basis, if any. In applications of finite cluster formation, all vectors of the cluster build up the corresponding individual. Depending on application, tradition, and taste, one may choose to represent the position vectors, which are variational quantities, either as real (decimal) numbers or in a binary representation, that is, as a succession of 1s and 0s. In the latter case, which is the one we adopt for the problem at hand, suitable generalized coordinates q_i that describe the unit cell are normalized to values

| 1 | 0 | 1 | 1 | 0 | 1 | 0 | 1 | 1 | 0 | 0 | 1 | 0 | ... | 0 | 1 | 0 | 0 | 1 |

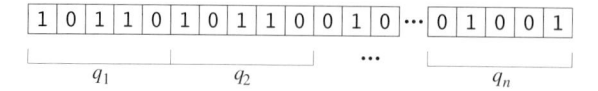

$\underbrace{\qquad\qquad}_{q_1}\ \underbrace{\qquad\qquad}_{q_2}\ \cdots\ \underbrace{\qquad\qquad}_{q_n}$

Figure 7.8 Schematic drawing of an individual representing the unit cell and possible basis vectors of a crystal. Each generalized coordinate q_i, $i = 1, 2, \ldots, n$ is assigned a fixed number of genes.

$0 \leq q_i \leq 1$ and are represented in the binary alphabet; for details, the reader may consult Ref. [72]. The number of genes (binary digits) devoted to each generalized coordinate q_i is fixed, and by concatenating successions of genes after one other, an individual is built that uniquely describes a crystal structure. An exemplary form of an individual is shown in Figure 7.8.

As a second ingredient of the GA, each individual $\mathcal{I}$ is assigned a *fitness value* derived by evaluation of the fitness function $f(\mathcal{I})$. The form of the fitness function must be chosen in such a way that the lower the (free) energy $F(\mathcal{I})$ of the crystal represented by the individual $\mathcal{I}$, the higher the (positive) value of its fitness $f(\mathcal{I})$. Though the choice of the form of $f(\mathcal{I})$ is *not* unique, we have empirically found that a suitable parametrization of the fitness is given by

$$f(\mathcal{I}) = \exp\{-[F(\mathcal{I}) - F(\mathcal{I}_{fcc})]/F(\mathcal{I}_{fcc})\} \tag{7.31}$$

where $F(\mathcal{I})$ is the free energy for a crystal structure represented by the individual I and I_{fcc} is the individual representing the fcc lattice; division of the exponent by $F(I_{fcc})$ renders all values of the former to be of order unity, avoiding in this way the technical difficulties associated with exponentiating very large numbers. The free energy of any given crystal structure can be calculated, for example, within a harmonic theory in the approximation of the Einstein model, or one can employ other approaches, such as phonon theory and density functional theory. However, we emphasize that the evaluation of $F(I)$ is the most severe bottleneck in this approach, thus economical ways of evaluating the free energy of crystals are crucial for convergence in reasonable amounts of time.

A GA starts with a large number of individuals, typically a few hundred to a few thousand ones, which are randomly generated, that is, as random successions of 1s and 0s. These form the first *generation* of the population, so that each individual $\mathcal{I}$ can be assigned a double index, $\mathcal{I}_j^{(\alpha)}$, with Latin subscripts denoting the individual *within* the generation and Greek superscripts the generation itself. Thus, we begin with a collection $\{\mathcal{I}_j^{(1)}\}$ $j = 1, 2, \ldots, j_{max}$. The next step is to choose two individuals as parents of two children for the next generation. The probability with which one individual is chosen as a parent is *proportional* to its fitness value. The combination between the mother and the father's genes is performed by cutting the two individuals at a randomly chosen place (the same for the two parents but different for different parent pairs) and by combining the first part of the mother with the second part of the father to create the first child and vice versa to create the second one. This *crossover operation* is performed $j_{max}/2$ times in each generation, creating thereby j_{max} children that form the next generation. The procedure is performed repeatedly until, after a number of generations (typically, a few hundred ones), the population has become fairly homogeneous and the fitness of the best individual of the generation does not improve any more. This fittest individual of the last generation is then taken as the optimal individual (crystal lattice) for the problem at hand.

Encoding of structures, evaluation of fitness, choice of the recombination probability, and the recombination algorithm are, thus, four fundamental ingredients of the GA. A fifth, crucial one, is mutations. Guided by the incessant mutations taking place in living organisms, and which play a key role in evolution and adaptation, we also introduce a mutation probability p_m, typically on the order of 0.1 %, into the algorithm to avoid complete and premature homogenization of the population toward a suboptimal individual. To this end, we randomly pick with probability p_m in a given generation an individual and flip the values of its genes from 1 to 0 and vice versa. In this way, we introduce fresh genetic material into the pool and "thermalize" the system, helping drive it out of local, metastable minima. It should be emphasized at this point that there is no guarantee that a GA will definitely converge to the absolute minimum; several runs of the GA should be performed at any rate and if these all converge to the same minimum, one gains confidence that the latter is indeed the best one. A decisive advantage of the GAs in comparison with other search techniques is that they can perform arbitrarily large "jumps" in the multidimensional search space through the recombination and mutation processes and, in this way, they are capable of escaping local minima in an efficient fashion.

Returning to the problem of ionic microgels, we have employed the effective interactions presented in Section 7.2. The GAs were employed at $T = 0$, that is, the free energy $F(\mathcal{I})$ of an individual with lattice vectors $\{\mathbf{R}_i\}$ is simply its lattice sum, namely,

$$F(\mathcal{I}) = \frac{1}{2} \sum_{\mathbf{R}_i \neq 0} \Phi_{\text{eff}}(\mathbf{R}_i; \varrho) \tag{7.32}$$

The $T = 0$-phase diagram resulting from the GA application is depicted in Figure 7.9, indicating which crystal structure represents for a given state of the system the stable one. It is drawn on the density–charge plane and for the *fixed* value of the microgel diameter $d = 100$ nm. At low concentrations $\varrho d^3 \lesssim 2.5$, the fcc and bcc lattices are stable because at these densities the average particle distances are such that the particles feel the Yukawa part of the interaction, which, as is well known from the case of charge-stabilized colloids [78–80] and block copolymer micelles [81–84], favors precisely one or the other of these two lattices, depending on the value of $\varkappa$.

Upon increasing the microgel concentration, open and anisotropic structures are stable, namely, hexagonal, trigonal with a bco section for higher charges, and hexagonal again. This is not unusual for ultrasoft potentials, as already indicated above. Similar structures also show up for star polymers [68] the bco and the diamond structures are encountered, while in polyelectrolyte star solutions the bco, the hexagonal, and the sc structures are the stable crystals [49]. At this point, it should be noted that the GA encounters convergence problems whenever the ground-state energies of two (or more) lattices are very close to one another, that is, in the

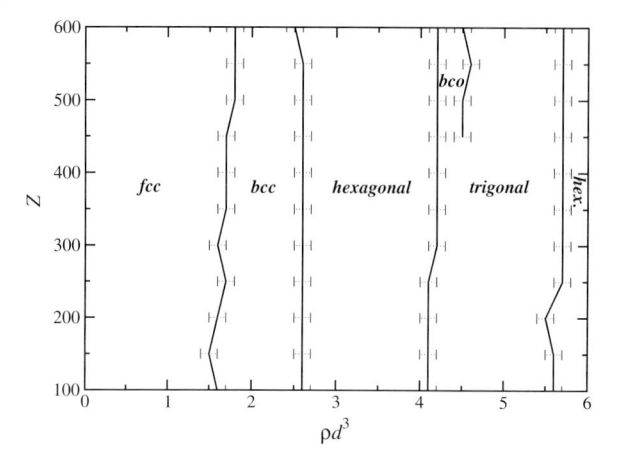

Figure 7.9 Ground-state configurations of microgel solutions (microgel diameter $d = 100$ nm) on the (ϱ, Z)-plane, as obtained by the genetic algorithm minimization of the lattice sums. Adapted with permission from Ref. [51], copyright 2005 American Institute of Physics.

neighborhood of the phase boundaries. There, the GA result "hops" rapidly between the two almost degenerate structures, an indication of an incipient phase transition. This is the reason for the existence of the error bars indicated in Figure 7.9, indicating the phase transition from one structure to the other. Finally, in Figure 7.10 we show representative sections of the hexagonal and the trigonal lattice, along with the respective conventional unit cells.

The five $T = 0$ crystal structures found by the GA comprise the set of candidate lattices taken into account in drawing the full phase diagram, the physical assumption being that $T \neq 0$ or no new candidates might emerge. These phase diagrams are now calculated as described in Ref. [49]: for the five candidate structures, we calculate the free energy within the approximate Einstein model [85, 86] and compare them with the free energies of the competing fluid phase (see preceding section). The resulting phase diagrams are presented and discussed in the following section.

7.5
Phase Behavior

Since the free energy of the fluid is readily available from the considerations discussed in Section 7.3 and those of the candidate crystal phases are calculated via a harmonic solid approximation, one can examine the phase diagram of concentrated microgel solutions at arbitrary temperatures and densities. In Figure 7.11, we show the phase diagram of ionic microgels with diameter $d = 100$ nm in the (ϱ, Z) plane, where the free energy of the fluid has been calculated by means of the λ-integration technique of Section 7.3 and the structural data, that is, the radial distribution function $g^{(\lambda)}(r)$, have been obtained by closing the Ornstein–Zernike equation with the RY-closure. As the closure has an effect on the quality of this structural input, we show in Figure 7.12 the

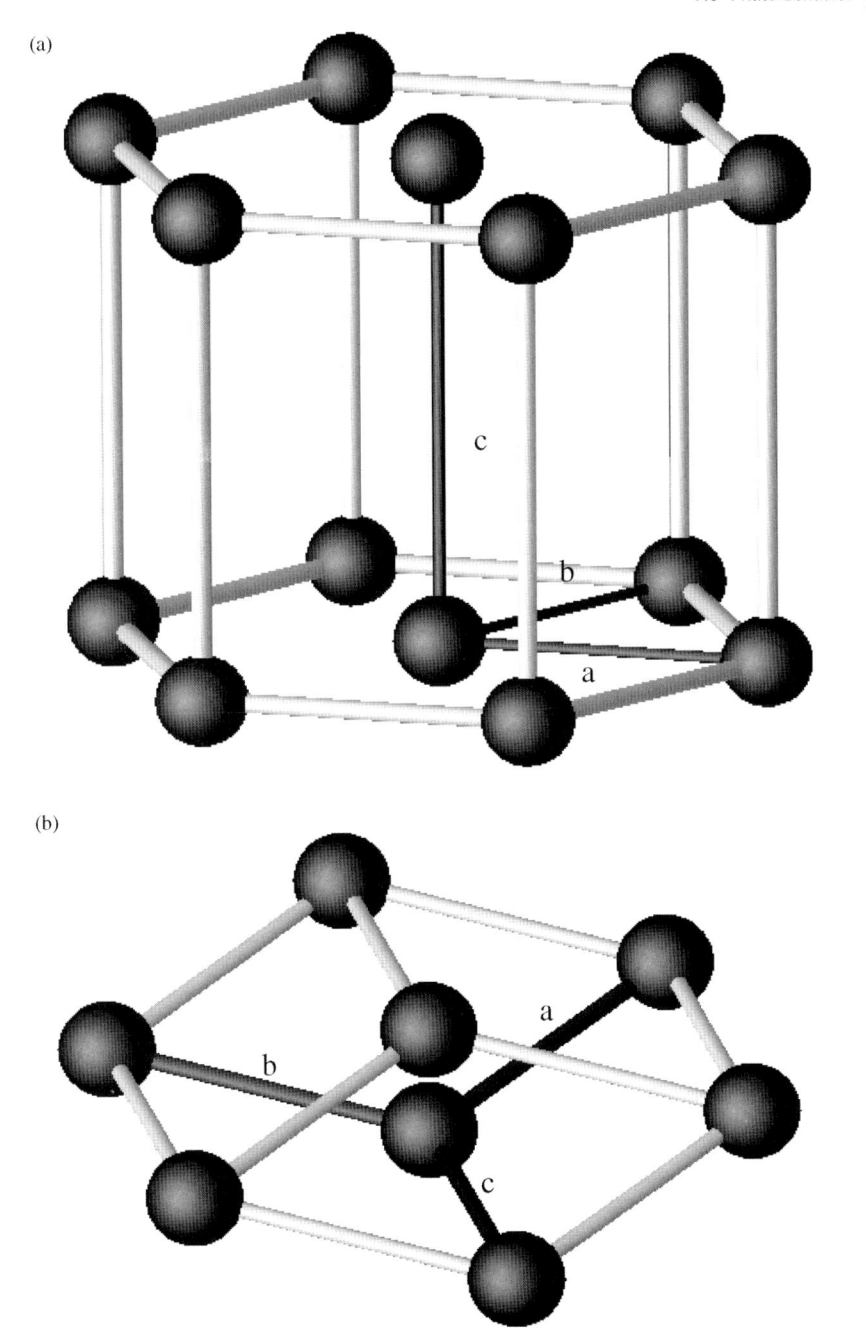

Figure 7.10 Unit cells of the hexagonal (a) and the trigonal (b) lattices. Adapted with permission from Ref. [51], copyright 2005 American Institute of Physics.

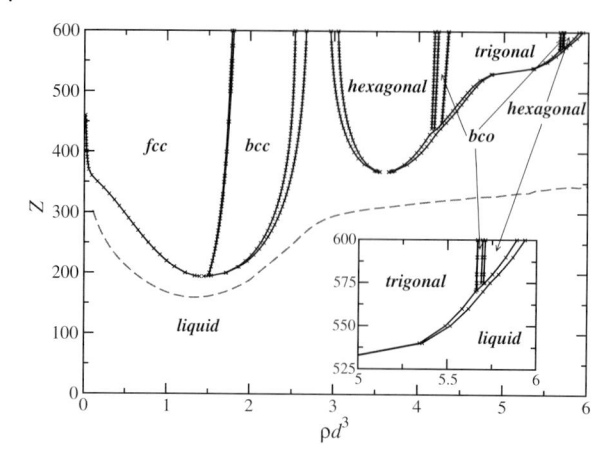

Figure 7.11 The phase diagram of ionic microgels with diameter $d = 100$ nm. The fluid free energy has been calculated using Rogers–Young structural input and the dashed line denotes the Hansen–Verlet locus (see the text). Adapted with permission from Ref. [51], copyright 2005 American Institute of Physics.

phase diagram of the same system for the case in which $g^{(\lambda)}(r)$ has been obtained via the HNC closure.

As mentioned in Section 7.2, the effective interaction potential $\Phi_{\text{eff}}(r)$ is bounded; moreover, it can be easily checked that its Fourier transform $\tilde{\Phi}_{\text{eff}}(k)$ is positive for all k. Thus, in the terminology of Ref. [87], it belongs to the Q^+ class and, according to the general criterion formulated there, its phase diagram is of the re-entrant melting type, a fact that has already been anticipated in Section 7.3 on the basis of the dependence of the structure factor $S(k)$ on density. Accordingly, there exists a "minimum crystallization charge," $Z_{\min}$, such that for $Z \leq Z_{\min}$ the system remains

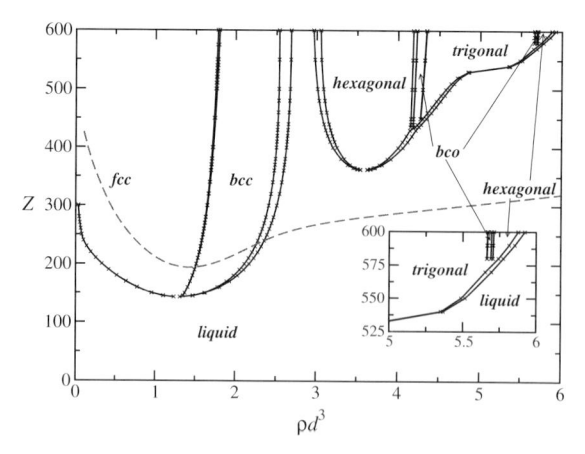

Figure 7.12 Same as Figure 7.11 but with the fluid free energy calculated through the HNC closure. Adapted with permission from Ref. [51], copyright 2005 American Institute of Physics.

fluid at all concentrations, a feature that microgels share with neutral [68] and charged [49] star polymers as well as with the Gaussian core model [60], for which cases the functionality f and the inverse temperature play a role analogous to Z.

Specifically for the ionic microgels, and referring now to Figure 7.11, we see that for microgel densities up to $\varrho d^3 \sim 3$ a first re-entrant melting process takes place: for $Z \gtrsim 200$, the liquid crystallizes into an fcc structure, which at higher densities undergoes a structural transformation into bcc structure – exactly the same feature as seen for other models mentioned above [49, 60, 68]. Further increase in ϱ leads to melting of the system. As far as this *generic* behavior is concerned, the RY and the HNC approximations give identical results and topologies of the phase diagrams, as a comparison between the results in Figures 7.11 and 7.12 demonstrates.

At higher densities, $3 \lesssim \varrho d^3 \lesssim 6$, we encounter, on the one hand, the occurrence of the exotic phases, hexagonal, bco, and trigonal, with structural phase transitions among them, and, on the other hand, the appearance of a *second* re-entrant melting scenario. The stability of the unusual crystal structures stems from the ultrasoft character of the potential [70]. Compared to the low-density part, this region of the phase diagram is considerably more complex and diversified (see insets in Figures 7.11 and 7.12): this points out that the energies of the competing crystal structures are very close to each other. At the same time, a comparison of the phase boundaries in Figures 7.11 and 7.12 shows that the agreement between RY and HNC results is *quantitative* in this density range. This is not surprising in view of the fact that at high concentrations the HNC closure becomes quasiexact [60, 88] and the RY closure reduces to the HNC. Free energy differences between the crystals and the fluid, showing the succession of the winning phases, are exemplarily shown in Figure 7.13 for the values $Z = 600$ and $d = 100$ nm.

Instead of performing full free energy calculations, which are expensive, one sometimes resorts to empirical criteria in order to estimate crystallization phase

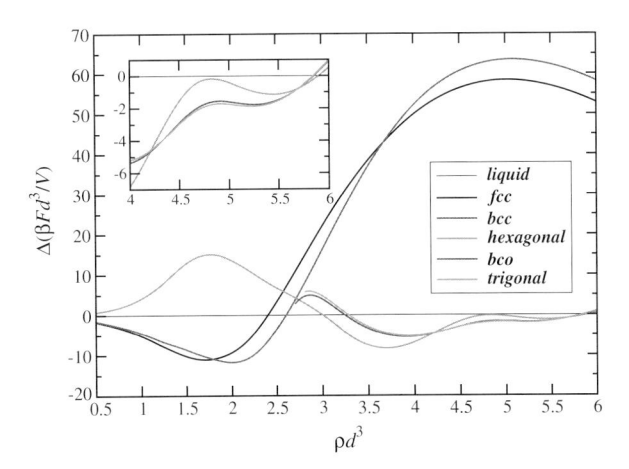

Figure 7.13 Free energy differences between the crystals and the fluid for $Z = 600$, $d = 100$ nm microgels, as functions of density. The inset shows a zoom at the high-density region. Adapted with permission from Ref. [51], copyright 2005 American Institute of Physics.

boundaries. One such criterion comes from the fluid side and is based on the height of the main peak of the fluid structure factor $S(k)$. This is the celebrated Hansen–Verlet rule [54, 89], stating that a liquid freezes whenever the peak height, $S_{\max}(k)$, exceeds the value 2.85. In order to test the applicability of this criterion to the system at hand, we draw in Figures 7.11 and 7.12 the locus of points $S_{\max}(k) = 2.85$ as a broken line; below this line, $S_{\max}(k) < 2.85$. It can be seen that the Hansen–Verlet rule serves as a rough indicator of the *topology* of the phase diagram, carrying the unmistakable signature of the re-entrant melting scenario. However, it cannot be employed for a *quantitative* estimation of the phase boundaries, especially at the high-density range, where it vastly overestimates the degree of stability of the (exotic) crystals, on the one hand, and it misses the second re-entrant melting on the other.

While the Hansen–Verlet criterion of *freezing* operates on the fluid side, another common empirical criterion comes from the crystal side of the phase diagram. This is the Lindemann criterion of *melting* [90]. It is based on the quantity L, which is defined as the ratio between the root mean square of the displacement of a particle around the lattice site (due to thermal oscillations) over the nearest-neighbor distance r_{nn}. For atomic systems, such as metals, melting occurs when L exceeds the typical value 10% [91], but in the case of soft particles things can be very different. In Figure 7.14, we display the Lindemann ratios L of the crystals that show up in the phase diagram for $Z = 600$ as functions of density (here $d = 100$ nm). At not too high concentrations, $\varrho d^3 \lesssim 3$, the crystals indeed have low Lindemann ratios but on the high-density side, where all the exotic lattices are also stable, L is considerably higher than 10% but the solids are still stable. Moreover, as can be seen in Figure 7.14, although for $4.5 \lesssim \varrho d^3 \lesssim 6$ the fcc and bcc lattices have the two lowest values of L, they are not thermodynamically stable. This demonstrates that, similar to the liquid-based Hansen–Verlet rule, the crystal-based Lindemann criterion fails at high concentrations. One should not, therefore, rely on empirical criteria; only a reliable calculation

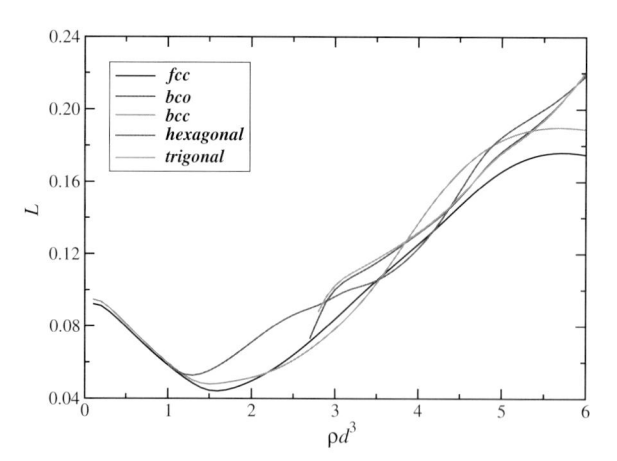

Figure 7.14 Lindemann parameters of microgel crystal structures drawn against the density, for the same Z- and d-values as in Figure 7.13. Adapted with permission from Ref. [51], copyright 2005 American Institute of Physics.

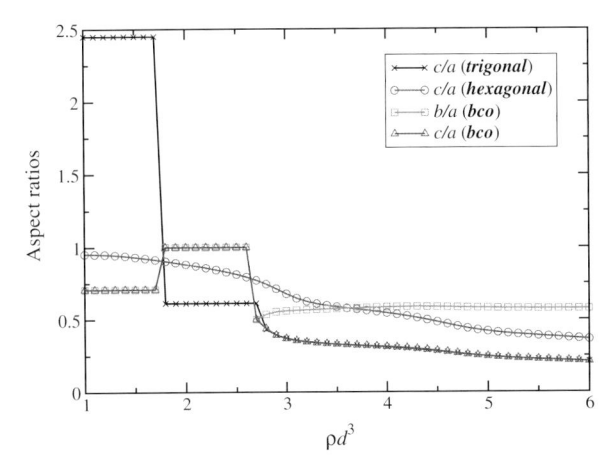

Figure 7.15 Aspect ratios of the crystals appearing in the phase diagram of microgels for Z- and d-values as in Figure 7.13. Adapted with permission from Ref. [51], copyright 2005 American Institute of Physics.

of the free energies of the involved phases can give information on the phase behavior of ultrasoft systems, such as the ionic microgels at hand.

A more detailed structural characterization on the noncubic crystals appearing in the phase diagram of ionic microgels can be obtained by considering the aspect ratios of the trigonal (c/a), of the hexagonal (c/a), and of the bco (b/a and c/a). Representative results for these quantities are shown in Figure 7.15 for $Z = 600$ and $d = 100$ nm. The trigonal lattice for densities up to ~ 1.8 with an aspect ratio of $c/a \equiv \sqrt{6}$ is equivalent to the fcc structure and that the same lattice for densities $1.8 \lesssim \varrho d^3 \lesssim 2.6$ with $c/a \equiv \sqrt{3/8}$ is equivalent to the bcc lattice. In Figure 7.15, the transition densities of fcc $\rightarrow$ bcc and bcc $\rightarrow$ hexagonal are clearly marked by the discontinuities in the aspect ratio curves. Note the agreement between the c/a ratio for the bco and the trigonal structure for $\varrho d^3 \gtrsim 2.7$: this reflects the fact that both structures have large similarities in their first coordination shells, as also seen for star polymers [68].

In order to investigate the effect of microgel size d on phase diagrams, we draw in Figure 7.16 phase diagrams for ionic microgels for two different values of Z in the $(d, \varrho d^3)$-plane; the properties of the liquid phase have been calculated here within the HNC approximation. For $Z = 200$, only freezing into an fcc and a bcc structure is observed, followed by a re-entrant melting process close to the overlap density $\varrho_\star$. The situation is considerably more complex for $Z = 400$, as could already be expected from Figure 7.12: in particular for small d-values (i.e., $d \lesssim 50$ nm) and intermediate-to-high densities, freezing into a hexagonal, a bco, and a trigonal structure with a subsequent re-entrant melting is observed. In general, it can be seen that d (for fixed Z) plays a role analogous to, roughly, $1/Z$ (for fixed d): indeed, as $d(1/Z)$ grows, crystallization disappears and there is a Z-dependent "maximum freezing diameter" d_{max}, so that for $d > d_{max}$ the system is fluid at all temperatures. The reason for this behavior can be traced back to the functional form of $\Phi_{eff}(r)$, see Equations 7.23–7.25.

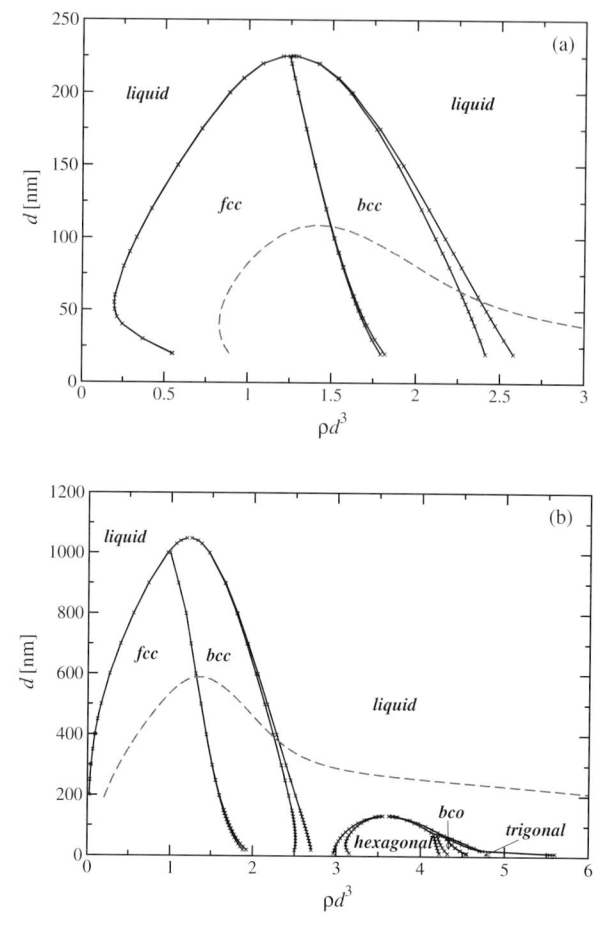

Figure 7.16 Microgel phase diagrams on the (ϱ, d)-plane, for fixed values of Z. (a) $Z = 200$; (b) $Z = 400$. The dashed line is the locus of Hansen–Verlet criterion values. Adapted with permission from Ref. [51], copyright 2005 American Institute of Physics.

Indeed, as d grows, $\Phi_{\text{eff}}(r)$ becomes weaker, so that beyond some $d_{\max}$ no crystals can be sustained. Physically, its origin is attributed to the fact that large microgels with fixed Z provide ample space in their interior for counterions to be absorbed, so that the free energy cost for two overlapping microgels is definitely smaller than that for smaller diameters. Concomitantly, the effective repulsive potential $\Phi_{\text{eff}}(r)$ weakens and crystals are destroyed by microgel swelling. This is the *opposite* effect from that occurring in the case of sterically interacting star polymers [92], for which case swelling leads to transition from a fluid into an arrested gel.

Finally, we turn our attention to the influence of steric interactions, which up to this point have been left out. We have assumed throughout that the contributions to $\Phi_{\text{eff}}(r)$ from charge (electrostatic energy of monomers and counterions as well as entropy of counterions) dominate the steric hindrance of the former. This assumption is justified as long as we are dealing with a solvent close to Θ-conditions, where the cross-linked polymers can be treated as ideal. Relaxing this condition and estimating the steric contribution $\Phi_{st}(r)$ to the interaction by the overlap volume of two spheres of diameter d results into the expression [51]

$$\Phi_{st}(r) = \begin{cases} V_{st} k_B T \left[1 - \frac{3}{2}\left(\frac{r}{d}\right) + \frac{1}{2}\left(\frac{r}{d}\right)^3 \right], & r \leq d \\ 0, & r > d \end{cases} \qquad (7.33)$$

where the strength V_{st} of the interaction can be expressed as

$$V_{st} = \frac{2 V_0}{v_c} k_B T \left(\frac{1}{2} - \chi\right) \phi^2 \qquad (7.34)$$

Here, $V_0 = \pi d^3/6$ denotes the volume of a microgel, v_c the typical volume occupied by a monomer and the strongly condensed counterions on it, ϕ is the typical internal monomer concentration of the microgel, and χ denotes the Flory–Huggins parameter that characterizes solvent quality ($\chi = 1/2$ for Θ-solvents and the steric contribution vanishes in this case).

We now consider the total interaction potential $\Phi_{tot}(r) = \Phi_{eff}(r) + \Phi_{st}(r)$, where $\Phi_{eff}(r)$ is given by Equations 7.23–7.25 and $\Phi_{st}(r)$ by Equation 7.33 above. As representative values for the prefactor α, we choose α = 50 and α = 100 and we redraw the phase diagrams following the same procedure described before. Results are shown in Figure 7.17. It can be seen that the steric interaction, since it enhances the repulsion between microgels, leads to a growth of the region of stability of the crystal phases and, provided it is strong enough (Figure 7.17b), it can even open a gap between the low-density and the high-density liquid phases, which are now separated by intervening fcc and bcc solids even for a vanishing value of Z. In this case, crystallization at low Z-values entirely stems from steric interactions; with growing Z, however, the electrostatic/entropic contributions of $\Phi_{eff}(r)$ dominate over steric ones and the phase boundaries converge to the ones calculated already on the basis of $\Phi_{eff}(r)$ alone.

Summarizing the results for the phase behavior of ultrasoft, ionic microgels, the following facts can be stated. First, ultrasoft microgel interactions give rise to open and noncubic crystal structures at the high-density region of the phase diagram. Second, intuition and sheer guessing are not reliable methods to find the correct crystal structures; unbiased search techniques, such as the genetic algorithms employed here, are necessary. And, finally, traditional and empirical criteria of melting and freezing break down in these regions.

7.6
Summary and Concluding Remarks

In this chapter, we have discussed the structure and phase behavior of ionic microgels by employing a coarse-graining procedure that allows us to bridge the length scales,

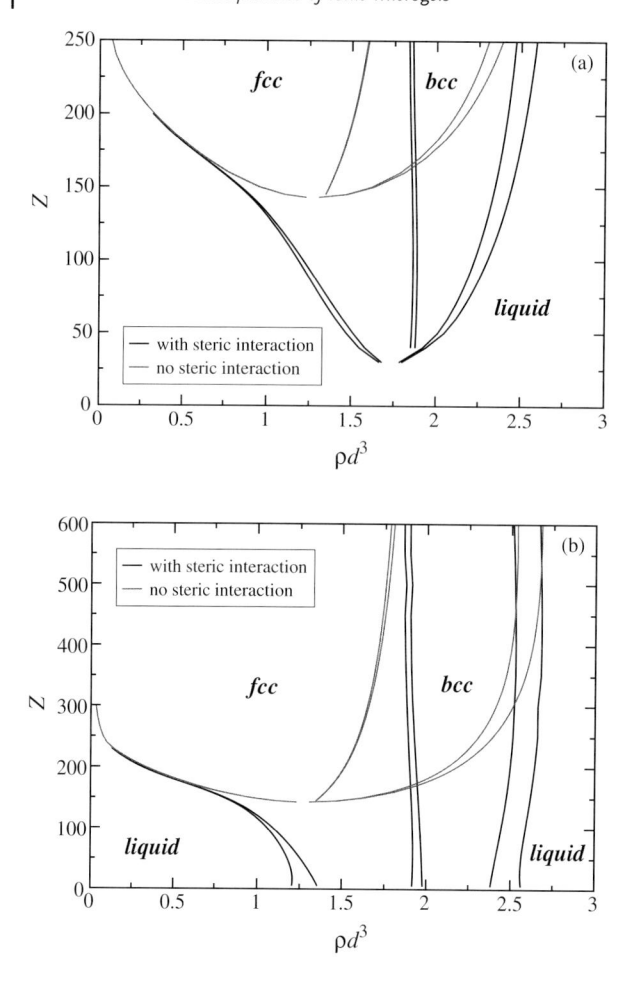

Figure 7.17 Phase diagrams of microgels with diameter $d = 100$ nm in the density–charge plane, drawn for an effective interaction potential that includes the steric contribution of Equation 7.33 and for two different values of the steric interaction strength V_{st} of Equation 7.34. (a) $V_{st} = 50$; (b) $V_{st} = 100$. Adapted with permission from Ref. [51], copyright 2005 American Institute of Physics.

starting from the monomers and ions and reaching the *mesoscopic* scale. In the latter, the whole microgel is modeled as a "point particle" that interacts with the others by means of an effective interaction potential. The power of the approach is demonstrated by its ability to, subsequently, bridge the next gap of length scales and make concrete predictions on the *macroscopic* phase behavior of the system, including some intriguing phase transitions to open crystals. It must be emphasized that, contrary to phenomenological approaches, the process of integrating microscopic degrees of freedom in the way of deriving effective interactions leaves the contact to the microscopic characteristics intact. In this way, the relevant microscopic parameters, such as the charge, the inner monomer and counterion concentrations, and so on, remain present in the effective interaction and provide a way to perform "inverse engineering" of the system. The physical origins of particular structural and phase characteristics of concentrated microgel solutions can be traced back to their microscopic origins.

The findings reported here are experimentally verifiable by employing cross-linked and highly charged microgel particles, whose synthesis is practicable with modern chemical techniques. It can also be anticipated that, in analogy with star polymer solutions that show a similar topology of their equilibrium phase diagram [68], ionic microgels may display unusual nonequilibrium glass behavior [93], including re-entrant liquification or melting by addition of free homopolymer chains [94].

Acknowledgment

I would like to thank Dieter Gottwald and Gerhard Kahl (Vienna University of Technology) for a very fruitful collaboration, on which this work has been based.

References

1. Flory, P.J. (1953) *Principles of Polymer Chemistry*, Cornell University Press, Ithaca.
2. Escobedo, F.A. and de Pablo, J.J. (1999) *Phys. Rep.*, **318**, 85.
3. Michaeli, I. and Katchalsky, A. (1955) *J. Polym. Sci.*, **15**, 69.
4. Katchalsky, A. and Michaeli, I. (1957) *J. Polym. Sci.*, **23**, 683.
5. Khokhlov, A.R., Starodubtzev, S.G., and Vasilevskaya, V.V. (1993) Conformational transitions in polymer gels: theory and experiment, in *Advances in Polymer Science*, vol. 109 (ed. K. Dušek), Springer-Verlag, New York, p. 123.
6. Schneider, S. and Linse, P. (2002) *Eur. Phys. J. E*, **8**, 457.
7. Schneider, S. and Linse, P. (2003) *J. Phys. Chem. B*, **107**, 8030.
8. Schneider, S. and Linse, P. (2004) *Macromolecules*, **37**, 3850.
9. Saunders, B.R. and Vincent, B. (1999) *Adv. Colloid Interface Sci.*, **80**, 1.
10. Pelton, P. (2000) *Adv. Colloid Interface Sci.*, **85**, 1.
11. Bromberg, L., Temchenko, M., and Hatton, T.A. (2002) *Langmuir*, **18**, 4944.
12. Antonietti, M., Bremser, W., and Schmidt, M. (1990) *Macromolecules*, **23**, 3796.
13. Dziechciarek, Y., van Soest, J.J.G., and Philipse, A.P. (2002) *J. Colloid Interface Sci.*, **246**, 48.
14. Eichenbaum, G.M., Kiser, P.F., Dobrynin, A.V., Simon, S.A., and Needham, D. (1999) *Macromolecules*, **32**, 4867.
15. Levin, Y., Diehl, A., Fernández-Nieves, A., and Fernández-Barbero, A. (2002) *Phys. Rev. E*, **65**, 036143.
16. Fernández-Nieves, A., Fernández-Barbero, A., Vincent, B., and de las Nieves, F.J. (2000) *Macromolecules*, **33**, 2114.
17. Fernández-Nieves, A., Fernández-Barbero, A., and de las Nieves, F.J. (2001) *J. Chem. Phys.*, **115**, 7644.
18. Varga, I., Gilányi, T., Mészáros, R., Filipscei, G., and Zrnyi, M. (2001) *J. Phys. Chem. B*, **105**, 9071.
19. Fernández-Barbero, A., Fernández-Nieves, A., Grillo, I., and López-Cabarcos, E. (2002) *Phys. Rev. E*, **66**, 051803.
20. Eckert, T. and Bartsch, E. (2002) *Phys. Rev. Lett.*, **89**, 125701.
21. Senff, H. and Richtering, W. (1999) *J. Chem. Phys.*, **111**, 1705.

22 Senff, H., Richtering, W., Norhausen, C., Weiss, A., and Ballauff, M. (1999) *Langmuir*, **15**, 102.
23 Senff, H. and Richtering, W. (2000) *Colloid Polym. Sci.*, **278**, 830.
24 Berndt, I. and Richtering, W. (2003) *Macromolecules*, **36**, 8780.
25 Stieger, M., Pedersen, J.S., Lindner, P., and Richtering, W. (2004) *Langmuir*, **20**, 7283.
26 Serpe, J.S., Kim, J., and Lyon, L.A. (2004) *Adv. Mater.*, **16**, 184.
27 Kim, J., Serpe, J.S., and Lyon, L.A. (2004) *J. Am. Chem. Soc.*, **126**, 9512.
28 Kim, J., Serpe, J.S., and Lyon, L.A. (2005) *Angew. Chem. Int. Ed. Engl.*, **44**, 1333.
29 Dong, L., Aggarwal, A.K., Beebe, D.J., and Jiang, H. (2006) *Nature*, **442**, 551.
30 Lahann, L., Mitragorti, S., Tran, T.-N., Kaido, H., Sundaram, J., Choi, I.S., Hoffer, S., Somorjai, G.A., and Langer, R. (2003) *Science*, **299**, 371.
31 Konieczny, M. and Likos, C.N. (2007) *Soft Matter*, **3**, 1130.
32 Brixner, T., Garca de Abajo, F.J., Schneider, J., and Pfeiffer, W. (2005) *Phys. Rev. Lett.*, **95**, 093901.
33 Aeschlimann, M., Bauer, M., Bayer, D., Brixner, T., Garca de Abajo, F.J., Pfeiffer, W., Rohmer, M., Spindler, C., and Steeb, F. (2007) *Nature*, **446**, 301.
34 Goychuk, I. and Hänggi, P. (2005) *Adv. Phys.*, **54**, 525.
35 Lewis, N.S. (2005) *Inorg. Chem.*, **44**, 6900.
36 Hellweg, T., Dewhurst, C.D., Brückner, E., Kratz, K., and Eimer, W. (2000) *Colloid Polym. Sci.*, **278**, 972.
37 Gröhn, F. and Antonietti, M. (2000) *Macromolecules*, **33**, 5938.
38 Fernández-Nieves, A., van Duijneveldt, J.S., Fernández-Barbero, A., Vincent, B., and de las Nieves, F.J. (2001) *Phys. Rev. E*, **64**, 051603.
39 Fernández-Barbero, A. and Vincent, B. (2000) *Phys. Rev. E*, **63**, 011509.
40 Likos, C.N. (2001) *Phys. Rep.*, **348**, 267.
41 Wu, J., Zhou, B., and Hu, Z. (2003) *Phys. Rev. Lett.*, **90**, 048304.
42 Wu, J., Huang, G., and Hu, Z. (2003) *Macromolecules*, **36**, 440.
43 Berli, C.L.A. and Quemada, D. (2000) *Langmuir*, **16**, 10509.
44 Denton, A.R. (2003) *Phys. Rev. E*, **67**, 011804; Erratum *ibid.* 68, 049904 (2003).
45 Baus, M. and Hansen, J.P. (1980) *Phys. Rep.*, **59**, 1.
46 Ichimaru, S., Iyetomi, H., and Tanaka, S. (1987) *Phys. Rep.*, **149**, 91.
47 Likos, C.N. and Ashcroft, N.W. (1992) *Phys. Rev. Lett.*, **69**, 316.
48 Hafner, J. (1987) *From Hamiltonians to Phase Diagrams*, Springer-Verlag, Berlin.
49 Hoffmann, N., Likos, C.N., and Löwen, H. (2004) *J. Chem. Phys.*, **121**, 7009.
50 van Roij, R., Dijkstra, M., and Hansen, J.-P. (1999) *Phys. Rev. E*, **59**, 2010.
51 Gottwald, D., Likos, C.N., Kahl, G., and Löwen, H. (2005) *J. Chem. Phys.*, **122**, 074903.
52 Jusufi, A., Likos, C.N., and Löwen, H. (2002) *Phys. Rev. Lett.*, **88**, 018301.
53 Jusufi, A., Likos, C.N., and Löwen, H. (2002) *J. Chem. Phys.*, **116**, 11011.
54 Hansen, J.-P. and McDonald, I.R. (1986) *Theory of Simple Liquids*, 2nd edn, Academic, New York.
55 Louis, A.A. (2002) *J. Phys.: Condens. Matter*, **14**, 9187.
56 Stillinger, F.H., Sakai, H., and Torquato, S. (2002) *J. Chem. Phys.*, **117**, 288.
57 Sakai, H., Stillinger, F.H., and Torquato, S. (2002) *J. Chem. Phys.*, **117**, 297.
58 Tejero, C.F. and Baus, M. (2003) *J. Chem. Phys.*, **118**, 892.
59 Rogers, F.A. and Young, D.A. (1984) *Phys. Rev. A*, **30**, 999.
60 Lang, A., Likos, C.N., Watzlawek, M., and Löwen, H. (2000) *J. Phys.: Condens. Matter*, **12**, 5087.
61 Löwen, H., Allahyarov, E., Likos, C.N., Blaak, R., Dzubiella, J., Jusufi, A., Hoffmann, N., and Harreis, H.M. (2003) *J. Phys. A: Math. Gen.*, **36**, 5827.
62 Watzlawek, M., Löwen, H., and Likos, C.N. (1998) *J. Phys.: Condens. Matter*, **10**, 8189.
63 Likos, C.N. (2006) *Soft Matter*, **2**, 478.
64 Likos, C.N., Rosenfeldt, S., Dingenouts, N., Ballauff, M., Lindner, P., Werner, N., and Vögtle, F. (2002) *J. Chem. Phys.*, **117**, 1869.
65 Götze, I.O., Harreis, H.M., and Likos, C.N. (2003) *J. Chem. Phys.*, **120**, 7761.

66 Ballauff, M. and Likos, C.N. (2004) *Angew. Chem. Int. Ed. Engl.*, **43**, 2998.
67 Evans, R. (1979) *Adv. Phys.*, **28**, 143.
68 Watzlawek, M., Löwen, H., and Likos, C.N. (1999) *Phys. Rev. Lett.*, **82**, 5289.
69 Ziherl, P. and Kamien, R. (2000) *Phys. Rev. Lett.*, **85**, 3528.
70 Likos, C.N., Hoffmann, N., Löwen, H., and Louis, A.A. (2002) *J. Phys.: Condens. Matter*, **14**, 7681.
71 Gottwald, D., Likos, C.N., Kahl, G., and Löwen, H. (2004) *Phys. Rev. Lett.*, **92**, 068301.
72 Gottwald, D., Kahl, G., and Likos, C.N. (2005) *J. Chem. Phys.*, **122**, 204503.
73 Goldberg, D.E. (1989) *Genetic Algorithms in Search, Optimization, and Machine Learning*, Addison-Wesley, Reading, MA.
74 Holland, J.H. (1975) *Adaptation in Natural and Artificial Systems*, The University of Michigan Press, Ann Arbor.
75 Deaven, D.M. and Ho, K.M. (1975) *Phys. Rev. Lett.*, **75**, 288.
76 Oganov, A.R. and Glass, C.W. (2006) *J. Chem. Phys.*, **124**, 244704.
77 Fornleitner, J., Lo Verso, F., Kahl, G., and Likos, C.N. (2008) *Soft Matter*, **4**, 480.
78 Sirota, E.B., Ou-Yang, H.D., Sinha, S.K., and Chaikin, P.M. (1989) *Phys. Rev. Lett.*, **62**, 1524.
79 Kremer, K., Robbins, M.O., and Grest, G.S. (1986) *Phys. Rev. Lett.*, **57**, 2694.
80 Robbins, M.O., Kremer, K., and Grest, G.S. (1988) *J. Chem. Phys.*, **88**, 3286.
81 McConnell, G.A., Gast, A.P., Huang, J.S., and Smith, S.D. (1993) *Phys. Rev. Lett.*, **71**, 2102.
82 Gast, A.P. (1996) *Langmuir*, **12**, 406.
83 Gast, A.P. (1997) *Curr. Opinion Colloid Interface Sci.*, **2**, 258.
84 McConnell, G.A., Lin, E.K., Gast, A.P., Huang, J.S., Lin, M.Y., and Smith, S.D. (1994) *Faraday Discuss. Chem. Soc.*, **98**, 121.
85 Ashcroft, N.W. and Mermin, N.D. (1976) *Solid State Physics*, Holt Saunders, Philadelphia.
86 Tejero, C.F., Daanoun, A., Lekkerkerker, H.N.W., and Baus, M. (1995) *Phys. Rev. E*, **51**, 558.
87 Likos, C.N., Lang, A., Watzlawek, M., and Löwen, H. (2001) *Phys. Rev. E*, **82**, 5289.
88 Likos, C.N., Mladek, B.M., Gottwald, D., and Kahl, G. (2007) *J. Chem. Phys.*, **126**, 224502.
89 Hansen, J.-P. and Verlet, L. (1969) *Phys. Rev.*, **184**, 151.
90 Lindemann, F.A. (1910) *Phys. Z.*, **11**, 609.
91 Shapiro, J.N. (1970) *Phys. Rev. B*, **1**, 3982.
92 Kapnistos, M., Vlassopoulos, D., Fytas, G., Mortensen, K., Fleischer, G., and Roovers, J. (2000) *Phys. Rev. Lett.*, **85**, 4072.
93 Foffi, G., Sciortino, F., Tartaglia, P., Zaccarelli, E., Verso, F.L., Reatto, L., Dawson, K., and Likos, C.N. (2003) *Phys. Rev. Lett.*, **90**, 238301.
94 Stiakakis, E., Vlassopoulos, D., Likos, C.N., Roovers, J., and Meier, G. (2002) *Phys. Rev. Lett.*, **89**, 208302.

8
Elasticity of Soft Particles and Colloids Near the Jamming Threshold
Matthieu Wyart

8.1
Introduction

Crystalline lattices are invariant under translation, which implies that vibrational modes are plane waves. From those excitations, one can build a theory of energy transport and a theory of elasticity. In amorphous solids, this symmetry breaks down. Although at large length scales a continuous (and translationary invariant) description is a good approximation, this fails at small-length scales, where the disorder has strong effects. At these scales, various properties of amorphous solids, such as energy transport, low-frequency excitations in glasses, or force propagation in granular matter, are both not satisfyingly understood so far and continue to be active fields of research. One inherent difficulty in the study of these phenomena is that the length scales at play are typically moderate, of the order of 10 particle sizes or less. This makes it harder to test and distinguish clearly the consequences of different theories. Finding a system where length scales can be large and controlled may therefore be extremely useful.

Emulsion experiments [1], followed with theoretical arguments [3–6] suggested that the "jamming transition" where repulsive, short-range particles are just in contact corresponds to a critical point. This idea was later substantiated by the findings that the elastic moduli [2], the vibrational spectrum [2], the microscopic structure [2, 7], and the force propagation [8] display scaling behavior near the jamming threshold. At that point, although the system is amorphous and isotropic, it cannot be described as a continuous elastic body on any length scale [8, 9]. Because the strong effects of disorder occurs already at large length scales near this critical point, this model system is a lens allowing to probe in a stringent manner the properties of amorphous solids, and the effects of disorder. It has enabled to build and test [9, 10] a theory for the excess low vibrational modes (the so-called Boson Peak) found in amorphous solids, which applies as well to covalent glasses [10] and to model systems of attractive glasses [10, 11]. This line of thought also permits to relate microscopic structure and some aspects of the dynamics near the glass transition of

Microgel Suspensions: Fundamentals and Applications
Edited by Alberto Fernandez-Nieves, Hans M. Wyss, Johan Mattsson, and David A. Weitz
Copyright © 2011 WILEY-VCH Verlag GmbH & Co. KGaA, Weinheim
ISBN: 978-3-527-32158-2

hard spheres [12], despite the fact that the glass transition occurs empirically at a packing fraction significantly smaller than those of the jammed configurations.

In this chapter, we will focus on the elastic moduli, as microgels are potentially a good system to vary the packing fraction around the jamming threshold to test predictions on elasticity. Near the jamming threshold, for particles interacting with either a harmonic potential (used to model emulsions) or a Hertzian potential (used to model elastic particles) [2], it is found numerically that the system is "almost" a liquid, its shear modulus becomes negligible in comparison with the bulk modulus:

$$\frac{G}{B} \sim \left(\frac{p}{B}\right)^{1/2} \tag{8.1}$$

where p is the pressure. In a gel, two different phenomena govern the shear and the bulk moduli: the elasticity of the polymeric network and the compressibility of the solvent, respectively. This can cause the two elastic moduli to be very different. Nevertheless this behavior is unusual for a solid made of identical particles interacting with a radial interaction, where both shear and bulk moduli are induced by local interactions, and are generally comparable in amplitude. The theoretical argument (see Section 8.3) will explain this observation and unravel a peculiar aspect of the elasticity near the jamming threshold: the elastic moduli can depend enormously on the stress applied on the system before the response is measured, the so-called prestress. Before discussing this argument, we will start by reviewing recent results on the geometry of the packings near the jamming threshold.

8.2
Structure and Mechanical Stability

Amorphous solids are typically out of equilibrium, and their structures *a priori* depend on their history. It may thus seem hard to infer their structure without a detailed description of the way they were made. There is, nevertheless, a limiting case that turns out to be conceptually important: when these systems are prepared via a very rapid quench from a fluid phase. In this situation, we expect that as soon as the liquid finds some metastable states, it remains in those states. For slower quenches the system will depart from such states, but there exists evidence that this effect is small, at least for hard particles and for quenches as slow as what is typically achieved numerically [12]. Thus, we are particularly interested in configurations that are marginally stable, that is, mechanically stable, but that are very close to yield. What are those configurations for an assembly of elastic particles?

Maxwell [13] studied mechanical stability in the context of engineering structures, and he found out that the key parameter is the coordination z, the average number of interactions per particle. For example, for a network of point particle connected via springs, he showed that stability requires $z \geq z_c = 2d$, where d is the spatial dimension of the system. His argument is explained as follows: consider a set of N

points interacting with N_c springs at rest of stiffness k. The expansion for the energy may be written as follows:

$$\delta E = \sum_{\langle ij \rangle} \frac{k}{2} [(\delta \vec{R}_i - \delta \vec{R}_j) \cdot \vec{n}_{ij}]^2 + o(\delta R^2) \qquad (8.2)$$

where the sum is made over all springs, $\vec{n}_{ij}$ is the unit vector going from i to j, and $\delta \vec{R}_i$ is the displacement of particle i. A system is floppy, that is, not mechanically stable, if it can be deformed without energy cost, that is if there is a displacement field for which $\delta E = 0$, or equivalently $(\delta \vec{R}_i - \delta \vec{R}_j) \cdot \vec{n}_{ij} = 0 \ \forall ij$. If the spatial dimension is d, this linear system has Nd degrees of freedom and $N_c \equiv Nz/2$ equations, and therefore there are always nontrivial solutions if $Nd > N_c$, that is, if $z < 2d \equiv z_c$. To be mechanically stable, a network must therefore have $z \geq 2d$.

It turns out that under compression the criterion of rigidity becomes more demanding. In this chapter, we shall derive this criterion in a simple model, as it yields the correct and more general result. The derivation for a generic amorphous packing is presented in Ref. [14]. Consider a square lattice made of springs of rest length l_0, which defines our unit length. It just satisfies the Maxwell criterion, since $z = 4$. Now, we add randomly a density δz of springs connecting second neighbors (Figure 8.1, diagonal lines), such that the coordination is $z = z_c + \delta z$. We add them in a rather homogeneous manner so that there are no large regions without springs (diagonal lines). The typical distance between two springs (diagonal lines) in a given row or column is of order $l_0/\delta z$. Dividing this length by the mesh size l_0, we define the dimensionless number $l^* = 1/\delta z$.

How much pressure can this system sustain before collapsing? For a system to be mechanically stable, all collective displacements need to have a positive energetic cost. It turns out that the first modes to collapse are of the type of the displacement mode (see Figure 8.1, gray arrows): they correspond to the longitudinal mode of wavelength $l_0 l^*$ of a segment of springs contained between two diagonal springs shown in Figure 8.1. These modes have a displacement field of the form $\delta \vec{R}_i = 2X \sin(\pi i/l^*)/\sqrt{l^*} \vec{e}_x$, where i denotes the particles along a segment and runs between 0 and l^*, $\vec{e}_x$ is the unit vector in the direction of the line, and X is the amplitude of the mode; $X = 1$ for a normalized mode. In the absence of pressure p, the energy of this mode comes only from the springs of the segment. In the limit of large l^*, a Taylor expansion of the displacement in Equation 8.2 gives the well-known result for the energy of the lowest frequency mode of a line of springs $\delta E \sim kX^2/l^{*2}$. When $p > 0$, all the springs now carry a force $f \sim pl_0$. The energy expansion contains other terms that are not indicated in Equation 8.2 [14], whose effect can be estimated quantitatively as follows. When particles are displaced along a longitudinal mode such as the one represented in Figure 8.1, the force of each spring directly connected and transverse to the segment considered (see Figure 8.1) now produces a work equal to f times the elongation of the spring. This elongation is simply $\delta R_i^2/l_0$ following Pythagoras' theorem. Summing on all the springs transverse to the segment leads to a work of order $fX^2/l_0 \approx pX^2$. This gives finally for the energy of the mode $\delta E \sim kX^2/l^{*2} - pX^2$, where numerical prefactors are omitted. Stability

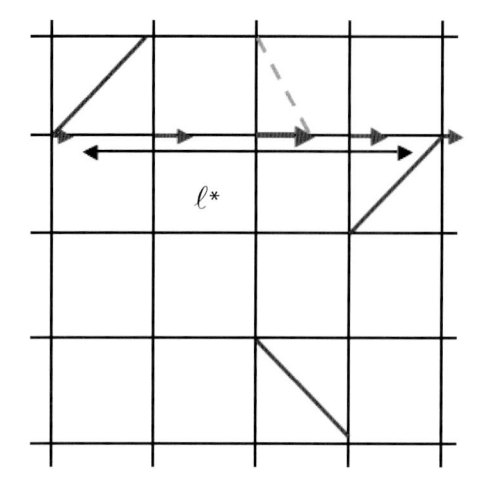

Figure 8.1 Square lattice with a density per particle δz of additional springs (diagonal lines). $l^* \sim 1/\delta z$ is the typical (dimensionless) distance of the segments contained between two springs on a given row or column. The gray arrows represent the longitudinal mode of wavelength $\sim l^*$ of such a segment: $\delta \vec{R}_i \propto \sin(\pi i/l^*)\vec{e}_x$, following the notation introduced in the text. The dashed line exemplifies the deformation of a spring transverse and directly connected to the segment considered; it is elongated by the longitudinal vibration of this segment. When the pressure is positive and contacts are under compression, this elongation lowers the energy those springs contain. This leads to an instability when δz becomes smaller than a quantity proportional to the square root of the contact strain, of order p/B.

requires $\delta E > 0$, implying that $k/l^{*2} > p$, or $\delta z > (p/k)^{1/2}$. As we shall see below, for repulsive particles the bulk modulus B simply follows $B \sim k$, and we can rewrite our result as

$$\delta z > (p/B)^{1/2} \tag{8.3}$$

Physically, these results signify that pressure has a destabilizing effect, which needs to be counterbalanced by the creation of more contacts to maintain elastic stability. The result is more general [14] and for sphere packing, one also gets $\delta z \equiv z - z_c > (p/B)^{1/2}$. p/B is a measure of the contact strain, and for all interaction potential of interest (approximately harmonic for emulsion or Hertzian for elastic body), one finds that $p/B \sim (\phi - \phi_c)$, where ϕ_c is the packing fraction at the jamming threshold (this comes from the fact that $B \equiv \partial p/\partial \phi$, leading to $p/B \sim (\phi - \phi_c)$ if power law behaviors are assumed) so that Equation 8.3 can be rewritten as $\delta z > (\phi - \phi_c)^{1/2}$. Packings generated numerically [2, 7] are consistent with the equality of this inequality, supporting that the packing lie indeed close to marginal stability: these systems have just enough contacts to counterbalance the destabilizing effect of compression.

Figure 8.2 Zigzag chain of springs. When pulled at the tips, such a chain yields without stiffness, until the line is straight and forces appears in the contacts. For larger strain, the stiffness jumps to a finite value. More generally weakly coordinated networks (with a moderate δz) are soft when an infinitesimal strain is imposed in a generic direction, but are stiff if this direction corresponds to the stress (called sometimes prestress) sustained by the system.

8.3
Elastic Moduli

Based on the above discussions, an assembly of repulsive particles close to the jamming threshold presents a vanishing excess coordination with respect to the jamming threshold: $\delta z \to 0$ as the strain $p/B \to 0$. Why the proximity of the Maxwell bound should impose that shearing becomes much softer than compressing is not obvious *a priori*. Our argument will show that this behavior is specific to purely repulsive systems, and that it is in fact possible to create spring networks where the behavior is opposite, for which $B/G \to 0$. More generally, we shall see that weakly connected systems (i.e., z close to z_c) are soft to most imposed strains, for which they display a dimensionless stiffness or elastic constant of order δz (or zero if the system is floppy, that is, $\delta z < 0$), with one exception: if the infinitesimal strain is imposed in the direction of the stress sustained by the system in its reference configuration, the "prestress." In this case, the response of the system is not soft and the elastic constant found is similar to the one of a well-connected solid. A simple example is given by the zigzag chain of springs of Figure 8.2. When no forces are applied, the chain is in a zigzag configuration. When pulled by the two ends, it yields freely up to the point where the line becomes straight and contact forces appear in the springs. At that point, the system does not yield freely anymore; it is stiff. In the following sections, we shall show how this simple idea applies to more complex networks.

8.3.1
Force Balance and Contact Deformation Operators

We will determine the elastic moduli of packing of a given coordination; let us introduce linear algebra. The argument proposed is a shortened version of Ref. [10]. Many of the properties discussed below concern the response of the system to external forces. It proves convenient to consider our system under the influence of an arbitrary set of forces $\vec{F}_i$ acting on all particles i. At equilibrium, forces are balanced on each particle i:

$$\sum_{\langle j \rangle} f_{ij} \vec{n}_{ij} = \vec{F}_i \qquad (8.4)$$

where f_{ij} is the compression in the contact $\langle ij \rangle$, the sum is on all the particles j in contact with particle i, and $\vec{F}_i$ the external force applied on i. $\vec{n}_{ij}$ is the unit vector going from i to j. This linear equation can be written as

$$\mathcal{T}|\mathbf{f}\rangle = |\mathbf{F}\rangle \tag{8.5}$$

where $|\mathbf{f}\rangle$ is the vector of contact tensions and has N_c components. F is the vector of external forces. Its dimension is[1] Nd, since there are d degrees of freedom for the external force on each particle. Therefore, $\mathcal{T}$ is an $Nd \times N_c$ matrix. In this chapter, we will use the notations in lowercase for the contact space of dimension N_c and uppercase for the particle's positions space of dimension Nd.

Another important linear operator describes the change of distances between particles for a given set of displacements $\delta \vec{R}_i$:

$$(\delta \vec{R}_i - \delta \vec{R}_j) \cdot \vec{n}_{ij} = \delta r_{ij} \tag{8.6}$$

It can be written as

$$\mathcal{S}|\delta \mathbf{R}\rangle = |\delta \mathbf{r}\rangle \tag{8.7}$$

where $\delta \mathbf{r} \equiv \{\delta r_{ij}\}$ is the set of distance changes for all contacts and $\mathcal{S}$ is an Nd by N_c matrix.[2]

As was shown, for example, in Ref. [5, 15], $\mathcal{T}$ and $\mathcal{S}$ are transpose of each other, as we now prove. At equilibrium, any force field applied should cost no energy at first order. This is the virtual work theorem, which is given as

$$\sum_i \delta \vec{R}_i \cdot \vec{F}_i - \sum_{ij} \delta r_{ij} f_{ij} \equiv \langle \delta \mathbf{R}|\mathbf{F}\rangle - \langle \delta \mathbf{r}|\mathbf{f}\rangle = 0 \tag{8.8}$$

where we used the scalar product notation $\langle \delta \mathbf{R}|\mathbf{F}\rangle \equiv \sum_i \delta \vec{R}_i \cdot \vec{F}_i$ and $\langle \delta \mathbf{r}|\mathbf{f}\rangle \equiv \sum_{ij} \delta r_{ij} f_{ij}$. Applying the definitions of $\mathcal{S}$ and $\mathcal{T}$ in Equation 8.8, we get

$$\langle \mathbf{f}|\mathcal{S}|\delta \mathbf{R}\rangle = \langle \delta \mathbf{R}|\mathcal{T}|\mathbf{f}\rangle \tag{8.9}$$

Introducing the transpose notation, this is equivalent to

$$\mathcal{S} = \mathcal{T}^t \tag{8.10}$$

8.3.2
Energy Expansion and Virtual Force Field

For simplicity of notation, we shall think about rather monodisperse emulsions, where the interaction potential is approximately harmonic of stiffness k. The energy expansion follows Equation 8.2,[3] which can be rewritten as

---

1) More precisely, its dimension is $(Nd - d(d+1)/2)$. The term $d(d+1)/2$ corresponds to the constraints on the total torques and forces that must be zero at equilibrium.
2) If one removes the global translations or rotations from the displacement fields, which obviously do not change any interparticle distance, $\mathcal{S}$ becomes an $(Nd - d(d+1)/2)$ by N_c matrix.
3) When forces are present in the contact, as is the case at finite pressure, another term enters the energy expansion, see Ref. [14]. It does not affect the scaling of the elastic moduli, and we shall ignore it.

$$\delta E = k \sum_{\langle ij \rangle} \frac{1}{2} \delta r_{ij}^2 = \frac{k}{2} \langle \mathcal{S}\delta\mathbf{R} | \mathcal{S}\delta\mathbf{R} \rangle \qquad (8.11)$$

Our system is equivalent to a set of point particles interacting with springs. In order to study its elasticity, it turns out to be convenient to consider the responses that follow arbitrary changes of rest of the length of these springs. This is in fact equivalent to imposing dipoles of force. As we shall see, the response to shear or compression can also be easily expressed in terms of changes of rest length. We impose an infinitesimal change of rest length on every contact $\mathbf{y} = \{y_{ij}\}$. The energy and the displacement field are given by the minimization of

$$\delta E = \frac{k}{2} \min_{\{\delta \mathbf{R}\}} (\delta r_{ij} - y_{ij})^2 = \frac{k}{2} \min_{\{\delta \mathbf{R}\}} \langle \mathcal{S}\delta\mathbf{R} - \mathbf{y} | \mathcal{S}\delta\mathbf{R} - \mathbf{y} \rangle \qquad (8.12)$$

Obviously if $\mathcal{S}$ was spanning its image space, we would have $\delta E = 0$: one could always find a displacement $|\delta \mathbf{R}\rangle$ that leads to a change of distances between particles in contact exactly equal to $|\mathbf{y}\rangle$. As we discussed, $\mathcal{S}$ is a $N_c \times Nd$ matrix. If $N_c < Nd$, $\mathcal{S}$ indeed spans its image space, and the energy associated with any strain $|\mathbf{y}\rangle$ is zero, which implies that the system is floppy. In the other case, if $N_c > Nd$, there are $N_c - Nd \equiv N\delta z/2$ relations of dependency among the columns of $\mathcal{S}$. One can choose a basis of $N\delta z/2$ vectors $|\mathbf{a}^p\rangle$, with $1 \leq p \leq N\delta z/2$, in the space of $|\delta r\rangle$ such that

$$\langle \mathbf{a}^p | \mathcal{S} = 0 \qquad (8.13)$$

All $|\mathbf{a}^p\rangle$ are orthogonal to all the vectors $\mathcal{S}|\delta\mathbf{R}\rangle$, for any displacement field $|\delta\mathbf{R}\rangle$. Transposing this relation, we have

$$\mathcal{T} |\mathbf{a}^p\rangle = 0 \qquad (8.14)$$

which indicates that all the vectors in the space of the $|\mathbf{a}^p\rangle$ satisfy force balance without external force Equation 8.4, but no others. The $|\mathbf{a}^p\rangle$ live in the contact force space, and in this chapter, we shall denote as $|\mathbf{a}^p\rangle \equiv |\mathbf{f}^p\rangle = \{f_{ij}^p\}$. In the following, we consider an orthogonal unit basis:

$$\langle \mathbf{f}^p | \mathbf{f}^{p'} \rangle \equiv \sum_{ij} f_{ij}^p f_{ij}^{p'} = \delta_{pp'} \qquad (8.15)$$

We can decompose any $|\mathbf{y}\rangle$ as

$$\mathbf{y} = \mathbf{y}^\perp + \sum_{p=1\ldots N\frac{\delta z}{2}} \langle \mathbf{f}^p | \mathbf{y} \rangle \mathbf{f}^p \qquad (8.16)$$

$\mathbf{y}^\perp$, the part of $|\mathbf{y}\rangle$ orthogonal to the $|\mathbf{f}^p\rangle$, is spanned by the matrix $\mathcal{S}$, and therefore does not contribute to the energy when the minimization of Equation 8.12 is performed. In other words, there exists a displacement field $\delta\mathbf{R}_o$ that leads to a strain $\mathbf{y}^\perp = \mathcal{S}\delta\mathbf{R}_o$. On the other hand, the strain field corresponding to the second term in Equation 8.16 is orthogonal to the space generated by $\mathcal{S}$, and cannot be accommodated by any displacements. The minimum of Equation 8.12 therefore occurs in $\delta\mathbf{R}_o$ and one gets the following equation for the energy:

$$\delta E = \frac{k}{2} \sum_{p=1,\ldots \frac{N\delta z}{2}} \langle \mathbf{f}^p | \mathbf{y} \rangle^2 \tag{8.17}$$

We now discuss properties of the contact force fields that we shall use to derive the scaling of the bulk and the shear moduli. Only one vector of the vector space of the force fields $|\mathbf{f}^p\rangle$ solutions of Equation 8.5 without external force is the real set of contact forces that supports the system, and that could be observed empirically. This vector is denoted by $|\mathbf{f}^1\rangle$. The rest of the basis $|\mathbf{f}^p\rangle$ with $p \neq 1$ are also solutions of Equation 8.5 without external force. Nevertheless, there are no "physical" solutions for the interaction potential chosen. Thus, we shall call them *virtual*. $|\mathbf{f}^1\rangle$ verifies the following properties: (i) In a system with repulsive interaction, as we consider in this chapter, all the contact forces are compressive and therefore $f_{ij}^1 > 0$ for all contacts and (ii) It is well known from simulations and experiments that the distribution of contact forces is roughly exponential, or compressed exponential (see, for example, Ref. [2] for simulations in the frictionless case). This implies that the fluctuations of the contact forces are of order of the average value, leading to $\langle f_{ij}^1 \rangle^2 \sim \langle (f_{ij}^1)^2 \rangle = 1/N_c$ for a normalized force field. Thus, we may introduce a constant c_0 such that

$$\langle f_{ij}^1 \rangle = c_0 \frac{1}{\sqrt{N_c}} \tag{8.18}$$

Now we turn to the properties of the virtual forces $|\mathbf{f}^p\rangle$: (i) There are no physical constraints on the sign of the contacts force for the virtual vectors. Furthermore, $|\mathbf{f}^p\rangle$ must be orthogonal to $|\mathbf{f}^1\rangle$, whose signs of contact forces are strictly positive, and where the fluctuations in the contact compression is small. Therefore, the virtual force fields have roughly as many compressive as tensile contacts.

8.3.3
Elastic Moduli

In our framework, it turns out to be convenient to study the response to shear or compression as they are generally implemented in simulations. When periodic boundaries conditions are used, an affine strain is first imposed on the system. Then the particles are let to relax. In general, the affine strain is obtained by changing the boundary condition. Consider a two-dimensional system with periodic boundary: it is a torus. For example, a shear strain can be implemented by increasing one of the principal radii of the torus and decreasing the other. Then the distance between the particles in contact increases or decreases depending on the direction of the contact. In fact, this procedure of change of boundary conditions is formally equivalent to a change of the metric of the system. If the metric is changed from identity I to the constant metric $G = I + U + U^t$, where U is the imposed infinitesimal global strain, the length of a vector $\vec{\delta l}_0$ becomes δl, such that $\delta l^2 = \vec{\delta l}_0 \cdot G \cdot \vec{\delta l}_0$. Using this expression with $\vec{\delta l}_0 = \vec{R}_{ij} \equiv \vec{R}_j - \vec{R}_i$, one can deduce the change of distance between the two particles as

8.3 Elastic Moduli

$$\delta r_{ij} = \vec{n}_{ij} \cdot \frac{U + U^t}{2} \cdot \vec{R}_{ij} \tag{8.19}$$

Near jamming, for monodisperse particle of diameter l_0, $\vec{R}_{ij} \approx l_0 \vec{n}_{ij}$ and therefore $\delta r_{ij} \approx l_0 \vec{n}_{ij} \cdot (U + U^t) \cdot \vec{n}_{ij}/2$. Formally, such a change of metric is strictly equivalent to a change of the rest length of the springs with $\gamma_{ij} = \delta r_{ij}$. Incidentally, Equation 8.17 can be used to compute the energy of such strain.

For a compression $(U + U^t)/2 = -\varepsilon I$, where I is the identity matrix and ε is the magnitude of the strain, Equation 8.17 becomes

$$\frac{\delta E}{k l_0^2} = \frac{1}{2}\left(\sum_{ij} -\varepsilon f_{ij}^1\right)^2 + \frac{1}{2}\sum_{p=2,\ldots\frac{N\delta z}{2}}\left(-\varepsilon \sum_{ij} f_{ij}^p\right)^2 \tag{8.20}$$

In the first sum, all the terms have the same sign for a purely repulsive system, and this term leads to the strongest contribution. We have

$$\frac{\delta E}{k l_0^2} \geq \left(\sum_{ij} -\varepsilon f_{ij}^1\right)^2 = \varepsilon^2 (N_c \langle f_{ij} \rangle)^2 = \varepsilon^2 c_0^2 N_c \tag{8.21}$$

On the other hand, δE is certainly smaller than an affine compression whose energy also goes as $k\varepsilon^2 N$. Therefore, we find that

$$\delta E \sim k N \varepsilon^2 l_0^2 \tag{8.22}$$

$$B \equiv \frac{\delta E}{V \varepsilon^2} \sim k l_0^{d-2} \tag{8.23}$$

which does not depend on δz. Here, $V \sim N l_0^d$ is the volume of the system. The bulk modulus of a harmonic system jumps from 0 in the fluid phase toward a constant when the system becomes jammed, as observed in the simulations. From this result follows that $p \sim B(\phi - \phi_c) \sim (\phi - \phi_c)$. Note that this result holds only for purely repulsive systems. If the potential has an attractive component, and if the pressure is set to zero, the real force field $|f^1\rangle$ presents as many negative contact forces as positive ones, and this term does not lead to a particularly large contribution in Equation 8.20. In this case, one expects generically to recover for the bulk modulus the result valid for the shear modulus, which we derive in the next section.

If a pure a shear strain is imposed, the tensor $(U + U^t)/2$ is traceless. Let ε be the largest eigenvalue (in absolute value). The change of distance of two particles in contact due to a pure shear δr_{ij} is a number of zero average if the system is isotropic, and fluctuates between $+\varepsilon$ and $-\varepsilon$ depending on the orientation of $\vec{R}_{ij}$. Equation 8.17 becomes

$$\delta E = \frac{k}{2}\sum_{p=1,\ldots\frac{N\delta z}{2}}\left(\sum_{ij} f_{ij}^p \delta r_{ij}\right)^2 \tag{8.24}$$

Each term in the summation gives the following on average:

$$\left\langle \left(\sum_{ij} f^p_{ij} \delta r_{ij} \right)^2 \right\rangle = \sum_{ij} \langle (f^p_{ij})^2 \delta r^2_{ij} \rangle + \sum_{mn \neq ij} \langle f^p_{ij} f^p_{mn} \delta r_{ij} \delta r_{mn} \rangle \qquad (8.25)$$

$$= \sum_{ij} \langle (f^p_{ij})^2 \rangle \langle \delta r^2_{ij} \rangle + \sum_{mn \neq ij} \langle f^p_{ij} f^p_{mn} \delta r_{ij} \delta r_{mn} \rangle \qquad (8.26)$$

For a generic imposed strain, δr_{ij} and δr_{mn} are not correlated so that the terms $\langle f^p_{ij} f^p_{mn} \delta r_{ij} \delta r_{mn} \rangle$ vanish in Equation 8.25. For a pure shear modulus, spatial correlations between δr_{ij} and δr_{mn} exist, and in order to estimate the shear modulus one must make an extra assumption on the nature of the disorder of the network that is being considered. For isotropic random sphere packings, one expects the contact network to be strongly disordered and the propagation of forces to be strongly scattered. We shall therefore assume that the contact forces fields only present weak spatial correlations, as is indeed observed numerically [2] (note that it is possible to build networks with reduced disorder where this assumption breaks down, for instance, in the example shown in Figure 8.1 where long straight lines are present in the microscopic structure. If this system is elongated in the direction of these lines, the restoring force will be large independently of the excess coordination δz). Following this assumption for packing of particles, we shall neglect the terms $\langle f^p_{ij} f^p_{mn} \delta r_{ij} \delta r_{mn} \rangle$ when $mn \neq ij$. Concerning the diagonal terms, one has $\delta r^2_{ij} \approx l^2_0 \varepsilon^2$ while $\sum (f^p_{ij})^2 = 1$ by construction. Thus, each term in the p summation is of the order $l^2_0 \varepsilon^2 \cdot 1$, and

$$\delta E \sim kN \delta z \varepsilon^2 l^2_0 \qquad (8.27)$$

Note that this estimation for the energy will apply for a generic imposed contact strain field $|y\rangle$, since in general such a field would only project weakly on the real force field $|\mathbf{f}^1\rangle$. Finally, Equation 8.27 implies

$$G \equiv \frac{\delta E}{V \varepsilon^2} \sim k \delta z l^{2-d}_0 \qquad (8.28)$$

Equations 8.28 and 8.22 lead to the result $G/B \sim \delta z$. In the context of frictionless particles, one may use Equation 8.3 to recover the observed result $G/B \sim (p/B)^{1/2}$.

8.4
Summary and Conclusion

We have shown that weakly coordinated elastic networks display an exotic elastic behavior: they are very soft and present an elastic constant proportional to δz (of order $\delta z k l^{d-2}_0$, rather than $k l^{d-2}_0$ expected for a usual solid) for generic deformations, but recover a usual, normally connected behavior when the strain is imposed along the prestress. These results apply to interacting particles, with the difference that the typical stiffness k now depends on compression. For emulsion the potential is nearly harmonic and the dependence of k with compression is negligible. For elastic particles interacting with a Herztian potential, one gets $k \sim p^{1/3}$. For a hard sphere glass, which lies below jamming $\phi < \phi_c$, the interaction is purely entropic, which is studied in Ref. [12]. One finds $k/k_B T \sim (p/k_B T)^2$.

A consequence of our analysis is that systems made of repulsive particles, which are always under pressure, are stiff to compression, with respect to the shear modulus that can be tiny if the contact strain is small. At a qualitative level, this has been observed empirically in emulsions [1], glass beads [16], and sand [17]. Nevertheless, as far as scaling exponents are concerned, this results has been backed only numerically, without [1, 2] and with [16, 18] friction. In order to observe this phenomenon quantitatively in nature, emulsions and microgel are presumably good candidates. Ideally one would like to have a system where (i) friction is small, as it limits how close one can get from criticality [19]; (ii) thermal effects are small near the jamming threshold, for the same reason. This requires to consider sufficiently large particles; and (iii) the osmotic pressure is controlled to vary the distance to threshold reliably.

Such an experimental system may enable to address additional intriguing questions; in particular, how amorphous solids yield under shear. From our analysis, we predict that as a shear strain is imposed, and as the shear stress builds up, the structure must stiffen. For example, if the shear stress can be made of the order of the compression near the jamming threshold – which remains to be established – the shear modulus must become of the order of the bulk modulus. Equivalently, as force chains orientate to hold the shear stress, the system becomes much stiffer in the directions where force chains align. At a qualitative level, such a stiffening has been observed numerically [20] with elastic particles, and we have discussed that this effect occurs more generally in various systems such as in gels of semiflexible polymers [21]. The small value of the shear modulus reflects the presence of soft collective modes in the amorphous structure [21], which stop coupling to the applied strain when stiffening occurs. It would be interesting to investigate if these modes, and the stiffening they generate, are causally related to the yielding of the structure. The scaling of the maximal strain with compression before yielding would be very informative as well to address this issue.

Acknowledgment

It is a pleasure to thank W.G. Ellenbroek, A. Kabla, S. Nagel, V. Vitelli, and T. Witten for helpful discussions.

References

1 Mason, T.G., Bibette, J., and Weitz, D.A. (1995) *Phys. Rev. Lett.*, **75**, 2051; Mason, T.G., Lacasse, M.D., Grest, G.S. *et al.* (1997) *Phys. Rev. E*, **56**, 3150–3166.

2 O'Hern, C.S., Silbert, L.E., Liu, A.J., and Nagel, S.R. (2003) *Phys. Rev. E*, **68**, 011306.
3 Alexander, S. (1998) *Phys. Rep.*, **296**, 65.
4 Tkachenko, A.V. and Witten, T.A. (1999) *Phys. Rev. E*, **60**, 687.
5 Tkachenko, A.V. and Witten, T.A. (2000) *Phys. Rev. E*, **62**, 2510.
6 Moukarzel, C.F. (1998) *Phys. Rev. Lett.*, **81**, 1634.
7 Durian, D.J. (1995) *Phys. Rev. Lett.*, **75**, 4780.
8 Ellenbroek, W.G., Somfai, E.K., van Hecke, M., and van Saarloos, W. (2006) *Phys. Rev. Lett.*, **97**, 258001.
9 Wyart, M., Nagel, S.R., and Witten, T.A. (2005) *Europhys. Lett.*, **72**, 486–492.
10 Wyart, M. (2005) *Annales de Physiques Fr.*, **30**, 1; arXiv 0512155.
11 Xu, N., Wyart, M., Liu, A.J., and Nagel, S.R. (2007) *Phys. Rev. Lett.*, **98**, 175502.
12 Brito, C. and Wyart, M. (2006) *Euro. Phys. Letters*, **76**, 149–155; Brito, C. and Wyart, M. (2007) *J. Stat. Mech. Theory Exp.*, **8**, L08003; Brito, C. and Wyart, M. (2009) *J. Chem. Phys.*, **131**, 024504, arXiv:0903.0148.
13 Maxwell, J.C. (1864) *Philos. Mag.*, **27**, 294–299.
14 Wyart, M., Silbert, L.E., Nagel, S.R., and Witten, T.A. (2005) *Phys. Rev. E*, **72**, 051306.
15 Roux, J.-N. (2000) *Phys. Rev. E*, **61**, 6802.
16 Agnolin, I. and Roux, J.-N. (2007) *Phys. Rev. E*, **76**, 061304.
17 Jacob, X., Aleshin, V., Tournat, V. et al. (2008) *Phys. Rev. Lett*, **100**, 158003; Bonneau, L., Andreotti, B., and Clement, E. (2008) *Phys. Rev. Lett.*, **101**, 118001.
18 Magnanimo, V., La Ragione, L., Jenkins, J.T., Wangand, P., and Makse, H.A. (2008) *EPL*, **81**, 34006.
19 Somfai, E., van Hecke, M., Ellenbroek, W.G., Shundyak, K., and van Saarloos, W. (2007) *Phys. Rev. E*, **75**, 020301.
20 Peyneau, P.-E. and Roux, J.-N. (2008) *Phys. Rev. E*, **78**, 041307.
21 Wyart, M., Liang, H., Kabla, A., and Mahadevan, L. (2008) *Phys. Rev. Lett.*, **101**, 215501.

9
Crystallization of Microgel Spheres
Zhibing Hu

9.1
Introduction

Colloidal suspensions exhibit fluid, crystal, and glassy phases similar to those observed in atomic systems [1–3]. Since colloidal particles have typical sizes in the range of a few microns and their movements are relatively slow, their phase behavior can be studied by optical microscopy and light scattering methods. Crystallization in colloidal systems have been intensely studied over the past decades not only due to the interest in reaching a better fundamental understanding of phase transitions [4–6] but also due to the large potential these materials offer for applications, such as in the fabrication of photonic materials [7, 8].

Recently, poly-*N*-isopropylacrylamide (PNIPAM) colloidal particles have attracted considerable attention [9–11]. These particles are cross-linked microgel spheres that are swollen in water, a good solvent, at room temperature but shrink and undergo a reversible volume-phase transition at around 33 °C [12]. This temperature is directly related to the lower critical solution temperature (LCST) of PNIPAM polymer chains in water [13]. PNIPAM-based microgel particle dispersions can form colloidal crystals as well as glasses and gels [14–21], and such dispersions have been used as a model to study both glass-formation [22] and colloidal gelation [23–26]. For crystallizing microgel dispersions, crystalline structures have been identified using small-angle neutron scattering techniques [27] and confocal microscopy [28], and a thermodynamic theory has been developed to describe the phase behavior of aqueous dispersions of PNIPAM microgels [29, 30]. Below the LCST, the neutral PNIPAM particles are hydrophilic. The interparticle interaction is repulsive and may be described with a soft sphere potential $U(r) = (1/r)^n$, where r is the separation between the particle centers and n is an exponent controlling the softness of the potential [14]. As the temperature is raised to near or above the LCST, the interparticle interaction becomes attractive due to the van der Waals attraction [29, 30]. It is also interesting that highly ionized PNIPAM microgels can be electrostatically stabilized and form colloidal crystal arrays at low polymer contents [31]. Moreover, it has been demon-

Microgel Suspensions: Fundamentals and Applications
Edited by Alberto Fernandez-Nieves, Hans M. Wyss, Johan Mattsson, and David A. Weitz
Copyright © 2011 WILEY-VCH Verlag GmbH & Co. KGaA, Weinheim
ISBN: 978-3-527-32158-2

strated that microgel glasses can be converted to crystalline lattices via a particle-based volume transition [32].

Microgels can also play important roles in applications. A fast responsive sensor has been invented by entrapping a microgel crystalline array into a hydrogel matrix [33]. Other applications have been achieved by self-assembly of PNIPAM microgels into a crystalline structure followed by covalent bonding to make the structure permanent [17, 20, 34]. Since the lattice spacing of such a structure is typically of the order of hundreds of nanometers, it leads to Bragg diffraction of visible light with colors that are tunable by controlling the microgel size, for instance, by controlling the temperature or pH [17, 20, 35]. Microgel networks have also been used for controlled drug release [36, 37].

PNIPAM microgels have a range of advantages in comparison, for example, with colloids made from polymethylmethacrylate, silica, or polystyrene: (i) since the microgels contain up to 97% water, their refractive index and density nearly match to the surrounding water and no refractive index matching or density matching of the liquid is required. (ii) The microgel size and the volume fraction of the microgel dispersions are tunable by external stimuli such as temperature and pH. Crystallization of the microgel dispersions can thus be probed by varying either temperature or particle concentration under essentially microgravity conditions (due to density matching). (iii) The interaction forces between microgel particles can be controlled by tuning the temperature or by introducing electric charges on the particles. (iv) PNIPAM microgels are soft and deformable and show physics that is different from that of the hard spheres. Therefore, it is not surprising that PNIPAM microgels have become one of the most important model systems in the study of colloidal physics and chemistry.

This brief review covers the basic aspects of crystallization of PNIPAM microgels including the sample preparation, the colloidal liquid-to-crystal transition, the phase behavior, the determination of volume fractions, crystalline structures, annealing, and aging effects, crystallization kinetics, and an outlook. The readers can find further information in several recent reviews on this topic [9–11, 38–41].

9.2
Synthesis and Characterization of PNIPAM Microgels

PNIPAM microgels have been synthesized using a precipitation polymerization method [42, Chapter 1 of this book]. Typically, NIPAM monomer and the cross-linking agent methylene-bis-acrylamide (BIS) are mixed in water under a nitrogen atmosphere [18]. Different amounts of sodium dodecyl sulfate in water are used as surfactant to control particle size. Potassium persulfate (KPS)–water solution is added to the reactor to initiate polymerization. The reaction is kept at 70 °C for 4 h. After cooling the solution to room temperature, the final reaction dispersion is exhaustively dialyzed in a dialysis tube to wash out unreacted chemicals and surfactant. Table 9.1 shows chemical compositions of two batches of typical PNIPAM microgel dispersions. Particle radii were 216 (batch 1) and 132 nm (batch 2) at 25 °C in water. To introduce electric charges to PNIPAM particles, one can make copolymer

Table 9.1 Dynamic and static light scattering results for the PNIPAM microgel spheres (batch 2) in water and composition of the pregel solutions for both batch 1 and 2.

R_h (nm)		R_g (nm)		R_g/R_h		M_w (10^8 g/mol)		A_2 (10^{-5} mol ml/g^2)		ϱ (g/cm^3)	
25 °C	40 °C	25 °C	40 °C	25 °C	40 °C	25 °C	40 °C	25 °C	40 °C	25 °C	40 °C
132	49.8	87	38	0.66	0.76	1.5	1.4	7.6	−3.5	0.026	0.45

	NIPA (g/ml)	BIS (g/ml)	KPS (g/ml)	SDS (g/ml)	Reaction temperature (°C)
Batch 1	1.536×10^{-2}	2.62×10^{-4}	6.24×10^{-4}	1.18×10^{-4}	68–70
Batch 2	1.538×10^{-2}	2.62×10^{-4}	6.24×10^{-4}	4.31×10^{-4}	68–70

microgels by incorporating different monomers. Typically, PNIPAM-co-acrylic acid (AAc) [9] microgels are negatively charged while PNIPAM-co-allylamine [20] microgels are positively charged in a neutral pH environment.

The average hydrodynamic radius and the size distribution of microgel particles are characterized using dynamic light scattering. The fluctuations of the scattered light intensity are measured as a function of time, t. These fluctuations are due to the fact that the dispersed microgels are undergoing Brownian motion [43, 44]. Figure 9.1 shows the determined distributions in hydrodynamic-radius for PNIPAM microgels (batch 2) in dilute dispersions at 25 and 40 °C, respectively [18].

In static light scattering (SLS), the angular dependence of the excess Rayleigh ratio $R_{vv}(q)$ of a dilute microgel dispersion is measured. $R_{vv}(q)$ is related to the weight average molar mass M_w, the second virial coefficient A_2, and the z-average root mean

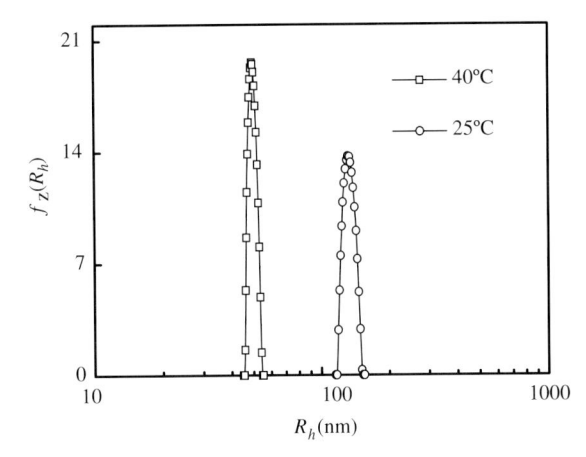

Figure 9.1 Hydrodynamic radius distributions ($f_z(R_h)$) of batch 2 PNIPAM microgel spheres in water at 25.0 (circles) and 40.0 °C (squares), respectively, where $C = 1.37 \times 10^{-5}$ g/ml and the scattering angle = 20°. Reproduced by permission of Gao and Hu [18]. Copyright 2007, American Chemical Society.

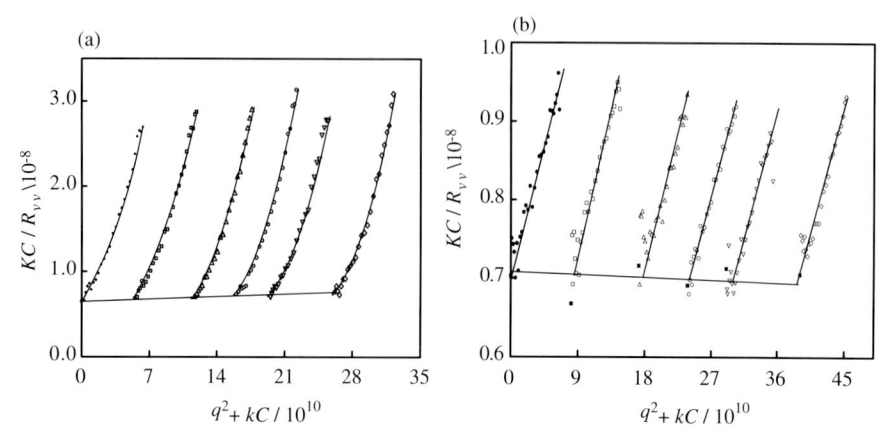

Figure 9.2 (a) Zimm plot of (batch 2) PNIPAM microgel spheres in water at 25.0 °C, where $C = 8.75 \times 10^{-7}$–4.2×10^{-6} g/g. (b) Zimm plot of (batch 2) PNIPAM microgel spheres in water at 40.0 °C, where $C = 8.75 \times 10^{-7}$–4.2×10^{-6} g/g. Reproduced with permission from Ref. [18], copyright 2007 American Chemical Society.

square radius of gyration $\langle R_g^2 \rangle^{1/2}$ by [45]

$$\frac{KC}{R_{vv}(q)} \cong \frac{1}{M_w}\left(1 + \frac{1}{3}\langle R_g^2 \rangle q^2\right) + 2A_2 C \tag{9.1}$$

where $K = 4\pi^2 n^2 (dn/dC)^2/(N_A \lambda_0^4)$ and $q = (4\pi n/\lambda_0)\sin(\theta/2)$. N_A is Avogadro's constant, n is the refractive index of the solvent, C is the solute concentration (g/g), λ_0 is the wavelength of light in vacuum, and θ is the scattering angle. Figures 9.2a and 9.2b show Zimm plots at 25 and 40 °C, respectively, where $dn/dC = 0.167$ cm^3/g was used. From the extrapolation of $KC/R_{vv}(q)$ in Equation 9.1 to the zero angle and zero concentration, the molar mass M_w, the second virial coefficient A_2, and the radius of gyration $\langle R_g^2 \rangle^{1/2}$ were obtained and are listed in Table 9.1.

By combining DLS and SLS results, many interesting features of the PNIPAM microgel dispersions can be revealed. For example, $R_g/R_h = 0.76$ for PNIPAM microgels at 40 °C is close to the theoretical value of $(3/5)^{1/2}$ for uniform hard spheres, indicating uniform dense spheres for PNIPAM microgel particles at temperatures higher than the critical temperature, T_c. At 25 °C, $R_g/R_h = 0.66$ is lower than $(3/5)^{1/2}$, which demonstrates that there should be pendant PNIPAM segments on the microgel particle surface that are not as dense as the PNIPAM networks in the center of the particles. The structure and swelling properties of microgels have been discussed in Chapters 5 and 4, respectively.

9.3
Phase Behavior of Dispersions of PNIPAM Microgels at Room Temperature

The phase behavior of PNIPAM microgel dispersions can be readily observed, as shown in Figure 9.3a. From this figure and from light scattering and turbidity

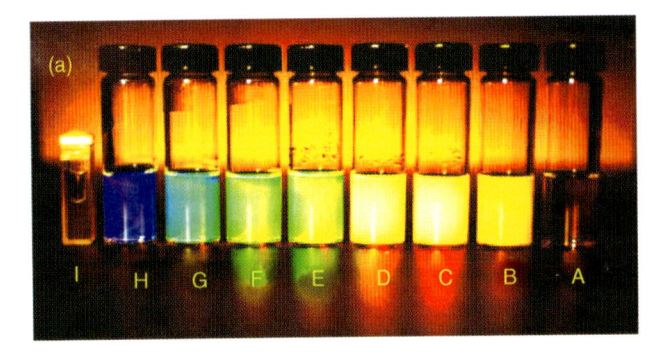

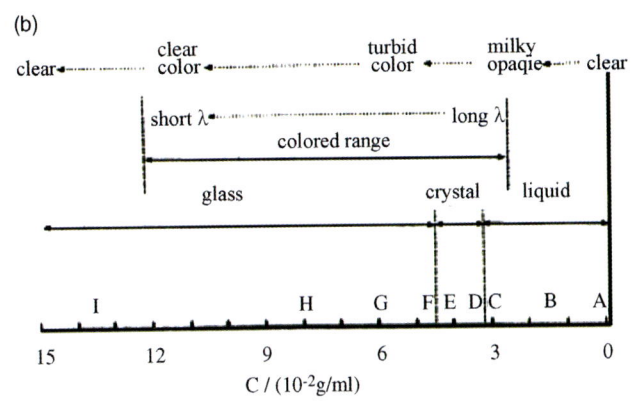

Figure 9.3 (a) Color pictures of batch 2 PNIPAM microgel dispersions at 21 °C. C/wt%: (A) 0.064, (B) 1.47, (C) 3.0, (D) 3.4, (E) 4.2, (F) 4.6, (G) 5.95, (H) 7.92, (I) 13.7 (samples A–I, right to left). The average hydrodynamic radius of PNIPAM microgel spheres in dilute water dispersions at 25 °C is 132 nm. (b) Schematic phase diagram of batch 2 microgel water dispersions as a function of C/wt% at room temperature. Reproduced with permission from Ref. [18], copyright 2007 American Chemical Society.

measurements, a schematic phase diagram is constructed for a very broad distribution of polymer concentrations (0.1–14 wt%), as shown in Figure 9.3b. Highly concentrated dispersions were obtained by evaporating the solutions at a temperature higher than 34 °C and then allowed them to reach a quasiequilibrium state at room temperature for 1 week.

The PNIPAM microgel dispersions with polymer concentrations ranging from ~3 to 4.5 wt% exhibit easily visible iridescent colors (samples E and D, Figure 9.3a) and UV–visible spectrum shows a sharp Bragg diffraction peak (that will be shown later), indicating that the particles self-assemble into an ordered arrangement. As the polymer concentration increases above 4.5% (samples F, G, H, and I), the color of the dispersion appears uniform and the UV–visible spectra of these samples do not have a sharp Bragg peak, indicating the dispersion enters a glass phase. When the concentration is below ~3 wt%, the dispersions become cloudy because the particles are well separated and scatter light strongly.

9.3.1
Characterization of Different Phases Using UV–Visible Spectroscopy

The various phases of the microgel dispersions can be characterized by UV–visible spectroscopy that is usually used to study the absorption by molecules undergoing electronic transitions. However, due to Bragg diffraction from an ordered colloidal array, the electromagnetic spectrum in the visible ranges exhibits an absorption peak. As a result, the information about crystalline formation can be obtained from the peak position and line width [46]. By measuring the UV–visible absorbance spectra on a diode array spectrometer, the turbidity of the samples was obtained from the ratio of the transmitted light intensity (I_t) to the incident intensity (I_0) as $\alpha = -(1/L) \ln(I_t/I_0)$, where L is the sample thickness.

Figure 9.4 shows UV–visible spectra for PNIPAM microgels with a radius of 216 nm at 25 °C (batch 1) for various polymer concentrations. The liquid phase is characterized by a monotonic decrease in absorption with wavelength (Figure 9.4a). At very low concentration ($C \approx 0.01$ wt%), the microgel spheres are well separated, and interactions between them are negligible. The dispersions are clear and colorless (see sample A in Figure 9.3a). As the concentration increases from 0.01 to 2 wt%, the dispersion becomes cloudy (Figure 9.3a) and turbidity increases across all wavelengths (Figure 9.4a) because of the increase in the scattering areas.

The crystalline phase is characterized by a sharp peak in the UV–visible spectra as shown in Figure 9.4b for polymer concentration between ~3 and ~4.5 wt%. These crystals are easy to observe because of their iridescent patterns, as shown in Figure 9.3a (samples D and E). The iridescent colors observed in the dispersions are due to diffraction from the ordered colloidal arrays with a lattice spacing of the order of the wavelength of visible light according to the Bragg's law: $2nd \sin\theta = m\lambda_c$, where n is the mean refractive index of the dispersion, θ is the diffraction angle, d is the lattice spacing, m is the diffraction order, and λ_c is the wavelength of diffracted light.

At even higher concentrations ($4.5 < C/\text{wt\%} < 14$), the microgel spheres are in a glass phase. In this concentration range, the dispersions look homogeneous, and their colors change from green to blue to blue purple to colorless, as shown in Figure 9.3a (samples F, G, H, and I). The turbidity of each dispersion also exhibits a shoulder-shape increase at a certain wavelength, λ_c (Figure 9.4c). This suggests that in the glass phase, there is a strong local order. As the concentration increases, the particle size decreases and λ_c shifts to shorter wavelengths.

The exact crystalline structure was determined by Hellweg et al. using small-angle neutron scattering (SANS) [27]. The scattering profiles were determined for PNIPAM microgels with 2% cross-linker for a dilute solution (0.85 wt% polymer), reflecting the particle form factor, and for a concentrated sample (8.5 wt% polymer), respectively, as shown in Figure 9.5. The effective structure factor obtained by division of the two measured scattering profiles shows several Bragg peaks, which are well described by scattering from an fcc lattice. Furthermore, the lattice parameter, a, calculated for an fcc structure is found to be close to the particle size, as predicted from the radius of gyration of the particles in dilute solution.

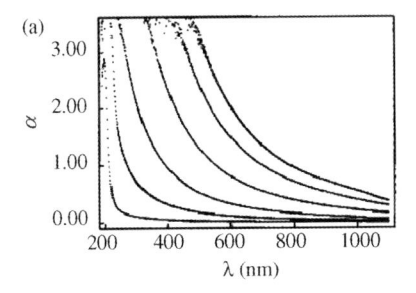

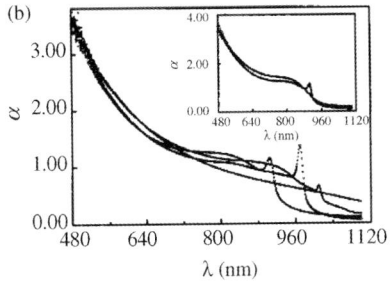

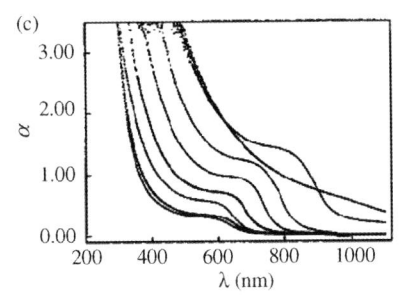

Figure 9.4 Turbidity versus wavelength for batch 1 PNIPAM microgel dispersions at various C, where the average hydrodynamic radius of PNIPAM microgel spheres in dilute water dispersions at 25 °C is 216 nm. (a) Dilute dispersions. The curves from left to right correspond to $C = 7.52 \times 10^{-5}$, 5.58×10^{-4}, 1.76×10^{-3}, 4.78×10^{-3}, 1.36×10^{-2}, and 1.99×10^{-2} g/g, respectively. (b) Concentrated dispersions that form crystals. The curves with a peak from left to right correspond to $C = 4.47 \times 10^{-2}$, 3.65×10^{-2}, and 3.21×10^{-2} g/g, respectively. The curve without a peak corresponds to $C = 1.99 \times 10^{-2}$ g/g. The inset compares the turbidity curves before (with a peak) and after shear melting (without a peak) for the sample of $C = 4.47 \times 10^{-2}$ g/g. (c) Highly concentrated dispersions that form a glassy state. The curves with a shoulder from left to right correspond to $C = 14.2 \times 10^{-2}$, 13.4×10^{-2}, 12.5×10^{-2}, 10.4×10^{-2}, 8.38×10^{-2}, 6.78×10^{-2}, and 4.47×10^{-2} g/g, respectively. The curve without a shoulder corresponds to $C = 1.99 \times 10^{-2}$ g/g. The unit of turbidity (R) is cm^{-1} for all turbidity figures. Reproduced with permission from Ref. [18], copyright 2007 American Chemical Society.

9.4
Temperature- and Polymer Concentration-Dependent Phases of the PNIPAM Microgel Dispersions

The phase behavior of PNIPAM microgels is not only dependent on polymer concentration but also on temperature. This is because the particle size decreases with temperature, as shown in Figure 9.6a for pure PNIPAM and PNIPAM-*co*-allylamine microgels [47]. The temperature at which the particle size undergoes a sharp decrease, the volume-phase transition temperature, corresponds to the LCST of the bulk polymers dissolved in water. The PNIPAM-*co*-allylamine particles showed a volume-phase transition similar to that of the pure PNIPAM gel with a slightly higher volume-phase transition temperature around 35 °C, indicating that this PNIPAM-*co*-allylamine microgel is only weakly charged at neutral pH. As a result, the phase

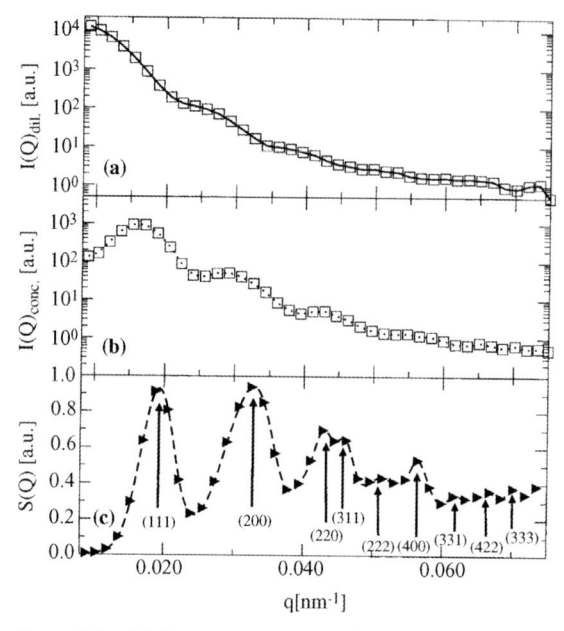

Figure 9.5 (a) Neutron scattering profile for a dilute solution of the microgel particles; (b) scattering profile for a concentrated crystallized sample; and (c) structure factor of the crystal. The Bragg peaks are interpreted according to the FCC lattice. Reproduced with permission from Ref. [27], copyright 2000 Springer.

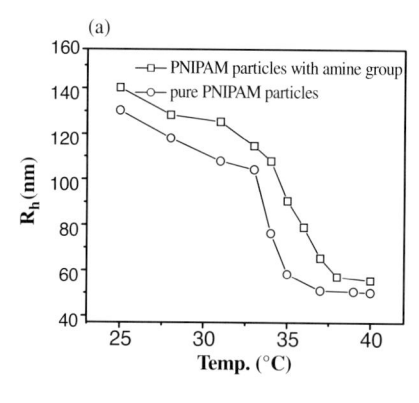

Figure 9.6 (a) Temperature-dependent hydrodynamic radii of PNIPAM-co-allylamine and PNIPAM microgels in water. (b) Phase behavior of the PNIPAM-co-allylamine microgel dispersions as functions of temperature and polymer concentration. Here, T_c (dashed line) is the volume-phase transition temperature of PNIPAM-co-allylamine particles. T_m (open circles) and T_g (open squares) are the melting and glass transition temperatures, respectively. T_m (solid circles) and T_g (solid squares) of microgel suspensions with same chemical compositions but prepared from different batches [20], included for comparison. Adapted from Ref. [47], copyright 2007 American Chemical Society.

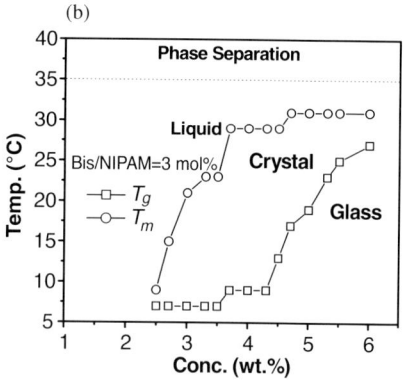

behavior for PNIPAM-*co*-allylamine microgels is observed to be similar to that of pure PNIPAM microgel systems [18, 23].

Figure 9.6b summarizes the phase behavior of the PNIPAM-*co*-allylamine microgel dispersions as a function of temperature and polymer concentration. Here, T_c (dashed line) is the volume-phase transition temperature of PNIPAM-*co*-allylamine particles. T_m (open circles) and T_g (open squares) are the melting and glass transition temperatures, respectively. At T_m, the UV–visible spectrum changes from exhibiting a Bragg peak from a crystalline phase to a featureless curve characteristic of a liquid phase. At T_g, a Bragg peak is replaced by a shoulder-like curve characterized by a glass phase as described in Figure 9.4c.

9.5
Theoretical Investigation of Phase Behavior

The dispersion of PNIPAM microgels undergoes two sequential transitions with temperature. Figure 9.7 shows that a colloidal crystalline phase with iridescent grains for a PNIPAM microgel dispersion at 21 °C (Figure 9.7a) is transferred into a homogeneous colloidal liquid at 26 °C (Figure 9.7b). When the temperature was raised further to 35 °C, the sample becomes white and opaque. This is caused by the increase in refractive index of microgels due to the sharp shrinkage of the particle size (Figure 9.7c) [29].

A quantitative phase diagram (Figure 9.8) for PNIPAM dispersions can be constructed by measuring the UV–visible absorbance spectra as a function of temperature and particle concentration. The melting temperature (solid circles in Figure 9.8) was determined by observing the disappearance of the turbidity peak due to Bragg diffraction. In contrast, as the temperature rises to the phase separation temperature (open circles in Figure 9.8), the turbidity increases sharply in the entire range of visible light wavelengths.

The phase behavior of microgel dispersions can be represented by a thermodynamic model. The pair potential between microgel particles includes a short-range repulsion that is similar to the interaction between two polymer-coated surfaces and a

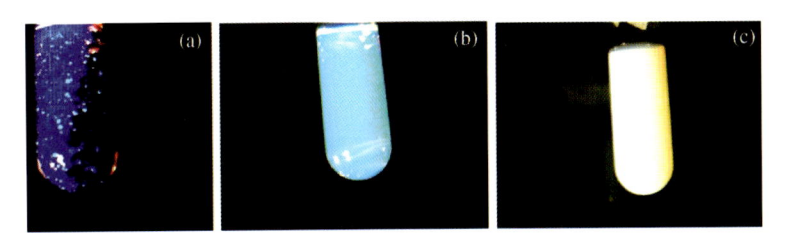

Figure 9.7 Color pictures of a PNIPAM microgel dispersion with a polymer concentration of 1.7 wt% at various temperatures: (a) 21, (b) 26, and (c) 35 °C. The average hydrodynamic radius of PNIPAM microgel spheres in water at 25 °C is 133 nm. The diameter of the vial is 1 cm. Adapted from Ref. [29], copyright 2003 American Physical Society.

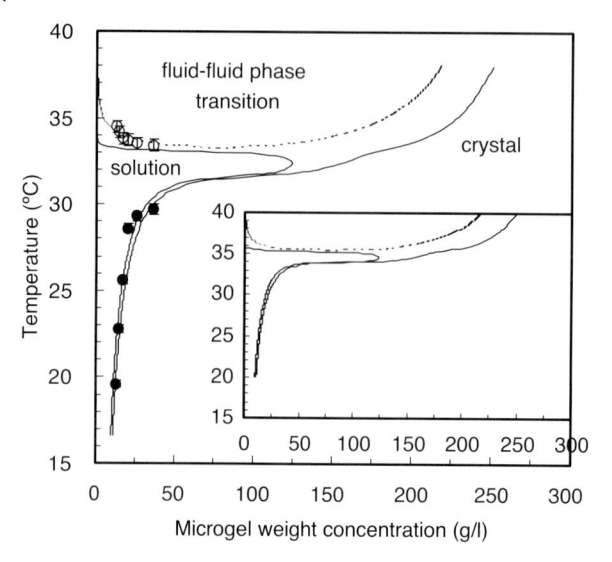

Figure 9.8 The phase diagram of aqueous dispersions of PNIPAM particles determined from turbidity measurements (symbols) and from the thermodynamic perturbation theory with an empirical correction of temperature (lines). The filled and open circles, respectively, represent the melting and the phase-separation temperatures. The inset shows the predicted phase diagram without any adjustable parameters. Reproduced with permission from Ref. [29], copyright 2003 American Physical Society.

longer range van der Waals-like attraction that arises from the difference in the Hamaker constants of the particle and the solvent [41, 48, 49]. The pair potential for microgel particles may be *effectively* represented by the Sutherland-like function [29, 30]:

$$\frac{u(r)}{kT} = \begin{cases} \infty & r < \sigma \\ -\frac{T_0}{T}\left(\frac{\sigma_0}{\sigma}\right)^{6+n}\left(\frac{\sigma}{r}\right)^n & r \geq \sigma \end{cases} \quad (9.2)$$

where σ is the effective diameter, r/σ denotes the reduced center-to-center distance, k is the Boltzmann constant, T is the absolute temperature, T_0 is an empirical proportionality constant that has the unit of temperature, and σ_0 is the particle diameter at a reference temperature where the conformation of the network chains is closest to that of unperturbed Gaussian chains. The introduction of T_0 and σ_0 in Equation 9.1 is solely for the purpose of dimensionality. It is assumed that $n = 8$ in Equation 9.1 in considering that the range of attraction between colloidal particles (relative to the particle size) is shorter than that between atomic molecules where $n = 6$ [41].

As the interaction potential between microgel particles is obtained, the phase diagram can be calculated using thermodynamic models for the fluid and the solid phases. Helmholtz free energy is determined for both the fluid and the solid crystalline colloidal phases. The chemical potential and osmotic pressure can then be determined [29, 30] and the resulting theoretical phase diagram is shown in Figure 9.8 as a solid line. The theoretical curve matches the experimental data well if the calculated temperatures are rescaled empirically as $T' = T \cdot (15/R_g)^{0.005}$, where R_g is the gyration radius of the microgel [29]. The discrepancy is probably because the slight size difference ($\sim$2 nm) between the microgel samples used in light scattering and in turbidity measurements.

9.6
Phase Diagram in Terms of Volume Fraction

In contrast to the case of hard spheres, it is difficult to determine the volume fraction of microgel suspensions. The origin of the difficulty is the deformable nature of microgels, which means that the volume, as measured, for instance, by light scattering in a dilute dispersion, might not be the same as that in a concentrated dispersion due to the increased osmotic pressure and/or the steric effects exerted between particles. To overcome this problem, a common method is to determine the effective volume fraction of microgel dispersions using the relation between the relative viscosity η_{rel} under dilute conditions and the effective volume fraction, φ_{eff} [14],

$$\eta_{rel} = 1 + 2.5\varphi_{eff} + 5.9\varphi_{eff}^2 \tag{9.3}$$

φ_{eff} is linearly related to the polymer concentration C in wt%, $\varphi_{eff} = kC$; the proportionality constant k thus converts between polymer mass concentration and effective volume fraction. It has been found [14] that by plotting the phase behavior versus the effective volume fraction for PNIPAM microgels suspended in water, φ_{eff} at the freezing and melting are 0.59 and 0.61, respectively, which differs considerably from the corresponding volume fractions of 0.494 and 0.55 for hard spheres. The fluid–crystal coexistence region in the PNIPAM systems is thus very narrow.

Recently, Eckert and Richtering revisited the phase behavior versus volume fraction experiment for PNIPAM microgels using simultaneous static and dynamic three-dimensional cross-correlated light scattering [50]. To avoid multiple scattering at high-volume fractions, they chose dimethylformamide (DMF) with refractive index of $n_D = 1.43$ as solvent. PNIPAM displays no temperature sensitivity in DMF, but DMF is a good solvent and the experimental data obtained in this study are thus relevant for microgel dispersions [50]. Figure 9.9 displays the phase behavior of concentrated microgel solutions in DMF. One notes that not only the freezing transition ($\varphi_{eff,\ freezing} = 0.494$) is mapped onto hard sphere behavior but also the width ($\varphi_{eff,\ melting} - \varphi_{eff,\ freezing} = 0.051$) of the coexistence region; in other words, the melting transition agrees with the predictions ($\varphi_{eff,\ melting} = 0.545$) from a hard sphere model.

An independent experiment also shows that φ_{eff} at the freezing and melting transitions are 0.494 and 0.545, respectively, for PNIPAM microgels by using optical microscopy for determining φ_{eff} [51]. By combining Refs [50] and [51], thus, it may be

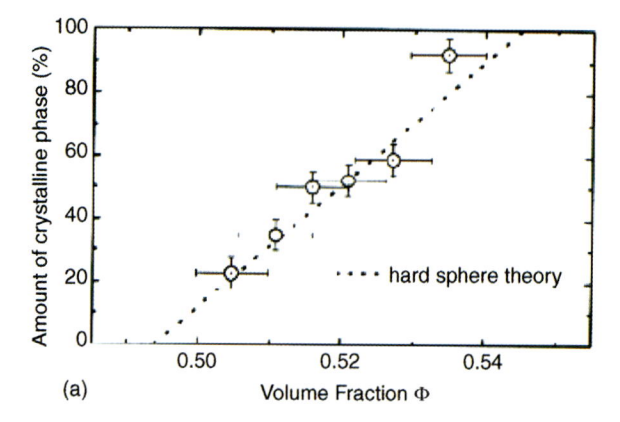

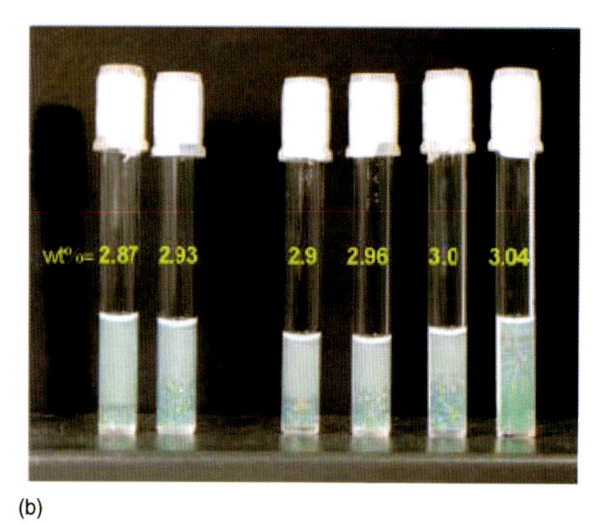

Figure 9.9 (a) Phase behavior of concentrated microgel solutions in DMF. The volume fractions are accurate to within ±1%. (b) Experimental phase behavior of concentrated microgel suspensions in DMF. Reproduced by permission of Eckert and Richtering [50], copyright 2008 American Chemical Society.

concluded that neutral homo-PNIPAM microgels at room temperature exhibit phase behavior similar to hard sphere systems.

By contrast, for PNIPAM-co-AAc microgels, the results are quite complex. For example, for crystals prepared from PNIPAM-co-AAc microgels at pH 3.8, the melting transition was observed at $\varphi_{eff} = 0.11$, thus far below the $\varphi = 0.494$ of hard spheres. The authors suggested that either an attractive enthalpic force or weak multibody interactions between particles could be involved in crystallization at such low particle concentrations [51]. In another experiment, the fluid–crystal coexistence region of the suspension of PNIPAM-co-acrylic acid microgels at pH 2.8 was measured by UV–visible spectroscopy [52]. The melting and freezing transition boundaries of the coexistence region were determined via a blueshift of the Bragg peak and the disappearance of the peak, respectively [52]. The polymer concentration was converted into the corresponding volume fraction for PNIPAM hard spheres at room temperature by setting the particle volume fraction at the freezing transition at 49.4%, as obtained for hard sphere suspensions. A melting transition was determined at 56.4%. The authors suggested that at low pH, the PNIPAM-*co*-acrylic acid microgel particles behave as thermosensitive hard spheres [52].

The complex behavior of the PNIPAM-*co*-acrylic acid microgels may be because acrylic acid contributes carboxyl (−COOH) functional groups. The pK_a value (an acid disscociation constant) for the carboxyl group is about pH 4.3, at which half the carboxyl groups are ionized. As a result, the PNIPAM-*co*-acrylic acid could be highly ionized at pH 7, or partially ionized at pH 4.3, or not ionized at pH 2. In addition to the electric interaction due to the presence of charges from carboxyl ionization, the unionized carboxyl groups can exhibit attractive interactions resulting from hydrogen bonding [51].

9.7
The Interparticle Potential

As discussed in Section 9.4, the interparticle interaction is often modeled by a combination of a repulsive hard sphere and an attractive temperature-dependent van der Waals potential. Far below the phase separation temperature (or LCST), the repulsive potential is dominant [23]. Alternatively, the interparticle potential of PNIPAM microgels has been modeled with a temperature-independent soft sphere potential (Ψ) [14],

$$\psi(r) \propto (1/r)^n \tag{9.4}$$

where r is the separation between the particle centers and n is an exponent controlling the softness of the potential.

To obtain the exponent n for PNIPAM microgel dispersions, Senff and Richtering have measured the frequency dependence of the storage and loss modulus, G' and G'', respectively, in low-amplitude oscillatory shear experiments [14]. At low temperatures, the sample behaved as a viscoelastic solid; G' was independent of frequency, thus corresponding to the plateau modulus G_p, and higher than G'' over the whole frequency range. G_p decreased with temperature and a master curve was obtained for G_p versus φ_{eff}, as shown in Figure 9.10.

Using Equation 9.4 and considering that $\varphi_{eff} \propto r^{-3}$, the relationship of G_p versus φ_{eff} is obtained [14]:

$$G_p \propto \frac{1}{r}\left(\frac{\partial^2 \psi}{\partial r^2}\right) \propto \varphi_{eff}^{(1+(n/3))} \tag{9.5}$$

Equation 9.5 was used to fit the data in Figure 9.10. The solid line corresponds to a power law fit of all data points and has the exponent $n = 9.5$; the dashed line describes

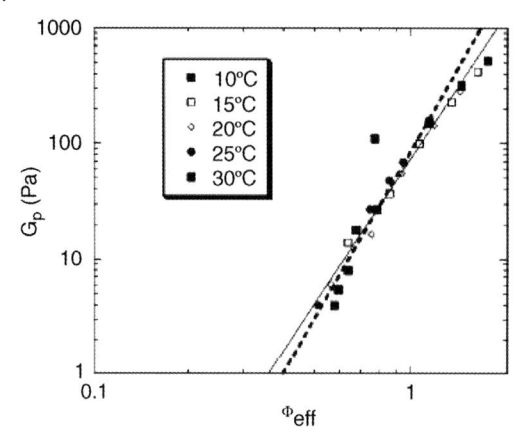

Figure 9.10 Master curve of the plateau modulus versus the effective volume fraction at different temperatures. The solid line is a fit for $G_p \propto \varphi_{eff}^m$ of all data points and the dashed line represents a fit at $\varphi_{eff} < 0.9$. Reproduced with permission from Ref. [14], copyright 2008 American Physical Society.

the behavior of volume fractions below 0.9 and has the exponent $n = 11.7$. These results indicate a soft sphere behavior.

It was demonstrated that the particle–particle interaction potential does not change significantly between 25 and 32 °C [53]. Even approximately 1 K below the LCST, the experimental scattering intensity distributions $I(q)/c$ are described very well by the hard sphere structure factor $S(q)$, using an equivalent hard sphere particle size and volume fraction. At temperatures well above the LCST, however, the interaction potential becomes strongly attractive and the collapsed microgel spheres form aggregates consisting of flocculated particles without significant long-range order. Heyes and Branka have written an excellent review on interactions between microgel particles, with particular focus on the current understanding and representations of the effective interactions between these particles [40].

9.8
Annealing and Aging Effects

At high particle concentrations, PNIPAM-based microgel suspensions, just as hard sphere suspensions, can enter a glassy state (see Figure 9.3b) and thus avoid crystallization. Interestingly, using the thermosensitive nature of PNIPAM-based systems, glasses based on such suspensions can be converted to ordered crystals through a temperature-induced particle volume contraction followed by reswelling, as demonstrated by the group of Lyon et al. [32]. A dilute solution of 260-nm diameter microgel particles was concentrated through centrifugation at a temperature below the phase transition temperature. This produced a glassy suspension; the UV–visible spectrum of this glassy sample is broad and largely featureless. However, upon cycling the suspension from 25 to 40 °C and back, the system spontaneously crystallizes into a brightly iridescent material, which displays a transmission spectrum with a sharp Bragg peak. Since the increase in temperature results in a 15-fold decrease in volume, the colloidal assembly goes from a close-packed, high-volume fraction system to a relatively low-volume fraction system where the particles are no longer touching. This transition increases the ability of the component particles to reorganize into the thermodynamically preferred crystalline array when the particles reswell upon cooling.

Moreover, it has been observed that unlike suspensions of simple repulsive hard spheres, PNIPAm–AAc microgel dispersions can evolve from a diffusive, fluid-like state characteristic of the suspensions immediately after introduction into capillary tubes, to crystalline or glassy phases on timescales of days or weeks [54]. In addition to this structural evolution, the free volume accessible to the microgels in the crystalline or glassy phases (i.e., the cage size) decreases with time indicating that the dispersion properties continue to evolve as the dispersions slowly proceed toward their equilibrium states.

The bright field micrographs and trajectories for a 2.0 wt% polymer in pH 3.5 buffer at different time points are displayed in Figure 9.11 (inset) [54]. Note that the inset trajectory maps are magnified by a factor of 4 relative to the bright-field images for easier visualization of the trajectories. The microgel dispersions are fluid-like immediately after sample preparation (1- and 5-day data), with diffuse trajectories and no evidence of spatial order. Eventually, the samples display crystalline order and caged trajectories, which are illustrated here by the 11- and 27-day data. Note that, even after formation of the ordered phase, the particle dynamics continues to evolve,

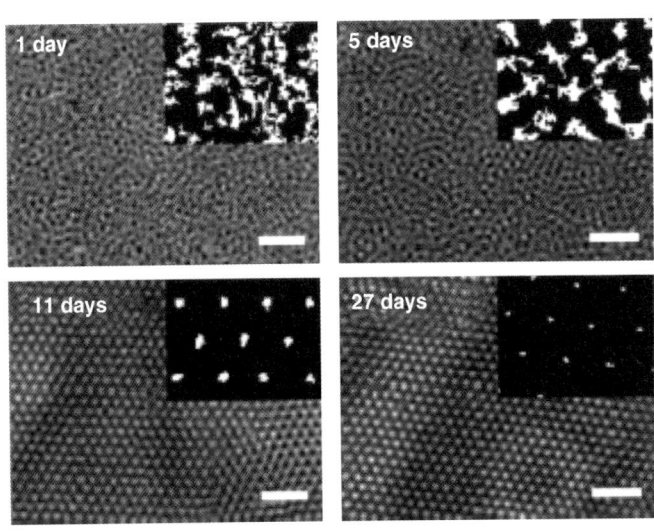

Figure 9.11 Microscopic images and trajectories (inset, ~10 s of observation) of PNIPAM–AAc microgel samples (2.0 wt% polymer in pH 3.5 buffer) at different ages (not in the same spot) and $20.0 \pm 0.1\,°C$; aging times are indicated on all microscopic images. Trajectories are magnified by a factor of 4 relative to the images for ease of trajectory visualization. Scale bar = 10 μm. Reproduced with permission from Ref. [54], copyright 2009 American Chemical Society.

with the particle trajectories becoming more constrained with aging (11- versus 27-day data).

The temperature dependence of PNIPAm–AAc microgel swelling and how it influences the colloidal assembly was evaluated during the aging process as well. These thermal melting experiments revealed an enhancement in the thermal stability of the assemblies during the aging process that was associated with an evolution of attractive interparticle interactions during aging. These effects were strongly pH dependent and it is at low pH that the attractive interactions leading to the strong aging behavior were found.

9.9
Kinetics of Crystallization

The colloidal crystallization process consists of nucleation and crystal growth. Nucleation is the step where the colloids dispersed in the solvent start to gather into clusters. The crystal growth is the subsequent growth of those nuclei that succeed in achieving the critical cluster size. The classical theory of nucleation and crystal growth has been adapted by Russel [55] to hard sphere colloids and extended and evaluated numerically [56]. Specifically, Ackerson and Schatzel [56] have compared the theory with data from small-angle scattering studies of nucleation and growth in suspensions of hard colloidal spheres and found that experimental nucleation rates are much larger than the theoretically predicted value.

According to classical nucleation theory (CNT) [56, 57], the crystal nucleation rate per unit volume I depends exponentially on the Gibbs energy barrier, ΔG_{crit}, which controls the formation of a critical nucleus,

$$I = \zeta \exp(-\Delta G_{crit}/k_B T) \tag{9.6}$$

with the kinetic prefactor ζ usually expressed as $A\phi^{5/3}D/a^5$, where A is a dimensionless factor, a is the particle radius, and D is the particle diffusivity. $D = (1-\phi/\phi_g)^{2.6}D_0$, where D_0 is the free particle diffusion constant and ϕ_g denotes the glass transition concentration. The nucleation rate is simply a Boltzmann probability for being at the barrier to nucleation, times a kinetic coefficient for crossing the barrier [56].

The critical Gibbs energy for nucleation is related to the difference in the chemical potentials of the solution and the crystal, $\Delta\mu$, the solution–solid interfacial tension, γ, and the number density of particles within the crystalline phase ϱ_s:

$$\Delta G_{crit} = \frac{16\pi}{3} \frac{\gamma^3}{[\varrho_s \Delta\mu]^2} \tag{9.7}$$

When analyzing crystal nucleation experiments, the parameter A in the kinetic prefactor and the interfacial tension are normally treated as adjustable variables.

The crystallization process has been quantitatively monitored by using UV–visible transmission spectroscopy [46]. After initialization of the suspensions by shear melting, the crystallization process can be followed by monitoring the Bragg peak of the UV–visible spectra as a function of time. I can then be obtained by analysis of the time-dependent area and position of the Bragg peak for various temperatures [48]. I obtained for the PNIPAM-co-AAc microgel dispersions [46] agrees well with CNT using $\log_{10}(A) = -5.96$ and $\gamma/k_B T = 0.45$.

It is suggested in Ref. [58] that in the bulk colloidal experiments the overall nucleation rates may have been overestimated as a result of the strong interactions between multiple nuclei. Using uniform microfluidic droplets, however, the interactions among the crystallites can be isolated. In this study, a colloidal PNIPAM microgel dispersion was injected into a microfluidic flow-focusing device constructed to generate monodisperse droplets in oil. As demonstrated in Figure 9.12, the nucleation process of crystals inside the droplets can be directly observed with an optical microscope due to the birefringence of the colloidal crystals. In a system with large, 500 μm, dispersion droplets (Figure 9.12a), crystallization of the droplets progressed through the formation of multiple nuclei, similar to what is observed in bulk dispersions [46]. No communication between different emulsion droplets was observed and droplets in the neighborhood of a crystallized droplet remained uncrystallized. In contrast, for a system consisting of smaller, 100 μm, PNIPAM droplets (Figure 9.12b), only one nucleus, or rarely two nuclei, triggered the crystallization in a single droplet. For both large and small emulsion droplet sizes,

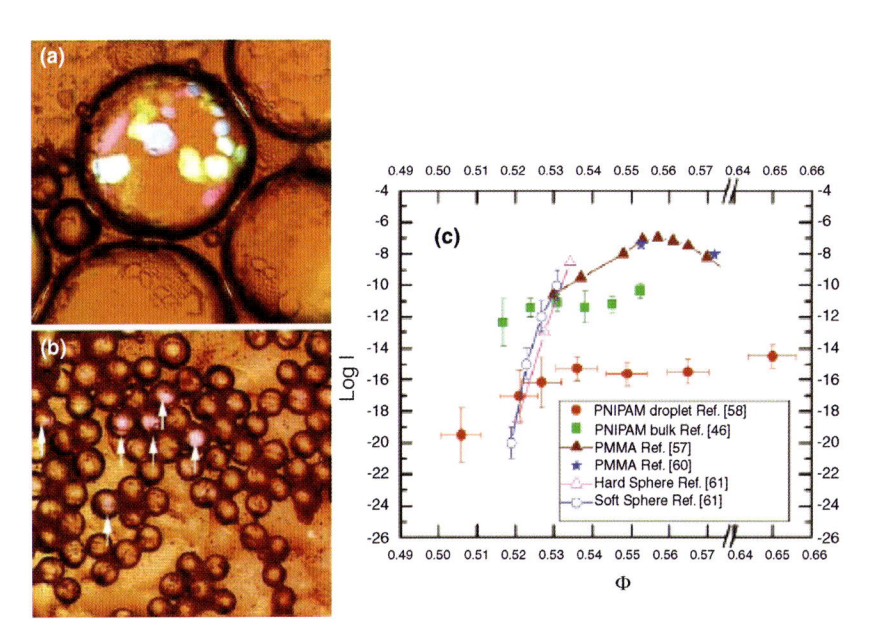

Figure 9.12 Emulsion crystallization observed between a pair of crossed polarizers on an optical microscope. (a) PNIPAM droplets (500 μm size) at 23.6 °C. (b) PNIPAM droplets (100 μm size) at 23.6 °C. The multiple crystallites in the larger droplet system indicate interactions among nuclei at high nucleation volume. Arrows indicate the crystallized droplets in the smaller droplet system.
(c) Normalized nucleation rate, I, comparison between PNIPAM spheres and hard sphere systems. Reproduced with permission from Ref. [58], copyright 2009 American Chemical Society.

the percentage of crystallized droplets among about 3000–4000 droplets was plotted versus time and the experimental result was fit to a model that provided the nucleation rate, I [58]. The effective volume fraction was estimated by assigning a value of 0.494 to 25.6 °C, which was the lowest temperature without any crystallized drops for more than 4 weeks. By comparing the temperature-dependent particle sizes, the effective volume fraction for other temperatures was estimated.

Figure 9.12c compares the kinetic nucleation data for a system of PNIPAM droplets [58] with that of PNIPAM bulk experiments [46] and with nucleation in PMMA hard sphere systems [57, 60]. The computer simulation results [61] on hard sphere nucleation rates are also included for comparison and have covered only part of the large-volume fraction regime that the colloidal data exist. The nucleation rate measured in microfluidic droplets is the lowest one in all experiments carried out so far. One reason may be that in bulk experiments the overall nucleation rates had been overestimated as a result of the strong interactions among multiple nuclei [58].

9.10
Crystallization Along a Single Direction

The single directional growth of hard sphere colloidal crystals has been studied using sedimentation [62] or diffusion-of-base methods [63]. If silica spheres are dispersed in an aqueous solution at volume fractions less than the freezing value [62] or in a pH gradient solution that results in highly charged silica spheres [63] and allowed to sediment onto a flat surface, they form crystals. For microgels, however, it is difficult to uniaxially grow crystals in water by natural sedimentation due to the good density matching between the microgels and the surrounding water.

Zhou *et al.* have, however, found a method to grow uniaxial crystals by mixing an aqueous microgel suspension with an organic solvent [64]. Typically, an aqueous dispersion of PNIPAM-*co*-allylamine microgels with a polymer concentration of 3.5 wt% was mixed with dichloromethane at 22 °C. After homogenization, the mixture was left to stand. This initial mixture (Figure 9.13a) appeared cloudy. Within about 4 h (Figure 9.13b), small crystals were observed growing from the top toward the bottom; this behavior is directly opposite to that observed for a hard sphere system, where crystals grow from the bottom toward the top because of sedimentation [62]. The crystals grew longer with time along the direction of gravity and reached about 1.5 cm in length after 82 h (Figure 9.13g). The dispersion can generally be divided into three portions: the top portion is the crystal phase, the bottom portion (cloudy) is a stable water–oil emulsion, and the middle portion is an unstable emulsion (cloudy and white).

Considering that PNIPAM particles have been used as emulsifiers [65], it was suggested that an unstable oil-in-water emulsion containing micelles of organic oil droplets coated with many microgels was initially formed. Using optical microscopy, the sizes of these micelles were found to range from 10 to 40 µm; these "micelles," which were heavier than water due to the higher mass density of the organic solvent (1.33 g/ml), gradually sank to the bottom of the cuvette. Due to coarsening, the

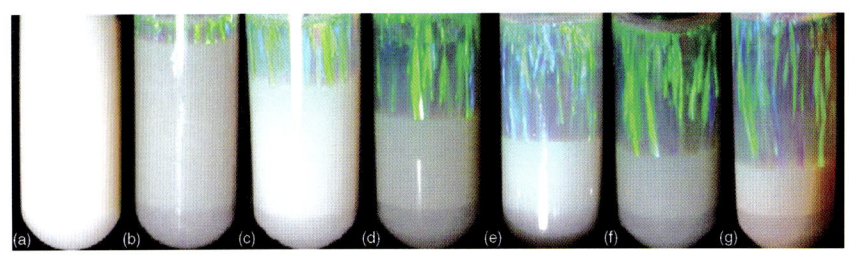

Figure 9.13 Time-dependent growth of columnar crystals in a mixture of an aqueous dispersion of PNIPAM-co-allylamine microgels and dichloromethane in a test tube with 10 mm diameter and 75 mm length. The time started after homogenization: (a) 0, (b) 4, (c) 33, (d) 43, (e) 55, (f) 72, and (g) 82 h. Reproduced with permission from Ref. [64], copyright 2006 American Chemical Society.

microgels at the micelle surfaces are released. The released microgels self-assemble into single directional crystals that originate at the interface between the dispersion mixture and air.

9.11
Summary and Outlook

A suspension of PNIPAM microgels provides a new model system for the study of crystallization. The PNIPAM microgels are soft and deformable, which may lead to new physics different from that of hard spheres. Although much recent work has been reported on the transition between fluid and solid states in microgel suspensions, there are many fundamental challenges ahead. For example, how can the microgel particle size be directly measured in a concentrated microgel dispersion and how can the elastic modulus, a direct measure of softness, of individual particles be determined? Chapter 12 of this book has addressed this question.

Recent work has been performed investigating the transition from a fluid to a crystal state, relating the behavior of neutral or weakly charged PNIPAM microgel suspensions to the behavior of hard sphere systems. However, crystallization in suspensions of highly charged ionic microgels is an interesting area [66–68]: an FCC crystalline order occurs at very low particle concentrations (0.03–0.3 wt%). Another interesting research avenue is the use of temperature-dependent microgels as depletants in colloidal suspension to tune colloidal attraction [69].

Moreover, it is known that hollow colloidal capsules, the so-called colloidosomes, can be formed by the formation of a layer of hard sphere particles around a liquid emulsion droplet [70]. In an analogy, microgels have been assembled on the surfaces of water droplets in oil [71, 72]. These experimental systems have a large potential for fundamental studies of crystallization in different geometries. Furthermore, the crystallization of microgels in spherical geometries may lead, for instance, to iridescent colloidosomes with temperature-tunable colors. Colloidosomes made from microgels coated with hard spheres have been reported [73].

Also, in addition to studies investigating the formation of ordered crystalline states, it is interesting to note the recent advances using microgels in investigating the formation of soft disordered glassy solids. Recent progress has been reported in the work using PNIPAM-based microgels [74], studying the transition from a fluid to a disordered "jammed" glassy state. The dispersions of microgels with tunable softness have been used as a model to understand the origins of dynamic processes of the glass formation [22]. Another important research topic for the future is to try and better understand the process of aging within the nonequilibrium glassy state [54, 75]. The current results suggested that attractive interparticle potential plays an important role in the aging process [54]. To understand how shear affects the structure and dynamics of soft disordered solids is a topic of large recent interest, where microgels could be very important as model systems. An important example is the recently proposed strain-rate frequency superposition method in rheology that may help further our understanding of structural relaxation in soft disordered solid states [76].

Many important investigations on the crystallization of microgels have not been discussed in this brief review due to space limitations including Refs [77–86]. Other aspects of microgel crystallization such as melting and applications are covered in chapters 10 and 14 of this book, respectively.

Acknowledgments

I thank J. Gao, G. Huang, S. J. Tang, B. Zhou, J. Zhou, T. Cai, M. Marquez, J. Z. Wu, and Z. D. Cheng for fruitful collaborations on microgel crystallization. This work is supported by the National Science Foundation (DMR-0 805 089).

References

1 Pusey, P.N. and van Megen, W. (1986) *Nature*, **320**, 340.
2 Russel, W.B., Saville, D.A., and Schowalter, W.R. (1989) *Colloidal Dispersions*, Cambridge University Press, Cambridge.
3 Weeks, E.R., Crocker, J.C., Levitt, C., Schofield, A., and Weitz, D.A. (2000) *Science*, **287**, 627.
4 Pusey, P.N. (1991) Colloidal suspensions, in *Liquid, Freezing, and the Glass Transition* (eds J.P. Hansen, D. Levesque, and J. Zinn-Justin), North-Holland, Amsterdam, pp. 763–942.
5 Gasser, U. (2009) *J. Phys. Condens. Matter*, **21**, 203101.
6 Okubo, T. (2008) *Polym. J.*, **40**, 882–890.
7 Xia, Y.N., Gates, B., Yin, Y., and Lu, Y. (2000) *Adv. Mater.*, **12**, 693.
8 Gast, A.P. and Russel, W.B. (1998) *Phys. Today*, **51**, 24.
9 Pelton, R.H. (2000) *Adv. Colloid Interface Sci.*, **85**, 1–33.
10 Saunders, B.R. and Vincent, B. (1999) *Adv. Colloid Interface Sci.*, **80**, 1.
11 Das, M., Zhang, H., and Kumacheva, E. (2006) *Annu. Rev. Mater. Res.*, **36**, 117–142.
12 Hirotsu, Y., Hirokawa, T., and Tanaka, T. (1987) *J. Chem. Phys.*, **87**, 1392.
13 Wu, C. (1998) *Polymer*, **39**, 4609.
14 Senff, H. and Richtering, W. (1999) *J. Chem. Phys.*, **111**, 1705.
15 Seneff, H. and Richtering, W. (1999) *Langmuir*, **15**, 102.
16 Debord, J.D. and Lyon, L.A. (2000) *J. Phys. Chem.*, **104**, 6327.

17 Hu, Z.B., Lu, X.H., and Gao, J. (2001) *Adv. Mater.*, **13**, 1708–1712.
18 Gao, J. and Hu, Z.B. (2003) *Langmuir*, **18**, 1360.
19 Tsuji, S. and Kawaguchi, H. (2005) *Langmuir*, **21**, 2434.
20 Hu, Z.B. and Huang, G. (2003) *Angew. Chem., Int. Ed.*, **42**, 4799.
21 Meng, Z., Cho, J.K., Debord, S., Breedveld, V., and Lyon, L.A. (2007) *J. Phys. Chem. B*, **111**, 6992–6997.
22 Mattsson, J., Wyss, H.M., Fernandez-Nieves, A., Miyazaki, K., Hu, Z.B., Reichman, D.R., and Weitz, D.A. (2009) *Nature*, **462**, 83–86.
23 Benee, L.S., Snowden, M.J., and Chowdhry, B.Z. (2002) *Langmuir*, **18**, 6025.
24 Zhao, Y., Cao, Y., Yang, Y.L., and Wu, C. (2003) *Macromolecules*, **36**, 855–859.
25 Hu, Z.B. and Xia, X.H. (2004) *Adv. Mater.*, **16**, 305–309.
26 Xia, X.H. and Hu, Z.B. (2004) *Langmuir*, **20**, 2094–2098.
27 Hellweg, T., Dewhurst, C.D., Bruckner, E., Kratz, K., and Eimer, W. (2000) *Colloid Polym. Sci.*, **278**, 972.
28 Alsayed, A.M., Islam, M.F., Zhang, J., Collings, P.J., and Yodh, A.G. (2005) *Science*, **309**, 1207–1210.
29 Wu, J.Z., Zhou, B., and Hu, Z.B. (2003) *Phys. Rev. Lett.*, **90**, 048304.
30 Wu, J.Z., Huang, G., and Hu, Z.B. (2003) *Macromolecules*, **36**, 440.
31 Weissman, J.M., Sunkara, H.B., Tse, A.S., and Asher, S.A. (1996) *Science*, **274**, 959.
32 Debord, J.D., Eustis, S., Debord, S.B., Lofye, M.T., and Lyon, L.A. (2002) *Adv. Mater.*, **14**, 658–662.
33 Reese, C., Mikhonin, A., Kamenjicki, M., Tikhonov, A., and Asher, S.A. (2004) *J. Am. Chem. Soc.*, **126**, 1493.
34 Hu, Z.B., Lu, X.H., Gao, J., and Wang, C.J. (2000) *Adv. Mater.*, **12**, 1173.
35 Garcia, A., Marquez, M., Cai, T., Rosario, R., Hu, Z.B., Gust, D., Hayes, M., Vail, S.A., and Park, C. (2007) *Langmuir*, **23**, 224–229.
36 Cai, T., Hu, Z.B., Ponder, B., St. John, J.V., and Moro, D. (2003) *Macromolecules*, **36**, 6559.
37 Huang, G., Gao, J., Hu, Z.B., St. John, J.V., Ponder, B., and Moro, D. (2004) *J. Control. Release*, **94**, 303–311.
38 Lyon, L.A., Debord, J.D., Debord, S.B., Jones, C.D., McGrath, J.G., and Serpe, M.J. (2004) *J. Phys. Chem. B*, **108**, 19099–19108.
39 Lyon, L.A., Meng, Z.Y., Singh, N., Sorrell, C.D., and St. John, A. (2009) *Chem. Soc. Rev.*, **38**, 865–874.
40 Heyes, D.M. and Branka, A.C. (2009) *Soft Matter*, **5**, 2681–2685.
41 Wu, J.Z. and Hu, Z.B. (2004) Microgel dispersions: colloidal forces and phase behavior, in *Encyclopedia of Nanoscience and Nanotechnology* (eds J. Schwarz, C. Contescu, and K. Putyera), Marcel Dekker, Inc., pp. 1967–1976.
42 Pelton, R.H. and Chibante, P. (1986) *Colloids Surf.*, **20**, 247.
43 Berne, B.J. and Pecora, R. (1976) *Dynamic Light Scattering*, John Wiley & Sons, Inc., New York.
44 Chu, B. (1974) *Laser Light Scattering*, Academic Press, New York.
45 Zimm, B.H. (1948) *J. Chem. Phys.*, **16**, 1099.
46 Tang, S.J., Hu, Z.B., Cheng, Z.D., and Wu, J.Z. (2004) *Langmuir*, **20**, 8858–8864.
47 Huang, G. and Hu, Z.B. (2007) *Macromolecules*, **40**, 3749–3756.
48 Berli, C.L.A. and Quemada, D. (2000) *Langmuir*, **16**, 10509.
49 Lowen, H. (1997) *Physica A*, **235**, 129.
50 Eckert, T. and Richtering, W. (2008) *J. Chem. Phys.*, **129**, 124902.
51 Debord, S.B. and Lyon, L.A. (2003) *J. Phys. Chem. B*, **107**, 2927–2932.
52 Clements, M., Pullela, S.R., Mejia, A.F., Shen, J.Y., Gong, T.Y., and Cheng, Z.D. (2008) *J. Colloid Interface Sci.*, **317**, 96–100.
53 Stieger, M., Pedersen, J.S., Lindner, P., and Richtering, W. (2004) *Langmuir*, **20**, 7283–7292.
54 Meng, Z., Cho, J.K., Breedveld, V., and Lyon, L.A. (2009) *J. Phys. Chem. B*, **113**, 4590–4599.
55 Russel, W.B. (1990) *Phase Transit.*, **21**, 27.
56 Ackerson, B.J. and Schatzel, K. (1995) *Phys. Rev. E*, **52**, 6448.
57 Harland, J.L. and van Megen, W. (1997) *Phys. Rev. E*, **55**, 3054–3067.
58 Gong, T., Shen, J., Hu, Z.B., Marquez, M., and Cheng, Z.D. (2007) *Langmuir*, **23**, 2919–2923.

59 Schatzel, K. and Ackerson, B.J. (1993) *Physica Scripta*, **T49**, 70–73.
60 Cheng, Z.D. (1998) Ph.D Dissertation, Princeton University, Princeton, NJ.
61 Auer, S. and Frenkel, D. (2001) *Nature*, **409**, 1020–1023.
62 Davis, K.E., Russel, W.B., and Glantschnig, W.J. (1989) *Science*, **245**, 507.
63 Yamanaka, J., Murai, M., Iwayama, Y., Yonese, M., Ito, K., and Sawada, T. (2004) *J. Am. Chem. Soc.*, **126**, 7156.
64 Zhou, J., Cai, T., Tang, S.J., Marquez, M., and Hu, Z.B. (2006) *Langmuir*, **22**, 863–866.
65 Ngai, T., Behrens, S.H., and Auweter, H. (2005) *Chem. Commun.*, **3**, 331.
66 Gottwald, D., Likos, C.N., Kahl, G., and Loewen, H. (2004) *Phys. Rev. Lett.*, **92**, 068301.
67 Mohanty, P.S. and Richtering, W. (2008) *J. Phys. Chem. B*, **112**, 14692–14697.
68 Gasser, U., Sierra-Martin, B., and Fernandez-Nieves, A. (2009) *Phys. Rev. E*, **79**, 051403.
69 Fernandes, G.E., Beltran-Villegas, D.J., and Bevan, M.A. (2008) *Langmuir*, **24**, 10776–10785.
70 Dinsmore, A.D., Hsu, M.F., Nikolaides, M.G., Marquez, M., Bausch, A.R., and Weitz, D.A. (2002) *Science*, **298**, 1006.
71 Lawrence, D., Cai, T., Hu, Z.B., Marquez, M., and Dinsmore, A.D. (2007) *Langmuir*, **23**, 395–398.
72 Brugger, B., Ruetten, S., Phan, K.H., Moeller, M., and Richtering, W. (2009) *Angew. Chem., Int. Ed.*, **48**, 3978–3981.
73 Cho, E.C., Kim, J.W., Fernández-Nieves, A., and Weitz, D.A. (2008) *Nano Lett.*, **318**, 1895–1899.
74 Zhang, Z., Xu, N., Chen, D.T.N., Yunker, P., Alsayed, A.M., Aptowicz, K.B., Habdas, P., Liu, A.J., Nagel, S.R., and Yodh, A.G. (2009) *Nature*, **459**, 230–233.
75 Purnomo, E.H., van den Ende, D., Mellema, J., and Mugele, F. (2007) *Phys. Rev. E*, **76**, 021404.
76 Wyss, H.M., Miyazaki, K., Mattsson, J., Hu, Z.B., Reichman, D.R., and Weitz, D.A. (2007) *Phys. Rev. Lett.*, **98**, 238303.
77 Motocarlo, D.A., Largo, J., and Solana, J.R. (2007) *J. Phys. Chem. B*, **111**, 10194–10201.
78 Wu, K.L. and Lai, S.K. (2007) *Colloids Surf. B*, **56**, 290–295.
79 Brijitta, J., Tata, B.V.R., and Kaliyappan, T. (2009) *J. Nanosci. Nanotech.*, **9**, 5323.
80 Brijitta, J., Tata, B.V.R., Joshi, R.G., and Kaliyappan, T. (2009) *J. Chem. Phys.*, **131**, 074904.
81 Mayorga, M., Osorio-Gonzalez, D., Romero-Salazar, L., Santamaria-Holek, I., and Rubi, J.M. (2009) *Physica A*, **388**, 1973–1977.
82 Yunker, P., Zhang, Z., Aptowicz, K.B., and Yodh, A.G. (2009) *Phys. Rev. Lett.*, **103**, 115701.
83 Sessoms, D.A., Bischofberger, I., Cipelletti, L., and Trappe, V. (2009) *Philos. Trans. R. Soc. A*, **367**, 5013.
84 Perro, A., Meng, G., Fung, J., and Manoharan, V.N. (2009) *Langmuir*, **25**, 11295.
85 Ueno, K., Sakamoto, J., Takeoka, Y., and Watanabe, M. (2009) *J. Mater. Chem.*, **19**, 4778.
86 Dai, S., Ravi, P., and Tam, K.C. (2009) *Soft Matter*, **5**, 2513.

10
Melting and Geometric Frustration in Temperature-Sensitive Colloids

Ahmed M. Alsayed, Yilong Han, and Arjun G. Yodh

10.1
Introduction

Colloid science explores the behavior of particles dispersed in a background fluid. Its phenomenology has captured the interest of many scientists across a wide range of disciplines for well over 100 years [1–7]. On the fundamental side, interest in colloids stems from a rich interplay of physical, chemical, and hydrodynamic mechanisms in suspension whose realization provides unique opportunities for the study of statistical mechanics and soft matter. On the more practical side, interest arises as a result of the demonstrated importance of colloids in conventional materials such as paints, motor oils, food, and cosmetics, and in high-tech problems such as photonics [8–16], lithography [17–21], biochemical sensing, and processing [22–40], and in the design of advanced composites [41–49].

Here, we are interested in fundamental science. Over the years, colloidal suspensions have proven to be elegant physical systems for tests of fundamental problems in statistical mechanics ranging from Brownian dynamics [50–77] to entropic phase transformations [78–95]. Their attraction as a model system arises from several factors. Chief among them is that colloids offer well-characterized "thermal" systems whose primary constituents (i.e., particles) can be readily observed and tracked by light scattering techniques or by video microscopy. In this context, colloids are often excellent models for traditional atomic materials with particles playing the role of atoms. In contrast to the atomic systems, however, experimental tools such as laser tweezers [95a], advanced optical microscopy [95b], and the CCD camera are readily used to manipulate particles in suspension [96], to create unusual potentials for the particles [89], and to follow and correlate their particle motions. At present, colloidal particles are used to explore a variety of new classes of interaction and self-assembly phenomenon [78, 97–132] and, more broadly, as a basis for novel measurement technologies [133–140].

The primary unifying feature of the research presented herein is the underlying colloidal system: ensembles of temperature-sensitive microgel particles. In particular, we have used microgel particles to create model lyotropic suspensions whose phase behavior can be tuned by small variations in temperature rather than by variation in mesogen concentration. The key ingredient in these samples is thermosensitive polymer, that is, NIPA polymer (poly(N-isopropylacrylamide)). The temperature-sensitive character of our samples stems from the temperature-dependent solubility of NIPA polymer in water. At relatively low temperature, water is a good solvent and NIPA polymer assumes a swollen coil form; in this regime, a small increase in temperature increases monomer–monomer attractions and the size of the isolated polymer coil decreases. Similar ideas apply to NIPA-based microgel particles [141]. In this case, "soft-repulsive" particles made from cross-linked NIPA polymer have diameters that can be tuned by temperature. By changing the particle diameter, one can vary particle volume fraction, and volume fraction is often the primary thermodynamic variable that drives melting and crystallization in colloidal systems [88].

The most significant advantage of these weakly temperature-sensitive suspensions is that changes in temperature enable us to prepare a "lyotropic" colloidal system in particular metastable or ordered states and then to study sample transformations (such as melting) *in situ*. The unique properties of NIPA, in both polymer and microgel particle forms, can therefore be used as a tool to more deeply explore classic physics problems. We have used NIPA *polymer*, for example (see Figure 10.1), to the control liquid crystal transitions of colloidal rods and to study the mechanisms by which lamellar phases melt into nematic phases [142], and we have employed chemically cross-linked NIPA polymer to gels to make new materials, nematic gels with carbon nanotubes, that are analogous to thermotropic liquid crystal elastomers [143, 144] (see Figure 10.2).

This chapter will describe recent research on microgel NIPA *particles* rather than NIPA polymer. In contrast to research oriented toward understanding particular chemical and physical phenomenology of NIPA microgel particles and NIPA polymer [145–158], our work takes advantage of the unique properties of microgel particles in order to explore classic physics problems in new ways. In particular, we will describe melting experiments in three-dimensional (3D) colloidal crystals that permit us to investigate the "nucleation" of fluid formation in crystals; these first fluidization events are observed near defects such as grain boundaries [141]. We will also discuss experiments that explore melting of microgel particles in two dimensions (2D); this work finds predicted hexatic, fluid, and crystal phases, and it introduces order parameter susceptibility as an experimental means to more clearly define transition points in colloid experiments [159]. Finally, we will describe experiments on geometric frustration [160]. In these measurements, self-assembled colloidal particles on a 2D triangular lattice behave like frustrated antiferromagnetic spins. In contrast to quantum spin systems, however, the NIPA colloidal systems offer the possibility to directly visualize an ensemble of particles (spins), to passively and actively observe their dynamics, and to use temperature to tune "antiferromagnetic" nearest-neighbor interaction strength. Thus, as in the other cases noted above, the microgel colloidal particle approach creates a fresh view of a classic problem. Finally, since our initial submission of this chapter, other recent experiments by us, which utilize the same unique features of NIPA particles, have

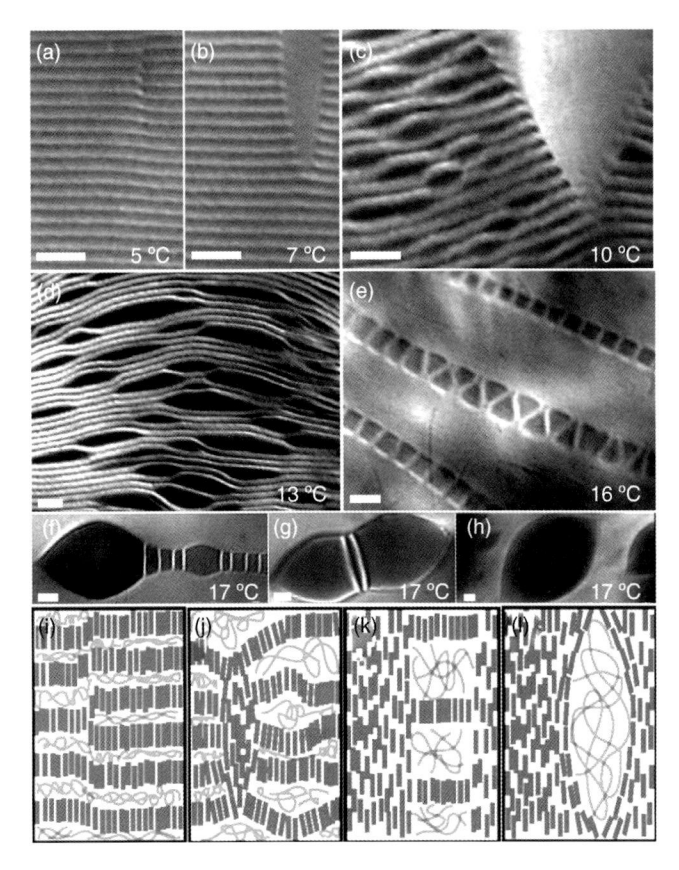

Figure 10.1 Melting behavior of sample of micron-long colloidal rods and NIPA polymers [142]. (a) The lamellar phase at low temperature, consisting of sheets of rods separated by polymer, exhibits visible dislocation defects. (b) At 7 °C, the dislocation defects act as a site for nucleation of the nematic phase. (c) Nematic domains grow, expelling NIPA polymer into lamellar phase, which leads to swelling of lamellar layers. (d) Swollen lamellar phase. (e) Coexistence between nematic phase and highly swollen lamellar phase. (f–g) Isolated monolayer-deformed isotropic tactoid expelling sheets of rods. (h) Isotropic-nematic coexistence observed at high temperature. (i–l) Illustrations of the proposed melting processes of the corresponding rod/polymer mixtures shown in the photographs (a, b, e, h). Scale bars are 5 μm.

proven useful for exploration of the behavior of disordered systems such as glasses [161, 162] and jammed matter [163, 163a], as well as the melting of thin crystalline and polycrystalline films [163b] and freezing criteria in two-dimensions [163c].

The remainder of the chapter is organized as follows. We first introduce the NIPA particle system, providing some discussion of synthesis and characterization. We also describe some of the general experimental and analysis tools that are used. The three sections that follow will describe, respectively, melting in 3D, melting in 2D, and "colloidal antiferromagnets." Finally, the chapter will close with a brief

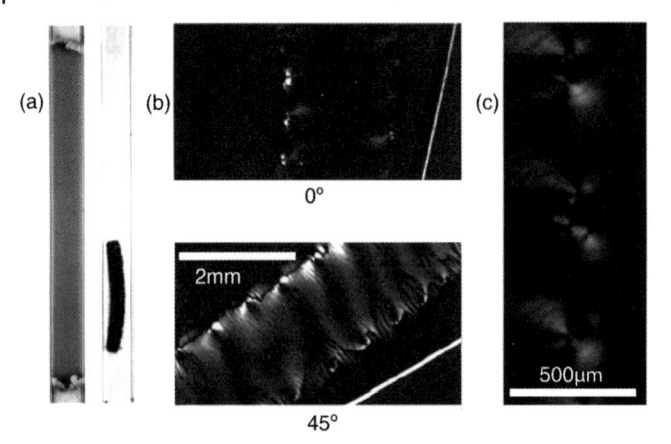

Figure 10.2 Nematic nanotube gel [143] containing isolated half-micron-long single-wall carbon nanotubes (SWNTs). Surfactant-coated SWNTs are homogenously dispersed in a thermally sensitive N-isopropylacrylamide gel. SWNT concentration is rapidly increased by shrinking the gel at high temperature. These shrunken gels exhibit hallmark properties of a nematic. (a) Capillary tube containing SWNT-NIPA gels before (left) and after (right) shrinking. (b) Birefringence images of the (shrunken) nematic nanotube gel in different orientations with respect to the input polarizer pass axis. (c) Liquid crystalline defects observed close to the sample edges.

discussion of future directions. Although significant effort will be made to touch on related work, the reader should note that the present chapter is intended to provide a snapshot of the field, rather than a comprehensive review.

10.2
The Experimental System

10.2.1
Synthesis of NIPA Microgel Particles

Poly(N-isopropylacrylamide) (NIPA) is a thermally sensitive polymer called PNIPA, PNIPAm, or NIPA, for short. NIPA polymer is known to undergo a coil–globule-type phase transition in water. As temperature is increased, NIPA changes from a predominantly hydrophilic molecule to a predominantly hydrophobic molecule at the lower critical solution temperature (LCST) of $\sim$32 °C [148–155].

Synthesis of microgel particles has been previously reported [145–147, 156]. In our lab, we use the dispersion polymerization technique to synthesize colloidal microgel particles from NIPA polymer.

The simplest scheme to make monodisperse 800-nm diameter NIPA particles (which we did not use for the research reported herein) is to dissolve NIPA monomer and cross-linker monomer (BIS) in water at 78 °C. Then the initiator APS (ammonium persulfate) is added to the solution. In 30 min the particles will grow to their maximum size. In our case, we employ slightly different procedures so we can attach fluorophores to the particles for microscopy or so we can modify the particle surfaces to have higher surface charges that prevent sticking to glass surfaces.

In order to prepare particles with larger negative surface charge, we add MMA (methyl-methacrylate) monomer and acrylic-acid monomer to the solution during the last stage of particle growth, that is, just before the particles reach their maximum size. This procedure effectively renders the surface hydrophilic at all temperatures. Thus, even above the LCST, only the interior of the particle will become hydrophobic.

The procedure employed for the case of added amine-reactive fluorophores is described below. Note, this method will make microgel particles with or without the added fluorophore. The fluorophore we utilize is TAMRA (5-(6)-carboxytetramethylrhodamine, succinimidyl ester). In this case, the initiator must make the particle surfaces positively charged; the initiator we use is azobis (2,2′-azobis(2-methylpropionamidine)). We also increase the number of amine groups per particle by copolymerizing NIPA with 2-aminoethylmethacrylate hydrochloride (AEMA). An AEMA-NIPA copolymer is expected to have higher LCST temperature than pure NIPA ($\sim$32 °C) since AEMA is a hydrophilic monomer. The stability of the beads is also improved by adding a cationic surfactant, didodecyldimethylammonium bromide (DTAB) to the suspension. We use BIS (methylene-bis-acrylamide) to cross-link the NIPA polymer within the particles.

The particles are synthesized as follows. First, 20 ml of 25 mM cationic surfactant DTAB, 300 mg of the cross-linker BIS (Polysciences, Inc.), 300 mg of AEMA, 30 gm of NIPA (Polysciences, Inc.), 100 mg of sodium chloride, and 375 ml of 50 mM acetic acid buffer solution, pH = 4.0, are loaded into a special three-neck flask equipped with a stirrer, thermometer, and a gas inlet (see Figure 10.3). The sodium chloride screens the particle surface charges, permitting us to produce bigger particles. In general, particle size can be changed from approximately 800 nm to 2 μm by varying sodium chloride concentration or by initiating NIPA-AEMA copolymerization using APS initiator. The resultant mixture is stirred, heated at 78 °C, and bubbled with dry nitrogen for 10 min to remove dissolved oxygen. (Dissolved oxygen reacts with the initiator, oxidizes the initiator, and it is therefore undesirable.) A solution of 500 mg of 2,2′-azobis(2-methylpropionamidine) dissolved in 10 ml of deionized water is then added to the mixture to start the polymerization reaction. The mixture is continuously stirred at 78 °C for 30 min and then allowed to cool down to room temperature.

The resultant particles are centrifuged and re-suspended in water a few times to remove unreacted monomer, homopolymers, and other salts. The particles are then centrifuged and re-suspended in a buffer solution (pH = 8.3, 0.1 M sodium bicarbonate) to enable the amine groups to react with the fluorophore TAMRA. Ten milligrams of TAMRA is dissolved in 300 μl of dimethylsulfoxide (DMSO) and then added slowly to 40 ml of particle solution. The solution is gently stirred for 24 h to permit the reaction of the fluorophore to take place.

Finally, the particles are cleaned cyclically, first concentrating them by centrifugation and then re-suspending them in a buffer solution (pH = 4.0, 20 mM acetic

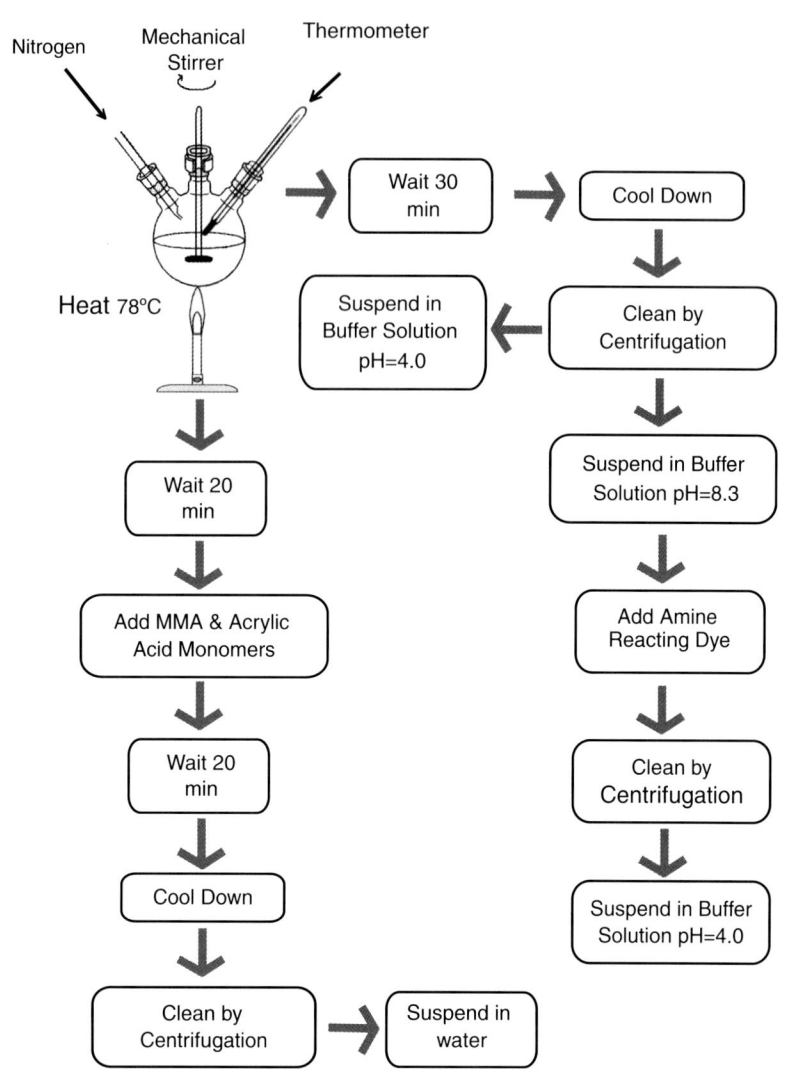

Figure 10.3 Two NIPA particle synthesis schemes, for preparation with negative surface charge (left) and for preparation with added fluorophore (right).

acid). In order to minimize particle aggregates, the suspensions are centrifuged for a few minutes, the supernatant is collected, and the process is repeated (approximately 10 times). For temperatures below 32 °C, NIPA solubility increases with decreasing temperature. Consequently, the NIPA-AEMA particle diameter varies in our experimental temperature regime (20–28 °C) as a result of water moving into and out of the microgel.

A confocal microscopy image of TAMRA-stained NIPA particles is shown in Figure 10.4. Note that particles made of NIPA-AEMA copolymer have higher surface

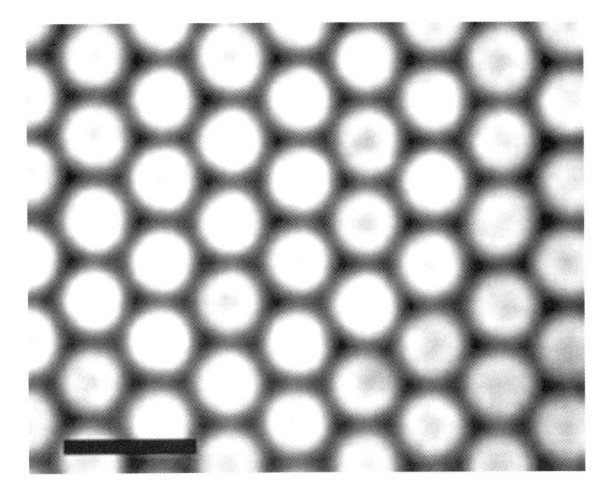

Figure 10.4 Confocal microscopy image of a slice within a NIPA microgel particle colloidal crystal. Scale bar is 3 μm.

charge and are more stable against aggregation in solution *during processing* than particles made of pure NIPA. Counter ions in the buffer solution ensure the charges on the particles are screened. In addition, the polymerization rate in our acidic solutions (pH 4.0) was slow, resulting in reduced size polydispersity. The presence or absence of TAMRA does not appreciably change the particle phase behavior.

10.2.2
Microscopy and Temperature Control

Particle motions were observed by microscope and recorded by CCD camera. In the simplest setup, an analogue CCD camera (Hitachi) is attached to the optical microscope (Leica), and 480×640 pixel images are captured with a video cassette recorder at a rate of 30 images per second. The videotape of particle dynamics is then digitized by a computer equipped with a frame grabber card. Digital images are analyzed to determine the particle positions in each frame (Figure 10.5). For 0.75-μm diameter NIPA spheres, the images are typically obtained with bright field microscopy using a $100\times$ oil objective with numerical aperture equal to 1.45. This gives a magnification of 128 nm/CCD pixel.

In order for particle tracking algorithms to accurately determine particle centers, the number of CCD pixels covered by the image of each particle must be four or more. The colloidal particle position is observed as a circularly symmetric Gaussian image with intensity profile centered about its geometrical center. In general, particle-tracking routines [96] locate particle positions with subpixel accuracy. At a magnification of $100\times$ (8 pixels/μm), subpixel accuracy is obtained, corresponding to spatial resolutions of ~ 20 nm. Thus, the experimental position resolution for particle tracking in our experiments is approximately one order of magnitude better than

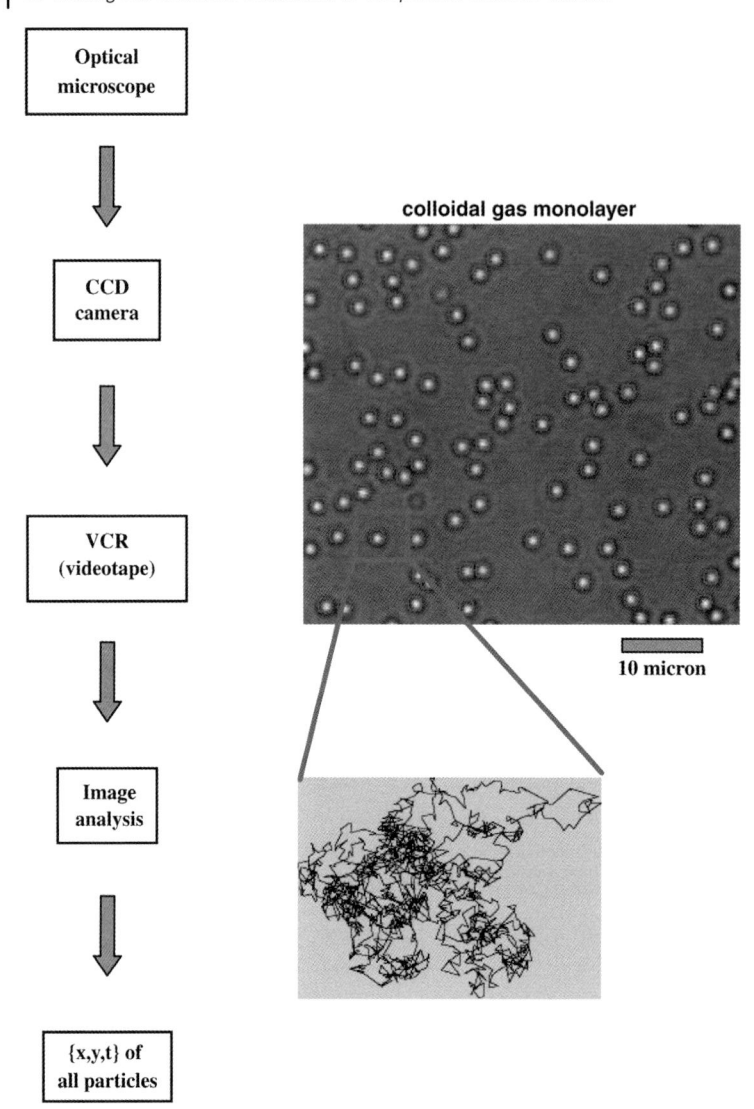

Figure 10.5 Schematic of particle tracking procedure. The image is a colloidal gas monolayer. Overlaid red circles show that the image analysis provides accurate (x,y) positions. The small image is a 1-min trajectory of a particle with 1/30 s per step.

the diffraction limit [96]. The shutter speed of the CCD camera sets an exposure time τ for a single image. If the exposure time is long enough to allow significant particle motion, then the microscope image of the particle will not be circularly symmetric, thus decreasing the ability to assign an accurate particle center. The shutter speed is therefore set so that a particle of radius a embedded in a fluid diffuses less than the spatial resolution, 20 nm, in time τ. The shutter speed in our experiment was adjusted from 1/500 to 1/2000s. Faster shutter speeds are feasible with our camera, however faster shutter speeds require higher light source intensity. In our experiment, we prefer to lower the light source intensity since too much light absorption can increase the temperature of the sample cell. A spatial "bandpass filter" routine, written by Crocker and Grier [96], was applied to each image to suppress noise and correct for background brightness variations. The particle positions in each frame were obtained using IDL software routines based on intensity weighting determination of the peak position [96].

We use a microscope objective heater to control sample temperature. The temperature control (Bioptechs) has 0.1 °C resolution (Figure 10.6). The small temperature difference (~0.3 °C) between the objective and the sample has been taken into account in calibration. With this device we were able to heat the sample very slowly, in 0.1 °C/steps; each temperature step typically equilibrated in a few minutes. Measurements were taken after the temperature had stabilized. Before starting the measurements, we cycled the samples one or two times near the melting point in order to relieve any possible shear stress built up during the particle loading process.

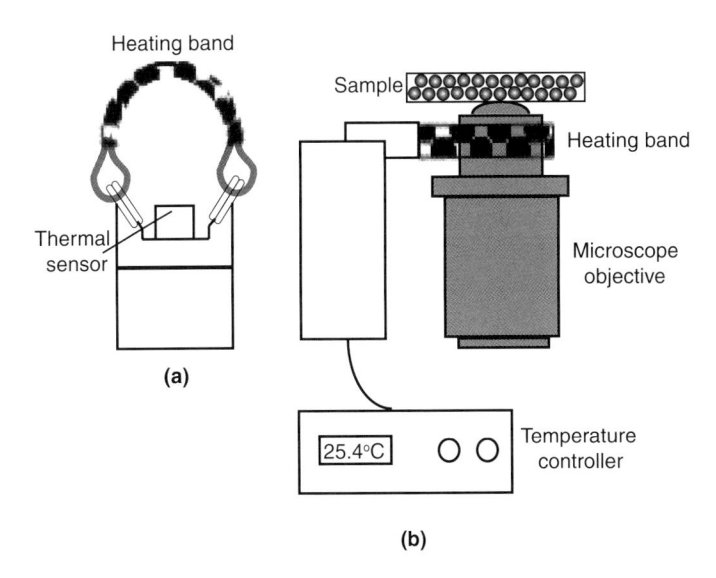

Figure 10.6 Schematic diagram shows microscope objective heating setup. (a) Heater and sensor. The heater is an adjustable thin-film heating band. A surface probe thermal sensor is designed to be in contact with the objective and measures temperature. (b) The objective heater is directly mounted onto the upper region of the microscope objective. The temperature is controlled via electronic temperature controller. This heater/sensor assembly is supported on an adjustable metal mounting to fit over microscope objectives. This configuration can be used on either upright or inverted microscopes.

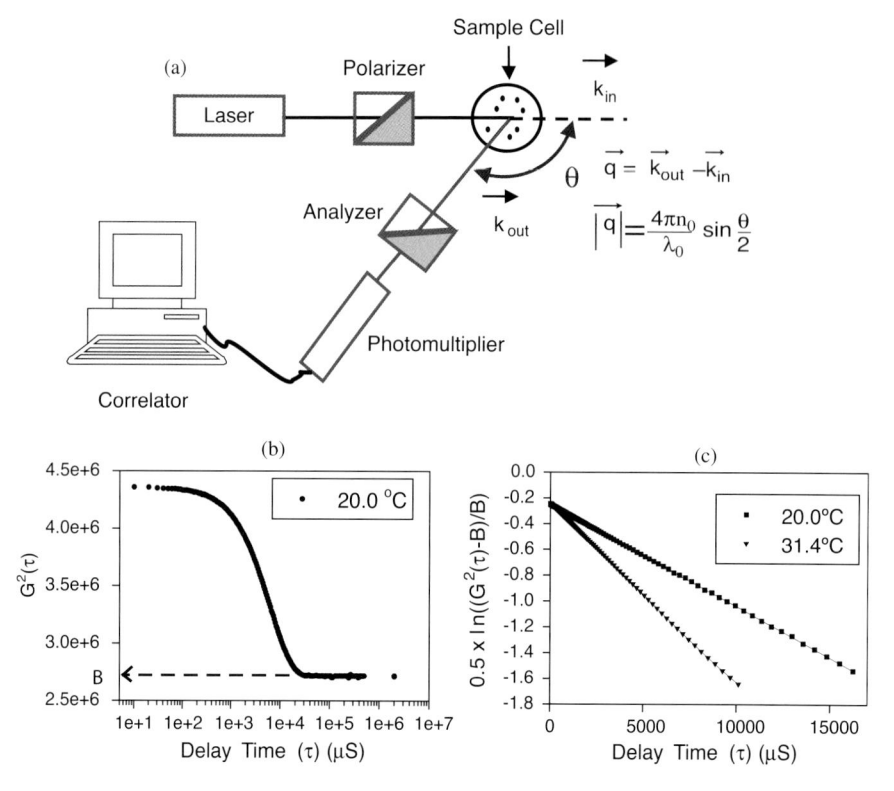

Figure 10.7 (a) Schematic of the dynamic light scattering (DLS) experimental setup. (b) The measured unnormalized intensity autocorrelation function as a function of delay time for a NIPA particle suspension at 20.0 °C. B is the baseline at large τ. (c) Processed data (i.e., electric field autocorrelation functions) at 20.0 and 31.4 °C. Notice that the temporal decay rate is larger at the higher temperature, indicating that the particles have smaller hydrodynamic radii at higher temperature.

10.2.3
Characterization: Dynamic Light Scattering

The hydrodynamic diameter of the microgel particles was measured by dynamic light scattering (DLS) [164, 165]. A schematic of the DLS setup is shown in Figure 10.7. The technique probes the Brownian motion of suspended particles in solution. The particles diffuse in water with diffusion constants that depend on the size of the particles. The light intensity scattered from particles will fluctuate in time as a result of particle diffusion. From the temporal intensity autocorrelation function of the scattered light, one can readily derive useful information about particle diffusion, which, in turn, can be related to particle "average" diameter and sample diameter distribution.

In practice, we measure the (unnormalized) temporal intensity autocorrelation function, $G^{(2)}(\tau) = \langle I(0)I(\tau)\rangle$, and we derive the average particle diffusion coefficient and the particle size distribution from this function. For a monodisperse suspension of particles, the quantity $(G^{(2)}(\tau)-B)/B$ decays exponentially; here, B is the asymptotic value of the intensity autocorrelation function as τ approaches infinity, and the quantity $(G^{(2)}(\tau)-B)/B$ is easily shown to be related to the temporal *electric field* autocorrelation function of the light scattered from the sample. For polydisperse suspensions, a distribution of "diffusion" decay rates will contribute to the autocorrelation signal. The field autocorrelation function will no longer decay purely exponentially. To second-order in the decay time τ, one can show that [164]

$$\frac{1}{2}\ln\left(\frac{G^{(2)}(\tau)-B}{B}\right) = A - \bar{\Gamma}\cdot\tau + \frac{\mu_2}{2}\cdot\tau^2 \qquad (10.1)$$

Here, $\bar{\Gamma}$ is the "average" exponential decay rate due to particle diffusion, μ_2 is the second-order coefficient of the decay due to particle polydispersity, and A is a constant.

For a dilute particle suspension, Γ depends on the Brownian diffusion of the particles,

$$\Gamma = D\cdot q^2 \qquad (10.2)$$

where $|\vec{q}|$ is the absolute value of the experimental momentum transfer, $|\vec{q}| = |\vec{K}_{out}-\vec{K}_{in}|$, and D is the average *particle* diffusion coefficient given by Stokes–Einstein relation,

$$D = \frac{k_B T}{6\pi\eta R_h} \qquad (10.3)$$

Here, k_B is the Boltzmann constant, T is temperature, η is the solvent viscosity, and R_h is the average particle hydrodynamic radius.

The particle size and the particle size distribution of our samples is obtained by fitting $0.5\ln\left((G^{(2)}(\tau)-B)/B\right)$ to a polynomial in τ. Then, the average hydrodynamic radius of the particles is obtained from Γ and the polydispersity is obtained from μ_2 (i.e., the variance of Γ). To first approximation, the variance of the hydrodynamic radius distribution is equal to the variance of Γ. This is a good approximation only for narrow particle size distributions, that is, $\Delta R/R \leq 0.30$. By fitting the data obtained from DLS to Equation 10.1, one obtains the polydispersity, P, of the particle hydrodynamic radius, that is,

$$P = \frac{\mu_2}{\bar{\Gamma}^2} \qquad (10.4)$$

Figure 10.8 shows the measured hydrodynamic diameter of a sample of microgel particles versus temperature. The DLS analysis shows that sphere polydispersity is lower than 3%. Such small polydispersity is not expected to appreciably affect the colloidal particle melting and crystallization processes [166].

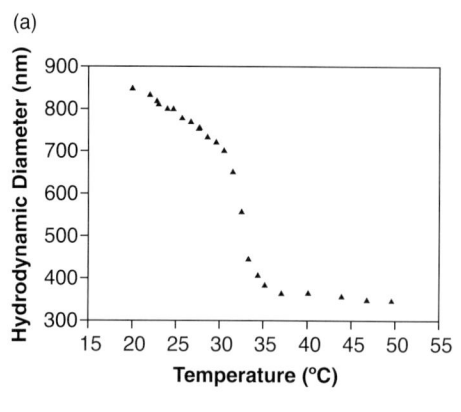

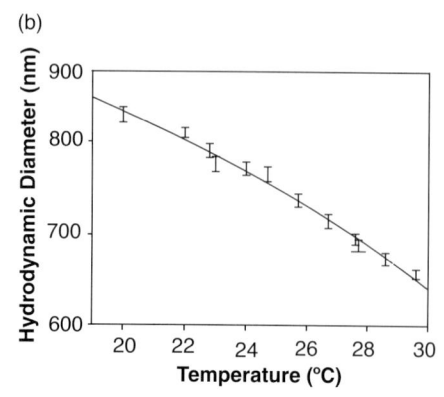

Figure 10.8 (a) Hydrodynamic diameter of NIPA particles as a function of temperature. The NIPA particle hydrodynamic diameter collapses at about 32.0 °C, corresponding to the coil–globule phase transition of the polymer. (b) A zoom in to the temperature range [20.0–30.0 °C] where the particle volume appears to linearly decrease with temperature (solid line fit).

10.2.4
Characterization: Video Microscopy Measurement of Interparticle Potentials

Particle interactions were derived from direct measurements of their radial distribution function (RDF), $g(r)$. The RDF describes how the density of surrounding particles varies as a function of the distance from a particle center placed at the origin. In a dilute (i.e., areal density ~10%) monolayer of spheres, $g(r = |\mathbf{r}|)$ is the azimuthal average of the pair correlation function

$$\tilde{g}(\mathbf{r}) = \frac{1}{n^2} \langle \varrho(\mathbf{r}' + \mathbf{r}) \varrho(\mathbf{r}') \rangle \tag{10.5}$$

Here, $\varrho(\mathbf{r}') = \sum_{j=1}^{N(t)} \delta(\mathbf{r} - \mathbf{r}_j(t))$ is the distribution of N particles in the field of view and n is the number density.

In bright field microscopy, though the centers of the individual spheres are characterized by brightness maxima, images will also contain weak alternating dark and bright diffraction fringes. Thus, when two particles are close enough for their fringes to overlap, the apparent (i.e., optically measured) pair separation $\tilde{r}$ deviates systematically from the true particle separation distance r by an amount $\Delta(r)$ [167]. It is important to correct for this effect. To this end, we employed the methods outlined in Ref. [167, 168] to quantify $\Delta(r)$. Briefly, the approach first finds an isolated particle in the colloidal gas monolayer and "cuts" a relatively large subarea around the particle so long as the image clip contains only one sphere. Then, the image clip is duplicated. The image clip and its duplicate are then placed side-by-side at a distance r, and a new "superimposed" intensity image of the "two" particles is thus derived. From the image analysis of this superposed image, we obtained a measured $\tilde{r}$ that

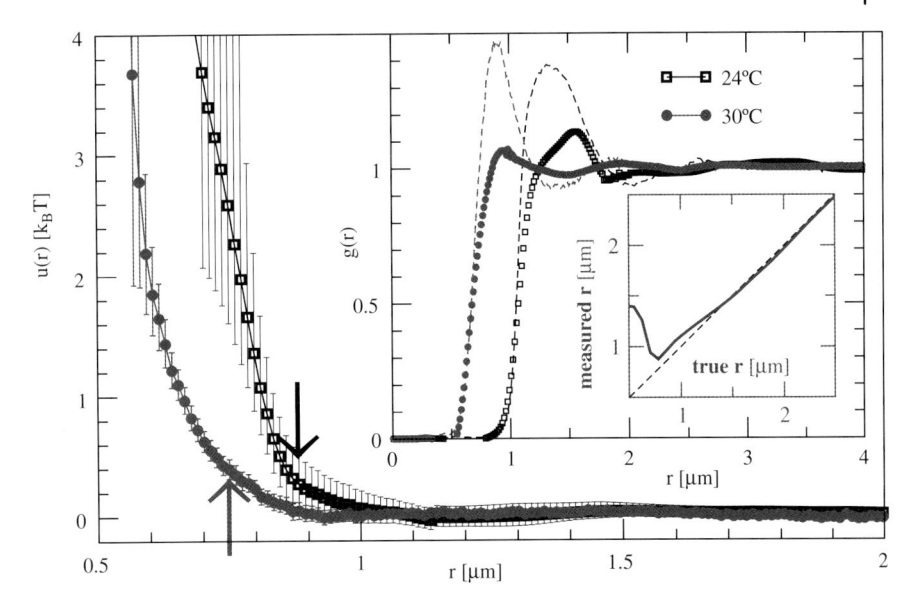

Figure 10.9 The pair potential $u(r)$ of NIPA spheres at 24 °C (square) and 30 °C (circle). Arrows indicate corresponding hydrodynamic diameter measured by dynamic light scattering. *Inset*: The raw $g(r)$ (dashed curve, with the ring artifact) and the corrected $g(r)$ (symbols) without the ring effect. *Small inset*: Measured pair separation with image artifact versus the true separation at 24 °C.

may slightly deviate from the true separation r due to the dark ring surrounded each sphere. We then repeat this process at various distances r, deriving quantitative information about how $\tilde{r}$ changes with r (see the small inset in Figure 10.9). To obtain good statistics, we repeat the above procedure for all isolated spheres in all frames. The undistorted pair correlation function $g(\mathbf{r})$ is then readily related to the directly measured $\tilde{g}(\tilde{\mathbf{r}})$ through conservation of probability [168],

$$g(\mathbf{r})\mathrm{d}r = \tilde{g}(\tilde{\mathbf{r}})\mathrm{d}\tilde{r} \tag{10.6}$$

As long as the sample's areal particle density is low enough to preclude substantial three-body overlap distortions, one can show that

$$g(\mathbf{r}) = \tilde{g}(\mathbf{r} + \Delta(\mathbf{r}))\left[1 + \frac{\mathrm{d}}{\mathrm{d}r}\Delta(\mathbf{r})\right] \tag{10.7}$$

where $\Delta(\mathbf{r})$ is the correction to the pair separation (see the large inset of Figure 10.9). Notice that this image artifact is a significant problem in bright field microscopy; its effect is much smaller in dark field microscopy and essentially disappears in fluorescence microscopy (i.e., using fluorescent spheres).

From $g(r)$, we apply liquid structure theory to extract the particle pair potentials [169, 170], $u(r)$, shown in Figure 10.9. We use the following relations [171]:

$$\frac{u(\mathbf{r})}{k_B T} = -\ln g(\mathbf{r}) + \begin{cases} nI & \text{(HNC)} \\ \ln[1 + nI(\mathbf{r})] & \text{(PY)} \end{cases} \qquad (10.8)$$

where n is the mean areal density of the particles.

The first term in the above equation is a Boltzmann relation that is valid for dilute systems where there are no many-body contributions to $g(\mathbf{r})$. For weak many-body contributions, the Boltzmann relation must be corrected. Here, we use the hypernetted chain (HNC) [172] or Percus–Yevick (PY) [172] approximations to correct for many-body effects. $I(r)$ is a convolution integral that can be solved iteratively starting with $I(r) = 0$,

$$I(\mathbf{r}) = \int_A [g(\mathbf{r}') - 1 - nI(\mathbf{r}')][g(|\mathbf{r}' - \mathbf{r}|) - 1] d^2\mathbf{r}' \qquad (10.9)$$

As shown in Figure 10.9, the potentials are short ranged and repulsive, which is consistent with a nearly hard-sphere description [173]. Different batches of microgels have similar $u(r)$. In liquid structure analysis, the HNC approximation is known to be accurate for "soft" potentials while the PY approximation is more accurate for short-ranged hardcore interactions. Here, we observed that both HNC and PY approximations yield almost the same $u(r)$, which suggests good accuracy. We quantitatively estimated other errors, including the statistical uncertainty, errors in image processing, the effects of 3% polydispersity, and the uncertainty stemming from the conversion from $g(r)$ to $u(r)$, which can be estimated as the difference between HNC and PY approximations [170]. The first two error sources dominate. The error bars in Figure 10.9 are total errors. Notice also that the effective particle diameter at $1\,k\,BT$ is $\sim$12% smaller than the hydrodynamic diameter measured by dynamic light scattering (see Figure 10.9). For soft spheres, the diameter is not clearly defined. Here, we use the hydrodynamic diameter R_h for defining the areal or volume fraction, which is the product of the hydrodynamic volume of the individual particle and the areal or volume number density. The areal number density can be measured by optical microscopy, while volume number density requires confocal microscopy.

10.3
"First" Melting in Bulk (3D) Colloidal Crystals

10.3.1
Background

Although melting and freezing are common phenomena in nature, the microscopic mechanisms involved in melting and freezing are still not well understood [174]. Scientists have speculated for more than a century about how crystalline solids melt [174–176], in the process generating microscopic models emphasizing the role of lattice vibrations [177, 178], dislocations [179, 180], grain boundaries [181, 182], surfaces [183–187], dimensionality [188], and combinations thereof.

Furthermore, experimental investigations to test underlying theoretical assumptions are difficult because they must track motions of individual atoms or defects within bulk crystals.

Microgel particle suspensions are especially suitable for the study of phase behavior in 3D. Their refractive index is very close to that of water; thus, both bright field and confocal microscopy can capture the behavior inside the bulk crystal without the detrimental effect of multiple scattering. Furthermore, the microgel particles are very nearly density matched to the solvent (water), so that sedimentation effects will be small. Finally, the microgel particles provide the unique opportunity to tune volume fraction and thus drive the phase transition within a single 3D sample. Compared to conventional colloids, wherein many samples with different particle concentrations must be prepared to study phase transitions, the temperature control characteristic of the NIPA microgel particle suspensions is attractive because it permits us to use one sample for all of the measurements; that is, it permits us to scan a range of volume fractions within essentially the same medium, while visualizing essentially the same sample volume.

A first-principles theory of the solid–liquid transition is not readily available. This theoretical situation contrasts significantly with that of the transitions that arise in ferromagnetism, two-dimensional melting, and liquid–vapor systems. Theoretical issues arise for conventional 3D melting from long-range many-body effects, symmetry, and a lack of universality. Recent experiments [189, 190] and theory [191] have shown that atomic crystal surfaces, at equilibrium below the bulk melting point, often form melted layers. This effect is sometimes referred to as premelting, the localized loss of crystalline order for temperatures below the bulk melting transition. It can be thought of as the nucleation of the melting process. "Premelting" lowers the energy barrier for liquid nucleation and effectively prevents superheating of the solid [189, 192].

Many theories have suggested that a similar "first" melting behavior should occur at defects within bulk crystals such as grain boundaries; however, these effects have not been observed. Simulations of grain boundaries [182, 193–197] have found that the free energy of the solid–solid interface can be larger than two solid–liquid interfaces, thereby favoring "premelting" near the grain boundary (GB). A hot-stage transmission electron microscopy (TEM) experiment [198] has also suggested that GB "premelting" may occur, but only at $0.999T_m$ for pure Al, where T_m is the bulk melting temperature.

In our experiment, we image the motions of particles in three-dimensional crystals during the melting process. These particles are micron-sized nearly hard-spheres [173]. The thermal response of the microgel permits precise control of particle volume fraction. At high volume fraction, these particles are driven entropically to condense into close-packed crystalline solids [88, 199], while at low volume fraction the particles are in the liquid state (see Figure 10.10). Thus by slightly changing sample temperature, we can precisely vary the volume fraction of particles in the crystal over a significant range, driving the crystal from close-packing toward its melting point at lower volume fraction. This general experimental approach enables us to learn about the nucleation of fluid phases inside bulk solids.

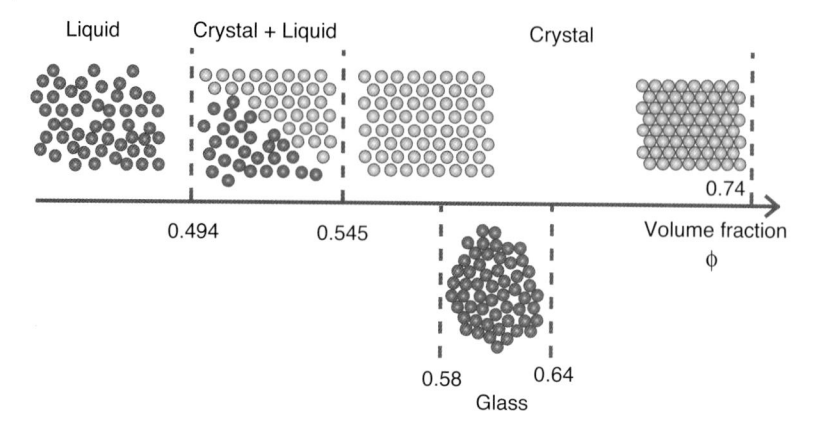

Figure 10.10 Phase behavior of monodisperse hard colloidal spheres.

Experimental images reveal "first" melting near grain boundaries and dislocations. Furthermore, particle tracking enables us to quantify the spatial extent of local particle fluctuations both near a variety of defects and within the more ordered parts of the crystal. Increased disorder and particle fluctuation is observed in regions bordering defects as a function of defect type (e.g., grain boundaries, dislocations, vacancies), distance from the defect, and particle volume fraction. These observations suggest that "first" melting of the grain boundaries is an important effect. Besides their intrinsic importance for colloid science and technology, all indications suggest interfacial free energy is the crucial parameter for melting. Thus, these results are also relevant for atomic-scale materials.

10.3.2
Sample Preparation and Imaging

The particle suspensions were loaded into the chamber using capillary forces at 28 °C, that is, just below the melting temperature. In this process, the suspension was sheared, giving the crystal a preferential orientation. Initially, we found that well-oriented fcc crystals grew from the glass coverslip surfaces and that the middle of the sample was glassy. After loading, we annealed the sample at 28 °C for 24 h during which the samples crystallized. Bragg diffraction (Figure 10.11a) from various parts of the annealed sample, measured using the microscope with a Bertran lens, exhibited no detectable change in peak positions. The colloidal crystal had very few defects close to the glass walls (see Figure 10.11c). We never observed melting near the walls; it is thus possible that the walls stabilized the crystal or that the (1 1 1) planes near the wall surfaces are intrinsically stable [192]. Interior crystalline regions had many more defects (see Figure 10.11d). A few defects in the sample interior are shown in Figures 10.14 and 10.16. Most of the defects observed were stacking faults, which caused the formation of partial dislocations (see Figure 10.16 [200]). Typically, the crystals lost their preferential orientation after melting and recrystallization,

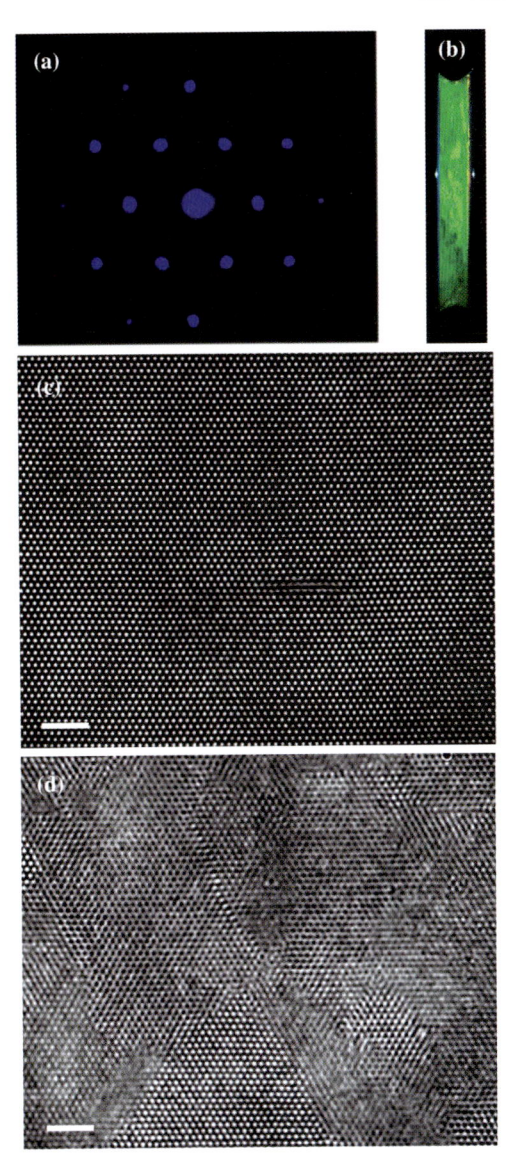

Figure 10.11 (a) Bragg diffraction (wavelength = 405 nm) of 0.75 μm diameter NIPA particle colloidal crystal. (b) Glass chamber containing NIPA particle suspension at high volume fraction. The color indicates that the suspension is in crystalline phase. (c) Bright field image of a layer in the crystal showing very few defects; the slice is of the seventh layer from the cover slip. Each bright spot corresponds to the central region of a 0.75 μm diameter particle. (d) Bright field image of the crystal showing many defects (part of the image is in focus while the other part is out of focus); the slice is of the 15th layer from the cover slip. Due to sample preparation and annealing, the primary defects are partial dislocations that exist in the interior of the crystal. Scale bars are 5 μm.

displaying large crystalline regions with different orientations separated by grain boundaries (Figure 10.14).

Experimental observations were made using an upright microscope (Leica DMRXA2) equipped with a 12-bit monochrome-cooled camera (QImaging RETIGA) and a motorized stage. The dimensions of the sample chamber were $18 \times 4 \times 0.1$ mm^3. The temperature of the sample and objective lens (100×1.4 NA) was controlled within 0.1 °C and was raised in 0.1 °C increments. Samples were left to equilibrate at each temperature for 1 h. In order to track melted regions and defects, we took bright field video images for 0.6 s in 100 nm intervals throughout the $\sim$100 μm thick chamber. In order to track individual particle movement, we employed a video shutter time of 2 ms. Image fields were chosen to contain $\sim$400 particles, and particle positions were determined at resolutions much smaller than the particle radius or crystal lattice constant [96]. Fifteen minutes of video were recorded at each temperature. (Note that, although bright-field images can potentially contain image artifacts, we have analyzed these effects [201] and are confident that none of the conclusions we make about the sample is very sensitive to the artifact.)

10.3.3
Positional Fluctuations and the Lindemann Parameter

Using the Lindemann parameter (L), a measure of the particle mean-square fluctuation, we have quantified one aspect of sample melting as a function of sample temperature and volume fraction. Figure 10.12a shows the time evolution of the particle mean square displacement (MSD),

$$\text{MSD}(\tau) = \langle |\vec{r}(t+\tau)-\vec{r}(t)|^2 \rangle \tag{10.10}$$

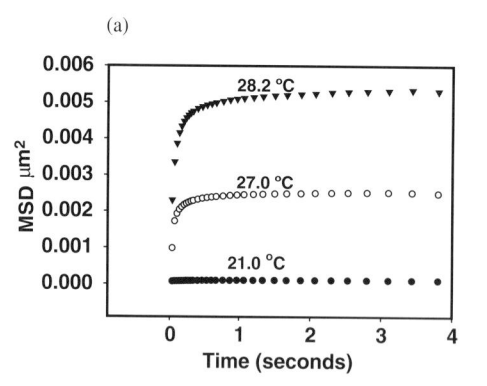

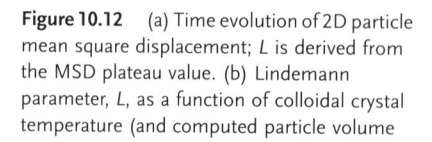

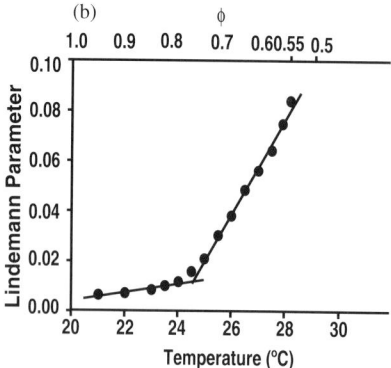

Figure 10.12 (a) Time evolution of 2D particle mean square displacement; L is derived from the MSD plateau value. (b) Lindemann parameter, L, as a function of colloidal crystal temperature (and computed particle volume fraction ϕ). These data are for regions far from defects. The curve exhibits a change in slope at 24.7 °C. The crystal melts at 28.3 °C, $\phi \sim 0.55$. The error bars for ϕ, L, and temperature are 0.02, 0.004, and 0.1 °C, respectively.

for particles in the bulk crystal at three different temperatures, where $\vec{r}(t)$ is the position of the particle center at time t measured from an arbitrary origin (usually the top left corner of the image). On short timescales, the MSD exhibits free particle diffusion; on long timescales, the particles are caged by nearest neighbors and the MSD asymptotically approaches a constant. From this plateau height of MSD measured in a 2D slice, we can calculate 3D Lindemann parameter (L_{3D}):

$$L_{3D} = \frac{1}{r_{nn}} \sqrt{\langle (r_{3D}(\tau \to \infty) - \bar{r}_{3D})^2 \rangle} = \frac{1}{r_{nn}} \sqrt{\frac{3}{2} \langle (r_{2D}(\tau \to \infty) - \bar{r}_{2D})^2 \rangle}$$
$$= \frac{1}{r_{nn}} \sqrt{\frac{3}{4} MSD_{2D}(\tau \to \infty)}$$
(10.11)

where r_{nn} is the lattice constant and $\bar{r}$ is particle's equilibrium position. Here, we assume that particle fluctuations are isotropic so that $MSD_{3D} = \frac{3}{2} MSD_{2D}$. The last step uses the fact that the asymptotic plateau height of MSD is twice the variance of the particle's displacement from its equilibrium position [202, 203].

10.3.4
Bulk Melting

Figure 10.12b shows the measured Lindemann parameter as a function of temperature. The Lindemann parameter experiences a change in slope at 24.7 °C. At this temperature, the hydrodynamic diameter of the particles measured by dynamic light scattering is ~754 nm, and the nearest neighbor distance derived from pair correlation functions measured by microscopy is ~750 nm (Figure 10.13). Thus at this temperature, it is reasonable to assume the particles are close-packed with a volume fraction $\Phi \sim 0.74$. Below this temperature, the particles "press" into one another and

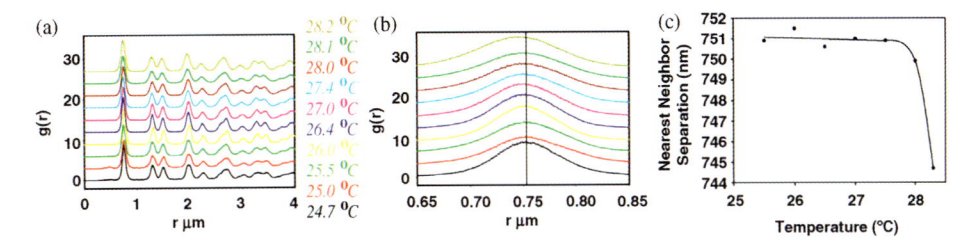

Figure 10.13 Radial distribution function of NIPA crystal at different temperatures. (a) First few peaks of g(r) versus r (b) The first peak of g(r) shows that the lattice constant changes very slightly as the temperature is increased up to the bulk melting temperature. The curves are displaced vertically for clarity. (c) The lattice constant (obtained from the position of the first peak of the pair correlation function) in the solid regions of a NIPA crystal as a function of temperature. Solid line is to guide your eye. Notice that at 28.2 °C the lattice constant starts to substantially decrease; this is approximately the same temperature that wetting appears at partial dislocation interfaces. Note that although the spatial resolution of the measurements is about 20 nm, the 5-nm peak position difference in (c) can still be well resolved if the statistics are good.

the particle motions are constrained; hence, L varies less strongly at low temperature. With this assumption we can deduce the particle number density from $\Phi(T) = n\frac{4}{3}\pi R^3(T)$, where n is the number density and $R(T)$ is particle radius as a function of the temperature (T). Putting $\Phi(24.7\,°C) = 0.74$ and $R(24.7\,°C) = (0.75/2)\,\mu m$ in the above formula, we can deduce particle number density per μm^3. This number density and the measured hydrodynamic radius determine the particle volume fraction (using the above formula), as a function of temperature (upper scale of Figure 10.12b). A particle volume fraction of ~ 0.54 corresponds to a temperature of 28.3 °C. Our *best* samples (i.e., samples with "lowest" disorder) contain very few grain boundaries and noticeably begin to melt at 28.3 °C. Nucleation of melting in these samples is at their center, where the largest concentration of partial dislocations exists. Upon melting, the sample is essentially composed of liquid regions and very ordered (nearly defect-free) crystalline regions. Interestingly, at 28.3 °C the particle volume fraction (based on the assumptions above) is ~ 0.54, close to the hard sphere melting prediction of 0.545 [130].

An alternative approach for establishing bulk melting may be derived from measurements of crystal lattice constant. The lattice constant in the solid regions is derived from our measurements of the particle pair correlation function. It is observed to decrease sharply near 28.3 °C (Figure 10.13c). Again, while our particles are not perfect hard spheres [173], the melting point suggests they may be approximated reasonably well as such.

10.3.5
"First" Melting Near Grain Boundaries

One of the common melting mechanisms exhibited by our colloidal crystals is illustrated in Figure 10.14. In crystals with grain boundaries, the grain boundary interfaces start to disorder at temperatures measurably below 28.3 °C. The figure shows a small angle (i.e., $\sim 13°$) grain boundary. The grain boundary is composed of an array of dislocations, one of which is shown in the inset of Figure 10.14a. Notice that the number of particle nearest neighbors along the grain boundary varies from five to seven (red and blue particles in the inset). These packing mismatches create stress in the crystal near the grain boundary. The dashed line in Figure 10.14a shows a Shockley partial dislocation that continues into the grain boundary. The region to the right (left) of the dashed line is out of focus (in-focus), and the particles in this portion of the image appear darker (whiter) than average.

Figure 10.14b shows the same region at higher temperature (i.e., lower particle volume fraction). In order to minimize the interfacial free energy caused by stress and surface tension, particles near the grain boundary start to melt. The inset of Figure 10.14b shows these particles rapidly jumping from one site to another. In contrast, melting is not observed near the partial dislocation (dashed line); its interfacial free energy is apparently less than that of the grain boundary. In Figure 10.14c, the temperature is slightly higher and melting has erupted along the grain boundary. At this stage, the sample volume fraction is higher than the bulk melting particle volume fraction, and the melted region has engulfed the partial

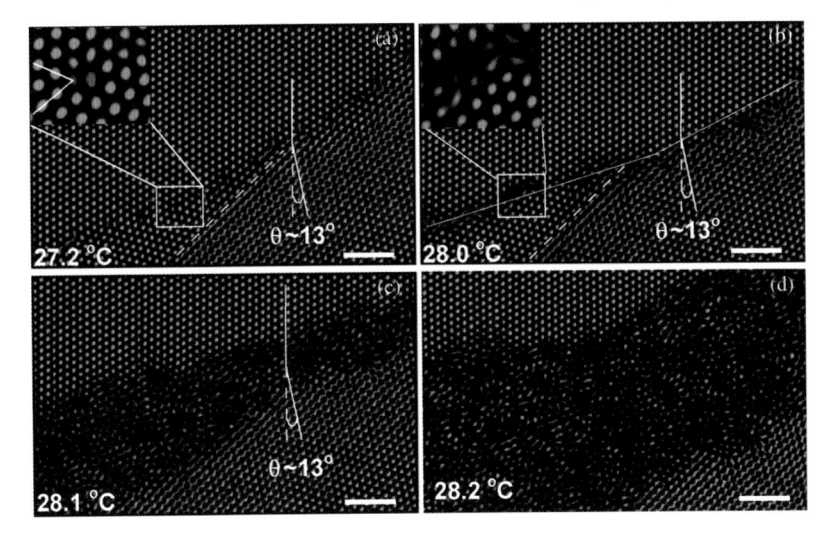

10.3 "First" Melting in Bulk (3D) Colloidal Crystals

Figure 10.14 Melting of the colloidal crystal at a grain boundary. The figure shows bright field images at different temperatures (i.e., particle volume fractions) of two crystallites separated by a grain boundary ($\theta\sim 13°$). (a) Sample at 27.2 °C. The solid and dashed lines show the grain boundary and a partial dislocation, respectively. The grain boundary cuts the two crystals along two different planes (yellow line has two slopes). It is composed of an array of dislocations; the two extra planes are indicated by lines in the inset. (b) Sample at 28.0 °C. The grain boundary starts to melt; nearby particles undergo liquid-like diffusion, inset. The partial dislocation, denoted by the dashed line, is not affected. (c and d) The same sample at 28.1 and 28.2 °C, respectively. The width of the melt region near the grain boundary increases. Scale bars are 5 µm.

dislocation. The width of the melted region continues to increase as the temperature is raised from 28.0 to 28.2 °C (see Figure 10.14b–d).

Melting is a heterogeneous process, that is, not all interfaces melt at the same temperature. In addition, the number of melted layers depends on the crystalline surface. Figure 10.15a shows a grain boundary that separates two crystallites (right and left). The boundary cuts the two crystallites at two different crystalline surfaces. Each cut has different interfacial energy. Below the bulk melting temperature of

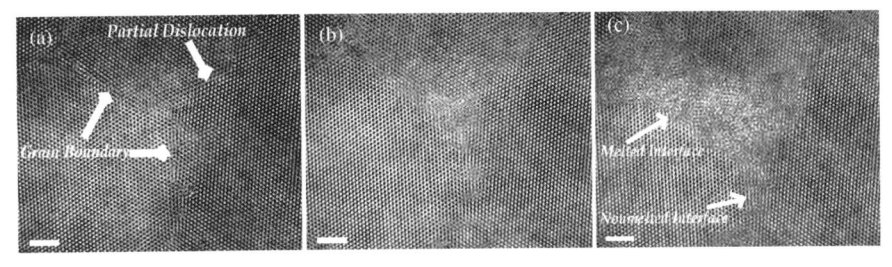

Figure 10.15 (a) Bright field image of a bulk layer in the NIPA particle colloidal crystal showing grain boundaries at 26.4 °C. (b and c) The same layer at 28.2 °C and at 28.3 °C. Scale bars are 5 µm.

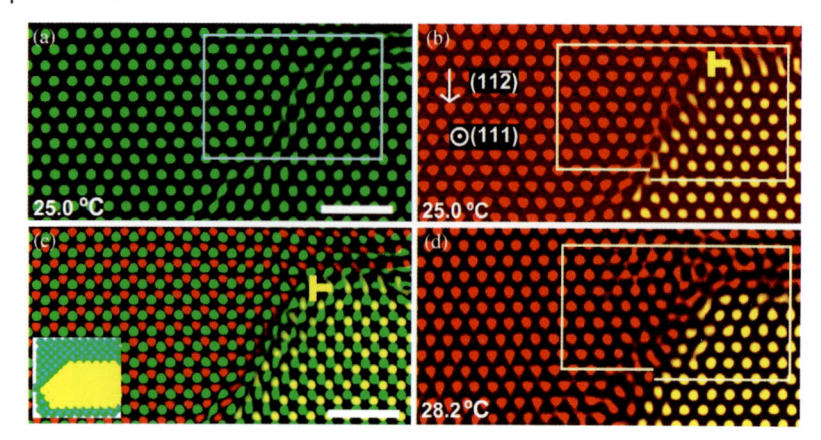

Figure 10.16 Melting of a colloidal crystal initiated at a Shockley partial dislocation in the absence of grain boundaries. (a) and (b) are bright field images of the 61st and 62nd layers at 25.0 °C, respectively. Colloidal particles fluctuate more in the 62nd layer due to the gap created by the dislocation. (c) Superposition of 61st (green) and 62nd (red and yellow) layers. The image shows particles in positions A (green), B (red), and C (yellow). Inset, 3D illustration of the 61st–64th layers (bottom to top) showing the displacement of the yellow spheres in the 62nd–64th layers. (d) 62nd layer at 28.2 °C where the crystal starts to melt at the dislocation. Scale bars are 3 μm.

28.3 °C, "first" melting nucleates at certain positions along the grain boundary (see Figure 10.15b). At higher temperatures (lower volume fraction), the number of melted layers increases at the crystalline interface, while remaining unmelted near other interfaces (see Figure 10.15c).

10.3.6
"First" Melting Near Dislocations

In addition to grain boundary melting, the colloidal crystals display melting from partial dislocations (Figure 10.16). This effect is more apparent when the grain boundaries are relatively far from the partial dislocations. Figure 10.16a and b show images of the 61st layer (green) and the 62nd layer (red and yellow) of the colloidal crystal at 25.0 °C, respectively. Figure 10.16c shows a superposition of these layers. Both of these layers represent (1 1 1) planes in the crystal. The Burger's circuit in the 61st layer (green) yields a zero Burger's vector, indicating no defect in the layer. Since a dislocation is present in the next layer, some of the particles are slightly out of focus. The Burger's circuit for the 62nd layer (yellow) reveals a Shockley partial dislocation with a Burger's vector of $\frac{1}{6}(\bar{1}\bar{1}2)$ [200]. The inset contains a three-dimensional illustration of the Shockley dislocation, showing the 61st layer and the undisplaced particles in the 62nd–64th layers in green, and the displaced particles in the 62nd–64th layers in yellow.

In monodisperse nearly hard-sphere colloidal crystals, the difference in energy between face-centered cubic (fcc) and hexagonal close packed (hcp) structures is very small [204, 205], and stacking faults are very common [88]. Shockley partial dislocations arise as a result of these stacking faults. Face-centered cubic crystals stack in the pattern ABCABC along the (1 1 1) direction, and hexagonal close-packed crystals stack in the pattern ABAB. The green particles in Figure 10.16a are in the A positions, while the red and yellow particles are in the B and C positions of the next layer, respectively. This stacking fault opens up gaps between the two close-packed structures within the crystal (two gaps are visible in the image and make an angle of 120° with respect to one another). Nearby particles fluctuate into and out of these gaps. The angle the gaps make with the (1 1 1) plane suggest the gaps cut the crystal along (1 0 0) planes as shown in the three-dimensional illustration. Finally, Figure 10.16b shows the 62nd crystal layer at 28.2 °C. At this temperature, which is higher than the grain boundary melting temperature, the crystal has begun to melt from the partial dislocation.

10.3.7
Positional and Angular Fluctuations Near Defects

The Lindemann melting criterion, which predicts melting for $L \sim 1/8$ [177], continues to provide a useful benchmark nearly 100 years after it was originally suggested. The data for L in Figure 10.12b are taken from deep within the crystalline regions of the sample, below the melting point. At 28.3 °C the sample begins to melt, and a coexistence of liquid and solid domains is readily observed. In Figure 10.17, we show local measurements of L near various crystalline defects and near the melt boundary just before bulk melting (28.3 °C). We find that the particle fluctuations in the proximity of these regions are measurably larger than in the bulk crystal. Furthermore, we find the magnitude of these fluctuations to decrease approximately

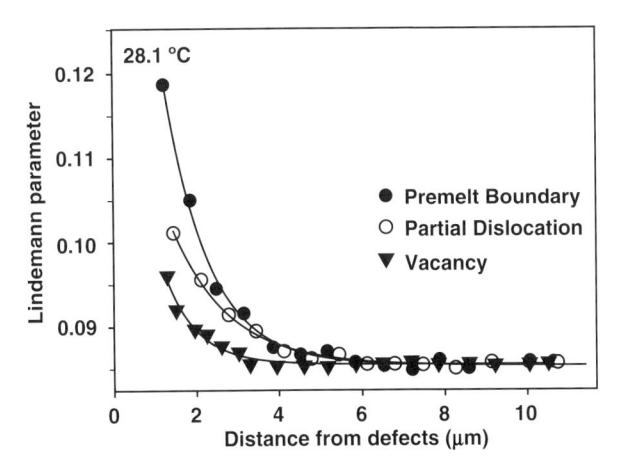

Figure 10.17 The local Lindemann parameter, L, as a function of distance from a vacancy, a partial dislocation, and a melt front. Within 1 μm of the defects, the particle motion was too rapid and calculation of L was unreliable.

exponentially as the measurement position is translated away from the melt region toward the interior of the bulk crystal. Extrapolation of our exponential fits of L to zero distance suggest that $L \sim 0.18$ in the melt region, twice its interior value of ~ 0.085 at the same temperature. Evidently, the greater number of vacancies in the melted region increases the free volume for particle movement so that the nearby particle fluctuations are large. Even the particles near isolated vacancies have large L, but the decay length of L to bulk values is shortest from isolated vacancies.

Orientational order is also useful for characterizing different phases in condensed matter systems [206]. For our sample with spherical particles, the orientational order can be measured from the bonds between nearest neighbors via Delaunay triangulation [206, 207]. Delaunay triangulation for a set of points in a plane is a triangulation such that there is no point inside the circumcircle of any triangle. It is the dual graph of the Voronoi diagram. Figure 10.18 shows the Delaunay triangulation near a dislocation.

Notice that instead of viewing the system as a set of vertexes, we can equivalently view it as a set of *bonds* and study the structure and dynamics of these bonds. The static structure of bonds can be quantified by the spatial correlation functions of their orientation [207], which are long ranged in a crystal. Bond dynamics can also be

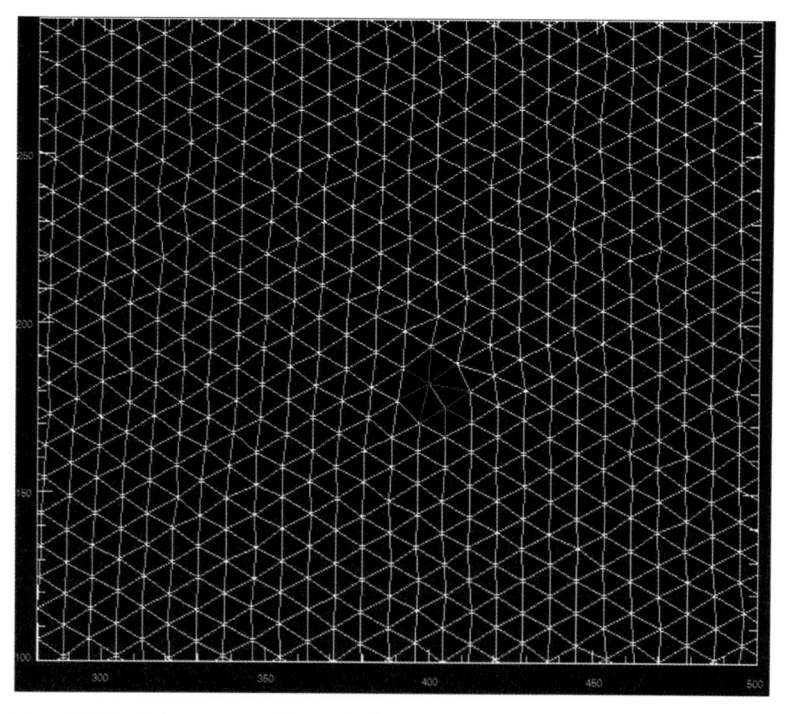

Figure 10.18 Delaunay triangulation of a layer in the bulk colloidal crystal. Colored bonds are associated with defect vortexes with non-six nearest neighbors: white 6–6; blue 7–6; yellow 5–7; red 5–6. Colored bonds show a 5–7 vortex pair, that is, a dislocation.

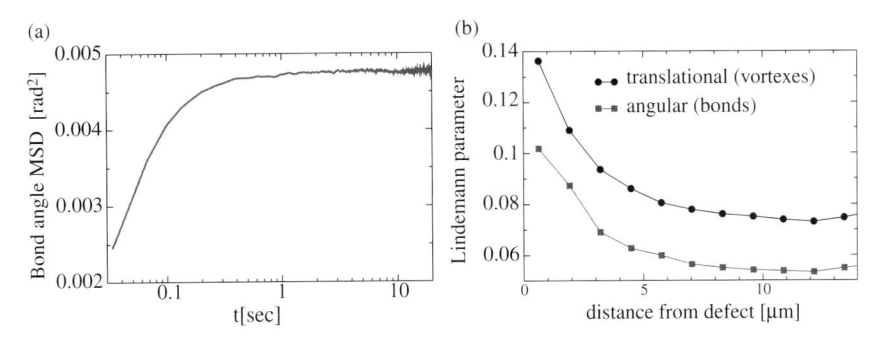

Figure 10.19 (a) Angular MSD of bonds in the crystalline region of Figure 10.18. (b) Translational and angular Lindemann parameters near the dislocation in Figure 10.18. Measurements were taken right before melting temperature. Both quantities decay to crystal values on the same length scale.

characterized by time correlations or bond angle MSDs. Here, we introduce the bond angle MSD, $\text{MSD}_\theta(t) = \langle (\theta(t+\tau)-\theta(\tau))^2 \rangle$, as a means to characterize bond dynamics. To our knowledge, the notion of bond angle MSD has barely been explored. As in the case of the translational MSD, the angular MSD diverges in the liquid phase and converges to a plateau in crystal phase, see Figure 10.19a.

Neighbor particles are defined from the Delaunay triangulation of the *first video frame* at $\tau = 0$, and we keep tracking the bonds of these neighbor pairs at $\tau > 0$. Thus in the liquid phase, the bond angle of two particles can change by more than 2π after long enough time and yield a diverging angular MSD. Furthermore, we can define an angular Lindemann parameter (L_θ) that is similar to the translational Lindemann parameter in Equation 10.11,

$$L_\theta = \frac{\sqrt{\frac{3}{4}\text{MSD}_\theta\,(t \to \infty)}}{\pi/3} \tag{10.12}$$

where the normalization factor $\pi/3$ is taken from the angle of two neighbor bonds at low temperature in a perfect triangular lattice. Figure 10.19b shows that the angular Lindemann parameter is about 30% lower than the translational Lindemann parameter. Interestingly, both parameters increase in parallel near the defect. Thus, the angular Lindemann parameter can also be used as a criterion for melting and is worthy of further exploration.

10.3.8
Summary

We have demonstrated that "first" melting occurs at grain boundaries and dislocations located within bulk colloidal crystals. The crystals are equilibrium close-packed three-dimensional colloidal structures made from thermally responsive microgel spheres. The thermal response of the microgel enables precision control of particle volume fraction. We used real-time video microscopy to track each particle. Our observations confirm an important mechanism for theories of melting. The amount of "first" melting depends on the nature of the interfaces and defects. Particle tracking has enabled us to study particle fluctuations both nearby and far from these defects in ways that are inaccessible to experimental probes of atomic crystals, revealing the excess free energy in these regions through higher values of the Lindemann parameter. Our observations suggest interfacial free energy is a crucial parameter for melting, in colloidal and atomic scale crystals.

10.4
Melting in Two Dimensions: The Hexatic Phase

10.4.1
Theoretical Background

Two-dimensional matter is qualitatively different from 3D matter. Questions about the existence of 2D crystals were first raised theoretically in the 1930s by Peierls [208] and Landau [209, 210]. They showed that thermal fluctuations should destroy long-range order, resulting in the melting of any 2D lattice at any finite temperature. Mermin and Wagner further proved that a long-range order in magnetic systems could not exist in both one and two dimensions [211], and they later extended the proof to crystalline order in 2D [212]. However, although true long-range translational order does not exist in 2D, the translational correlations can be quite extended, that is, extended over the finite sample size. In this case, the translational order is said to be quasi-long-range. In addition, 2D systems can have another long-ranged order called *orientational order* [207]. The behaviors of these features distinguish 2D "crystals" from 3D crystals and liquids (see Table 10.1).

The most popular theory for melting in 2D is Kosterlitz, Thouless, Halperin, Nelson, and Young (KTHNY) theory [207, 213–215] that predicts two-stage melting from a 2D crystal to hexatic phase and then from hexatic to liquid phase via two continuous KT transitions (see Figure 10.20). This theory has been confirmed in simulation and experiment for particles with long-range interactions, but is not as

Table 10.1 Translational and orientational orders of different phases.

	Translational order	Orientational order
3D crystal	Long	Long
2D crystal	Quasi-long	Long
Hexatic phase	Short	Quasi-long
Liquid	Short	Short
Buckled 2D crystal	Short	Long

Orders are characterized by correlation functions of the order parameters as shown in Table 10.3. The correlation functions defined in Table 10.2 can have three types of behaviors: approaching a finite constant (long-range order), power law decay (quasi-long-range order), and exponential decay (short-range order).

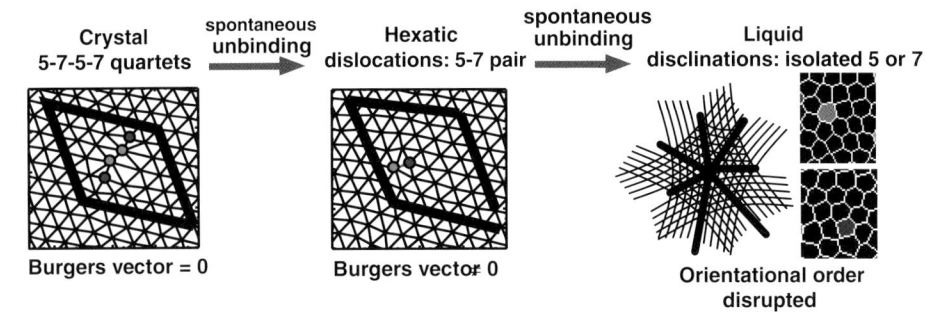

Figure 10.20 Schematic of 2D phases based on KTHNY theory.

well understood in systems with short-range interactions. Here, we study the melting of 2D microgel colloidal crystals confined between two glass walls. KTHNY predictions are thus quantitatively tested in this *short-range* interaction system.

The intermediate hexatic phase of KTHNY theory has short-range translational and quasi-long-range orientational order, and the two transitions from crystal to hexatic phase and then from hexatic to liquid phase are characterized by topological defects, see Figure 10.20. For 2D triangular lattices, nearest neighbors can be characterized by the Delaunay triangulation or its dual structure, the Voronoi diagram. Particles with nearest neighbor $n_n \neq 6$ are considered to be defects. Isolated $n_n = 5$ or 7 defects are called disclinations that can disrupt both translational and orientational order. Isolated 5–7 pairs are dislocations that disrupt the translational order by producing a nonzero Burgers vector, while still preserving the orientational order by keeping the lattice orientation unchanged. KTHNY theory suggests that the creation of free dislocations drives the system from crystal to hexatic phase, and the creation of free disclinations drives the transition from hexatic to liquid phase.

The phases and transitions are characterized by different behaviors of the translational and orientational order parameters. Their fluctuations and correlations are defined in Table 10.2 and its caption. In the local translational order parameter, **G** is the optimal vector that maximizes the order parameter. For crystals, it is simply a reciprocal lattice vector. Usually, **G** is chosen to be a primary reciprocal lattice vector derived from the peak of the structure factor. In the liquid phase, reciprocal lattice vectors are not defined, so we use **G** of the crystal phase to compute ψ_T in the liquid and hexatic phases. This approach has been employed previously [216–218]. To assign accurately **G**, we maximized the sample's translational order parameter ψ_T at each temperature (including the liquid phase), by iteratively varying **G** around the initial estimate.

Orientational order is measured from the "bonds" between nearest-neighbor particles. ψ_{6j}^b in Table 10.2 is a complex local bond orientational order parameter for systems with sixfold symmetry. It is calculated from the position and orientation of the bond j between two nearest-neighbor particles. The triangular crystal's spatial orientational order is reflected by $g_6^b(\mathbf{r}_{jk}) = \langle e^{i6(\theta_j - \theta_k)} \rangle$ between bonds j and k with bond center separation $\mathbf{r}_{jk}$. Alternatively, the local orientational order parameter can

Table 10.2 Translational and orientational order parameters, susceptibilities, and correlations.

	Translational order	Sixfold orientational order
Local order parameter	$\psi_{Tj}(\mathbf{r}) = e^{i\mathbf{G}\cdot\mathbf{r}_j}$	Bond: $\psi^b_{6j}(\mathbf{r}) = e^{i6\theta_j(\mathbf{r})}$
		Particle: $\psi_{6j}(\mathbf{r}) = \left(\sum_{k=1}^{n_n} e^{i6\theta_{jk}}\right)/n_n$
Global order parameter	$\psi_T = |\langle\psi_{Tj}\rangle|$	$\psi_6 = |\langle\psi_{6j}\rangle|$
Susceptibility (fluctuation)	$\chi_T = \lim_{A\to\infty} A(\langle|\psi_T^2|\rangle - \langle|\psi_T|\rangle^2)$	$\chi_6 = \lim_{A\to\infty} A(\langle|\psi_6^2|\rangle - \langle|\psi_6|\rangle^2)$
Spatial correlation	$g_T(r = |\mathbf{r}_i - \mathbf{r}_j|) = \langle\psi^*_{Ti}(r_i)\psi_{Tj}(r_j)\rangle$	$g_6(r = |\mathbf{r}_i - \mathbf{r}_j|) = \langle\psi^*_{6i}(r_i)\psi_{6j}(r_j)\rangle$
Time autocorrelation	$g_T(t) = \langle\psi^*_{Ti}(t_0)\psi_{Ti}(t_0 + t)\rangle$	$g_6(t) = \langle\psi^*_{6i}(t_0)\psi_{6i}(t_0 + t)\rangle$

G is the optimal vector that maximizes the order parameter. For crystal, **G** is a primary reciprocal lattice vector. Orientational order can be equivalently represented by the particle orientational order parameter ψ_{6j} and by the bond orientational order parameter ψ^b_{6j}. n_n is the number of nearest neighbors of particle j at position $\mathbf{r}_j = (x_j, y_j)$; θ_{jk} is the angle of the bond between particle j and its neighbor k; for ψ^b_{6j}, θ_j is the orientation of bond j and $\mathbf{r}$ is the position of the bond center; $\langle\rangle$ is the ensemble average over all particles or bonds; A is the system area; and t is the time.

be defined based on particles as shown in Table 10.2. It is simply the average of ψ^b_{6j} over all bonds of particle j. Both ψ_{6j} based on particles and ψ^b_{6j} based on bonds yield the same global order parameter, susceptibility, and correlation functions. We calculated both to double check our data analysis. The global order parameters range from 0 (i.e., totally disordered) to 1 (i.e., perfectly ordered).

The fluctuation of a global order parameter is called its susceptibility χ [219, 220]. For example, χ_T measures the response of the translational order parameter to sinusoidal density fluctuations with periodicity characterized by the primary reciprocal lattice constant **G**. The divergence or discontinuity of order parameter fluctuations, that is, the divergence or discontinuity of its susceptibility, signatures a phase transition. We found that such divergences or discontinuities can be used to determine the phase transition points accurately, avoiding ambiguities inherent to many other analyses. The traditional way to distinguish different phases is from the shapes of the order parameter correlation functions. Correlating single-particle ψ_6 or ψ_T in space or in time yields four correlation functions shown in Table 10.2. The predicted functional forms of these correlation functions in KTHNY theory are listed in Table 10.3. Note, $g_T(r)$ and $g_6(r)$ are two-body quantities, and $g_T(t)$ and $g_6(t)$ are one-body quantities.

Table 10.3 KTHNY predictions for order parameter correlations in three phases.

	Solid	Hexatic	Liquid
$g_6(r)$	Nonzero constant at $r \to \infty$	$\sim r^{-\eta_{6r}}$, $0 < \eta_{6r} \leq 1/4$	$\sim e^{-r/\xi_{6r}}$
$g_6(t)$	Nonzero constant at $t \to \infty$	$\sim r^{-\eta_{6t}}$, $0 < \eta_{6t} = \eta_{6r}/2 \leq 1/8$	$\sim e^{-t/\tau}$
$g_T(r)$	$\sim r^{-\eta_{Tr}}$, $1/4 < \eta_{Tr} \leq 1/3$	$\sim e^{-r/\xi_{Tr}}$	$\sim e^{-r/\xi_{Tr}}$

10.4.2
Experimental Background

Experimenters have sought out KTHNY predictions across a wide range of materials including monolayers of molecules and electrons [221], liquid crystals [222], vortex lattices in superconductors [223], diblock copolymers [224], and colloidal suspensions [225–231]. Compared to other materials, colloids have the advantage of measurable single-particle dynamics in real space. Some experiments and simulations have demonstrated substantial agreement with KTHNY theory, but others exhibit deviations and ambiguities possibly due to finite size effects [221], interaction range and form [227], and out-of-plane fluctuations [216, 220]. Charged colloids with screened Coulomb repulsion were the first colloidal systems used in the search for the hexatic phase. In these experiments [225, 232, 233], charge stabilized colloidal suspensions are filled in chambers with two smooth glass walls. The two glass walls form a wedge with a small angle. The spheres repel each other and are repelled by the glass plates due to surface charges, so that squeezing the plates closer together lowers the areal density of the spheres in between. Thus, the gradual change in plate separation induces gradual density gradient along the wedge. This system exhibited the hexatic phase, but the topological defects found in the system were complex and were not consistent with the simple KTHNY picture [225]. Nearly hard spheres with very short-range *attraction* [226] have been studied in a series of samples with different packing fractions; in this case, two first-order transitions and a middle hexatic phase were reported [226]. However, similar colloids with purely short-range *repulsion* did not exhibit a hexatic phase during melting [227]. Arguably the most well defined system to investigate the hexatic phase are monolayers of magnetic spheres with long-range (tunable) dipole repulsion [228, 229]. Many KTHNY predictions have been confirmed within this system class including the structural properties [228], the dynamics [229], and the crystal's elasticity [230]. Thus, the preponderance of simulation and experimental evidence clearly points to the validity of the KTHNY scenario in 2D systems with long-range interaction potentials [228–230, 234]. The evidence is less convincing, however, in systems with *short-range interactions* [219, 235–238a]. Note that KTHNY theory does not guarantee a scenario with a middle hexatic phase, and different 2D systems may have different melting paths as shown in the phase diagrams of Figure 2.21 in Ref. [207].

The NIPA microgel particles are ideal to study 2D melting. The pair potential between particles is short ranged and repulsive, and temperature tuning can be used to vary sample volume fraction and drive the melting transition. Furthermore, temperature-sensitive particles enable us to follow the spatiotemporal evolution of the *same* sample area through the entire sequence of transitions. This feature is attractive and was not realized in previous colloidal samples that employed charged spheres in the wedge geometry with density gradients [225] or in the more hard-sphere-like systems that employ a series of concentration-dependent sample cells [226]. In the wedge geometry, different densities are achieved at different wall separations that can affect correlations and the phase behavior. The same problem exists when making a series of samples; the samples are hard to make uniformly and it is also difficult to accurately measure and control wall separations in different cells [226]. Finally, the nontunable systems are more easily trapped in metastable glassy states [226]. By contrast, our temperature-sensitive samples start in the equilibrium 2D crystal phase and re-equilibrate rapidly after each tiny (0.2 °C) temperature step. They are, therefore, far less likely to be trapped in metastable glassy states during melting. Again, although the recent experiments using magnetic spheres with tunable dipole–dipole interactions [228–230] share some of these advantages, they also differ from the microgel experiments in a complementary way as a result of their *long-range* dipolar interactions.

10.4.3
2D Samples

Samples consisted of a monolayer of NIPA spheres confined between two glass coverslips (see Figure 10.21). Dynamic light scattering measurements found the NIPA sphere hydrodynamic diameter to vary from 950 nm at 20 °C to 740 nm at 30 °C. Note that we conducted this experiment with larger NIPA particles than the particles that are used in the 3D melting experiment (see Figure 10.8); however, the behavior of both samples is similar as a function of temperature. The cleaned glass surfaces were coated with a layer of 100 nm diameter NIPA spheres to prevent particle sticking. A simple geometric calculation showed that 100 nm close-packed spheres on the surface give rise to 3 nm surface roughness for the 800 nm diameter spheres. This surface roughness is negligible compared to sphere polydispersity and wall separation fluctuations. In addition, our observations of the large sphere motions at lower concentrations did not find evidence for preferential spatial locations, that is, significant surface potentials. We increased the temperature from 26.5 to 28.5 °C in 0.2 °C steps and recorded 5 min of video at each temperature.

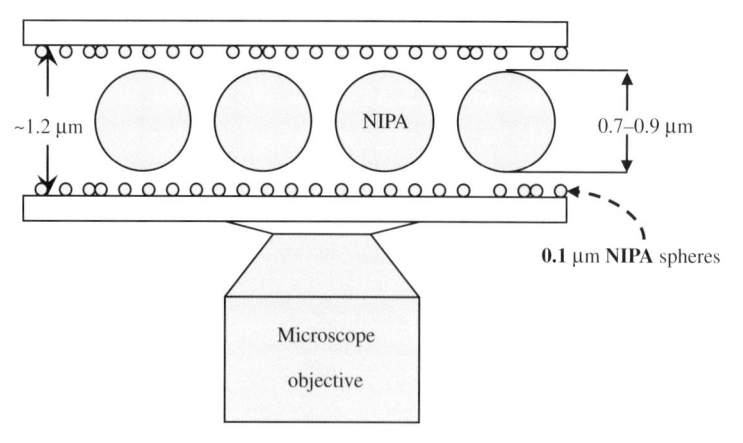

Figure 10.21 Schematic of 2D sample cell. A uniform layer of 0.1 μm diameter NIPA spheres is coated on the inner surfaces in order to avoid large NIPA particles sticking on glass walls. (The microscope objective is not drawn to scale.)

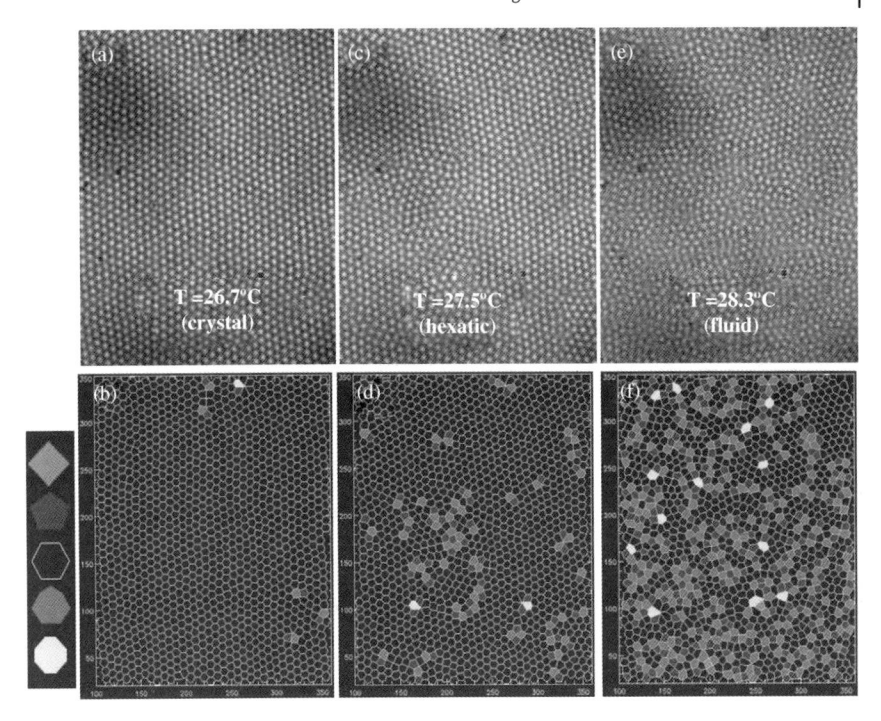

Figure 10.22 Real-space images of 2D NIPA microgel crystals at (a) crystal, (b) hexatic, and (c) liquid phases. (b,d,f) are the corresponding Voronoi diagrams of (a,c,e), respectively. Particles without six nearest neighbors are labeled by dark gray polygons.

The dense monolayers of 800 nm spheres formed crystal domains within the sample cell of typical size $(40\,\mu m)^2$, corresponding to $\sim$3000 particles. Measurements were carried out on a $(20\,\mu m)^2$ central area, well separated from the grain boundaries, see Figure 10.22a, c, and e. In practice, we found that grain boundaries affected only a few neighboring lines of particles and that melting started almost simultaneously throughout the crystal, both from inside the crystal domains and at the grain boundaries. This behavior differs qualitatively from grain-boundary melting in 3D [141] and edge melting in 2D [174], wherein the melting starts from grain boundaries or edges and then propagates into the crystal. Our observations suggest that the interfacial energies for liquid nucleation from within crystal domains might be similar to those of liquid nucleation starting from grain boundaries.

10.4.4
Data Analysis

Figure 10.23 shows typical particle trajectories in the three phases. From the $g_6(r)$ shown in Figure 10.24a, we can semiquantitatively distinguish three regimes corresponding to crystal, hexatic, and liquid as predicted by KTHNY theory: $g_6(r) \sim$ constant (long-range orientational order) for 26.5–26.9 °C, $g_6(r) \sim r^{-\eta_{6r}}$

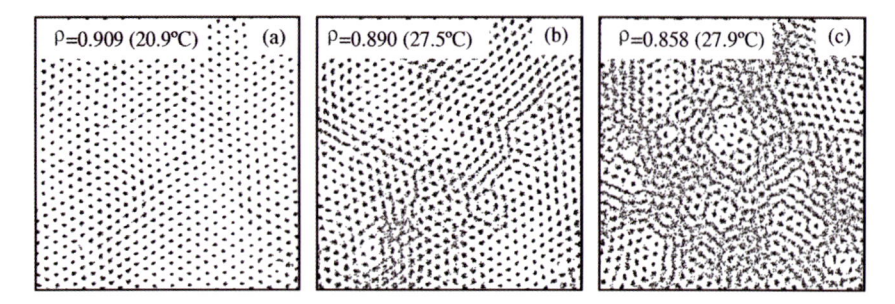

Figure 10.23 Typical 10-s particle trajectories in the (a) crystal, (b) hexatic, and (c) liquid phases, respectively.

(quasi-long-range order) for 27.1–27.5 °C, and $g_6(r) \sim e^{-r/\xi_6}$ (short-range order) for 27.7–28.5 °C. These three regions are more clearly resolved over three decades of dynamic range in Figure 10.24c and d, which plots the dynamic quantity $g_6(t)$. Comparing Figure 10.24a and c, we confirm KTHNY predictions [207] that the power

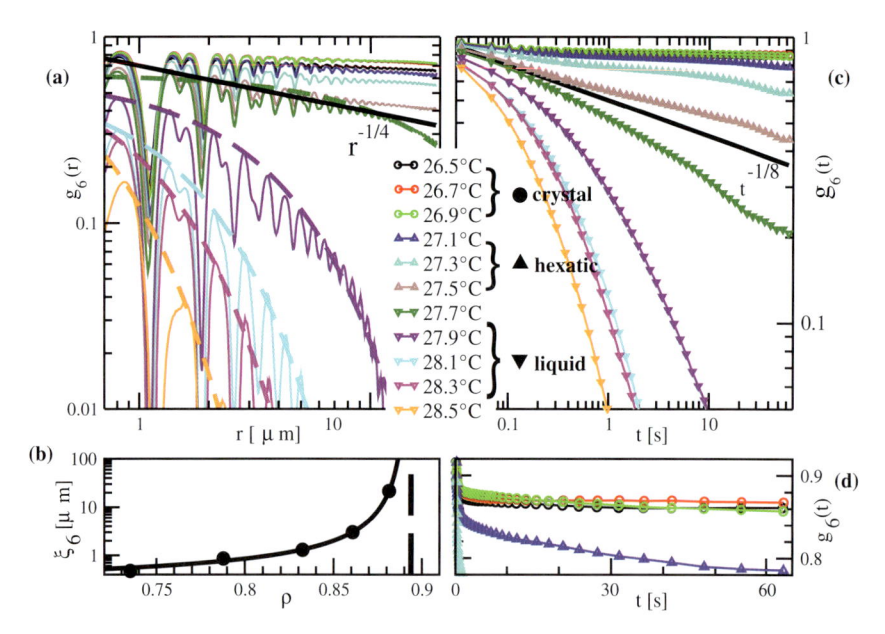

Figure 10.24 (a) Orientational correlation functions $g_6(r)$. Minima in the oscillations are associated with off-lattice site particles. The five dashed curves are fits of $g_6(r)$ to $e^{-r/\xi_6} \cdot r^{-1/4}$ is the KTHNY prediction at hexatic–liquid transition point. (b) Circles are the orientational correlation lengths ξ_6 obtained from the fits in (a). The solid curve is a fit to the KTHNY prediction $\xi_6(\varrho) \propto e^{-b_\xi/\sqrt{\varrho_i - \varrho}}$ with $b_\xi = 0.566$ and $\varrho_i = 0.894$. These fit values, however, are prone to systematic error as a result of finite size effects [239]. (c) The orientational correlation function $g_6(t)$ in time. $t^{-1/8}$ is the KTHNY prediction at hexatic–liquid transition point. (d) Expanded version of (c) that more clearly exhibits the transition from long-range to quasi-long-range order. The 11 temperatures correspond to the 11 densities in Figure 10.25.

law decay of $g_6(t)$ is two times slower than that of $g_6(r)$, and $2\eta_{6t} = \eta_{6r} = 1/4$ at the hexatic–liquid transition point. $g_T(t)$ and $g_T(r)$ yielded consistent results. We observed that $g_T(t) \sim t^{-\eta}$ (crystal) for $T < 27$ °C and $g_T(t) \sim e^{-t/\tau}$ (hexatic and liquid) for $T > 27$ °C. The KTHNY prediction that $\eta_{Tr} = 1/3$ [207] at the crystal hexatic transition point was also confirmed. The oscillations in $g_6(r)$ and $g_T(r)$ correspond to the oscillations in radial distribution function $g(r)$: the off-lattice particles usually have lower probability in $g(r)$, lower $|\psi_6|$, and less correlation with other particles.

Despite substantial agreement with the KTHNY model, two major ambiguities arise in the traditional correlation function analysis: (i) the power law decay of g_6 can reflect crystal–liquid coexistence rather than the hexatic phase and (ii) finite size and finite time effects introduce ambiguities into the correlation function curve shapes near transition points. For example, the $T = 27.7$ °C curve in Figure 10.24c appears to decay algebraically over the finite measured timescale, but it could also decay exponentially at longer times. Since the curve appears below the theoretical $t^{-1/8}$ transition curve, we assigned the system to the liquid phase.

Other correlation functions, such as the spatial density autocorrelation function $g(r)$, that is the radial distribution function, and the 2D structure factor $s(k)$ can also be used to determine the phase transition point [225, 226]. In our experiments, however, these methods appeared to have more ambiguities [159] because the theoretical functional forms of different phases are very similar close to the phase transition points.

Another function of interest is the Lindemann parameter L that is a measure of the particle mean-square fluctuation. It has been used as a traditional criterion of melting. For 2D melting, however, L diverges slowly even in the crystal phase due to strong long-wavelength fluctuations in 2D. To avoid such divergences, we calculated the *dynamic* Lindemann parameter L in a local coordinate system based on the neighbor positions [229, 239a]. It is defined as a bond length fluctuation, that is,

$$L^2 = \frac{\langle (\Delta \mathbf{r}_{\text{rel}}(t))^2 \rangle}{2a^2} = \frac{\langle (\Delta \mathbf{u}_i(t) - \Delta \mathbf{u}_j(t))^2 \rangle}{2a^2} \qquad (10.13)$$

where $\Delta \mathbf{r}_{\text{rel}}$ is the relative neighbor–neighbor displacement, $\Delta \mathbf{u}_i$ is the displacement of particle i, and particles i and j are nearest neighbors. As shown in Figure 10.25, L^2 converges below 27 °C; in this case, particles remain close to their lattice sites. Divergence of L^2 is found above 27 °C; in this case, particles can more readily exchange positions with their neighbors via the gliding and climbing of dislocations [207]. This transition at 27 °C is also consistent with our direct measurement of dislocation densities (see discussion below) in Figure 10.26a.

Defect densities are helpful for distinguishing different phases. The Voronoi diagram in Figure 10.22b, d, and f shows typical defects in the three phases. Particles with $n_n \neq 6$ are considered to be defects. KTHNY theory suggests that the creation of *free* dislocations (isolated 5–7 pairs) drives the system from crystal to hexatic phase, and the creation of *free* disclinations (isolated $n_n = 5$ or 7 defects) drives the transition from hexatic to liquid phase. We measured defect concentrations as a function of

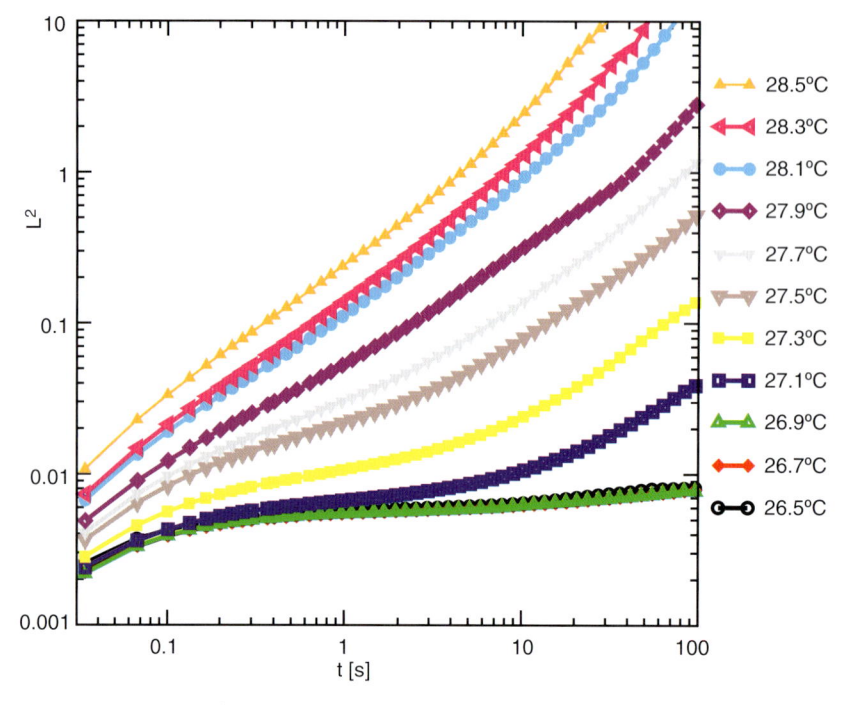

Figure 10.25 Square of dynamic Lindemann parameters at 11 temperatures.

temperature. Figure 10.26a shows that dislocations start to appear for $T > 27\ °C$ ($\varrho_m = 0.905$), and disclinations start to appear for $T > 27.7\ °C$ ($\varrho_i = 0.875$). Although defect density measurements are less sensitive to finite size effects than the correlation functions [239], the assignment of melting volume fraction ϱ_m based on defect density (Figure 10.26a) is somewhat problematic too. Problems can arise because (1) the data (Figure 10.26a) inevitably includes dislocations that are not completely "free", for example, dislocation pairs in Figure 10.27a that are nearly adjacent to one another, point in opposite directions, and thus give zero Burgers vector for a large Burgers circuit; (2) the data (Figure 10.26a) are susceptible to other systematic errors, for example, miscounting large defect clusters equivalent to a free dislocation, for example, a "free" 6-mer 5-7-5-7-5-7. In fact, Figure 10.26a very likely overestimates ϱ_m because sufficient numbers of "nonfree" dislocations are needed before the dislocation chemical potential reaches zero and free dislocations are produced. Consequently, a dislocation precursor stage in the crystal phase might be expected.

Note that the densities of $5n_n$ and $7n_n$ are the same only in perfect crystals with periodic boundary conditions if neglecting $n_n = 3, 4, 8 \ldots$ defects [207]. In the experiment, we observed a density imbalance of $5n_n$ and $7n_n$ in Figure 10.26a because (1) our polycrystals have "free" boundary conditions, thus $5n_n$ and $7n_n$ can be generated not only by pair but also by diffusing from boundaries [159, 207]; (2) any deviation from the strict monolayer limit can produce a concentration

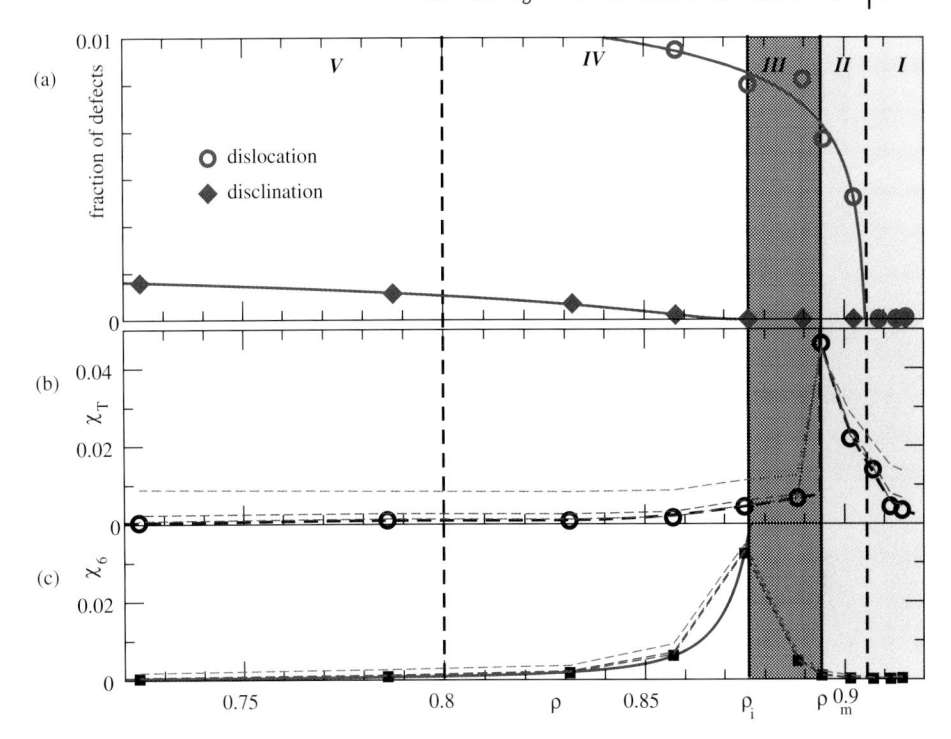

Figure 10.26 (a) Thick dashed curve: $n_n = 5$ disclination density. Thin dashed curve: $n_n = 7$ disclination density. Diamonds: net disclination density; circles: dislocation density fit by $e^{-2b_m/(\varrho_m-\varrho)^{0.36963}}$ [207]. (b) Translational and (c) orientational susceptibilities. Dashed curves: χ_L derived from subbox sizes $L = 5, 10, 20\,\mu m$ from top to bottom. Symbols: χ_∞ extrapolated from dashed curves. The solid curve in (c) is a fit to the KTHNY prediction [237] $\chi_6(\varrho) \propto e^{-b_\chi/\sqrt{\varrho_i-\varrho}}$ with $\varrho_i = 0.901$ and $b_\chi = 1.14$. Vertical solid lines partition crystal (regions I and II), hexatic (region III), and liquid (regions IV and V) phases as determined from susceptibilities in (b) and (c). Region II is a "dislocation precursor stage" of crystal with dislocations. Region IV is a prefreezing stage [242] of liquid with ordered patches.

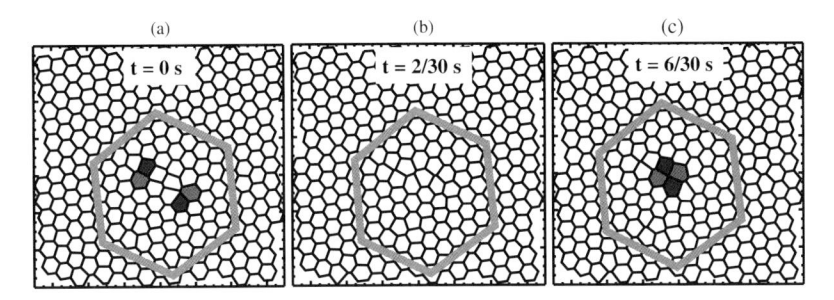

Figure 10.27 Voronoi diagram of the time evolution of a nonfree dislocation pair at 27.1 °C. Dark and light polygons represent particles with 5 and 7 nearest neighbors, respectively. (a–c) All yield zero Burgers vector as shown by the closed hexagonal loop. Dislocations can rapidly form and annihilate in pairs if they are in the same lattice line.

asymmetry [207]; (3) the observed higher density of $n_n = 8, 9$ than $n_n = 3, 4$ defects partially compensated the imbalance. The higher density of $5n_n$ reflects their lower free energy compared to that of $7n_n$. Besides the static properties noted above, we observed some interesting defect dynamics. For example, dislocations often dissociated from larger defect clusters (e.g., 6-mer 5-7-5-7-5-7) rather than from isolated pairs of dislocations (5-7-5-7 quartet), perhaps because the energy change for such disassociation is small.

In order to avoid the ambiguities outlined above and determine the true ϱ_m, we explored the utility of using the order parameter susceptibility, χ, for finding phase transition points. χ is a measure of the fluctuations of the order parameter in 9000 frames. To ameliorate finite size effects, we calculated χ_L in different size subboxes within the sample (dashed curves in Figure 10.26b and c)) and then extrapolated to χ_∞, thus attaining the thermodynamic limit (see Ref. [159] for details). The sharp divergence/discontinuity of $\chi_{T\infty}$ and $\chi_{6\infty}$ in Figure 10.26b and c clearly indicates the two transitions of the melting process. Although the magnitude of χ suffered from size effects, the diverging point of χ was robust to box size. Thus, the susceptibility method avoided finite-size ambiguities. The divergence of χ also avoids ambiguities arising from the similar functional forms of other measures (e.g., correlation functions) near transition points. Theoretically, we expect the divergence of χ to have better statistics than correlation function shape because χ is essentially an integral of the correlation function. We also observed that the diverging points of χ_T were robust to small uncertainties in **G**, though the exact magnitude of χ_T was somewhat sensitive to **G**.

10.4.5
The Hexatic Phase and Other Features of the Phase Diagram

We clearly observed the middle hexatic phase that has short-ranged translational order, quasi-long-range orientational order, zero disclination density, and finite dislocation density. Furthermore, we resolve five regimes that are marked off in Figure 10.26 based on the various analyses we have carried out. Region I is crystalline with few dislocations (Figure 10.26a), convergent dynamic Lindemann parameters over the measured timescales (Figure 10.25), constant $g_6(r), g_6(t)$ (Figure 10.24), and algebraic decay of $g_T(t)$. We take region II to be a "dislocation precursor stage" in the crystal because dislocations have started to appear, but their density is not high enough for the system to reach the hexatic phase, wherein the chemical potential of dislocation reaches zero. In other words, the observed dislocations in region II are not "free." This gas of nonfree dislocations causes a softening of the crystal, an effect that has been observed in the crystal phase [240, 241]. The dynamic Lindemann parameter is divergent in region II, a direct consequence of the nonzero dislocation density that permits particles near dislocations to diffuse out of their cages via the gliding and climbing [207] of dislocations. The correlation function, $g_6(t)$, has finite-size ambiguity in region II. For example, the $T = 27.1$ °C curve in Figure 10.24d appears to have lost orientational order over the measured timescale, but could become constant at longer times. Region III is the hexatic phase as determined from the χ measurements and other analyses. In region IV, disclinations start to appear (Figure 10.26a), and $g_6(r)$, $g_6(t)$ decay exponentially (Figure 10.24). We take region IV to be a "prefreezing" liquid [242] because it has visible ordered patches. The nonzero ψ_6, the splitting of the second peak in $g(r)$, and the hexagonal shape of the structure factor $s(k)$ in region IV are also indicative of the presence of ordered patches. Region V is the liquid phase.

10.4.6
The Order of the Phase Transitions

The order of the phase transition can, in principle, be deduced from the shape of susceptibility curves. If the curve on the left of the diverging point and the curve on the right of the diverging point have the same asymptotic ϱ, then the transition is second order; otherwise, it is first order [243]. The curve shape in Figure 10.26b and c are consistent with second-order transitions. However, the χ_T curve shape is sensitive to the choice of **G** even though the diverging point is quite robust. For the liquid hexatic transition in Figure 10.26c, when we fit the left part (liquid regime) of the curve with the KTHNY prediction, we obtained an unreasonably high asymptotic transition density $\varrho_i = 0.901$ as obtained in Ref. [219]. This discrepancy suggests the hexatic–liquid transition may be more first-order-like. In addition, the continuous phase transitions must satisfy universality relations, while first-order transitions need not. Our $b_{6\xi} = 0.566$, from Figure 10.24b, and $b_{6\chi} = 1.14$, from Figure 10.26c, do not completely satisfy the universality [237] $b_{6\chi} = (2-\eta_6)b_{6\xi}$ where $\eta_6 = 1/4$. This failure could be viewed as further evidence of a first-order transition; however, when we forced $b_{6\chi}$ and $b_{6\xi}$ to satisfy the universality relation, they still gave somewhat reasonable (albeit worse) fitting curves because other fitting parameters were adjustable too. For example, the five data points in Figure 10.24b can be fit well by the other two free parameters when $b_{6\xi}$ is fixed. In total, the evidence leans slightly to favor a first-order liquid hexatic transition, but is not sufficient to unambiguously exclude a second-order transition. Future work with finer control of the approach to the phase transition should enable us to pin down the order of two transitions more precisely.

10.4.7
Summary

In summary, we used the divergence of susceptibilities to determine the phase transition points of a 2D microgel suspension during the melting process. This approach avoided ambiguities from finite-size effects, and the diverging points were robust. We clearly observed the hexatic phase in a system of particles interacting via short-range soft repulsion potentials. Five regimes were assigned to the phase diagram in Figure 10.26. A number of KTHNY predictions were quantitatively confirmed, especially near the hexatic–liquid transition, but the order of two phase transitions was not unambiguously resolved due to our limited temperature resolution.

10.5
Geometric Frustration in Colloidal "Antiferromagnets"

10.5.1
Background

Geometric frustration is a phenomenon that arises when lattice structure prevents simultaneous minimization of local interactions. Arguably the most famous example of geometric frustration arises in the context of antiferromagnetic (AF) materials. AF Ising spins on a 2D triangular lattice have strong geometric frustration. The problem was famously studied by Wannier in 1950 [244, 245]. Consider three spins on a triangle shown in Figure 10.29a. When two spins are arranged to be antiparallel to satisfy their antiferromagnetic interaction, the third one has no way to be antiparallel to both of the other spins. Thus, it is impossible to simultaneously satisfy all nearest-neighbor interactions on a triangular lattice. In contrast, antiferromagnetic Ising spins on a square lattice have no geometric frustration because every spin can be antiparallel to all four of its nearest neighbors (see Figure 10.28b). Frustration leads to materials with many degenerate ground states. Figure 10.28c is one possible ground state with a striped configuration wherein each triangular plaquette has two satisfied bonds and one frustrated bond, and Figure 10.28d and e denotes more general ground-state configurations. By removing all frustrated bonds and drawing in all satisfied bonds only in the latter configuration, the ground state can be viewed as a stack of cubes (see Figure 10.28f and g) [246]. The ground states and cube stacks have one-to-one correspondence, thus there are many possible ground states because there are many ways to pack cubes.

For the 2D triangular lattice, there are $W \simeq e^{0.3231N}$ ground states [244, 245], and thus the system has an extensive entropy at zero temperature, that is, $S \simeq 0.3231 N k_B$, where N is the number of spins in the system. The triangular lattice antiferromagnet is the only geometrically frustrated 2D Bravais lattice, and it therefore plays an important role within the theory of cooperative phenomenon in 2D. However, although the model outlined above is a prototypical frustrated magnet, progress studying it has suffered from a lack of experimental realization.

Broadly speaking, geometric frustration arises in many physical and biological systems [247] ranging from water [248] and spin ice [249] to magnets [250–252], ceramics [253], and high-T_c superconductors [254]. Traditionally, these phenomena have been explored in atomic materials by ensemble averaging techniques such as neutron and X-ray scattering, muon spin rotation, nuclear magnetic resonance, and heat capacity and susceptibility measurements [252, 253]. Artificial arrays of mesoscopic constituents have also been fabricated to probe geometric frustration at the single-"particle" level. Examples of the latter include Josephson junctions [255, 256], superconducting rings [257, 258], ferromagnetic islands [259], and recent simulations [260] of charged colloids in optical traps. Observations in these model systems, however, have been limited to the static patterns into which these systems freeze when cooled. Thus, many questions about frustrated systems remain unexplored, particularly those associated with single-particle dynamics.

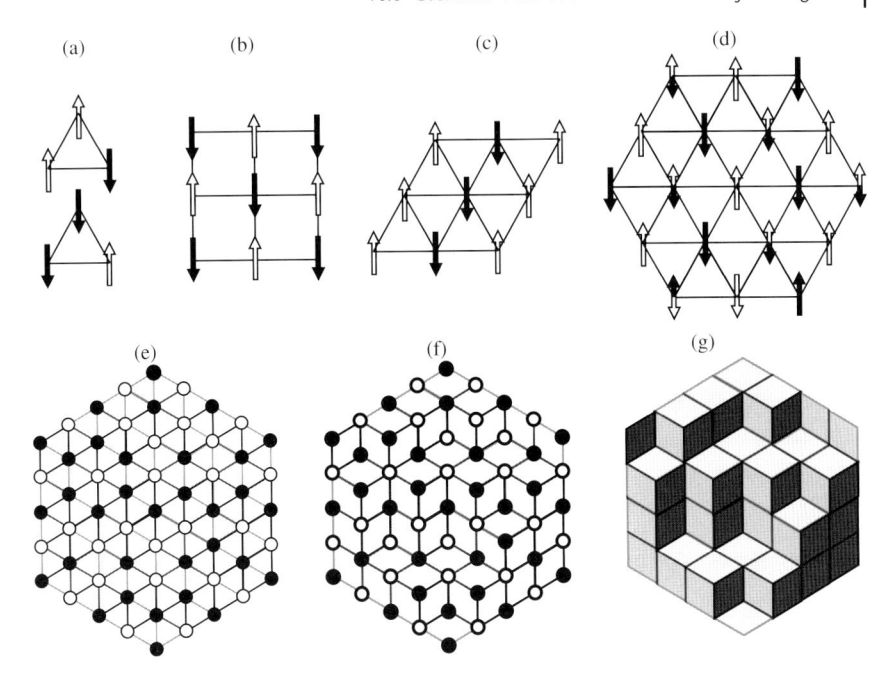

Figure 10.28 Schematics of ground states antiferromagnetic Ising spin 2D lattices. (a) The ground state of AF spins in triangular unit lattices, each triangle has two satisfied and one frustrated bonds. (b) The ground state of AF spins in a square lattice. All bonds (nearest-neighbor interactions) are satisfied. (c) One possible ground state of triangular AF spin lattice with a striped configuration. (d) A more general ground state, the up–down arrows represent spins that are surrounded by three up and three down spins, thus each of them can be up or down. The configuration shows that the ground state is highly degenerate. (e) Another ground-state configuration, each triangular plaquette has one frustrated and two satisfied bonds. Here, open (closed) circle denotes as up (down) state. (f) The same configuration of (e) where only satisfied bonds are shown. (g) Rhombuses in (f) with three orientations are painted by three gray scales so that the ground state can be better viewed as a cube stack.

The buckled colloidal monolayer provides an elegant model system for measuring the *single-spin* static and *dynamical* properties of a geometrically frustrated system. It is readily constructed by confining colloidal spheres between two parallel walls. When the wall separation is about 1.5 sphere diameter, the particles assemble (at high packing fraction) into a buckled triangular lattice with either up or down displacements analogous to an antiferromagnetic Ising model on a triangular lattice. Buckling minimizes system free energy $F = U - TS$, where U is the internal energy, T is temperature, and S is the entropy; spheres move apart to lower their repulsive interaction potential energy U and to increase their free volume V, which in turn leads to an entropy increase with $S \propto \ln V$. The effective repulsion causes spheres to move to the top or bottom sample wall. Nearest neighbors maximize free volume by moving to opposite walls (see Figure 10.29b). Furthermore, by using microgel spheres, the effective antiferromagnetic interactions can be tuned by

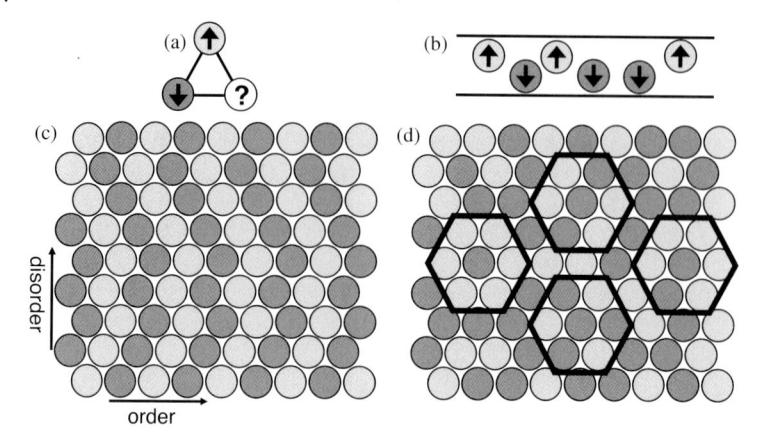

Figure 10.29 (a) Three spins on a triangular plaquette cannot simultaneously satisfy all AF interactions. (b) For colloids confined between walls separated of order 1.5 sphere diameters (side view), particles move to opposite walls in order to maximize free volume. (c and d) Ising ground-state configurations wherein each triangular plaquette has two satisfied bonds and one frustrated bond. (c) Zigzag stripes generated by stacking rows of alternating up/down particles with random side-wise shifts; all particles have exactly two frustrated neighbors. (d) Particles in disordered configurations have zero, one, two, or three frustrated neighbors (gray hexagons).

changing microgel particle diameter, and therefore the samples can be driven from spin liquid to glassy states. Once again, this class of colloid experiment provides access to new physics, in this case bridging the fields of frustrated magnetism and soft matter.

Buckled colloidal monolayers were first observed more than two decades ago [233, 261, 262], and the AF analogy was then suggested [262, 263]. On the basis of Koshikiya and Hachisu's colloidal monolayer image [263], Ogawa suggested the AF analogy and estimated the probabilities of different local configurations in ground state. However, image analysis was not readily available in the 1980s limiting the quantitative study of the experiment. To date, few quantitative measurements have been performed on this system class, and the themes explored by most of the early work centered largely on structural transitions exhibited by colloidal thin films as a function of increasing sample thickness [233, 262, 264, 265], rather than their connection to frustrated antiferromagnets. In a wedged sample, the following sequence of crystal phases has been observed as a function of increasing wall separation [262]: monolayer triangular lattice, buckled monolayer (i.e., our configuration), two-layer square lattice, two-layer triangular lattice, three-layer square lattice, three-layer triangular lattice, and more. Simulations studied the nature of these solid–solid phase transitions and provided quantitative phase diagrams for hard spheres [220, 266, 267]. Usually, the crystal domain size formed by traditional colloids is not large, however [233, 262, 263, 265]. The diameter tunable spheres permit us to anneal the buckled crystals near the melting point and form larger crystals with better quality.

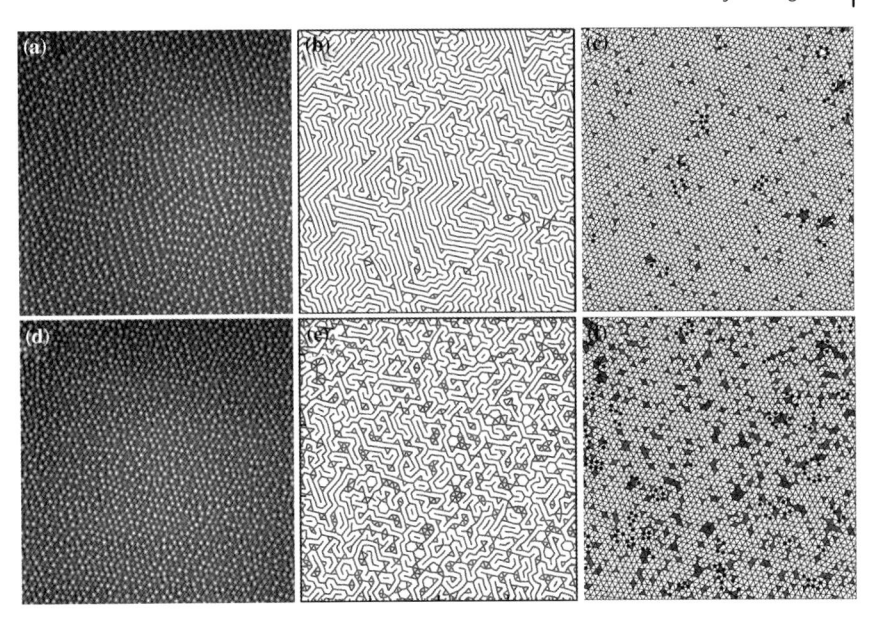

Figure 10.30 Buckled monolayer of colloidal spheres. $(32\,\mu m)^2$ area at $T = 24.7\,°C$ (a–c) and $27.1\,°C$ (d–f). Bright spheres: up; dark spheres: down. (b and e) Labyrinth patterns obtained by drawing in only the frustrated up–up (dark gray) and down–down (light gray) bonds. (c and f) Corresponding Delaunay triangulations. Black dots mark defects in the triangular lattice, that is, particles without six nearest neighbors. Thermally excited triangles with three spheres up/down are labeled by dark gray/light gray.

10.5.2
The Experimental System

The experiments employ densely packed spheres confined by parallel glass walls. Microgel spheres are cross-linked with PMMA (polymethyl methacrylate) at their surfaces to prevent sticking to glass walls. For walls separated by a distance of about 1.5 sphere diameter, the particles maintain in-plane triangular order but buckle out of plane (see Figure 10.30a and d). Samples were equilibrated at low volume fraction near the melting point to produce 2D crystal domains with $\sim 10^4$ spheres covering an area of order $(60\,\mu m)^2$. Video microscopy measurements were carried out far from grain boundaries on a $\sim (32\,\mu m)^2$ central area ($\sim$2600 spheres) within the larger crystal domain. Below $24\,°C$, the system is jammed and no dynamics is observed and above $27.5\,°C$ the in-plane crystals melt. Our primary measurements of the frustrated states probe five temperatures in between, from 24.7 to $27.1\,°C$ in $0.6\,°C$ steps. In this range, the hydrodynamic diameter of the particles decreases linearly with temperature from $0.89\,\mu m$ to $0.76\,\mu m$, while the average in-plane particle separation remains constant. We slowly cycled through this temperature range and hysteresis was not observed.

10.5.3
Antiferromagnetic Order

Typical frustrated samples are illustrated in Figure 10.30. Figure 10.30a and d shows roughly half of the spheres as bright because they are in the focal plane of the microscope; the other half, located close to the bottom plate, are slightly out of focus and appear darker. To analyze these images, we discretize the continuous brightness profile of the particles into two Ising states with $s_i = \pm 1$ (see Figure 10.30).

The nature of the frustrated states can be exhibited in different ways in processed images based on the data in Figure 10.30a and d. One way focuses on the "bonds" between particles. We refer to the line connecting a pair of neighboring particles in opposite states ($s_i s_j = -1$) as a satisfied bond, that is, a bond that satisfies the effective AF interaction, and we refer to the line connecting up–up or down–down pairs (with $s_i s_j = 1$) as a frustrated bond. Images of these bonds (Figure 10.30b and e) show that the frustrated bonds form an almost single-line labyrinth (Figure 10.30b) at low temperature that then form small domains (Figure 10.30e) at high temperature. AF order can also be characterized by the average number, $\langle N_f \rangle$, of frustrated bonds per particle. In the limit of weak interactions, that is, $T \to \infty$, an Ising system will choose a completely random configuration with half of the six bonds satisfied and half frustrated, leading to $\langle N_f \rangle = 3$. In the limit of strong interactions, that is, $T = 0$, on the other hand, each triangular plaquette will have one frustrated bond (see Figure 10.29a), a third of the bonds are frustrated, and therefore $\langle N_f \rangle = 2$. $\langle N_f \rangle$ is essentially a linear rescaling of the density of excited triangles (3 up or 3 down) in Figure 10.30c and f, which ranges from 0 in the AF Ising ground state to 0.5 for a random configuration. By analyzing the experimental movies, we found that $\langle N_f \rangle$ decreased from approximately 2.5 to 2.1 in the temperature interval 27.1–24.7 °C (see Figure 10.31).

10.5.4
Stripes and the Zigzagging Ground State

Ideal geometrically frustrated systems are highly degenerate with extensive entropy at zero temperature. However, the third law of thermodynamics dictates that system

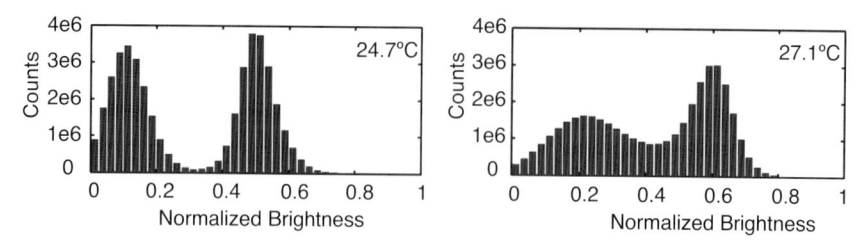

Figure 10.31 Histograms of particle brightness normalized to [0,1]. The histogram has a bimodal distribution where ∼50% of the particles were darker/brighter than the central minimum point for all temperatures.

entropy vanish as $T \to 0$. This suggests a problem with the Wannier model. Of course, the model is an idealization using a rigid lattice and including only nearest-neighbor interactions. In real materials, subtle effects, for example, anisotropic interactions [268], long-range interactions [269], boundary conditions [270], and lattice distortions [271, 272] can relieve frustration.

The partially ordered zigzagging striped sample observed at high volume fraction in Figure 10.30a and b is an example of a frustration relief by symmetry-reducing lattice distortions. In the colloidal monolayer, the triangular packing is self-assembled, and (like atoms in real solids) the particles are not forced to remain at fixed positions on the lattice [273]. This deformability, and the fact that the free volume of the system is a collective function of all particle positions, breaks the mapping to simple Ising models with pair-wise additive nearest-neighbor interactions. In short, the Ising ground state is a single-line labyrinth with $N_f = 0, 1, 2, 3$ and $\langle N_f \rangle = 2$, while the colloidal spheres' real ground state appears to be a subset of the Ising ground states, including nonbranching single-line labyrinths with $N_f = 2$ in the bulk.

We can understand zigzagging stripes as a ground state by viewing them as random stacks of ordered lines of alternating up and down states, see Figure 10.29c. Notice that the straight and zigzagging stripes are essentially a 2D analogy of the face-centered cubic and randomly hexagonal close packed (rhcp) [274] structures in 3D. Equal-sized spheres can be most efficiently packed in 3D by stacking hexagonal close-packed 2D layers, that is, fcc or rhcp. Ordered stacking as ABCABC... (fcc) or random stacking such as ABACB... (rhcp) has the same volume fraction, simply because nonneighbor layers cannot affect each other. For a similar reason, zigzagging stripes have the same high closed packing area fraction as the straight stripes. In Figure 10.29c, the disordered stack sequence is random along the vertical direction and hence yield $\sim 2^{\sqrt{N}}$ configurations. Consequently, the ground-state entropy $S \sim \sqrt{N}$ is subextensive.

10.5.5
Dynamics

Real-time videos permit direct visualization of "spin flipping", as well as the motions of thermal excitations and defects, in frustrated systems for the first time. The ability to track particle dynamics at the "single-spin" level will likely be the most important contribution of the colloid experiments to the frustration subfield. Thermal excitations, for example, labeled as colored triangles in Figure 10.30c and f were typically found to be generated/annihilated in pairs due to the flipping of a particle shared by the two triangles; isolated thermal excitations, on the other hand, appear to be quite stable. Similarly, if we simply follow the up–down spin trajectory of single particles, then we should obtain a range of interesting dynamics. An example of spin trajectory as a function of time is shown in Figure 10.32.

As the first step toward quantifying these effects, we first extract the full time spin trajectory, $s_i(t)$, of each particle i. In Figure 10.33b, the ensemble-averaged single-particle temporal spin autocorrelation function, that is,

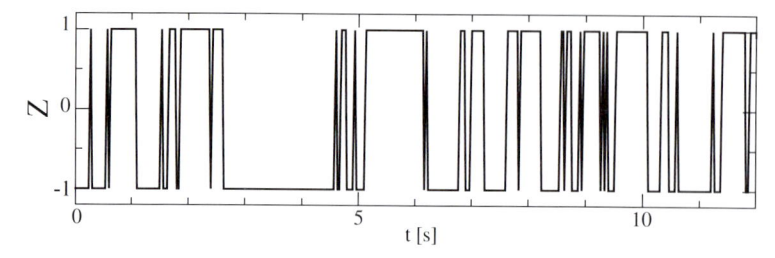

Figure 10.32 A typical trajectory of a single "spin" flip. $+$ $(-)$ 1 denotes an up (down) Ising state.

$$C(t) = [\langle s_i(t)s_i(0)\rangle - \langle s_i\rangle^2]/[\langle s_i^2\rangle - \langle s_i\rangle^2] \quad (10.14)$$

is plotted as a function of temperature. The function is averaged over all particles that are not located at lattice defects. The correlation function cannot be well fitted by a power law or an exponential, but it can be fit to a stretched exponential form, that is, $C(t) = \exp\left[-(t/\tau)^\beta\right]$. The measured relaxation time τ exhibits a dramatic increase as the particles swell at low temperature, while the extracted stretching exponent β decreases.

This behavior is suggestive of dynamics similar to those found in glasses. We speculate that particles in the frustrated system experience a complex energy landscape wherein transitions between different local configurations have different

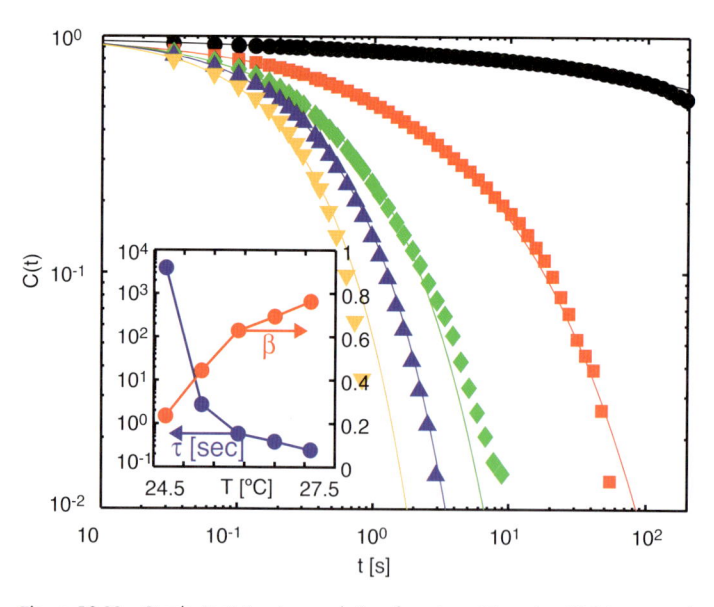

Figure 10.33 Single "spin" autocorrelation functions (Equation 10.14 averaged over all particle trajectories. Lines are fits to stretched exponentials $C(t) = \exp[-(t/\tau)^\beta]$, with τ and β given in the inset.

energy barriers and decay rates. It is not clear why the final averaged autocorrelation function can be fitted so economically by a simple stretched exponential form with only two free parameters. The stretched exponential behavior has been referred to as "one of Nature's best-kept secrets [275]," and it is possible that these geometrically frustrated colloids, with measurable single-particle dynamics, could once again provide a fresh platform from which to tackle this interesting challenge. For example, it should be possible to measure the correlation functions and flip rates of spins sitting in different local environments. We have begun computations along these lines [160].

Finally, in a different vein, defects in the underlying lattice can strongly affect the properties of frustrated systems. However, detailed knowledge about the role of defects in frustrated systems is very limited. Our experiments permit us to directly visualize defects nucleating, annihilating, and diffusing, see Figure 10.30c and f. By comparing trajectories containing different numbers and types of defects, our initial studies suggest that defect particles have faster in-plane diffusion and slower flipping dynamics than the average of the particles with six nearest neighbors.

10.5.6
Summary

We have demonstrated two-dimensional colloidal frustrated antiferromagnets. Colloidal microgel spheres with tunable diameter self-assemble to buckled monolayer crystals and form a system analogous to the triangular lattice antiferromagnetic Ising model. By tuning volume fraction, we found that at high compaction, in-plane lattice deformation relieves most frustration and yields a zigzag stripe ground state with subextensive entropy. We measured spatial correlations and the statistics of various local configurations as well as their flipping rates and found strong dependences on arrangements of neighboring particles. As the glassy phase is approached, we observed dramatic slowing of the dynamics and formation of stretched exponential correlation functions. Single-defect dynamics were directly visualized and measured for the first time. The new system opens the door for the study of detailed single-particle dynamics in frustrated systems and begins an exploration of the connections between the frustrated soft materials and the more studied frustrated magnetic and related materials.

10.6
Future

Many more directions for fundamental physics experiments with NIPA microgel particles should be explored. Interesting new model systems will be created in the future. For example, very recent experiments have shown that NIPA particles are ideal for investigating thin-film melting [163b], 2D freezing [163b], jamming [163] and glassy behaviors [161, 162], dynamic heterogeneity, and even the crystal-to-glass transition. In principle, the microgel systems permit creation of frustration in other contexts; for example, by filling thin cylinders with spheres, one can mimic a 1D chain of xy-spins [276]. Another exciting experimental arena, which need not even involve the development of new systems, concerns the perturbation and active manipulation of existing samples using laser tweezers and other tools. For example, potential energy landscapes for the particles can be created using laser tweezers of varying strength and periodicity (including rigid lattices), enabling experimenters to explore the role of lattice deformability on the dynamics and the creation of structure. Optical tweezers or magnetic traps can also be used to flip and to move individual spins, and video microscopy can be used to probe the resulting system's responses. Light beams can even be used to locally heat a region within the sample, causing the energy landscape to "reset" and permitting experimenters to understand different classes of response.

Indeed, research in this subfield of soft matter appears promising for years to come.

Acknowledgments

The research we have discussed in this chapter has been made possible, in part, through interactions with many outstanding colleagues. In some cases, we benefited from discussion, in some cases we benefited from detailed assistance with experiment, and in many cases these colleagues coauthored papers with us. We particularly want to thank Peter Collings, Mohammad Islam, Zvonimir Dogic, Na Young Ha, Yair Shokef, Randy Kamien, David Nelson, Kevin Aptowicz, Piotr Habdas, Zexin Zhang, Peter Yunker, Anindita Basu, Matt Lohr, and Tom Lubensky for their collaboration and their helpful discussions. This work was supported by the National Science Foundation (NSF) through the MRSEC program and, partially, through individual NSF investigator grants; it was also supported partially by the National Aeronautics and Space Administration (NASA).

References

1 Hunter, R.J. (2001) *Foundations of Colloid Science*, 2nd edn, Oxford University Press.

2 Israelachvili, J.N. (1992) *Intermolecular and Surface Forces*, 2nd edn, Academic Press, London.

3 Daoud, M. and Williams, C.E. (1999) *Soft Matter Physics*, 1st edn, Springer, Germany.

4 Russel, W.B., Saville, D.A., and Schowalter, W.R. (1991) *Colloidal Dispersions*, 1st edn, Cambridge University Press, Cambridge.

5 Morrison, I.D. and Ross, S. (2002) *Colloidal Dispersions*, 1st edn, John Wiley & Sons, Inc., New York.

6 Larson, R.G. (1998) *The Structure and Rheology of Complex Fluids (Topics in Chemical Engineering)*, 1st edn, Oxford University Press, Oxford, New York.

7 Witten, T.A. and Pincus, P.A. (2004) *Structured Fluids: Polymers, Colloids, Surfactants*, 1st edn, Oxford University Press, New York.

8 Joannopoulos, J.D., Johnson, S.G., Winn, J.N., and Meade, R.D. (2008)

Photonic Crystals: Molding the Flow of Light, 2nd edn, Princeton University Press.
9 Holland, B.T., Blanford, C.F., and Stein, A. (1998) *Science*, **281**, 538.
10 Imhof, A. and Pine, D.J. (1997) *Nature*, **389**, 948.
11 Subramania, G., Constant, K., Biswas, R., Sigalas, M.M., and Ho, K. (1999) *Appl. Phys. Lett.*, **74**, 3933.
12 Velev, O.D., Tessier, P.M., Lenhoff, A.M., and Kaler, E.W. (1999) *Nature*, **401**, 548.
13 Vlasov, Y.A., Yao, N., and Norris, D.J. (1999) *Adv. Mater.*, **11**, 165.
14 Wijnhoven, J.E.G.J. and Vos, W.L. (1998) *Science*, **281**, 802.
15 Yablonovitch, E. (1999) *Nature*, **401**, 539.
16 Lopez, C. (2003) *Adv. Mater.*, **15**, 1679.
17 Xia, Y.N. and Whitesides, G.M. (1998) *Annu. Rev. Mater. Sci.*, **28**, 153.
18 Burmeister, F., Schafle, C., Keilhofer, B., Bechinger, C., Boneberg, J., and Leiderer, P. (1998a) *Adv. Mater.*, **10**, 495.
19 Hulteen, J.C. and Vanduyne, R.P. (1995) *Abstr. Pap. Am. Chem. Soc.*, **210**, 25.
20 Winzer, M., Kleiber, M., Dix, N., and Wiesendanger, R. (1996) *Appl. Phys. A: Mater.*, **63**, 617.
21 Burmeister, F., Schafle, C., Keilhofer, B., Bechinger, C., Boneberg, J., and Leiderer, P. (1998b) *Adv. Mater.*, **10**, 495.
22 Bhave, R.R. (1991) *Inorganic Membranes: Synthesis, Characteristics and Applications*, Van Nostrand Reinhold, New York.
23 Senkevich, J.J. and Desu, S.B. (1998) *Appl. Phys. Lett.*, **72**, 258.
24 Seino, H., Haba, O., Mochizuki, A., Yoshioka, M., and Ueda, M. (1997) *High. Perform. Polym.*, **9**, 333.
25 Kong, J., Franklin, N., Zhou, C., Chapline, M., Peng, S., Cho, K., and Dai, H. (2000) *Science*, **287**, 622.
26 Tanev, P.T., Chibwe, M., and Pinnavaia, T.J. (1994) *Nature*, **368**, 321.
27 Tennikov, M.B., Gazdina, N.V., Tennikova, T.B., and Svec, F. (1998) *J. Chromatogr. A*, **798**, 55.
28 Xie, S.F., Svec, F., and Frechet, J.M.J. (1997) *J. Chromatogr. A*, **775**, 65.
29 Lewandowski, K., Murer, P., Svec, F., and Frechet, J.M.J. (1998) *Anal. Chem.*, **70**, 1629.
30 Litovsky, E., Shapiro, M., and Shavit, A. (1996) *J. Am. Ceram. Soc.*, **79**, 1366.
31 Maquet, V. and Jerome, R. (1997) *Mater. Sci. Forum*, **250**, 15.
32 Palm, A. and Novotny, M.V. (1997) *Anal. Chem.*, **69**, 4499.
33 Peters, M.C. and Mooney, D.J. (1997) *Mater. Sci. Forum*, **250**, 43.
34 Deleuze, H., Schultze, X., and Sherrington, D. (1998) *Polymer*, **39**, 6109.
35 Akolekar, D.B., Hind, A.R., and Bhargava, S.K. (1998) *J. Colloid Interface Sci.*, **199**, 92.
36 Schugens, C., Maquet, V., Grandfils, C., Jerome, R., and Teyssie, P. (1996) *Polymer*, **37**, 1027.
37 Durkop, T., Getty, S.A., Cobas, E., and Fuhrer, M.S. (2004) *Nano Lett.*, **4**, 35.
38 Storhoff, J.J., Elghanian, R., Mucic, R.C., Mirkin, C.A., and Letsinger, R.L. (1998) *J. Am. Chem. Soc.*, **120**, 1959.
39 Bancel, S. and Hu, W.S. (1996) *Biotechnol. Prog.*, **12**, 398.
40 Elghanian, R., Storhoff, J.J., Mucic, R.C., Letsinger, R.L., and Mirkin, C.A. (1997) *Science*, **277**, 1078.
41 Lewis, J.A. (2000) *J. Am. Ceram. Soc.*, **83**, 2341.
42 Ray, S.S. and Bousmina, M. (2005) *Prog. Mater. Sci.*, **50**, 962.
43 Ray, S.S. and Okamoto, M. (2003) *Prog. Polym. Sci.*, **28**, 1539.
44 Tang, Z.Y., Kotov, N.A., Magonov, S., and Ozturk, B. (2003) *Nat. Mater.*, **2**, 413.
45 Baughman, R.H., Zakhidov, A.A., and de Heer, W.A. (2002) *Science*, **297**, 787.
46 Vigolo, B., Penicaud, A., Coulon, C., Sauder, C., Pailler, R., Journet, C., Bernier, P., and Poulin, P. (2000) *Science*, **290**, 1331.
47 Haggenmueller, R., Gommans, H.H., Rinzler, A.G., Fischer, J.E., and Winey, K.I. (2000) *Chem. Phys. Lett.*, **330**, 219.
48 Schadler, L.S., Giannaris, S.C., and Ajayan, P.M. (1998) *Appl. Phys. Lett.*, **73**, 3842.
49 Biercuk, M.J., Llaguno, M.C., Radosavljevic, M., Hyun, J.K.,

Johnson, A.T., and Fischer, J.E. (2002) *Appl. Phys. Lett.*, **80**, 2767.
50 Einstein, A. (1905) *Ann. Phys. Berlin*, **17**, 549.
51 Perrin, F. (1908) *C. R. Acad. Sci.*, **146**, 967.
52 Langevin, P. (1908) *C. R. Acad. Sci.*, **146**, 530.
53 Kubo, R. (1966) *Rep. Prog. Phys.*, **29**, 255.
54 Alder, B.J. and Wainwright, T.E. (1970) *Phys. Rev. A*, **1**, 18.
55 Mazur, P. and Oppenheim, I. (1970) *Physica*, **50**, 241.
56 Zwanzig, R. and Bixon, M. (1975) *J. Fluid Mech.*, **69**, 21.
57 Hinch, E.J. (1975) *J. Fluid Mech.*, **72**, 499.
58 Felderhof, B.U. (1978) *J. Phys. A: Math. Gen.*, **11**, 929.
59 Ermak, D.L. and Mccammon, J.A. (1978) *J. Chem. Phys.*, **69**, 1352.
60 Doi, M. and Edwards, S.F. (1978) *J. Chem. Soc. Faraday Trans. II*, **74**, 1789.
61 Jones, R.B. (1979) *Physica A*, **97**, 113.
62 Mazur, P. (1982) *Physica A*, **110**, 128.
63 Beenakker, C.W.J. and Mazur, P. (1983) *Physica A*, **120**, 388.
64 Beenakker, C.W.J. and Mazur, P. (1984) *Physica A*, **126**, 349.
65 Brady, J.F. and Bossis, G. (1988) *Annu. Rev. Fluid Mech.*, **20**, 111.
66 Granick, S. (1991) *Science*, **253**, 1374.
67 Jones, R.B. and Pusey, P.N. (1991) *Annu. Rev. Phys. Chem.*, **42**, 137.
68 Kaplan, P.D., Dinsmore, A.D., Yodh, A.G., and Pine, D.J. (1994) *Phys. Rev. E*, **50**, 4827.
69 Segre, P.N., Meeker, S.P., Pusey, P.N., and Poon, W.C.K. (1995) *Phys. Rev. Lett.*, **75**, 958.
70 Bedeaux, D. and Mazur, P. (1974) *Physica*, **76**, 247.
71 Batchelor, G.K. (1976) *J. Fluid Mech.*, **74**, 1.
72 Kao, M.H., Yodh, A.G., and Pine, D.J. (1993) *Phys. Rev. Lett.*, **70**, 242.
73 Zhu, J.X., Durian, D.J., Muller, J., Weitz, D.A., and Pine, D.J. (1992) *Phys. Rev. Lett.*, **68**, 2559.
74 Weitz, D.A., Pine, D.J., Pusey, P.N., and Tough, R.J.A. (1989) *Phys. Rev. Lett.*, **63**, 1747.
75 Qiu, X., Wu, X.L., Xue, J.Z., Pine, D.J., Weitz, D.A., and Chaikin, P.M. (1990) *Phys. Rev. Lett.*, **65**, 516.
76 Han, Y., Alsayed, A.M., Nobili, M., Zhang, J., Lubensky, T.C., and Yodh, A.G. (2006) *Science*, **314**, 626.
77 Astumian, R.D. (1997) *Science*, **276**, 917.
78 Anderson, V.J. and Lekkerkerker, H.N.W. (2002) *Nature*, **416**, 811.
79 van der Kooij, F.M., Kassapidou, K., and Lekkerkerker, H.N.W. (2000) *Nature*, **406**, 868.
80 van Bruggen, M.P.B., van der Kooij, F.M., and Lekkerkerker, H.N.W. (1996) *J. Phys. Condens. Matter*, **8**, 9451.
81 Buining, P.A., Philipse, A.P., and Lekkerkerker, H.N.W. (1994) *Langmuir*, **10**, 2106.
82 Lekkerkerker, H.N.W. (1992) Disorder-to-Order Phase Tranitions in Concentrated Colloidal Dispersions, in *Structure and Dynamics of Strongly Interacting Colloids and Supramolecular Aggregates in Solution* (eds S. Chen, J. Huang, and P. Tartaglia), Kluwer Academic Publishers, p. 97.
83 Stroobants, A., Lekkerkerker, H.N.W., and Odijk, T. (1986) *Macromolecules*, **19**, 2232.
84 Lekkerkerker, H.N.W., Coulon, P., Haegen, V.D., and Deblieck, R. (1984) *J. Chem. Phys.*, **80**, 3427.
85 Lekkerkerker, H.N.W. and Stroobants, A. (1993) *Physica A*, **195**, 387.
86 Lekkerkerker, H.N.W., Buining, P., Buitnhuis, J., Vroege, G.J., and Stroobants, A. (1995) Liquid Crystal Phase Transitions in Dispersions of Rodlike Colloidal Particles in *Observation, Prediction and Simulation of Phase Transitions in Complex Fluids* (eds M. Baus, L.F. Rull, and J.P. Ryckaert), Kluwer Academic Publishers, pp. 53–112.
87 Poon, W.C.K. and Pusey, P.N. (1995) Phase Transitions of Spherical Colloids in *Observation, Prediction, and Simulation of Phase Transitions in Complex Fluids* (eds M. Baus, L.F. Rull, and J.P. Ryckaert), Kluwer Academic Publishers.
88 Pusey, P.N. and VanMegen, W. (1986) *Nature*, **320**, 340.
89 Crocker, J.C., Matteo, J.A., Dinsmore, A.D., and Yodh, A.G. (1999) *Phys. Rev. Lett.*, **82**, 4352.

90 Dinsmore, A.D., Warren, P.B., Poon, W.C.K., and Yodh, A.G. (1997) *Europhys. Lett.*, **40**, 337.
91 Adams, M., Dogic, Z., Keller, S.L., and Fraden, S. (1998) *Nature*, **393**, 349.
92 Dogic, Z. and Fraden, S. (1997) *Phys. Rev. Lett.*, **78**, 2417.
93 Dogic, Z., Frenkel, D., and Fraden, S. (2000) *Phys. Rev. E*, **62**, 3925.
94 Dogic, Z. and Fraden, S. (2000) *Langmuir*, **16**, 7820.
95 Dogic, Z., Purdy, K.R., Grelet, E., Adams, M., and Fraden, S. (2004) *Phys. Rev. E*, **69**, 051702. (a) Grier, D.G. (2003) *Nature*, **424**, 810. (b) Prasad, V., Semwogerere, D., and Weeks, E.R. (2007) *J. Phys. Cond. Matt.*, **19**, 113102.
96 Crocker, J.C. and Grier, D.G. (1996) *J. Colloid Interface Sci.*, **179**, 298.
97 Lau, A.W.C., Lin, K.H., and Yodh, A.G. (2003a) *Phys. Rev. E*, **66**, 020401.
98 Dinsmore, A.D., Yodh, A.G., and Pine, D.J. (1995) *Phys. Rev. E*, **52**, 4045.
99 Berthier, L., Biroli, G., Bouchaud, J.P., Cipelletti, L., Masri, D.E., L'Hote, D., Ladieu, F., and Pierno, M. (2005) *Science*, **310**, 1797.
100 Leunissen, M.E., Christova, C.G., Hynninen, A.P., Royall, C.P., Campbell, A.I., Imhof, A., Dijkstra, M., van Roij, R., and van Blaaderen, A. (2005) *Nature*, **437**, 235.
101 Yethiraj, A. and van Blaaderen, A. (2003) *Nature*, **421**, 513.
102 Nikolaides, M.G., Bausch, A.R., Hsu, M.F., Dinsmore, A.D., Brenner, M.P., Weitz, D.A., and Gay, C. (2002) *Nature*, **420**, 299.
103 Dinsmore, A.D., Hsu, M.F., Nikolaides, M.G., Marquez, M., Bausch, A.R., and Weitz, D.A. (2002) *Science*, **298**, 1006.
104 Pham, K.N., Puertas, A.M., Bergenholtz, J., Egelhaaf, S.U., Moussaid, A., Pusey, P.N., Schofield, A.B., Cates, M.E., Fuchs, M., and Poon, W.C.K. (2002) *Science*, **296**, 104.
105 Likos, C.N. (2001) *Phys. Rep.*, **348**, 267.
106 Trappe, V., Prasad, V., Cipelletti, L., Segre, P.N., and Weitz, D.A. (2001) *Nature*, **411**, 772.
107 Gasser, U., Weeks, E.R., Schofield, A., Pusey, P.N., and Weitz, D.A. (2001) *Science*, **292**, 258.
108 Lin, K.H., Crocker, J.C., Prasad, V., Schofield, A., Weitz, D.A., Lubensky, T.C., and Yodh, A.G. (2000) *Phys. Rev. Lett.*, **85**, 1770.
109 Xia, Y.N., Gates, B., Yin, Y.D., and Lu, Y. (2000) *Adv. Mater.*, **12**, 693.
110 Velev, O.D., Lenhoff, A.M., and Kaler, E.W. (2000) *Science*, **287**, 2240.
111 Weeks, E.R., Crocker, J.C., Levitt, A.C., Schofield, A., and Weitz, D.A. (2000) *Science*, **287**, 627.
112 Kegel, W.K. and van Blaaderen, A. (2000) *Science*, **287**, 290.
113 van Megen, W., Mortensen, T.C., Williams, S.R., and Muller, J. (1998) *Phys. Rev. E*, **58**, 6073.
114 Zhu, J.X., Li, M., Rogers, R., Meyer, W., Ottewill, R.H., Russell, W.B., and Chaikin, P.M. (1997) *Nature*, **387**, 883.
115 Harland, J.L. and van Megen, W. (1997) *Phys. Rev. E*, **55**, 3054.
116 van Blaaderen, A., Ruel, R., and Wiltzius, P. (1997) *Nature*, **385**, 321.
117 Dinsmore, A.D., Yodh, A.G., and Pine, D.J. (1996) *Nature*, **383**, 239.
118 Trau, M., Saville, D.A., and Aksay, I.A. (1996) *Science*, **272**, 706.
119 van Blaaderen, A. and Wiltzius, P. (1995) *Science*, **270**, 1177.
120 Mason, T.G. and Weitz, D.A. (1995) *Phys. Rev. Lett.*, **75**, 2770.
121 Segre, P.N., Behrend, O.P., and Pusey, P.N. (1995b) *Phys. Rev. E*, **52**, 5070.
122 Ilett, S.M., Orrock, A., Poon, W.C.K., and Pusey, P.N. (1995) *Phys. Rev. E*, **51**, 1344.
123 van Megen, W. and Underwood, S.M. (1994) *Phys. Rev. E*, **49**, 4206.
124 van Megen, W. and Underwood, S.M. (1993) *Phys. Rev. Lett.*, **70**, 2766.
125 van Blaaderen, A. and Vrij, A. (1992) *Langmuir*, **8**, 2921.
126 Bartsch, E., Antonietti, M., Schupp, W., and Sillescu, H. (1992) *J. Chem. Phys.*, **97**, 3950.
127 Schatzel, K. and Ackerson, B.J. (1992) *Phys. Rev. Lett.*, **68**, 337.
128 van Megen, W., Underwood, S.M., and Pusey, P.N. (1991) *Phys. Rev. Lett.*, **67**, 1586.
129 van Megen, W. and Pusey, P.N. (1991) *Phys. Rev. A*, **43**, 5429.

130 Pusey, P.N., van Megen, W., Bartlett, P., Ackerson, B.J., Rarity, J.G., and Underwood, S.M. (1989) *Phys. Rev. Lett.*, **63**, 2753.

131 Pusey, P.N. and van Megen, W. (1987) *Phys. Rev. Lett.*, **59**, 2083.

132 Pusey, P.N. (1987) *De Physique*, **48**, 709.

133 Chen, D.T., Lau, A.W.C., Hough, L.A., Islam, M.F., Lubensky, T.C., and Yodh, A.G. (2007) *Phys. Rev. Lett.*, **99**, 148302.

134 Lau, A.W.C., Hoffman, B.D., Davies, A., Crocker, J.C., and Lubensky, T.C. (2003b) *Phys. Rev. Lett.*, **91**, 198101.

135 Levine, A.J. and Lubensky, T.C. (2000) *Phys. Rev. Lett.*, **85**, 1774.

136 Crocker, J.C., Valentine, M.T., Weeks, E.R., Gisler, T., Kaplan, P.D., Yodh, A.G., and Weitz, D.A. (2000) *Phys. Rev. Lett.*, **85**, 888.

137 MacKintosh, F.C. and Schmidt, C.F. (1999) *Curr. Opin. Colloid Interface. Sci.*, **4**, 300.

138 Palmer, A., Mason, T.G., Xu, J.Y., Kuo, S.C., and Wirtz, D. (1999) *Biophysical*, **76**, 1063.

139 Mason, T.G., Ganesan, K., van Zanten, J.H., Wirtz, D., and Kuo, S.C. (1997) *Phys. Rev. Lett.*, **79**, 3282.

140 Gittes, F., Schnurr, B., Olmsted, P.D., MacKintosh, F.C., and Schmidt, C.F. (1997) *Phys. Rev. Lett.*, **79**, 3286.

141 Alsayed, A.M., Islam, M.F., Zhang, J., Collings, P.J., and Yodh, A.G. (2005) *Science*, **309**, 1207.

142 Alsayed, A., Dogic, Z., and Yodh, A. (2004) *Phys. Rev. Lett.*, **93**, 057801.

143 Islam, M.F., Alsayed, A.M., Dogic, Z., Zhang, J., Lubensky, T.C., and Yodh, A.G. (2004) *Phys. Rev. Lett.*, **92**, 088303.

144 Islam, M., Nobili, M., Ye, F., Lubensky, T., and Yodh, A. (2005) *Phys. Rev. Lett.*, **95**, 148301.

145 Pelton, R. (2000) *Adv. Colloid Interface Sci.*, **85**, 1.

146 Senff, H. and Richtering, W.J. (1999) *Chem. Phys.*, **111**, 1705.

147 Stieger, M., Richtering, W., Pedersen, J.S., and Lindner, P. (2004) *J. Chem. Phys.*, **120**, 6197.

148 Shibayama, M. and Tanaka, T. (1993) *Adv. Polym. Sci.*, **109**, 1.

149 Kokufuta, E., Zhang, Y., Tanaka, T., and Mamada, A. (1993) *Macromolecules*, **26**, 1053.

150 Shibayama, M., Tanaka, T., and Han, C. (1992) *J. Chem. Phys.*, **97**, 6829.

151 Shibayama, M., Tanaka, T., and Han, C. (1992) *J. Chem. Phys.*, **97**, 6842.

152 Li, Y. and Tanaka, T. (1992) *Annu. Rev. Mater. Sci.*, **22**, 243.

153 Tokuhiro, T., Amiya, T., Mamada, A., and Tanaka, T. (1991) *Macromolecules*, **24**, 2936.

154 Hirotsu, S., Hirokawa, Y., and Tanaka, T. (1987) *J. Chem. Phys.*, **87**, 1392.

155 Otake, K., Inomata, H., Konno, M., and Saito, S. (1990) *Macromolecules*, **23**, 283.

156 Debord, J.D., Eustis, S., Debord, S.B., Lofye, M.T., and Lyon, L.A. (2002) *Adv. Mater.*, **14**, 658.

157 Wu, J., Huang, G., and Hu, Z. (2003a) *Macromolecules*, **36**, 440.

158 Wu, J., Zhou, B., and Hu, Z. (2003b) *Phys. Rev. Lett.*, **90**, 048304.

159 Han, Y., Ha, N.Y., Alsayed, A.M., and Yodh, A.G. (2008a) *Phys. Rev. E*, **77**, 041406.

160 Han, Y., Yair, S., Alsayed, A.M., Yunker, P., Lubensky, T.C., and Yodh, A.G. (2008b) *Nature*, **456**, 898.

161 Yunker, P., Zhang, Z., Aptowicz, K.B., and Yodh, A.G. (2009) *Phys. Rev. Lett.*, **103**, 115701.

162 Yunker, P., Zhang, Z., and Yodh, A.G. (2010) *Phys. Rev. Lett.*, **104**, 015701.

163 Zhang, Z., Xu, N., Chen, D.T.N., Yunker, P., Alsayed, A.M., Aptowicz, K., Habdas, P., Liu, A.J., Nagel, S., and Yodh, A.G. (2009) *Nature*, **459**, 230. (a) Chen, K., Ellenbroek, W.G., Zhang, Z.X., Chen, D.T.N., Yunker, P.J., Henkes, S., Brito, C., Dauchot, O., van Saarloos, W., Liu, A.J., and Yodh, A.G. (2010) *Phys. Rev. Lett.*, **105**, 025501. (b) Peng, Y., Wang, Z., Alsayed, A.M., Yodh, A.G., and Han, Y. (2010) *Phys. Rev. Lett.*, **20**, 205703. (c) Wang, Z,-R., Alsayed A.M., Yodh, A.G. and Han, Y. (2010) *J. Chem. Phys.*, **132**, 154501.

164 Berne, B.J. and Pecora, R. (2000) *Dynamic Light Scattering: With Applications to Chemistry, Biology, and Physics*, Dover Publications, Inc., Mineola, NY.

165 Johnson, C.S. and Gabriel, D.A. (1994) *Laser Light Scattering*, Dover Publications, Inc., Mineola, NY.
166 Pronk, S. and Frenkel, D. (2004) *Phys. Rev. E*, **69**, 066123.
167 Baumgartl, J. and Bechinger, C. (2005) *Europhys. Lett.*, **71**, 487.
168 Polin, M., Grier, D.G., and Han, Y. (2007) *Phys. Rev. E*, **76**, 041406.
169 Behrens, S.H. and Grier, D.G. (2001) *Phys. Rev. E*, **64**, 050401.
170 Han, Y. and Grier, D.G. (2003) *Phys. Rev. Lett.*, **91**, 038302.
171 Chan, E.M. (1977) *J. Phys. C*, **10**, 3477.
172 Hansen, J.P. and McDonald, I.R. (1986) *Theory of Simple Liquids*, 2nd edn, Academic, New York.
173 Stieger, M., Pedersen, J.S., Lindner, P., and Richtering, W. (2004) *Langmuir*, **20**, 7283.
174 Dash, J.G. (1999) *Rev. Mod. Phy.*, **71**, 1737.
175 Löwen, H. (1994) *Phys. Rep.*, **237**, 249.
176 Dash, J.G., Fu, H., and Wettlaufer, J.S. (1995) *Rep. Prog. Phys.*, **58**, 115.
177 Cahn, R.W. (2001) *Nature*, **413**, 582.
178 Lindemann, F.A. (1910) *Z. Phys.*, **11**, 609.
179 Edwards, S.F. and Warner, M. (1979) *Philos. Mag.*, **40**, 257.
180 Burakovsky, L., Preston, D.L., and Silbar, R.R. (2000) *Phys. Rev. B*, **61**, 15011.
181 Lipowsky, R. (1986) *Phys. Rev. Lett.*, **57**, 2876.
182 Ciccotti, G., Guillope, M., and Pontikis, V. (1983) *Phys. Rev. B*, **27**, 5576.
183 Curtin, W.A. (1989) *Phys. Rev. B*, **39**, 6775.
184 Pluis, B., Frenkel, D., and van der Veen, J.F. (1990) *Surf. Sci.*, **239**, 282.
185 Ohnesorge, R., Löwen, H., and Wagner, H. (1994) *Phys. Rev. E*, **50**, 4801.
186 Lipowsky, R., Breuer, U., Prince, K.C., and Bonzel, H.P. (1989) *Phys. Rev. Lett.*, **62**, 913.
187 Cahn, R.W. (1989) *Nature*, **323**, 668.
188 Pettersen, M.S., Lysek, M.J., and Goodstein, D.L. (1989) *Phys. Rev. B*, **40**, 4938.
189 Dahmen, U., Hagege, S., Faudot, F., Radetic, T., and Johnson, E. (2004) *Philos. Mag.*, **84**, 2651.
190 Frenken, J.W.M. and van der Veen, J.F. (1985) *Phys. Rev. Lett.*, **54**, 134.
191 van der Veen, J.F. (1999) *Surf. Sci.*, **433–435**, 1.
192 Pluis, B., van der Gon, A.W.D., Frenken, J.W.M., and van der Veen, J.F. (1987) *Phys. Rev. Lett.*, **59**, 2678.
193 Broughton, J.Q. and Gilmer, G. (1986) *Phys. Rev. Lett.*, **56**, 2692.
194 Phillpot, S., Lutsko, J.F., Wolf, D., and Yip, S. (1989) *Phys. Rev. B*, **40**, 2831.
195 Kikuchi, R. and Cahn, J. (1980) *Phys. Rev. B*, **21**, 1893.
196 Nguyen, T. and Yip, S. (1989) *Mater. Sci. Eng. A: Struct.*, **107**, 15.
197 Ho, P.S., Kwok, T., Nguyen, T., Nitta, C., and Yip, S. (1985) *Scripta. Math.*, **19**, 993.
198 Hsieh, T.E. and Balluffi, R.W. (1989) *Acta Meter.*, **37**, 1637.
199 Zhu, J., Li, M., Rogers, R., Meyer, W., Ottewill, R.H., Crew, S.-S.S., Russel, W.B., and Chaikin, P.M. (1997b) *Nature*, **387**, 883.
200 Hirth, J.P. and Lothe, J. (1982) *Theory of Dislocations*, 2nd edn, John Wiley & Sons, Inc., New York.
201 Alsayed, A.M., Han, Y., Aptowicz, K., and Yodh, A.G., *Phys. Rev. E*, in press.
202 Bongers, J. and Versmold, H. (1996) *J. Chem. Phys.*, **104**, 1519.
203 Ohshima, Y.N. and Nishio, I. (2001) *J. Chem. Phys.*, **114**, 8649.
204 Schall, P., Cohen, I., Weitz, D.A., and Spaepen, F. (2004) *Science*, **305**, 1944.
205 Pronk, S. and Frenkel, D. (1999) *J. Chem. Phys.*, **110**, 4589.
206 Strandburg, K. (1992) *Bond-Orientational Order in Condensed Matter Systems*, Springer, New York.
207 Nelson, D.R. (2002) *Defects and Geometry in Condensed Matter Physics*, Cambridge University Press, Cambridge.
208 Peierls, R.E. (1934) *Helv. Phys. Acta*, **7**, 81.
209 Landau, L.D. and Lifshitz, E.M. (1980) *Statistical Physics*, Part I, Pergamon, Oxford.
210 Landau, L.D. (1937) *Phys. Z. Sowjetunion*, **11**, 26.
211 Mermin, N.D. and Wagner, H. (1966) *Phys. Rev. Lett.*, **17**, 1133.
212 Mermin, N.D. (1968) *Phys. Rev.*, **176**, 250.
213 Kosterlitz, J.M. and Thouless, D.J. (1973) *J. Phys. C: Solid State Physics*, **6**, 1181.

214 Nelson, D.R. and Halperin, B.I. (1979) *Phys. Rev. B*, **19**, 2457.
215 Young, A.P. (1979) *Phys. Rev. B*, **19**, 1855.
216 Li, D. and Rice, S.A. (2005) *Phys. Rev. E*, **72**, 041506.
217 Pang, H., Pan, Q., and Song, P.H. (2007) *Phys. Rev. B*, **76**, 064109.
218 Chekmarev, D.S., Oxtoby, D.W., and Rice, S.A. (2001) *Phys. Rev. E*, **63**, 051502.
219 Weber, H., Marx, D., and Binder, K. (1995) *Phys. Rev. B*, **51**, 14636.
220 Zangi, R. and Rice, S.A. (1998) *Phys. Rev. E*, **58**, 7529.
221 Strandburg, K.J. (1988) *Rev. Mod. Phys.*, **60**, 161.
222 Chou, C.F., Jin, A.j., Hui, S.W., Huang, C.C., and Ho, J.T. (1998) *Science*, **280**, 1424.
223 Grier, D.G., Murray, C.A., Bolle, C.A., Gammel, P.L., Bishop, D.J., Mitzi, D.B., and Kapitulnik, A. (1991) *Phys. Rev. Lett.*, **66**, 2270.
224 Angelescu, D.E., Harrison, C.K., Trawick, M.L., Register, R.A., and Chaikin, P.M. (2005) *Phys. Rev. Lett.*, **95**, 025702.
225 Murray, C.A. and Winkle, D.H.V. (1987) *Phys. Rev. Lett.*, **58**, 1200.
226 Marcus, A.H. and Rice, S.A. (1996) *Phys. Rev. Lett.*, **77**, 2577.
227 Karnchanaphanurach, P., Lin, B.H., and Rice, S.A. (2000) *Phys. Rev. E*, **61**, 4036.
228 Zahn, K., Lenke, R., and Maret, G. (1999) *Phys. Rev. Lett.*, **82**, 2721.
229 Zahn, K. and Maret, G. (2000) *Phys. Rev. Lett.*, **85**, 3656.
230 von Grünberg, H.H., Keim, P., Zahn, K., and Maret, G. (2003) *Phys. Rev. Lett.*, **93**, 255703.
231 von Grünberg, H.H., Keim, P., and Maret, G. (2007) *Soft Matter*, **3**, 41.
232 Murray, C.A. and Grier, D.G. (1996) *Annu. Rev. Phys. Chem.*, **47**, 421.
233 van Winkle, D.H. and Murray, C.A. (1986) *Phys. Rev. A*, **34**, 562.
234 Lin, S.Z., Zheng, B., and Trimper, S. (2006) *Phys. Rev. E*, **73**, 066106.
235 Zollweg, J.A. and Chester, G.V. (1992) *Phys. Rev. B*, **46**, 11186.
236 Fernandez, J.F., Alonso, J.J., and Stankiewicz, J. (1995) *Phys. Rev. Lett.*, **75**, 3477.
237 Jaster, A. (1998) *Europhys. Lett.*, **42**, 277.
238 Jaster, A. (2004) *Phys. Lett. A*, **330**, 120. (a) Mak, C.H. (2006) *Phys. Rev. E*, **73**, 065104.
239 Celestini, F., Ercolessi, F., and Tosatti, E. (1997) *Phys. Rev. Lett.*, **78**, 3153. (a) Bedanov, V.M., Gadiyak, G.V., and Lozovik, Y.E. (1985) *Phys. Lett. A*, **109**, 289.
240 von Grünberg, H.H., Keim, P., Zahn, K., and Maret, G. (2004) *Phys. Rev. Lett.*, **93**, 255703.
241 Keim, P., Maret, G., and von Grünberg, H.H. (2007) *Phys. Rev. E*, **75**, 031402.
242 Ubbelohde, A.R. (1965) *Melting and Crystal Structure*, Clarendon, Oxford.
243 Binder, K. (1987) *Rep. Prog. Phys.*, **50**, 783.
244 Wannier, G.H. (1950) *Phys. Rev.*, **79**, 357.
245 Wannier, G.H. (1973) *Phys. Rev. B: Erratum*, **7**, 5017.
246 Blote, H.W.J. and Nienhuis, B. (1994) *Phys. Rev. Lett.*, **72**, 1372.
247 Moessner, R. and Ramirez, A.R. (2006) *Phys. Today*, **59**, 24.
248 Pauling, L. (1935) *J. Am. Chem. Soc.*, **57**, 2680.
249 Harris, M.J., Bramwell, S.T., McMorrow, D.F., Zeiske, T., and Godfrey, K.W. (1997) *Phys. Rev. Lett.*, **79**, 2554.
250 Nakatsuji, S., Nambu, Y., Tonomura, H., Sakai, O., Jonas, S., Broholm, C., Tsunetsugu, H., Qiu, Y., and Maeno, Y. (2005) *Science*, **309**, 1697.
251 Bramwell, S.T. and Gingras, M.J.P. (2001) *Science*, **294**, 1495.
252 Moessner, R. (2001) *Can. J. Phys.*, **79**, 1283.
253 Ramirez, A.R. (2003) *Nature*, **421**, 483.
254 Anderson, P.W. (1987) *Science*, **235**, 1196.
255 Davidovic, D., Kumar, S., Reich, D.H., Siegel, J., Field, S.B., Tiberio, R.C., Hey, R., and Ploog, K. (1996) *Phys. Rev. Lett.*, **76**, 815.
256 Davidovic, D., Kumar, S., Reich, D.H., Siegel, J., Field, S.B., Tiberio, R.C., Hey, R., and Ploog, K. (1997) *Phys. Rev. B*, **55**, 6518.
257 Hilgenkamp, H., Ariando, A., Smilde, H.H., Blank, D.H.A., Rijnders, G., Rogalla, H., Kırtley, J.R., and Tsueiet, C.C. (2003) *Nature*, **422**, 50.

258 Kirtley, J.R., Tsuei, C.C., Ariando, A., Smilde, J.J.H., and Hilgenkamp, H. (2005) *Phys. Rev. B* **72** 214521.
259 Wang R.W. *et al.* (2006) *Nature*, **439**, 303.
260 Libal, A., Reichhardt, C., and Reichhardt, C.J.O. (2006) *Phys. Rev. Lett.*, **97**, 228302.
261 Koshikiya, Y. and Hachisu, S. (1982) Proceedings of the Colloid Symposium of Japan (in Japanese, 1982).
262 Pieranski, P., Strzlecki, L., and Pansu, B. (1983) *Phys. Rev. Lett.*, **50**, 900.
263 Ogawa, T. (1983) *J. Phys. Soc. Jpn.* (Suppl.), **52**, 167.
264 Chou, T. and Nelson, D.R. (1993) *Phys. Rev. E*, **48**, 4611.
265 Weiss, J.A., Oxtoby, D.W., Grier, D., and Murray, C.A. (1995) *J. Chem. Phys.*, **103**, 1180.
266 Schmidt, M. and Lowen, H. (1996) *Phys. Rev. Lett.*, **76**, 4552.
267 Schmidt, M. and Lowen, H. (1997) *Phys. Rev. E*, **55**, 7228.
268 Houtappel, R.M.F. (1950) *Physica*, **16**, 425.
269 Melko, R.G., den Hertog, B.C., and Gingras, M.J.P. (2001) *Phys. Rev. Lett.*, **87**, 067203.
270 Millane, R.P. and Blakeley, N.D. (2004) *Phys. Rev. E*, **70**, 057101.
271 Chen, Z.Y. and Kardar, M. (1986) *J. Phys. C: Solid State Physics*, **19**, 6825.
272 Gu, L., Chakraborty, B., Garrido, P.L., Phani, M., and Lebowitz, J.L. (1996) *Phys. Rev. B*, **53**, 11985.
273 Osterman, N., Babic, D., Poberaj, I., Dobnikar, J., and Ziherl, P. (2007) *Phys. Rev. Lett.*, **99**, 248301.
274 Mau, S.C. and Huse, D.A. (1999) *Phys. Rev. E*, **59**, 4396.
275 Phillips, J. (1996) *Rep. Prog. Phys.*, **59**, 1133.
276 Lohr, M.A., Alsayed, A.M., Chen, B.G., Zhang, Z., Kamien, R.D., and Yodh, A.G. (2010) *Phys. Rev. E*, **81**, 040401.

Part Four
Mechanical Properties

11
Yielding, Flow, and Slip in Microgel Suspensions: From Microstructure to Macroscopic Rheology
Michel Cloitre

11.1
Introduction

Microgels are hybrid particles consisting of an intramolecular cross-linked polymeric micronetwork swollen by a good solvent. Because of this architecture, they are partially impenetrable just like colloids but at the same time inherently soft and deformable like polymers. Most of their properties result from this subtle interplay between colloid-like and polymer-like features. On the one hand, the dispersion and the degree of swelling of microgels in a solvent are mainly governed by polymer–solvent interactions. On the other hand, microgel suspensions exhibit the main characteristic features of colloidal behavior for extremely low solid contents. Dilute suspensions are weakly elastic viscous fluids similar to conventional particulate suspensions. At higher concentrations, microgels come into contact and pack into a continuous elastic network that resists deformation. Upon application of a sufficiently high stress, the particles can flow past one another appreciably and the elastic network is disrupted making the suspensions strongly shear thinning. This unique property is widely exploited to impart solid-like behavior to formulations used in a variety of industries for coatings, ceramics, inks, personal care products, and foods.

Over the years, the rheology of microgel suspensions has stimulated a lot of work both for industrial applications and for fundamental science. *On the applied level*, it is well established that rheology plays a major role in determining the appearance and many of the performance properties of microgel-based products. At the same time, conferring the desired rheology on a formulation is often a difficult task. Although a rich literature describes how the composition and the architecture of microgels (monomer composition, cross-link density, particle size, surface charge, and introduction of functional groups) can be customized to meet the requirements of specific situations, many features of the rheology of these materials remain to be rationalized before it will be possible to tune their rheology at will. *On the fundamental level*, microgel suspensions have recently attracted renewed interest as model systems for colloidal gels and glasses. Microgel suspensions share common features with many other soft particle dispersions such as emulsions, multilamellar vesicles, block

Microgel Suspensions: Fundamentals and Applications
Edited by Alberto Fernandez-Nieves, Hans M. Wyss, Johan Mattsson, and David A. Weitz
Copyright © 2011 WILEY-VCH Verlag GmbH & Co. KGaA, Weinheim
ISBN: 978-3-527-32158-2

copolymer, or star micelles, which all form soft glasses at high packing density. Since the softness of microgels can be finely adjusted through their architecture and external parameters, they constitute an exquisite system that has permitted definite progress in the understanding of the structure, dynamics, and flow behavior of soft particle glasses.

In this chapter, we review several generic features of the rheology of microgel dispersions with the ambition to draw a bridge between the local dynamics at the particle scale and the macroscopic rheology. Several recent advances in this direction have been stimulated by the development of novel nonintrusive techniques that are able to probe the dynamics of soft materials over a broad range of length scales and timescales. We present two of these techniques in Section 11.2: diffusive wave spectroscopy (DWS), which appears to be an outstanding method to probe the local dynamics of microgel dispersions, and particle-tracking velocimetry techniques, which allow visualizing the flow of suspensions in real space. Section 11.3 deals with the near-equilibrium properties and the linear viscoelasticity of microgel suspensions both in the dilute and in the concentrated regimes. In Section 11.4, we discuss how these materials yield, flow, and ultimately age upon flow cessation. Section 11.5 is devoted to the flow and slipping properties of microgel suspensions near confining surfaces.

11.2
Advanced Techniques for Microgel Rheology

11.2.1
Macroscopic Shear Rheology

Conventional shear rheology remains the most popular technique devoted to the characterization of microgel suspensions in academy and industry. Recent advances have led to the development of commercial rheometers capable of performing most of the controlled stress and controlled strain rheological tests using a single instrument. Most often, the microgel dispersions involved in real applications are highly elastic solids for which cone and plate or parallel plate geometries are recommended. To characterize the rheological properties of a newly formulated product, it is crucial to determine the linear viscoelastic properties, the yield stress, and the flow curves. This requires correlating the results obtained from frequency sweep, strain sweep, and creep measurements. Generally, measurements are straightforward once a number of experimental issues have been seriously considered.

Wall slip is the most serious difficulty experienced when testing microgel dispersions. Like many other soft materials, microgel dispersions are prone to slip when they are sheared between solid surfaces [1, 2]. A direct manifestation of wall slip is that apparent motion can be detected below the bulk yield stress. This greatly affects the apparent yield stress values, the shape of the flow curves, and the linear storage and loss moduli [3]. Wall slip is therefore a major concern for rheologists interested in complex formulations involving microgel particles. Over the years, various solutions have been proposed to suppress slip. This is achieved by modifying the surfaces of the tools either by roughening [2], by sticking a rough coating such as solvent proof sandpaper [4, 5], or by using specific serrated tools [6]. On the fundamental level, wall slip is a key feature of microgel tribology that will be studied in Section 11.5.

Wall slip is often associated with edge flows and instabilities. These phenomena are very common in viscoelastic materials such as polymer solutions and melts, and concentrated suspensions. Generally, a crack develops at the free surface and propagates into the rheometer gap, which seriously affects the rheological behavior and can ultimately lead to the loss of cohesion of the studied material [7]. Although frequently observed, this phenomenon is not well understood yet. While for polymeric materials edge fracture occurs at high shear rates when the second normal stress difference exceeds a critical value, in pastes it generally takes place at low shear rates close to yield stress when the material becomes solid. The onset of edge banding often coincides with the development of inhomogeneous bulk flow associated with wall slip or shear banding [5]. The most detrimental manifestation of edge fracture is the formation of a highly irregular paste distribution at the periphery of the shearing surfaces, which affects the quality of the rheological signals.

The vane tool is a very useful and simple means to measure the rheology of microgel suspensions and other solid-like materials without artifacts associated with wall slip or elastic instabilities [8]. The vane geometry is similar to the Couette geometry with the inner cylinder replaced by a vane; in general, the outer cup is profiled or roughened. The vane avoids the wall slip problem since the shear surface is within the test material itself. It enables to reach higher rates than cone and plate or parallel plate geometries because ejection of the sample is restricted. It also minimizes structure breakdown during sample loading particularly when the measurement takes place inside the container where the sample has been prepared.

Another difficulty encountered in probing microgel pastes is the apparent lack of reproducibility of measurements. The rheological properties of concentrated pastes often vary in time and strongly depend on sample preparation and measurement history. This affects the determination of the linear viscoelastic moduli and yielding properties. The solution consists in controlling the initial preparation and the mechanical history. To prepare microgel pastes in a reproducible state, a preshear stress much larger than the yield stress, resulting in high shear rates, is applied to the material prior to each measurement. Upon flow cessation, the material slowly recovers the strain accumulated during preshearing. This recovery is very slow and persists up to the longest time accessible, indicating that mechanical equilibrium is never reached and that subsequent measurements probe out-of-equilibrium situations. Nevertheless, valuable information is obtained when the time elapsed after preparation is considered as an experimental variable. In Section 11.4, we shall show that these history-dependent phenomena exhibit the characteristic features of physical aging and that they can be rationalized in terms of concepts borrowed from the physics of disordered materials [9].

11.2.2
DWS-Based Microrheology

Microrheology represents a class of techniques in which the properties are inferred from the response of colloidal tracers to an applied force, which can arise from thermal fluctuations (passive methods), or which can be an external force of magnetic or optical origin (active methods) [10]. The tracers can be either colloidal particles artificially embedded in the system under study or can be part of it. Since microrheology measures properties at the microscopic scale, it provides information that could not be obtained by other conventional shear rheology. The window of accessible timescales and frequencies is extraordinarily wide, ranging from Brownian timescales to days. Microrheology also provides unique information about dynamics at the particle scale. Finally, it requires much less product than conventional rheology so that it can be used whenever small volumes of materials are available. While the development and the use of microrheology have initially been restricted to the area of fundamental science, it has begun to be increasingly attractive to engineers and applied researchers.

Experimentally, microrheology infers the linear viscoelastic properties of a material from the mean square displacement (MSD), $\langle \Delta r^2(t) \rangle$, of the tracers. For a purely elastic solid-like material, the particle motion is constrained at long time so that the MSD reaches an average plateau value that is a function of elastic modulus $G'(\omega)$ and particle radius R through $\langle \Delta r^2(t) \rangle = k_B T / \pi G'(\omega) R$. If the material is purely viscous, the probe particles diffuse through it and the MSD increases linearly with time according to $\langle \Delta r^2(t) \rangle \sim k_B T t / \pi \eta R$. The mean square displacement can be measured using a number of tehniques such as fluorescence microscopy, photon correlation spectroscopy (PCS), or diffusive wave spectroscopy. Since DWS has an excellent temporal and spatial resolution, it has a definite advantage in probing highly elastic suspensions where the thermally induced motion can be extremely small.

Diffusive wave spectroscopy is a recent extension of conventional photon correlation spectroscopy to the multiple scattering regimes that characterize turbid materials [11]. The implementation of a typical DWS experiment in transmission geometry is shown in Figure 11.1. An intense expanded laser beam illuminates a multiple scattering medium contained in a transparent parallelepiped cell of thickness L. Photons propagate in the medium, emerge from the surface, and interfere with the detector where they form a speckle pattern. The propagation of photons in the medium can be described as a random walk with persistence length ℓ^*, also called the transport mean path. ℓ^* is a function of the nature of the tracers, of their size, and of their concentration and it can be measured independently. In addition to the transmission geometry, many other configurations have been proposed to satisfy various experimental constraints [11].

DWS techniques probe a wide range of timescales from Brownian timescales to days depending on the characteristics of the detection system (Figure 11.1). The first method uses a monomode optical fiber with low acceptance angle, which collects light from a single speckle spot and sends it to a photomultiplier. The signals are then analyzed by means of a commercial high-speed correlator. This method gives access

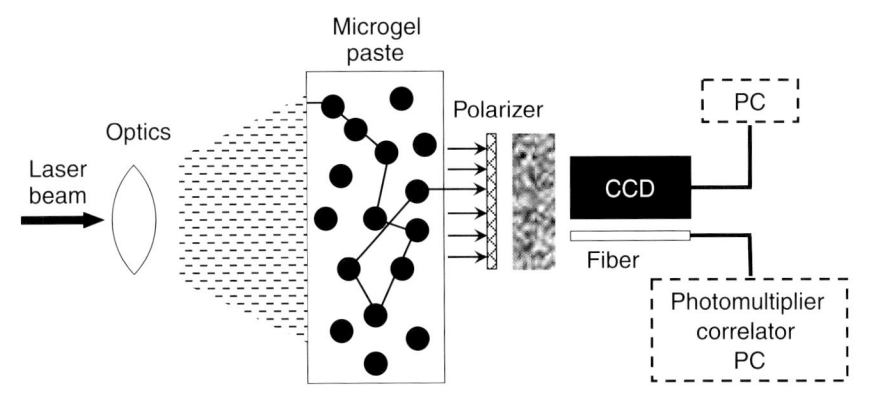

11.2 Advanced Techniques for Microgel Rheology

Figure 11.1 Experimental setup for diffusive wave spectroscopy investigations of microgel suspensions (transmission geometry).

to timescales ranging from 10^{-6} s or less to 10^2 s. A major difficulty inherent to this method is that concentrated microgel suspensions like many other soft solids are nonergodic media [12]. The probe particles are trapped near fixed average locations around which they execute spatially limited excursions. As a result, the time average intensity correlation function, which is measured at one particular location of the optical fiber, does not coincide with the ensemble-average correlation function that ultimately characterizes the dynamics. Experimentally different methods have been developed to account for nonergodicity [13].

Another outstanding development is the so-called multispeckle scheme that uses a CCD camera connected to a computer that numerically computes intensity–intensity correlation functions [14, 15]. Instead of analyzing a single speckle spot, one records a large area of the speckle pattern, which is equivalent to performing a large number of experiments in parallel. The acquisition time does not have to be larger than the relaxation time of the material (if any) like in conventional DWS, which has definite advantages in probing time-evolving materials. Moreover, the difficulties arising from nonergodicity disappear. The drawback is the limitation of the experimental timescale window at short times due to the low time resolution of conventional CCD cameras. Typically, it is possible to access timescales ranging from 10^{-2} to 10^5 s, which is perfectly adapted to investigate slow dynamics and aging phenomena in microgel dispersions.

Once ensemble-averaged correlations are obtained, the MSD of the scatterers, $\langle \Delta r^2(t) \rangle$, is easily calculated using the analytical expressions provided by the general DWS formalism for different experimental configurations [11]. The full frequency dependence of the storage and loss moduli are determined from the MSD using a generalized Stokes–Einstein equation, assuming that the stress response follows the fluctuations that determine the local dynamics of the tracers [16]. The medium must be incompressible and inertia negligible [17]. Different efficient schemes have been proposed to analyze the data in practical situations [18]. To fully assess the validity of the results, it is useful to check that the data are insensitive to the size of the probe particles. Strong discrepancies indicate the presence of spatial heterogeneities or of specific interactions between the microgels and the tracers [19].

By combining high-frequency detection and multispeckle DWS, it is possible to cover a range of frequencies from 10^{-7} to 10^5 rad/s. Due to its extreme sensitivity, DWS-based microrheology can easily probe microgel pastes with a shear modulus of about 100 Pa or larger, but it is somewhat limited when the shear modulus is much lower. The smallest accessible shear modulus ($G_{\min}$) is set by the maximum excursion of the probe particles that can be detected by the experimental setup, due to the finite value of the correlation base line. A useful estimate of $G_{\min}$ is given by the expression $G_{\min} \cong k_B T (L/\ell^*)^2 / \lambda^2 R$ (λ is the wavelength of light and R is the probe radius). The various parameters involved in this analysis are not all independent. For instance, the validity of the approximation of strong multiple scattering requires a ratio L/ℓ^* at least larger than 5 and preferentially greater than 10. Similarly, varying the radius of the probe particle affects the transport mean free path ℓ^*. This shows that the technique requires a careful definition of the experimental configuration prior to measurements. Recent applications of DWS-based microrheology to probe the local dynamics of microgel suspensions will be discussed in Section 11.3 [20].

11.2.3
Real Space Particle-Tracking Techniques

This class of techniques is based on real space imaging of a great number of Brownian tracer particles embedded in the studied material using conventional video microscopy, epifluorescence microscopy, or confocal microscopy [21]. The locations of the probes and their MSD are determined using image processing techniques [22]. The viscoelastic moduli are then extracted from the MSD just as described previously. Due to the finite spatial resolution associated with optical microscopy, the minimum detectable excursion is larger than that in DWS-based microrheology, which in turn restricts the applicability of the technique in its passive version to weakly elastic suspensions. As for DWS-based microrheology, possible sources of attractive or repulsive interactions between the tracers and the studied material must be seriously considered since they can affect the local mobility. However, these techniques have the unique advantage of performing spatially resolved measurements of rheological particles on micron length scales. This feature combined with the ability to perform many local measurements at the same time provides a quantitative understanding of the local elasticity and heterogeneities of soft materials. In its passive version, particle tracking microrheology has been successfully applied to various microgel systems [23–25].

Active particle tracking microrheology differs from its passive counterpart in that the probes are actively forced to move through the material under an external field, rather than simply allowed to diffuse. A promising application in the field of rheology is the measurement of the local velocity in complex fluids subject to shear flows in various geometries [5, 26–28]. Figure 11.2 illustrates the practical implementation of particle tracking velocimetry to characterize microgel flow in a cone and plate rheometer [5]. The microgel suspension is seeded with tiny hollow glass spheres

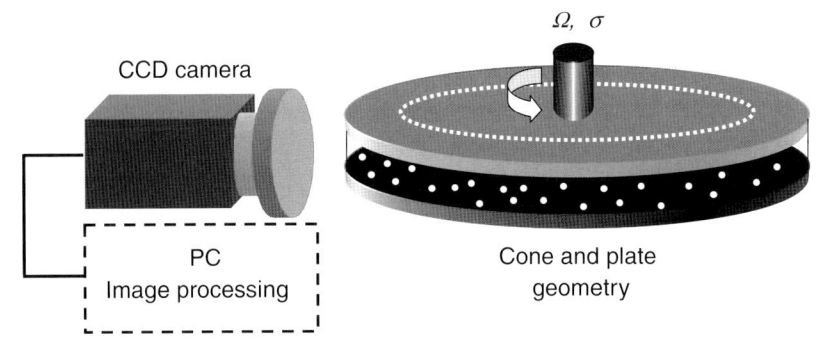

Figure 11.2 Particle tracking velocimetry for measuring shear profiles of microgel suspensions in a cone and plate geometry.

that reflect light when illuminated. The sample is then imaged onto a CCD camera at high magnification using a zoom lens focused at a radial position inside the cone and plate device. To avoid optical distortion by the outer meniscus, a thin transparent film is placed at the sample periphery. Flow profiles are obtained by tracking the successive positions of the tracers in the course of time. This technique has proven very efficient to probe the flow properties of concentrated microgel suspensions near confining surfaces [2, 5]. These aspects will be presented in Section 11.5.

11.3
Near-Equilibrium Properties and Linear Rheology of Microgel Suspensions

11.3.1
Dilute Regime and Paste Formation

At low concentrations, the linear rheology of microgel suspensions is that of dilute colloidal suspensions. The linear viscoelastic moduli show terminal regime with $G''(\omega) \sim \omega$, $G'(\omega) \sim \omega^2$, and $G''(\omega) \gg G'(\omega)$, allowing to define a Newtonian viscosity in the limit of low frequencies and low shear rates (η_0). At high frequencies and shear rates, the viscosity generally exhibits shear-thinning behavior. Qualitatively, the variation of the zero-shear viscosity η_0 with concentration is reminiscent of that measured in other particulate dispersions: η_0 first increases smoothly before rising sharply at a relatively well-defined concentration C_m the value of which depends on the degree of swelling. Figure 11.3a presents typical results obtained for alkali-swellable microgels made from ethyl-acrylate (EA) and methacrylic acid (MAA) [29]. Similar results have been reported for commercial microgel systems [30], neutral PMMA microgels [31], various polyelectrolyte microgels [32], thermosensitive microgels, and core–shell particles [33, 34].

A quantitative comparison between microgel dispersions and conventional colloidal suspensions requires conversion of the concentration C, which is the natural experimental control variable, into the volume fraction, Φ, which is normally used for

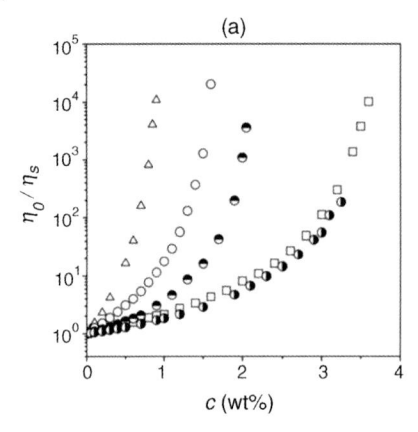

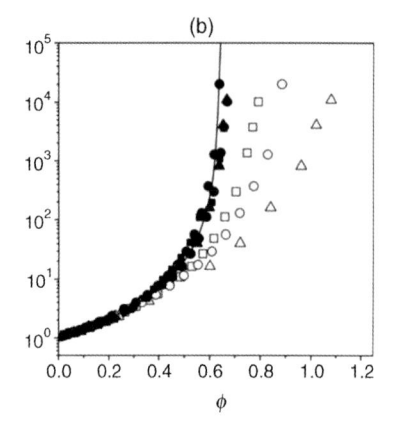

Figure 11.3 Viscosity of dilute suspensions of alkali-swellable microgel particles. The microgels are prepared by standard polymerization techniques using ethyl acrylate, methacrylic acid, and a difunctional cross-linking agent [37]. The cross-link density is characterized by the average number of monomers between cross-links (N_x). Graph (a) shows the increase in the viscosity with polymer concentration with (●: 0.01 mol/l and $N_x = 70$; ◐: 0.1 mol/l and $N_x = 70$) and without (△: $N_x = 140$; ○: $N_x = 70$; □: $N_x = 28$) added sodium chloride. Graph (b) shows the variation of the viscosity of the salt-free solutions versus the effective volume fraction before (open symbols) and after (full symbols) accounting for deswelling effects (same symbols as in a)). The continuous line represents the Krieger-Dougherty equation: $\eta_0/\eta_s = (1 - \Phi/\Phi_C)^{-2}$ where $\Phi_C = 0.64$. The data are replotted from Ref. [29].

colloids. This is generally achieved by postulating a linear relation of the type $\Phi = kC$, where the proportionality coefficient k is obtained from measurements of the hydrodynamic radius at infinite dilution using capillary viscometry or photon correlation spectroscopy [35]. Although very appealing, this procedure must be applied with caution since it can lead to erroneous conclusions. Complications arise from the very nature of microgels that are soft polymeric networks subject to osmotic deswelling when the total concentration is increased. Osmotic deswelling is particularly significant in polyelectrolyte microgels where the distribution of counterions at the origin of swelling is a function of concentration [29, 36, 37]. Neutral thermo-sensitive microgels also shrink at large volume fraction [38]. In ideal situations, it is possible to account for osmotic deswelling after adequate modeling. Once the volume fraction is correctly determined and/or eventually corrected from deswelling effect, it is found that the zero-shear viscosity diverges at C_m and that the variations of the viscosity with volume fraction are well described by the semiempirical Krieger-Dougherty relation standing for hard sphere dispersions (Figure 11.3b). More accurate conclusions can be drawn when microgels are sufficiently monodisperse to crystallize providing an unambiguous determination of the relation between Φ and C at high concentrations since the onset of crystallization is known to occur at $\Phi = 0.494$ [39, 40].

Above C_m, the rheological properties of microgel dispersions drastically change. The storage modulus $G'(\omega)$ becomes larger than the loss modulus $G''(\omega)$ over the entire range of experimentally accessible frequencies. This is the signature of solid-like behavior. In some fairly monodisperse model systems, the transition occurring at C_m has been shown to exhibit all the hallmarks of glass formation ($\Phi_g = 0.58$) and has been described in the framework of mode-coupling theory [40, 41]. For polydisperse samples, the transition occurring at C_m is generally associated with close packing ($\Phi_C = 0.64$). Yet, there is a fundamental difference between hard sphere suspensions and soft microgel suspensions since the latter can be concentrated well above the glass transition, microgels being inherently soft and deformable. In the following sections, we shall refer to these highly concentrated microgel suspensions as pastes.

11.3.2
Linear Viscoelasticity of Microgel Pastes

Although extremely useful, conventional rheological techniques do not provide much information about the dynamics of concentrated pastes due to their limited frequency window. Valuable information about the high-frequency viscoelastic behavior has been obtained using DWS-based microrheology [20]. A typical result is reproduced in Figure 11.4a, which shows the time variation of the mean square displacement of strongly scattering tracers embedded in an EA/MAA microgel paste. The technique gives access to the dynamics of the material over many decades of timescales. At short times, the motion of the tracers is subdiffusive ($\langle \Delta r^2(t) \rangle \sim t^\beta$ with $\beta < 1$). At long times, the mean square displacement reaches a nearly constant plateau that expresses dynamical arrest.

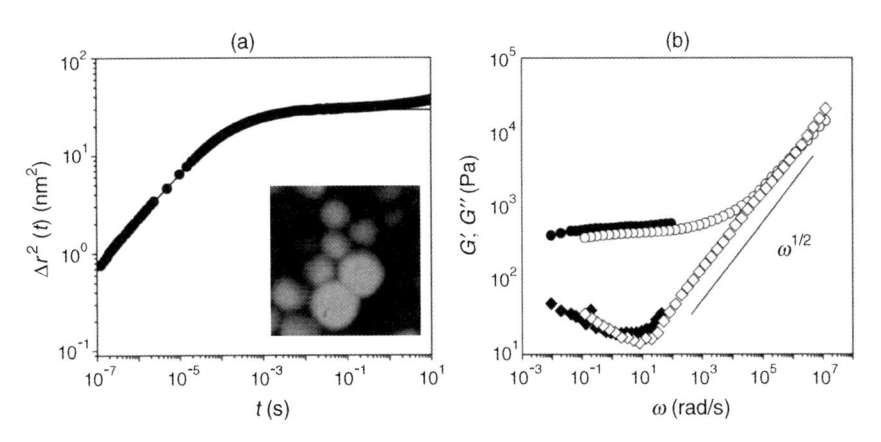

Figure 11.4 DWS microrheology of concentrated microgel suspensions. Graph (a) shows the mean square displacement of polystyrene beads ($R = 85$ nm) embedded in a suspension of EA/MAA microgels ($N_X = 140$, $C = 2$ wt%). The data are well described by a stretched exponential (continuous line) [20]. The inset is an AFM picture of a concentrated suspension showing the dense amorphous structure of microgel particles [37]. The polyhedron shape of the particles and the contacting facets are clearly visible. Graph (b) shows the variations of the storage (G': ○) and loss modulus (G'': ◇) extracted from the MSD. The full symbols represent data measured using conventional rheology.

This behavior can be understood in relation to the amorphous close-packed structure of the pastes at high-volume fraction (inset of Figure 11.4a). Each tracer particle is locally trapped in a cage formed by its many neighbors so it cannot move over large distances. The short-time diffusive motion is due to the local motion of the particles inside their cages. When a particle moves, it deforms elastically its neighbors that subsequently push it back inside the cage. The maximum excursion of a particle is reached when the elastic restoring forces exerted by the cages become larger than Brownian forces. This provides a direct way to estimate the shear modulus of the cage from the plateau value of the mean square displacement: $G_0 = k_B T / \pi R \langle \Delta r^2(t) \rangle_\infty$. For sufficiently dense pastes, the time variations of the mean square displacement are well described by stretched exponentials, which indicates the presence of a wide spectrum of relaxation times. This description establishes a direct analogy between the rheology of microgel pastes and the dynamics of glasses [20]. The local motion of particles inside their cages can be viewed as analogous to the β-relaxation processes existing in colloidal glasses. The crossover between short time subdiffusive motion and dynamical arrest occurs at a characteristic relaxation time that is governed by the cage elasticity (G_0) and the local viscous friction (η_S) only: $\tau_\beta \sim \eta_S / G_0$.

The time variations of the mean-square displacement also give access to the full frequency dependence of the storage and loss moduli as explained in Section 11.2.2. The data presented in Figure 11.4b clearly show that microgel pastes behave like elastic solids at low frequencies with $G'(\omega) \gg G''(\omega)$ being nearly independent of frequency. $G''(\omega)$ exhibits a minimum that results from an increase in $G''(\omega)$ at low and high frequencies. The rise at low frequencies suggests the existence of relaxation processes associated with very slow structural rearrangements. The rise at high frequency, where $G''(\omega)$ becomes larger than the shear modulus, is well described over several decades by a robust power-law variation with an exponent close to 0.5. A very similar behavior has been found in thermosensitive microgel suspensions using a high-frequency squeeze flow rheometer [42]. In Figure 11.4b, we have also plotted the variations of the storage and loss moduli measured using conventional shear rheology for the same sample. We observe a perfect agreement between the macroscopic elastic modulus and the local cage modulus. This result indicates that the tracer particles all experience the same local environment in this system. This result is not general and large discrepancies between microrheology and conventional rheology are found in heterogeneous systems [23, 24].

11.3.3
Elastic Properties of Concentrated Microgel Pastes

The value and the prediction of the low-frequency plateau modulus G_0 have attracted a lot of attention since this quantity is central to many applications. In general, it is observed that the plateau shear modulus rises sharply above close packing before reaching a slower variation at high concentration (Figure 11.5a). In most systems, the initial rise in the shear modulus with concentration can be empirically described by a power law variation of the form $G_0 \sim C^n$, where the value of the exponent n varies

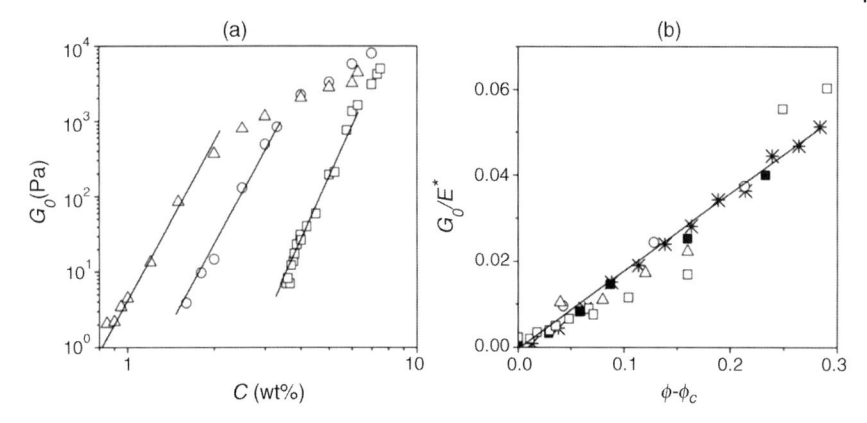

Figure 11.5 Elasticity of concentrated EA/MAA microgel suspensions. Graph (a) shows the increase with concentration of the plateau shear modulus G_0 for different cross-link densities ($\triangle$: $N_X = 140$; $\bigcirc$: $N_X = 70$; $\square$: $N_X = 28$). The initial rise of G_0 is well described by power laws of the form $G_0 \sim \Phi^n$ with $n \cong 7–8$ (continuous lines); at high concentration, G_0 becomes independent of the cross-link density and increases more slowly. Graph (b) compares the predictions of the micromechanical model ($*$) and the experimental data (E^* is the contact Young modulus; Φ is the effective volume fraction; $\Phi_C = 0.64$). The data are replotted from Ref. [48].

between 5 and 10 [33, 34, 38, 43, 44]. At large concentrations, G_0 increases linearly with the concentration $G_0 \sim C$ [37]. There have been many attempts to relate the power law variation observed near C_m to the form of the interaction potential between soft spheres [33, 34, 38, 43, 45]. Several authors have also tried to rationalize the concentration dependence of the shear modulus above close packing using the Zwanzig and Mountain formalism initially developed for conventional colloids [33, 46, 47]. This technique requires analytical expressions both for the interparticle potential and for the pair correlation function. It is not clear, however, whether the Zwanzig and Mountain formalism normally intended to predict the high frequency shear modulus yields satisfactory predictions of the low-frequency modulus G_0 currently measured in experiments. Moreover, these analyses rely on the assumption that the effective volume is estimated with reasonable accuracy. As discussed previously, this is a difficult task especially at high solid content where swelling equilibrium is generally not reached and osmotic deswelling plays a significant role.

A micromechanical model accounting explicitly for the disordered microstructure of microgel pastes and the existence of repulsive interactions has been proposed recently [48]. Pastes are modeled as disordered arrays of compressed elastic spheres interacting through a pairwise Hertzian potential. Beyond close packing, the particles adapt their volume and shape to packing constraints by developing flat facets at contact. Because the number and the size of the facets increase with volume fraction, the particles are expected to progressively take the shape of rounded polyhedrons, just like microgels in pastes (inset of Figure 11.4a). When an external deformation is applied, particles resist by exerting repulsive Hertzian forces through the contacting facets. Simulations have been performed wherein the packing is subject to isochoric uniaxial extension, allowing computing of the total energy stored during deformation and the full stress tensor, from which the high- and low-frequency storage moduli and the osmotic pressure are obtained. These quantities are found to rise by more than two orders of magnitude when the volume fraction is increased above Φ_C, just as observed experimentally [48]. This behavior is associated with the fact that there are more and more contacting facets per particle and that they are increasingly deformed. The increase in the low-shear modulus above close packing can also be represented empirically by a power law variation with an exponent of about 6 just as discussed above for experiments [48]. This is thus strong evidence that the model captures the essential physical features that determine microgel paste elasticity. Interestingly, the low-frequency modulus is found to be proportional to $\Phi_C - \Phi$ (Figure 11.5b). This result can be rationalized in terms of rigidity percolation by considering that the facets transmitting the stress form a disordered network of springs, the critical volume fraction being that of random close packing where the facets first appear. Similar arguments have been used to describe the elasticity of concentrated emulsions [49].

The numerical predictions shown in Figure 11.5b involves the so-called contact modulus E^* of individual particles. In the framework of Hertz theory, E^* is a function of the bulk Young modulus E of the particles and of the Poisson ratio: $E^* = E/2(1-\nu^2)$ [48]. While the value of E^* does not affect the variations of the shear modulus with volume fraction, it sets its absolute value. This explains why pastes made of stiffer particles have larger elastic properties [36, 37, 47]. It is therefore crucial to have good estimates of the elastic modulus of individual microgels. Like their macroscopic counterparts, microgels swell up to the point where their modulus becomes equal to the difference between the osmotic pressure inside the polymer network and the osmotic pressure of the solution. Simple relations exist between the elastic modulus and the swelling equilibrium both for neutral and charged gels [50, 51]. Any parameter that acts to increase or decrease the swelling will affect the elastic modulus in the opposite direction. For neutral gels, the osmotic pressure inside the polymer network results from the solvent–polymer mixing free energy so that the particle elasticity depends essentially on the solvent quality and on the cross-link density [50, 52]. For polyelectrolyte gels, the osmotic pressure is dominated by the counterions associated with the fixed charges borne by the polymer network. The gel elasticity strongly depends on the degree of ionization and on the ionic strength [53]. While providing useful guides, this approach neglects the fact that the shear moduli of gels are often extremely dependent on the conditions of synthesis [53] and preparation [24, 47]. Moreover, it is not obvious that predictions holding for macroscopic gels can be transposed to microgels of micron size that have a heterogeneous structure and where finite size effects play an important role [54]. Advanced micromanipulation techniques capable of measuring directly the swelling behavior and the elastic properties of micron-size microgels appear very attractive in that respect [55].

The importance of preparation conditions is particularly well exemplified by the case of microstructured materials prepared by shear gelation. Shear gelation is a standard technique to make microgel-like materials from thermosensitive biopolymers such as agar or gellan gum [24]. The microgels are prepared by cooling a polymer solution initially heated at high temperature and applying a constant shear either during cooling or after cooling. The microgels obtained when the shear is applied after cooling behave much like conventional microgel systems. Those prepared by shearing during cooling exhibit surprising differences. The elastic modulus still increases with the polymer concentration as a power law, but the exponent ($\cong 2$) is much weaker than that classically measured in other systems ($\cong 7$). This probably reflects the fact that the individual particle moduli and the phase morphology change with polymer concentration [56].

11.4
Yielding, Flow, and Aging

11.4.1
Yielding

While concentrated microgel pastes behave like weak elastic solids at rest, they yield very much like liquids upon the application of a large enough stress. This property is of great technological importance in the ink or coating industry where microgels are often used as rheological additives to impart solid-like properties to complex formulations. Conventional rheology tests such as oscillatory strain sweeps are particularly well adapted to probe the yielding of microgel pastes. The sample is sheared at fixed oscillatory frequency but variable stress or strain amplitude. Figure 11.6 illustrates the generic yielding behavior of microgel pastes. At low strain

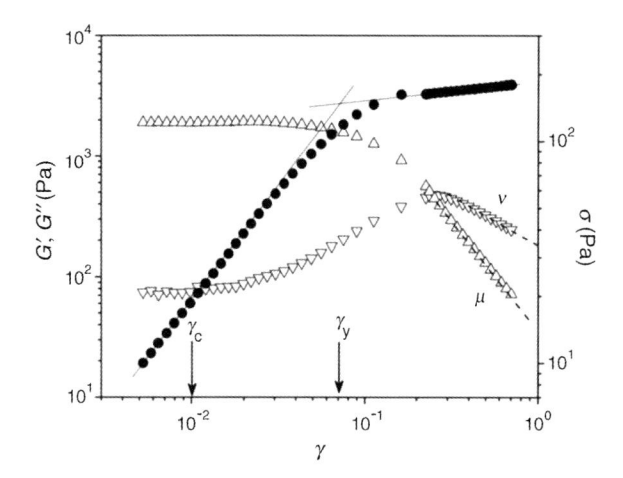

Figure 11.6 Characterization of the yielding properties of a concentrated EA/MAA microgel suspension ($N_x = 140$; $C = 4$ wt%) using the oscillatory strain sweep method ($\triangle$: storage modulus G'; $\triangledown$: loss modulus G''; ●: stress amplitude). The yield point is obtained from the variations of the stress at low and large strain amplitudes. γ_y is the yield strain; γ_c marks the limit of the regime of linear response. The dashed lines show that G' and G'' follow power law variations at large strains ($\mu \cong 1.78$; $\nu \cong 0.75$).

and stress, both the storage and the loss moduli are constant and the stress–strain relationship is perfectly linear. At higher strain and stress ($\gamma > \gamma_C$; $\sigma > \sigma_C$), the storage modulus is still constant but the loss modulus rises toward a pronounced peak before decreasing again larger strains. At the largest strains and stresses ($\gamma > \gamma_y$; $\sigma > \sigma_y$), the storage and loss moduli decay rapidly, the loss modulus being larger than the storage modulus, and the stress–strain relationship is no longer linear but instead varies like a power law: the paste yields. In the yielding regime, both moduli follow power law variations: $G'(\omega) \sim \omega^{-\mu}$ and $G''(\omega) \sim \omega^{-\nu}$. The yield point ($\gamma_y$, σ_y) can be determined from the intersection of the low-strain linear power law and the high-strain sublinear power law.

It is useful to connect the different behaviors observed during yielding to the local structure of microgel pastes. In the linear viscoelastic domain ($\gamma < \gamma_C$), the dynamics is dominated by the equilibrium microstructure, elastic forces associated with cage elasticity, and inherent dissipation of thermal fluctuations. The strain γ_C marking the upper limit of linear behavior is reached when the elastic energy stored during deformation exceeds the thermal energy: $\gamma_C \cong (kT/G_0 V)^{1/2}$ where G_0 is the plateau shear modulus and V is the volume of a microgel. This relation predicts that γ_C first drops above C_m before reaching a constant value at higher concentrations where $G_0 \sim C$ and $V \sim C^{-1}$ (Figure 11.5a). This is in good agreement with experimental data [36]. The pronounced increase of $G''(\omega)$ in the range $\gamma_C < \gamma < \gamma_y$ is due to the existence of structural plastic rearrangements that cause intense dissipation although the material keeps its integrity. When the mechanical strain or strain reaches its yield value, cages maintaining microgel particles locally trapped break and large-scale motions become possible. In view of this, we expect the yield stress to be proportional to the plateau modulus G_0 through the yield strain γ_y: $\sigma_y = G_0 \gamma_y$. This is in good agreement with experimental observations [20]. The yield strain γ_y, which is generally on the order of 0.05, is essentially constant and varies only very slightly with concentration, cross-link density, and polydispersity.

The results described above are general, and a very similar behavior has been observed in many different systems [57–60]. The case of shear gel materials [24] already discussed in Section 11.3.3 constitutes one notable exception. The yield strain is now proportional to the polymer concentration instead of being simply constant. Yielding occurs at a constant $\gamma_y \sigma_y$ value suggesting that the system flows at a critical energy density. Again, these results highlight the importance of the local microstructure with respect to the yielding properties of microgel materials.

The rheological response shown in Figure 11.6 is ubiquitous in many soft disordered systems such as emulsions, star solutions, block copolymer micelles, and onion phases [61]. This suggests that common behaviors may be well governed by the same underlying mechanisms. A fruitful approach establishes an analogy between the close-packed amorphous structure of microgel and that of supercooled liquids [62]. Within picture, the peak in $G''(\omega)$ before yielding is directly related to a decrease in the structural relaxation time under shear. If this assumption were true, it would provide a way to connect linear and nonlinear rheology because slow relaxation modes normally undetected in the linear regime might become accessible under shear. This new technique called strain-rate frequency superposition gives access to the viscoelastic behavior of soft solids over an extended range of timescales in the low-frequency domain. It has been successfully applied to many soft matter systems including PNIPAm–polyacrylic acid microgel systems in water [63].

11.4.2
Flow of Microgel Pastes

Above yielding microgel pastes flow very much like viscoelastic liquids and their rheological properties can be characterized by flow curves giving the shear rate response to the stress excitation. The following experimental protocol generally yields reliable results. A preshear stress resulting in high shear rates is first applied prior to each measurement. The stress is then quenched to a lower value and the steady-state shear rate is measured. The time required to reach the steady state becomes longer and longer as the applied stress gets closer to the yield stress. Figure 11.7 shows typical flow curves for ethyl-acrylate/methacrylic acid polyelectrolyte microgel pastes [20]. This representation shows that the stress first increases very slowly over several decades of shear rates and then finally rises up at higher shear rates. There is a definite advantage of using the $\sigma(\dot{\gamma})$ representation instead of the plot of the effective viscosity versus shear rate, $\eta(\dot{\gamma})$, since the latter does not generally capture deviations from a simple shear-thinning power law with exponent -1 [30, 31, 57]. The flow curve $\sigma(\dot{\gamma})$ is well represented by the Herschel–Bulkley equation: $\sigma = \sigma_y + a\dot{\gamma}^n$, where a and n are fitting parameters. Similar results have been reported in the literature for many types of microgel systems. Figure 11.7 also highlights the

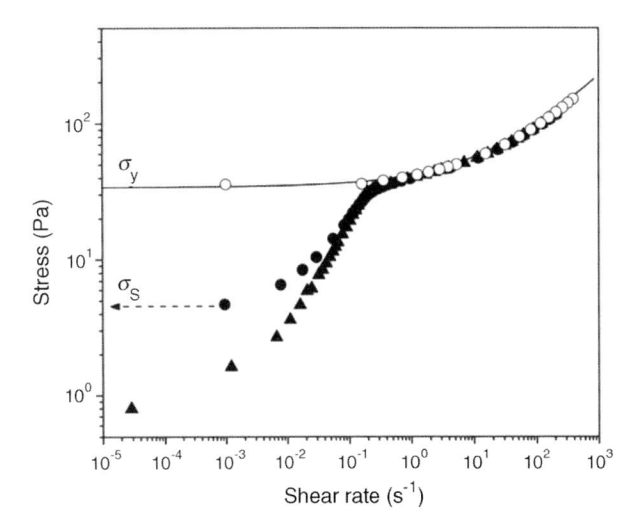

Figure 11.7 Flow curves of an EA/MAA microgel suspension sheared along different surfaces ($N_X = 28$, $C = 5.5$ wt%). The flow curve measured with rough surfaces (○) is not affected by wall slip and is well described by the Herschel–Bulkley equation (continuous line). When the shearing surfaces are smooth (●: surface treated with octadecyltrichlorosilane; ▲: silicon wafer), slip occurs and apparent motion is detected below the bulk yield stress σ_y.

importance of controlling the nature of the shearing surfaces. With smooth surfaces, apparent flow continues to be detected below the bulk yield stress, which obviously leads to erroneous determinations of σ_y. The apparent motion below the yield stress is due to wall slip (see Section 11.5).

Although the flow behavior depicted in Figure 11.7 is remarkably universal, the connection between the macroscopic flow rheology of microgel pastes and their microscopic structure remains an open issue. Recently, the nonlinear rheology of thermosensitive core–shell microgels has been discussed in the framework of an extension of mode coupling theory (MCT) to dense colloidal suspensions subject to an external drive [64]. In the MCT microscopic equations, competition arises from the structural arrest induced by the caging of the particles at high densities and shear advection that speeds up the structural relaxation. Interestingly, the theory predicts a nonequilibrium transition between a shear-thinning fluid and a yielding solid. The resulting stress simply reflects the slowing down of particle rearrangements as the external drive is reduced. Interestingly, the predictions of the theory capture the main characteristic features observed in experiments; in particular, the predicted flow curves nicely match the Herschel–Bulkley variation found in experiments [40, 65]. This analysis, however, is restricted to a narrow range of volume fraction situated between glass transition ($\Phi_g = 0.58$) and close packing ($\Phi_C = 0.64$). In addition, the exact comparison with experimental data involves adjustable parameters that do not seem to be directly related to materials properties.

The MCT approach focuses on the glass-like structure of the dispersions and neglects all physicochemical aspects. A recent study clearly demonstrates that the nonlinear rheology of highly concentrated microgel is largely influenced by the presence of the continuous phase [20]. The model is again based on the amorphous, jammed structure of pastes (inset of Figure 11.4). At rest, the microgels are trapped in cages from which they can escape and relax back to another position when the stress is large enough. The duration of a rearrangement is set by the competition between the elastic restoring forces that push back the microgels in their cage and the viscous forces due to interparticle friction: $\tau = \eta_S/G_0$ (η_S is the solvent viscosity, G_0 is the low-frequency plateau modulus). At low shear rates, the microgels relax to their equilibrium position because the duration of a rearrangement is shorter than the advection time ($\dot{\gamma}\tau \ll 1$): the stress remains on the order of the yield stress. At high shear rates, the driving flow can induce a continuous sequence of rearrangements ($\dot{\gamma}\tau \gg 1$): the stress increases due to viscous dissipation. Interestingly, the model predicts that the flow curves should be described by a universal curve of the form $\sigma/\sigma_y = 1 + f(\dot{\gamma}\eta_S/G_0)$. This prediction has been successfully tested for the case of microgel pastes of various compositions [20]. This result gives evidence for the importance of contact dynamics and interparticle friction in the flow of soft wet particulate systems such as microgel pastes. It also provides a unique way to control the nonrheology of microgel pastes, once the viscosity of the suspending medium and the plateau shear modulus are known. The former can be adjusted either by changing the temperature or by adding rheological additives, while the latter can be controlled by the molecular architecture and the composition of the microgels (Section 11.3.3).

11.4.3
Slow Dynamics and Aging of Microgel Pastes

Below the yield stress, it is technically difficult to perform rheological measurements: the properties seem to vary endlessly in time and to depend strongly on sample preparation and measurement history. While these observations have been attributed for a long time to experimental artifacts, there is now clear evidence that they exhibit all the hallmarks of the aging phenomena commonly encountered in glassy systems [9, 66]. A general protocol for investigating rheological aging involves sample loading, rejuvenation, waiting, and measurement of the rheological properties. Rejuvenation is intended to erase all internal stresses stored during the preparation and loading of the rheometer. This is generally achieved by applying a high stress resulting in high shear rates prior to any measurement. The end of the rejuvenation step sets the time origin; the paste is then kept at rest during a time interval t_W called waiting time. Upon flow cessation, the paste solidifies and the strain recovers very slowly. Strain recovery is logarithmic in time and persists up to extremely long times without any evidence of arrest. Similarly, during the waiting time, the viscoelastic moduli strongly depend on the time elapsed after flow cessation; the storage modulus $G'(\omega, t)$ increases in time while the loss modulus $G''(\omega, t)$ decreases, both quantities following logarithmic functions. Such logarithmic variations, which are not found in usual viscoelastic materials, are the signature of aging phenomena [9, 36]. Since the paste is still evolving slowly, mechanical equilibrium is not reached and the properties are function of the waiting time t_W.

Most conventional rheological tests such as creep, step strain, or step rate measurements can been used to investigate aging [36]. Figure 11.8a shows the creep response of a microgel paste to a stress smaller than the yield stress applied at time t_W. Clearly, the strain following the application of the stress depends on the time the paste has been kept at rest. The longer the waiting time before applying the probe stress, the slower the overall response, the smaller initial elastic jump, and the slower creep: the paste becomes stiffer. This dynamics is nonstationary: the response at time t to an excitation at t_W depends on the independent timescales t and t_W, indicating that time translational invariance is broken and aging is taking place. Nonstationary creep is observed when the probe stress is in the range $\sigma_C < \sigma < \sigma_y$ (according to Section 11.3.3, σ_C marks the limit of the regime of linear response). Below σ_C, a new phenomenon appears: the material first creeps and then at some moment it begins to recover in the direction opposite to that of the applied stress. This effect is due to the fact that the sample continues to feel the effect of the internal stress stored during rejuvenation ("memory effect") [9, 67]. Above σ_y, stationary flow is possible, which suppresses aging.

Similar aging phenomena have been observed in the field of polymer mechanics [68], spin glass magnetization [69], and other concentrated suspensions [70]. A way to rationalize the response of such aging systems is to rescale the experimental time t by a scaling variable λ that is a function of t and t_W. A simple functional form for λ is obtained by scaling t with an effective relaxation time that is proportional to t_W^μ, where $0 < \mu < 1$ is a phenomenological exponent: $\lambda = t/t_W^\mu$. This type of scaling has

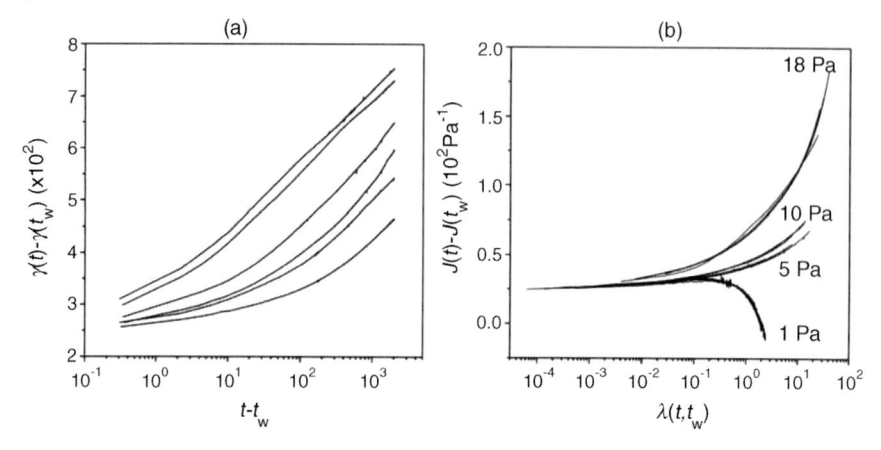

Figure 11.8 Creep properties of aging EA/MAA microgel suspensions ($N_x = 140$; $C = 2$ wt %). Graph (a) shows the strain response to an applied stress of 10 Pa for different waiting times (from top to bottom: $t_w = 15, 30, 300, 1000, 2000$, and $10\,000$ s). Graph (b) shows the collapse of the compliance data measured for different applied stresses when the data are plotted against the scaling variable $\lambda(t, t_w)$ ($J(t) = \gamma(t)/\sigma$) (see Equation 11.1). Exponent μ takes different values depending on the stress amplitude: $\mu \cong 1, 0.85, 0.8, 0.6$, respectively, for $\sigma = 1, 5, 10$, and 18 Pa. The yield stress is $\sigma_y = 19$ Pa.

been successfully applied to characterize the creep response of various microgel pastes [9]; μ is found to decrease from 1 to 0 when the amplitude of the probe stress is increased from σ_C to σ_y. The simple form $\lambda = t/t_w^\mu$ may be not valid when the duration of a measurement exceeds the waiting time t_w since then the material ages during testing. It is then interesting to consider that the effective relaxation time is proportional to $(t + t_w)^\mu$, which leads to the alternative form [68]:

$$\lambda(t, t_w) = \frac{t_w^{\mu-1}}{1-\mu}\left[\left(\frac{t}{t_w}\right)^{1-\mu} - 1\right] \quad \text{for} \quad \mu \neq 1 \tag{11.1}$$

$$\lambda(t, t_w) = \log\left(\frac{t}{t_w}\right) \quad \text{for} \quad \mu \neq 1 \tag{11.2}$$

Figure 11.8b shows that this scaling variable succeeds in collapsing the creep responses measured at different waiting times onto a single curve.

This scaling is very useful to rationalize the long-term behavior of microgel pastes. The μ exponent parameterizes the subaging deviations from a full t/t_w scaling when the probe stress is not vanishingly small. This provides a simple method to predict the long-time behavior of aging microgel pastes from a limited set of data. There is yet a major difference between rheological aging in pastes and aging phenomenon in polymer or structural glasses. In glasses, a metastable state is reached when the system is brought below the glass transition temperature and rejuvenation happens when the temperature is increased. In pastes, rejuvenation is due to macroscopic flow, whereas metastability appears upon flow cessation. The connection between this phenomenological description and the microscopic dynamics remains an open issue. The absence of intrinsic relaxation time suggests that rheological aging is closely associated with the slow cooperative relaxation of stains and stresses accumulated during the initial preparation and trapped inside interlocked neighboring particles. Microscopic evidence for such a scenario of rheological aging is still missing.

11.5
Slip and Flow of Microgel Suspensions Near Confining Surfaces

11.5.1
Wall Slip

The rheological behavior of concentrated microgel suspensions depends not only on bulk properties but also on the nature of the confining surfaces. In real situations, the motion of concentrated microgel suspensions is often dominated by wall slip. Wall slip is a very common phenomenon that has been reported in systems as different as particulate dispersions, flocculated suspensions, colloidal gels, emulsions, and foams [71]. Generally, slip arises from a depletion of particles adjacent to the surfaces, resulting in the presence of a thin layer of solvent between the particles and the wall. Since the local shear rate at the wall is much greater than that in the bulk material, the apparent flow can be markedly different from that of the bulk. This has crucial implications with respect to the rheological characterization, the transport, the storage, and the processing of concentrated suspensions. While this picture appears to be very general, quantitative predictions concerning the wall slip of colloidal dispersions have remained surprisingly scarce and fragmented until recently. In this context, microgel suspensions constitute an exquisite system that has permitted definite progress in the understanding of the slip behavior of soft particle pastes.

The presence of wall slip greatly affects the flow properties of microgel suspensions. This is shown in Figure 11.7, which represents the apparent flow curves for polyelectrolyte microgel suspensions measured using different shearing surfaces. When the suspension is sheared with rough surfaces, the flow curve is well described by the Herschel–Bukley variation that has been shown to characterize the bulk flow properties of this class of materials (Section 11.4.2). This behavior is dramatically changed when at least one of the shearing surfaces is smooth. While the flow curve is still superimposed on the bulk flow curve at high shear rates, it deviates at lower shear rates and apparent motion continues to be detected well below the yield stress σ_y. Apparent motion stops at a much lower stress σ_s, which is termed the sticking yield stress. The resulting shape of the flow curve, which displays a sharp kink at a stress close to the yield stress, can be considered as an unambiguous signature of wall slip occurrence. Figure 11.7 also shows that the apparent flow curve below the yield stress is extremely sensitive to the nature of the shearing surface. The sticking yield stress is obviously much larger for hydrophobic surfaces than for hydrophilic ones. This shows that short-range forces and possible adhesion of microgels onto the shearing surfaces have a great influence on the slip properties.

11.5.2
Direct Measurements of Slip Velocity

Until recently, the occurrence of wall slip has been inferred indirectly from rheological measurements. Indeed, in the presence of wall slip results, the rheological measurements depend on the measuring geometry. This is because the apparent shear rate is a combination of the bulk shear rate associated with material deformation and slip effects. Assuming that the slip velocity depends only on the shear stress and not on the gap h between shearing surfaces, the apparent shear rate is $\dot{\gamma}_{app} = \dot{\gamma} + V_S/h$ ($\dot{\gamma}$ is the shear rate associated with bulk flow; V_S is the slip velocity). This provides a simple way to extract slip velocities from measurements performed using different gaps [72].

Recently, the possibility to directly observe the flow of concentrated microgel suspensions in a rheometer and to measure velocity profiles using particle tracking velocimetry (Section 11.2.3) has stimulated important progress with respect to the detection, understanding, and curing of wall slip [2, 5]. Figure 11.9 shows the velocity profiles of an alkali-swellable EA/MAA microgel paste sheared under different

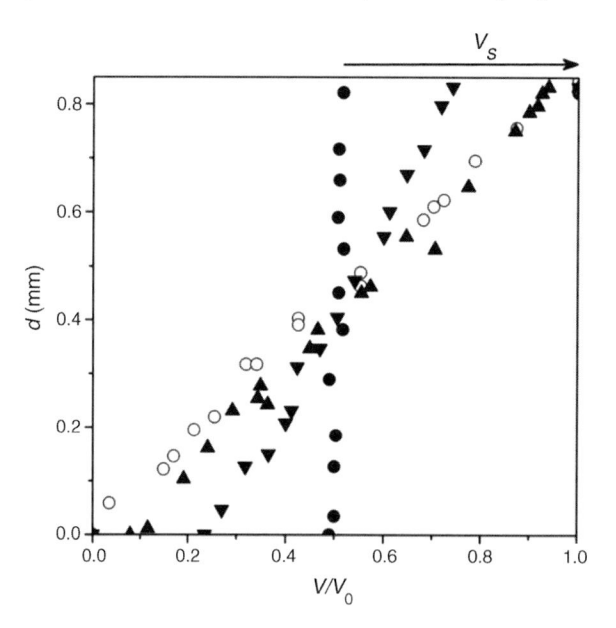

Figure 11.9 Velocity profiles measured by particle-tracking velocimetry for a concentrated EA/MAA microgel suspension ($N_x = 140$, $C = 2$ wt%). The suspension is sheared in a cone and plate geometry. The velocity profile measured with rough surfaces is linear without any evidence of slip (○). With smooth surfaces, slip can be partial (▲: ▼) or total (●) depending on the applied stress: ▲: $\sigma/\sigma_y = 1.7$; ▼: $\sigma/\sigma_y = 1.3$; ●: $\sigma/\sigma_y = 0.9$. V_S denotes the slip velocity. The data are replotted from Ref. [2].

conditions. When sheared between rough surfaces, the pastes flow homogeneously in the whole range of accessible shear stresses and shear rates. When one or both of the shearing surfaces are smooth, wall slip occurs. The relative importance of slip depends on the shear stress and three flow regimes have been identified. At high shear stresses ($\sigma > \sigma_y$), the apparent motion due to slip is negligible compared to the displacement associated with bulk flow. As the stress becomes comparable to the yield stress ($\sigma \cong \sigma_y$), wall slip becomes increasingly significant so that the total deformation results from a combination of bulk flow and slip. At stresses at and below the yield stress ($\sigma < \sigma_y$), bulk flow is negligible and slip totally dominates paste motion. In this regime, the slip velocity is simply the velocity of the mobile surface for slip at one wall or half of it for slip at both walls. This provides a unique and convenient way to determine the slip velocity from rheology.

Experimentally, the variations of the slip velocity are found to be very sensitive to the wetting properties of the shearing surfaces. This is illustrated in Figure 11.10a and b where we compare the variations of the slip velocity with stress when the nature of the confining surfaces is changed. For hydrophobic surfaces that are not wetted by the continuous phase of the paste, the dependence of V_S on the applied stress σ is nearly quadratic. Slip stops at a finite value σ_S of the sticking value stress. For hydrophilic wetting surfaces, σ_S is much lower than for the nonwetting surface and the slip velocity follows a quadratic dependence only in the vicinity of the bulk yield stress. At small stresses, the slip velocity increases linearly with stress.

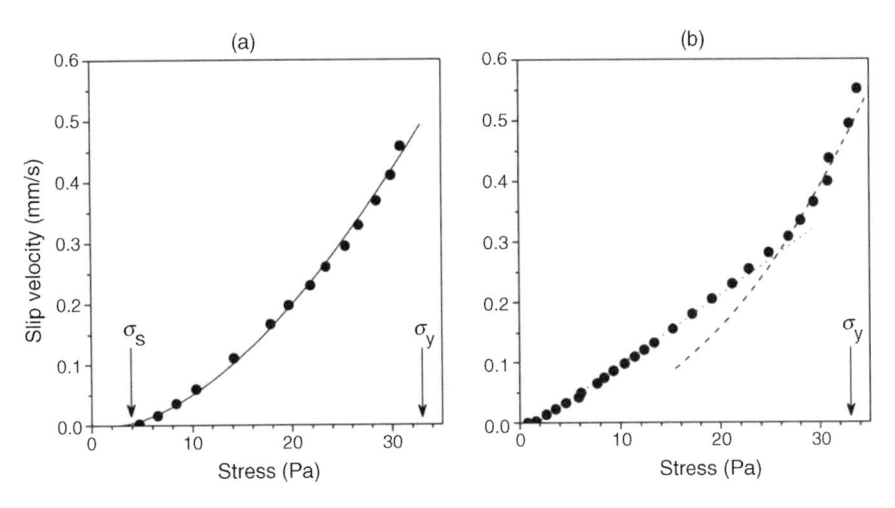

Figure 11.10 Variations of the slip velocity with the applied stress for a concentrated EA/MAA microgel suspension ($N_X = 28$, $C = 5.5$ wt%). In graph (a), the suspension is sheared along a hydrophobic surface consisting of a glass substrate coated with octadecyltrichlorosilane. The continuous line shows the quadratic variation predicted by the elastohydrodynamic theory. In graph (b), the suspension is sheared along a hydrophilic silicon wafer. The variation of the slip velocity is linear at small stresses (short-dashed line) and quadratic at higher stresses (long-dashed line).

11.5.3
Elastohydrodynamic Lubrication as the Origin of Wall Slip

In order to account for the unusual quadratic variation of the slip velocity with the applied stress, Meeker *et al.* have proposed a noncontact elastohydrodynamic lubrication theory that explains wall slip of soft and deformable particles such as microgels, particle gels, and emulsions [2, 5]. Elastohydrodynamic slip can be described qualitatively as follows. At rest, the osmotic pressure of the compressed suspension forces the particle to come into contact through facets with the bounding surfaces. These facets are somewhat analogous to those existing in the bulk between squeezed particles (inset of Figure 11.4). If these contacts were to persist during flow, no-slip behavior would be expected. It turns out that due to their softness and deformability, any relative motion between the microgels and the wall deforms asymmetrically the contacting facets. The resulting pressure field generates a coupled lift force that pushes the particles away from the wall and maintains a lubricated film that causes slip. The balance between the lift force and the osmotic force acting on microgels determines the thickness of the lubricating film, and ultimately the drag between the particle and the smooth surface. According to this model, the high-pressure lubricating film does not form when the surface is rough, which suppresses slip just as observed experimentally. Similarly, if the adhesion forces due to short-range attraction are large enough to oppose the lift force, slip also disappears.

The elastohydrodynamic formalism involves a set of coupled nonlinear equations that can be solved exactly using numerical techniques or analyzed in terms of scaling arguments [5]. The main prediction of the model provides an expression of the slip velocity as a function of the applied stress:

$$V_S = \gamma_y^2 \left(\frac{G_0 R}{\eta_S}\right) \left(\frac{\sigma}{\sigma_y}\right)^2 \tag{11.3}$$

The predicted slip velocity involves the main rheological parameters we have analyzed in the previous sections: the shear modulus G_0, the particle radius R, the solvent viscosity η_S, the yield stress σ_y, and the yield strain γ_y. The quadratic variation of the slip velocity with the applied stress comes from the fact that the thickness of the lubricating film situated between the microgels and the shearing surface increases linearly with V_S due to the elastohydrodynamic coupling. A recent version of this model incorporates the existence of attractive and repulsive interactions of different origins between the slipping microgel particles and the shearing surfaces with the objective of providing guides to chemically manipulate surfaces in order to promote or suppress wall slip [73].

Interestingly, the elastohydrodynamic lubrication model quantitatively captures the main features found experimentally for the case of the EA/MAA microgel pastes slipping along nonwetting substrates (Figure 11.10a). The slip velocity–shear stress curve is very close to the predicted variation. Below a well-defined shear stress, short-range forces such as Van der Waals dispersive forces provoke the dewetting of the lubricating film, which makes the microgels snap into contact with the substrate and suppresses slip. This model also accounts for the quadratic variation that is observed at high stresses when the paste is sheared along a hydrophilic surface resulting in net repulsion (Figure 11.10b). It is important to note, however, that the paste can slip at much lower stresses without sticking because a lubricating film of solvent always exists even in the absence of elastohydrodynamic phenomena. The slip velocity–stress relationship is then linear because the thickness of the lubricating film is set by the short-range forces and not by the hydrodynamics [73].

11.6
Outlook

In this chapter, we have shown that many microgel suspensions share generic rheological properties, although the structure, the composition, and even the architecture of individual particles exhibit a great variability. Concentrated suspensions behave like weak elastic solids at rest but yield and flow under high enough solicitation. They are also subject to slip near solid surfaces with dramatic effect on the apparent rheology. The close-packed amorphous structure of concentrated microgel suspensions lies at the heart of their properties. The particles are highly compressed by osmotic forces and forced to develop flat facets at contact that exert central repulsive forces of elastic origin. These forces are at the origin of the bulk elasticity of concentrated suspensions. The lubrication of the contacting facets by the solvent forming the continuous phase determines the local dynamical processes, the flow properties, and the behavior of the suspensions near confining surfaces.

Acknowledgments

The author is indebted to Roger T. Bonnecaze (University of Texas at Austin) with whom many theoretical topics described in this chapter were developed. He is very grateful to his many collaborators who, over the years, contributed to a better understanding of these fascinating systems: Régis Borrega, Steven P. Meeker, Fabrice Monti, and Jyoti Seth. The author thanks Dr Brian Erwin for a critical reading of the manuscript. Last but not least, he is indebted to Ludwik Leibler for his stimulating and enthusiastic support.

References

1 Barnes, H.A. (1995) *J. Non-Newtonian. Fluid Mech.*, **56**, 221.
2 Meeker, S.P., Bonnecaze, R.T., and Cloitre, M. (2004) *Phys. Rev. Lett.*, **92**, 198302.
3 Walls, H.J., Brett Caines, S., Sanchez, A.M., and Khan, S.A. (2003) *J. Rheol.*, **47**, 847.
4 Khan, S.A., Schnepper, C.A., and Armstrong, R.C. (1988) *J. Rheol.*, **1**, 69.

5. Meeker, S.P., Bonnecaze, R.T., and Cloitre, M. (2004) *J. Rheol.*, **48**, 1205.
6. Nickerson, C.S. and Kornfield, J.A. (2005) *J. Rheol.*, **49**, 865.
7. Keentok, M. and Xue, S.-C. (1999) *Rheol. Acta*, **38**, 321.
8. Stokes, J.R. and Telford, J.H. (2004) *J. Non-Newtonian. Fluid Mech.*, **124**, 137; Davies, G.A. and Stokes, J.R. (2008) *J. Non-Newtonian. Fluid Mech.*, **148**, 73.
9. Cloitre, M., Borrega, R., and Leibler, L. (2000) *Phys. Rev. Lett.*, **95**, 4819.
10. There is a rich literature on the subject. The interested reader can also refer to several reviews. A nonexhaustive list includes Gilser, T. and Weitz, D.A. (1998) *Cur. Opin. Colloid Interface Sci.*, **3**, 586; MacKintosh, F.C. and Schmidt, C.F. (1998) *Cur. Opin. Colloid Interface Sci.*, **3**, 391; Solomon, M.J. and Lu, Q. (2001) *Cur. Opin. Colloid Interface Sci.*, **6**, 2001; Harden, J.L. and Viasnoff, V. (2001) *Cur. Opin. Colloid Interface Sci.*, **6**, 438; Breedveld, V. and Pine, D.J. (2003) *J. Mater. Sci.*, **38**, 4461; Cicuta, P. and Donald, A.M.M. (2007) *Soft Matter*, **3**, 1449.
11. Weitz, D.A. and Pine, D.J. (1993) Diffusive-wave spectroscopy, in *Dynamic Light Scattering: The Method and Some Applications* (ed. W. Brown), Clarendon Press, Oxford.
12. Pusey, P.N. and Van Megen, W. (1989) *Physica A*, **157**, 705.
13. Joosten, J.G.H., Geladé, E.T.F., and Pusey, P.N. (1990) *Phys. Rev. A*, **42**, 2161; Xue, J.-Z., Pine, D.J., Milner, S.T., Wu, X.-I., and Chaikin, P.M. (1992) *Phys. Rev. A*, **46**, 6550; Schätzel, K. (1993) *Appl. Optics*, **32**, 3880; Scheffold, F., Skipetrov, S.E., Romer, S., and Schurtenberger, P. (2001) *Phys. Rev. E*, **63**, 061404.
14. Cipelletti, L. and Weitz, D.A. (1999) *Rev. Sci. Instrum.*, **70**, 3214.
15. Viasnoff, V., Lequeux, F., and Pine, D.J. (2002) *Rev. Sci. Instrum.*, **73**, 2336.
16. Mason, T.G. and Weitz, D.A. (1995) *Phys. Rev. Lett.*, **1250**, 74.
17. Gittes, F., Schnurr, B., Olsmsted, P.D., Mackintosh, F.C., and Schmidt, C.F. (1997) *Phys. Rev. Lett.*, **79**, 3286.
18. Mason, T.G. (2000) *Rheol. Acta*, **39**, 371; Dasgupta, B.R., Tee, S.-Y., Crocker, J.C., Frisken, B.J., and Weitz, D.A. (2002) *Phys. Rev. E*, **65**, 051505.
19. Liu, Q. and Solomon, M.J. (2002) *Phys. Rev. E*, **66**, 061504.
20. Cloitre, M., Borrega, R., Monti, F., and Leibler, L. (2003) *Phys. Rev. Lett.*, **90**, 068303.
21. Valentine, M.T., Kaplan, P.D., Thota, D., Crocker, J.C., Gisler, T., Prud'homme, R.K., Beck, M., and Weitz, D.A. (2001) *Phys. Rev. E*, **64**, 061506.
22. Crocker, J.C. and Grier, D.A. (1996) *J. Colloid Interface Sci.*, **179**, 298.
23. Oppong, F.K., Rubatat, L., Frisken, B.J., Bailey, A.E., and de Bruyn, J.R. (2006) *Phys. Rev. E*, **73**, 041405; Oppong, F.K. and de Bruyn, J.R. (2007) *J. Non-Newtonian. Fluid Mech.*, **142**, 104.
24. Caggioni, M., Spicer, P.T., Blair, D.L., Lindberg, S.E., and Weitz, D.A. (2007) *J. Rheol.*, **51**, 851.
25. Ashlee, J., Breedveld, V., and Lyon, L. (2007) *Phys. Chem. B.*, **19**, 17579396.
26. Degré, G., Joseph, P., Tabeling, P., Lerouge, S., Cloitre, M., and Ajdari, A. (2006) *Appl. Phys. Lett.*, **89**, 024104.
27. Tapadia, P. and Wang, S.-Q. (2006) *Phys. Rev. Lett.*, **96**, 016001.
28. Isa, L., Besseling, R., and Poon, W.C.K. (2007) *Phys. Rev. Lett.*, **98**, 198305.
29. Borrega, R., Cloitre, M., Betremieux, I., Ernst, B., and Leibler, L. (1999) *Europhys. Lett.*, **47**, 729.
30. Boggs, L.J., Rivers, M., and Bike, S.G. (1996) *J. Coat. Tech.*, **68**, 63.
31. Wolfe, M.S. and Scopazzi, C. (1989) *J. Colloid Interface Sci.*, **133**, 265; Wolfe, M.S. (1992) *Prog. Org. Coat.*, **20**, 487.
32. Tan, B.H., Tam, K.C., Lam, Y.C., and Tan, C.B. (2004) *J. Rheol.*, **48**, 915; Tan, B.H., Tam, K.C., Lam, Y.C., and Tan, C.B. (2005) *Adv. Colloid Interface Sci.*, **113**, 111.
33. Senff, H. and Richtering, W. (1999) *J. Chem. Phys.*, **111**, 1705.
34. Senff, H., Richtering, W., Norhausen, Ch., Weiss, A., and Ballauff, M. (1999) *Langmuir*, **15**, 102.
35. Buscall, R. (1994) *Colloids Surf. A*, **83**, 33.
36. Borrega, R. (2000) Suspensions de microgels polyélectrolytes: propriétés physico-chimiques, rhéologie, écoulement PhD Thesis, University of Paris VI.

37 Cloitre, M., Borrega, R., Monti, F., and Leibler, L. (2003) *C. R. Physique*, **4**, 221.
38 Stieger, M., Pedersen, J.S., Lindner, P., and Richtering, W. (2004) *Langmuir*, **20**, 7283.
39 Lyon, L.A., Debord, J.D., Debord, S.B., Jones, C.D., McGrath, J.G., and Serpe, M.J. (2004) *J. Phys. Chem. B*, **108**, 19099.
40 Crassous, J.J., Siebenbürger, M., Ballauff, M., Drechsler, M., Henrich, O., and Fuchs, M. (2006) *J. Chem. Phys.*, **125**, 204906.
41 Bartsch, E., Frenz, V., Baschnagel, J., Schartl, W., and Sillescu, H. (1997) *J. Chem. Phys.*, **106**, 3743.
42 Crassous, J.J., Régissier, R., Ballauff, M., and Willenbacher, N. (2005) *J. Rheol.*, **49**, 851.
43 Paulin, S.E., Ackerson, B.J., and Wolfe, M.S. (1996) *J. Colloid Interface Sci.*, **178**, 251; Paulin, S.E., Ackerson, B.J., and Wolfe, M.S. (1997) *Phys. Rev. E*, **55**, 5812.
44 Fridrikh, S., Raquois, C., Tassin, J.F., and Rezaiguia, S. (1996) *J. Chem. Phys.*, **93**, 941.
45 Berli, C.L.A. and Quemada, D. (2000) *Langmuir*, **16**, 7968.
46 Zwanzig, R. and Mountain, R.D. (1965) *J. Chem. Phys.*, **43**, 4464.
47 Adams, S., Frith, W.J., and Stokes, J.R. (2004) *J. Rheol.*, **48**, 1195.
48 Seth, J., Cloitre, M., and Bonnecaze, R.T. (2006) *J. Rheol.*, **50**, 353.
49 Mason, T.G., Bibette, J., and Weitz, D.A. (1995) *Phys. Rev. Lett.*, **75**, 2051; Lacasse, M.-D., Grest, G.S., Levine, D., Mason, T.G., and Weitz, D.A. (1996) *Phys. Rev. Lett.*, **76**, 3448.
50 Obukhov, S.P., Rubinstein, M., and Colby, R.H. (1994) *Macromolecules*, **27**, 3191.
51 Rubinstein, M., Colby, R., Dobrynin, A.V., and Joanny, J.-F. (1996) *Macromolecules*, **29**, 398.
52 Rubinstein, M. and Colby, R. (2003) *Polymer Physics*, Oxford University Press.
53 Skouri, R., Schosseler, F., Munch, J.P., and Candau, S.J. (1995) *Macromolecules*, **28**, 197.
54 Rofriguez, B.E., Wolfe, M.S., and Fryd, M. (1994) *Macromolecules*, **27**, 6642.
55 Eichenbaum, G.M., Kiser, P.F., Dobrynin, A.V., Simon, S.A., and Needham, D. (1999) *Macromolecules*, **32**, 4867.
56 Stokes, J.S. and Frith, W. (2008) *Soft Matter*, **4**, 1133.
57 Ketz, R.J., Prud'homme, R.K., and Graessley, W.W. (1988) *Rheol. Acta*, **27**, 531.
58 Raquois, C., Tassin, J.F., Rezaiguia, S., and Gindre, A.V. (1995) *Prog. Org. Coat.*, **26**, 239.
59 Islam, M.T., Rodriguez-Hornedo, N., Ciotti, S., and Ackermann, C. (2004) *Pharm. Res.*, **21**, 1192.
60 Kaneda, I. and Sogabe, A. (2005) *Colloids Surf. A*, **270**, 163.
61 Hyun, K., Kim, S.H., Ahn, K.H., and Lee, S.J. (2002) *J. Nonnewton. Fluid Mech.*, **107**, 51.
62 Miyazaki, K., Wyss, H.M., Weitz, D.A., and Reichman, D.R. (2006) *Europhys. Lett.*, **75**, 915.
63 Wyss, H.M., Miyazaki, K., Mattsson, J., Hu, Z., Reichman, D.R., and Weitz, D.A. (2007) *Phys. Rev. Lett.*, **98**, 238303.
64 Fuchs, M. and Cates, M.E. (2002) *Phys. Rev. Lett.*, **89**, 248304.
65 Crassous, J.J., Siebenbürger, M., Ballauff, M., Drechsler, M., Hajnal, D., Henrich, O., and Fuchs, M. (2008) *J. Chem. Phys.*, **128**, 204902; Ballauff, M. and Fuchs, M. (2005) *J. Chem. Phys.*, **122**, 094707.
66 Purnomo, E.H., van den Ende, D., Mellema, J., and Mugele, F. (2007) *Phys. Rev. E*, **76**, 021404.
67 Weitz, D.A. (2001) *Nature*, **410**, 32.
68 Struik, L.C.E. (1978) *Physical Aging in Amorphous Polymers and Other Materials*, Elsevier, Amsterdam.
69 Vincent, E., Hammann, J., Ocio, M., Bouchaud, J.-P., and Cugliandolo, L.F. (1997) Slow Dynamics and Aging in Spin Glasses, in *Complex Behaviour of Glassy Systems* (ed. M. Rubi), Lecture Notes in Physics, Springer Verlag.
70 Derec, C., Ajdari, A., Ducouret, G., and Lequeux, F. (2000) *C. R. Acad. Sci. IV*, **1**, 1115; Ramos, L. and Cipelletti, L. (2002) *Phys. Rev. Lett.*, **89**, 065701.
71 Barnes, H.A. (1995) *J. Non-Newtonian. Fluid Mech.*, **56**, 221.
72 Yoshimura, A. and Prud'homme, R.K. (1988) *J. Rheol.*, **32**, 53.
73 Seth, J., Cloitre, M., and Bonnecaze, R.T. (2008) *J. Rheol.*, **52**, 1241.

12
Mechanics of Single Microgel Particles

Hans M. Wyss, Johan Mattsson, Thomas Franke, Alberto Fernandez-Nieves, and David A. Weitz

Microgel suspensions are important examples of systems where subtle differences in the chemistry and structure at small length scales can lead to dramatic effects at the macroscopic scale. One of the most important macroscopic properties of these systems is their mechanical behavior. The mechanical properties of a single microgel particle can be systematically controlled during synthesis by variation of its polymer concentration, cross-link density, or stiffener content. In turn, variations in the mechanical properties at the single-particle level can lead to dramatic changes in the macroscopic mechanics of the resulting microgel suspensions. This ease of control is an important reason why microgels have such a wide range of important applications as additives to control the flow behavior of industrial products. A fundamental understanding of the macroscopic behavior of microgel suspensions is, however, still lacking.

Most studies on suspension rheology have focused on studying hard colloidal particles, where a good understanding of flow behavior has been established from experiments [1–4], theory, and simulations [5, 6]. To extend our understanding of hard particle suspensions to describe the wealth of behaviors found in microgel suspensions will clearly require new approaches.

In contrast to hard colloidal particles, microgels are soft and can change both their shape and their size in response to changes in mechanical stress, temperature, or chemical environment. The phase behaviors and mechanical responses of microgel systems are therefore much richer than those observed for hard sphere suspensions [7–17]. To reach an understanding of the mechanics of the single microgel particle is an essential prerequisite for understanding the macroscopic mechanics of microgels in suspension. Measurements of the macroscopic behavior of these systems can be interpreted in a meaningful way only if the behavior of a single microgel particle can also be experimentally accessed.

Unfortunately, measuring the properties of submicrometer-sized particles is extremely challenging; it typically requires the measurement of forces on the order pN for particles remaining immersed in their surrounding fluid. While the feasibility of such measurements has recently been demonstrated using atomic force microscopy (AFM) on poly-NIPAM microgels [18, 19], these measurements are difficult to carry out for many microgel systems.

In this chapter, we review experimental methods that have recently been developed to experimentally characterize the behavior of microgels at the single-particle level. We focus on simple techniques that yet offer a surprising degree of accuracy in characterizing the mechanical response of microgel spheres. We will also discuss the effects of the single-particle elastic properties on the macroscopic mechanics of dense suspensions of microgel spheres. Suspensions of soft, compressible microgel particles clearly exhibit a mechanical behavior that is qualitatively different from that of hard spheres. We show some examples where this link between the macroscopic mechanics and the properties at the single-particle level is important. These examples serve as an illustration to show that the characterization of single-particle mechanics is an important prerequisite for gaining a better understanding of the macroscopic mechanics of microgel.

12.1
Compressive Measurements by Variation of the Osmotic Pressure

A microgel particle can be viewed as a porous bead where the pores are filled with liquid and the typical pore size corresponds to the mesh size of the polymer network. If molecules or particles that are larger than this typical pore size are added to the background liquid, they will not be able to penetrate the pores. This results in an unbalanced osmotic pressure between the inside and the outside of the microgel particle, resulting in a uniform compressive stress that compresses the particle. This compressive stress acting on the particle is equal to the osmotic pressure Π of the background fluid. As a result of this externally applied stress, the particle volume decreases as Π increases, as shown schematically in Figure 12.1a.

The application of an external osmotic pressure can thus be used to directly characterize the compressive elastic modulus K of bulk hydrogels [20, 21] or of microgel particles [12, 16] by measuring the relative volume change of the polymer network in response to an increase in osmotic pressure. Depending on the typical mesh size and on the elastic properties of the gels to be characterized, macromolecules such as dextran or polyethylene oxide are often used to induce the osmotic pressure. The molecular weight and correspondingly the size of the used species should be sufficiently large to prevent penetration into the polymer network of the microgel particles. For cases where a smaller species is used, it has been observed that an increase in osmotic pressure can cause the particles to initially swell before finally deswelling, as Π is further increased [22]; in this case, the species can penetrate into the network and thereby cause the swelling of the microgel particles [23]. Thus, the optimal choice of osmotic species will vary with the properties of the specific microgel particles to be characterized. Moreover, the concentration dependence of the osmotic pressure $\Pi(c)$ for the chosen species is usually characterized in a separate set of measurements, for instance, by the use of a membrane osmometer.

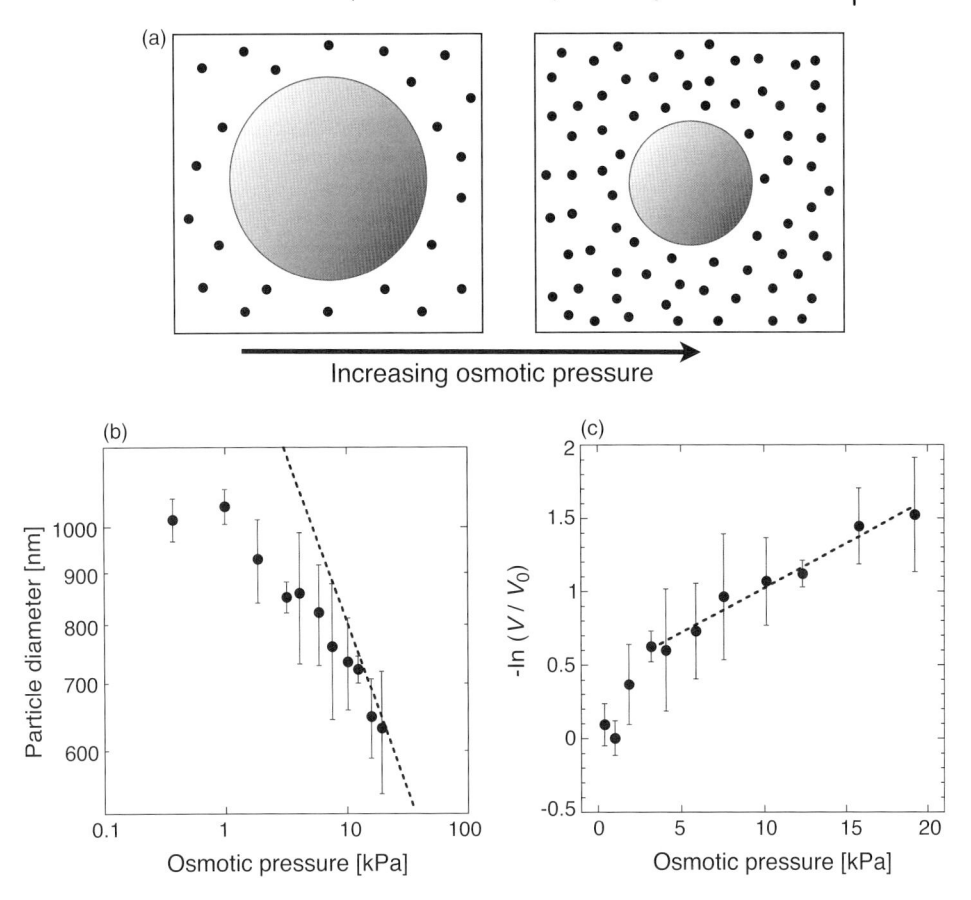

Figure 12.1 Osmotic compression of microgel particles. (a) Schematic representation of osmotic compression measurements. As the osmotic pressure in the background fluid is increased, the microgel particles are compressed isotropically. The compressive modulus of the particles is characterized by measuring the particle size as a function of the applied osmotic pressure. (b) Example of an osmotic compression measurement [12]. The osmotically induced deswelling of ionic microgel particles is characterized by measuring the diameter d of the particles as a function of the osmotic pressure Π with dynamic light scattering. The particle size decreases with increasing Π; at large Π the diameter d of the particles approaches the expected $d \propto \Pi^{1/3}$ behavior indicated by the dashed line. (c) Evaluation of the overall compressive modulus K from the same data. We plot $-\ln(V/V_0)$ as a function of Π, where the local slope corresponds to the inverse of the compressive modulus, $1/K$. The dashed line is a linear fit to the data at $\Pi > 3$ kPa corresponding to a compressive modulus $K = 17$ kPa.

Osmotic compression measurements are performed by exposing the microgels to increasing levels of osmotic pressure and characterizing their equilibrium size at each level of Π. For large enough particles the resulting particle sizes can be measured directly by imaging the particles in an optical microscope. For submicron particles, the size change and thus the microgel elastic properties are usually determined by dynamic light scattering (DLS) measurements. Here, the diffusion coefficient of the microgel particles is characterized; if the viscosity of the background fluid is known, the particle size can be determined directly from the Stokes–Einstein relation. Thus, in order for DLS to be applicable as a means to quantify the particle response, the dependence of viscosity on the solution concentration also needs to be characterized in separate measurements.

To derive the compressive modulus K from the Π-dependence of the particle size, we have to bear in mind that K is strictly defined as

$$K = -V \frac{d\sigma}{dV} \qquad (12.1)$$

where $\sigma = \Pi$ is the externally applied stress and V is the particle volume. For many materials, where the changes in volume are relatively small, this definition can be approximated as $K = -V_0(\Delta\sigma/\Delta V)$, where V_0 is the initial volume without applied stress and $\Delta V = V - V_0$ is the volume difference as a result of the applied stress difference $\Delta\sigma = \sigma - \sigma_0$. However, for microgel particles the volume changes can be as large as several orders of magnitude, hence this approximation is no longer valid. Moreover, generally the compressive modulus will depend sensitively on the volume of the microgel particle. The compressive modulus is determined by the internal osmotic pressure of the microgel particle,

$$\Pi_{int} = \Pi_m + \Pi_e + \Pi_i \qquad (12.2)$$

where Π_m, Π_e, and Π_i are the mixing, elastic, and ionic contributions, respectively, to the osmotic pressure. In the context of Flory theory [24],

$$\Pi_m = \frac{N_a k_B T}{V_s}\left[\phi + \ln(1-\phi) + \chi\phi^2\right] \qquad (12.3)$$

$$\Pi_e = \frac{N_c k_B T}{V_0}\left[\frac{\phi}{2\phi_0} - \left(\frac{\phi}{\phi_0}\right)^{1/3}\right] \qquad (12.4)$$

where N_a is the Avogadro number, k is the Boltzmann constant, T is the temperature, v_s is the molar volume of the solvent, ϕ and ϕ_0 are the volume fractions of polymer within the particle in the collapsed and swollen state, respectively. Furthermore, V_0 and d_0 are the particle volume and diameter in the collapsed state, respectively, and χ is the Flory mixing parameter, which describes the interactions between the solvent and the polymer chains. Finally, N_c is the effective number of chains within one particle.

For ionic microgels, Π_i also has to be considered. If the salt concentration in the continuous phase is smaller than the counterion concentration inside a microgel particle, this contribution can be attributed to the osmotic pressure of the counterions [12, 25], expressed as

$$\Pi_i = k_B T \frac{f}{V} \qquad (12.5)$$

where V is the particle volume and f is the number of mobile counterions inside the gel, which for an electrically neutral gel has to match the number of network charges inside the gel.

In equilibrium, the total internal osmotic pressure Π_{int} is in equilibrium with the externally applied osmotic pressure, $\Pi_{int} = \Pi$. Balancing the applied osmotic pressure with the size-dependent internal osmotic contributions allows a theoretical prediction of the osmotically induced deswelling of microgels, which has been found to be in agreement with the experimentally observed behavior both for neutral [23] and for ionic [12] microgels.

A typical deswelling curve taken from Fernandez-Nieves et al. [12] for ionic microgel particles is shown in Figure 12.1b, where the particle diameter is plotted as a function of the applied osmotic pressure. As Π increases, the particle diameter decreases significantly. At the highest osmotic pressures, a limiting behavior of $d \propto \Pi^{1/3}$ is expected; indeed, at the highest osmotic pressures accessed, the experimental values approach such a power law behavior, shown in the figure as a dashed line.

To evaluate the compressive modulus K of the particles from such data, we follow Equation 12.1 with $\sigma = \Pi$ and substitute $x = V/V_0$, which for constant K yields

$$\frac{d\Pi}{dx} = -\frac{K}{x} \tag{12.6}$$

As a consequence,

$$\Pi = -K \ln\left(\frac{V}{V_0}\right) \tag{12.7}$$

and thus in a plot of Π as a function of $-\ln(V/V_0)$, the local slope represents the inverse of the compressive modulus, $1/K$, at the corresponding level of deformation. In Figure 12.1c, we replot the data from Figure 12.1b in this way. Even though the scatter in the data is considerable, we clearly see that the local slope of this curve decreases with increasing osmotic pressure, representing an increase in the compressive modulus as the particle is compressed. However, at $\Pi > 3$ kPa a nearly linear behavior is observed. The dashed line is a fit to the data in this regime, which yields a slope that corresponds to a compressive modulus of $K = 17$ kPa.

This example illustrates that osmotically induced deswelling is an effective and simple method for measuring the elastic response of microgel particles under compression. However, the method allows only a measurement of the compressive elastic modulus K. Since the mode of deformation is limited to isotropic compression, it cannot provide information on the elastic shear modulus or on the Poisson ratio of the material.

12.2
Capillary Micromechanics: Full Mechanical Behavior of a Single Microgel Particle

To access the full elastic behavior of microgel particles, a different mode of deformation, in addition to isotropic compression, is thus required. A full characterization can be given in terms of a set of two independent moduli, such as the compressive modulus K and the shear modulus G or, alternatively, the Poisson ratio ν and the Young's modulus E. If any two of these are known, all the other quantities follow and the elastic properties of the material are fully characterized within the limit of linear elasticity. As the osmotic compression measurement allows only the characterization of K, at least one complementary test is necessary to fully characterize the properties of a microgel particle.

Here, we report on a recently developed technique termed *capillary micromechanics* [26], which is based on the pressure-induced deformation of soft objects in tapered microcapillaries. Because the resulting deformation is a combination of compression and shear, by quantifying the two modes of deformation, the method enables access to both the compressive elastic modulus K and the shear elastic modulus G in one single experiment. This local *squeezing experiment*, while simple, is thus surprisingly powerful. However, it can be applied only to particles that are large enough in size to allow a direct observation of their deformation in an optical microscope. This restricts the size of the objects that can be measured to a few micrometers or a larger scale.

The capillary micromechanics technique is based on a simple microfluidic device, shown schematically in Figure 12.2. It consists of a glass capillary that at the back is connected to a flexible tube and at the front is drawn to a tapered tip. In a typical experiment, a dilute suspension of microgel particles is pumped through the device by application of an overpressure p in the flexible tube at the inlet of the device. To prevent the surface tension between the fluid and the air to influence the flow, the outlet of the device, the tapered tip, is led into a fluid reservoir that contains the same fluid as is used as the continuous phase of the particle suspension. The pressure p is applied via an air pressure regulator that is connected to the tube; alternatively, to achieve small values of p, the hydrostatic pressure in the tube can be directly used by variation of the filling height.

If the applied pressure is small enough, the first particle that approaches the tapered tip will become trapped near the tip and block further flow of fluid through the device. As a result of the pressure-induced external stresses acting on it, the microgel particle will deform elastically. In this situation, the internal stresses in the particle due to the elastic deformation have to balance the pressure-induced external stresses acting on the particle. Thus, by monitoring the particle deformation as a result of the applied pressure difference, the mechanical properties of a single particle can be directly characterized.

To illustrate the experimental procedure, we use poly-acrylamide particles, which were synthesized by polymerization of aqueous drops in a water-in-oil emulsion; the drops contained 10% weight fraction of acrylamide monomer as well as the cross-linker BIS-acrylamide at a cross-linker to monomer weight ratio of 5%. The pressure-induced deformation of such a poly-acrylamide particle is illustrated in Figure 12.3, where we show a series of micrographs of a particle at applied pressure levels increasing from $p = 6.9\,\text{kPa}$ to $p = 62\,\text{kPa}$. The microgel particle clearly deforms as the pressure is increased, changing both its shape and its volume.

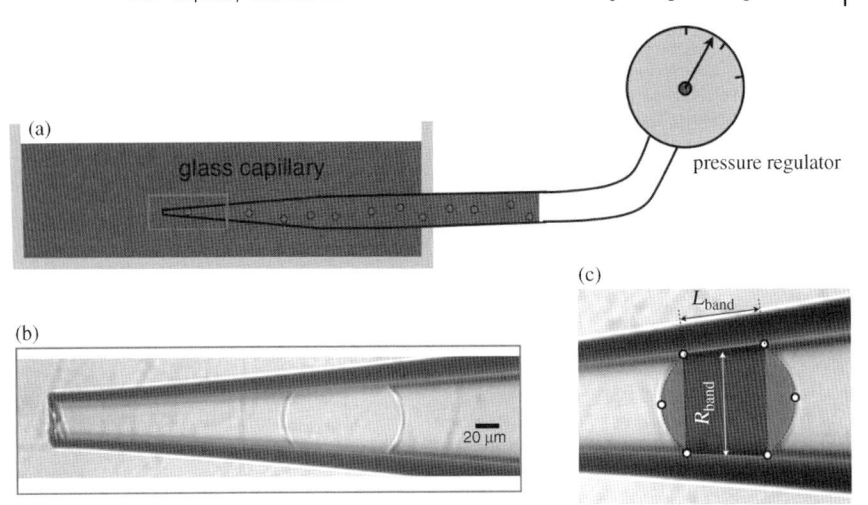

12.2 Capillary Micromechanics: Full Mechanical Behavior of a Single Microgel Particle

Figure 12.2 Experimental setup for squeezing in tapered capillaries. (a) Schematic of basic experimental setup. A dilute suspension of particles is pumped through a glass capillary; the end of the capillary is tapered, narrowing down toward the tip with a typical diameter of around 5–20 µm, which is smaller than the diameter of the particles. The pressure at the inlet of the capillary is controlled by a pressure regulator. Eventually, a single particle will block the flow and the entire pressure gradient is balanced by the elastic stress response of the microgel particle. (b) Typical microscope image of a soft particle during a capillary micromechanics experiment. (c) Data analysis: 6 characteristic points on the particle are selected to characterize its geometry at each level of applied pressure. The particle shape is approximated as the sum of three bodies with circular cross section: one cylinder-cut volume (central dark gray area) and two sphere-cut volumes (front and rear light gray areas).

To quantify this deformation, we analyze the shape of the particle as a function of the applied pressure using digital image analysis. We parameterize the shape by identifying six characteristic points on the images, as shown in Figure 12.2c. Four of these points are at the triple boundary between the glass surface, the aqueous phase, and the microgel particle. Two more points indicate the front and the rear end of the particle with respect to the flow direction in the capillary. Using these characteristic points, the particle shape is approximated as the sum of three bodies with circular cross sections: a cylinder-cut volume at the center and two spherical cuts at the front and at the rear of the particle, as shown schematically in Figure 12.2c. From this approximation, we obtain the particle volume V directly from the coordinates of the six characteristic points. To further characterize the shape, we focus on the contact surface between the particle and the glass wall. This contact surface has the shape of a tapered circular band; we denote the length of this band along the flow direction as L_{band} and the average radius of this band as R_{band}, as shown in Figure 12.2c. In our simple model description of the particle deformation, these parameters are sufficient to describe the shape and volume change of the particle in order to analyze its elastic properties. In equilibrium, the externally applied stress on the particle is balanced by the internal elastic stresses in the particle. This balance allows us to quantify the

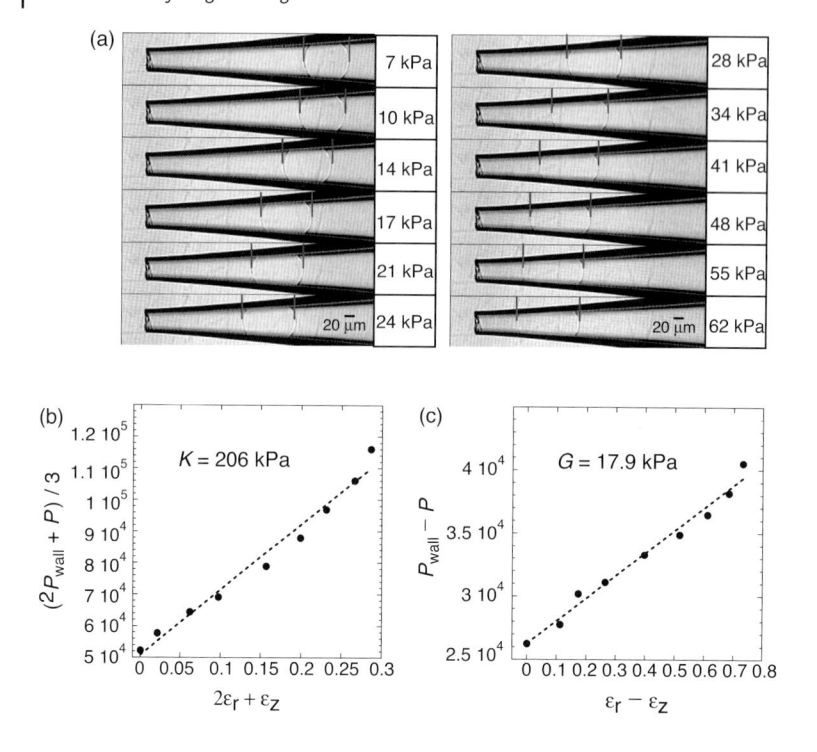

Figure 12.3 Example of micromechanical testing in capillaries. (a) Geometry of a microgel particle as the pressure difference between inlet and outlet is gradually increased. Each picture represents an equilibrium situation, where the particle has been allowed to equilibrate at constant p for at least 10 min. The dark gray marks at the edges of the particles are drawn as a guide to the eye to illustrate the deformation of the particle with increasing pressure. (b) Plot of σ_{compr} as a function of $\Delta V/V$; the slope of this curve corresponds to a compressive modulus $K = 206$ kPa. (c) Plot of σ_{shear}s as a function of ε_{shear}; the slope of this curve corresponds to an elastic shear modulus $G = 17.9$ kPa.

elastic properties of the material in terms of the compressive elastic modulus K and the shear elastic modulus G.

The externally applied stresses are directly proportional to the applied pressure difference p; because the background fluid is incompressible, the absolute pressure has no effect on the stress exerted on the particle. Due to the porous nature of the particle network, the background fluid will continue to flow through the particle, thereby exerting viscous drag forces on the network in a homogeneous manner throughout the particle. We therefore equate the stress σ_z along the longitudinal direction with the applied pressure difference, $\sigma_z = p$.

In order to quantify the stress exerted on the particle by the walls of the glass capillary, we assume the absence of friction between the particle and the wall. Previous studies have shown that the static friction of microgel particles at a wall is generally very low, which, for instance, leads to wall slip in rheological measurements of microgel suspensions [13]; thus, neglecting static friction should be a reasonable approximation for the microgel particles studied. In the absence of static friction, the longitudinal component of all forces from the wall must be equal to the force exerted on the particle directly through the applied pressure difference. As a result, the average wall pressure is given by [26]

$$p_{wall} = \frac{1}{2\sin(\alpha)} \frac{R_{band}}{L_{band}} p \qquad (12.8)$$

where α is the taper angle of the capillary, while R_{band} and L_{band} are the average radius and the length of the band, respectively, around the particle that is in contact with the glass wall. Thus, if we assume a uniform stress distribution within the particle, the stress in the radial direction is given by the wall pressure, $\sigma_r = p_{wall}$, and the stress in the longitudinal direction is given by the applied pressure difference, $\sigma_z = p$.

The internal elastic stress of the material as a result of the particle deformation can be written as a function of K, G, as well as the three-dimensional strain deformations. This internal elastic stress is given by $\sigma_{r,elast.} = 2G\varepsilon_r + l(2\varepsilon_r + \varepsilon_z)$ for the radial direction and $\sigma_{z,elast.} = 2G\varepsilon_z + l(2\varepsilon_r + \varepsilon_z)$ for the longitudinal direction, where $l = K - (2/3)G$ is the Lamé parameter and $\varepsilon_r \approx \Delta R_{band}/R_{band}$ and $\varepsilon_z \approx \Delta L_{band}/L_{band}$ are the radial and longitudinal strain deformations, respectively.

By balancing these internal elastic stresses with the externally applied stresses, we arrive at the following expressions for the elastic moduli:

$$K = \frac{(1/3)(2p_{wall} + p)}{2\varepsilon_r + \varepsilon_z} \qquad (12.9)$$

$$G = \frac{(1/2)(p_{wall} - p)}{\varepsilon_r - \varepsilon_z} \qquad (12.10)$$

Both expressions can be rationalized in terms of an elastic stress in response to a strain deformation, characteristic of the mode of deformation probed. The compressive modulus K characterizes resistance of the material to a volume change. The characteristic strain for this deformation is thus the volumetric strain, $\Delta V/V$, and the relevant stress is the average of the stresses along the principal axes, thus $\sigma_{compr.} = (2p_{wall} + p)/3$.

The shear modulus G characterizes the resistance of the material to a pure shape deformation. The characteristic strain here characterizes the shape change: it is the difference of the strains in the longitudinal and the radial direction $\varepsilon_{shear} = \varepsilon_r - \varepsilon_z$. The relevant stress is the difference of the stresses in the longitudinal and the radial directions, $\sigma_{shear} = (p_{wall} - p)/2$.

We thus obtain both the compressive and the shear modulus of our particles directly from one experiment, as shown in Figure 12.3a. The compressive modulus is determined from the slope of σ_{compr} as a function of $\Delta V/V$; for the particle shown in Figure 12.3a, we obtain $K = 206$ kPa, as shown in Figure 12.3b. The shear modulus is determined from the slope of σ_{shear} as a function of ε_{shear}; here, we obtain $G = 17.9$ kPa, as shown in Figure 12.3c.

It may seem surprising that we are able to significantly compress a particle with a modulus of 200 kPa by applying typical pressure differences on the order of only tens of kPa. However, we have to keep in mind that the wall pressure p_{wall} can become much larger than p for small enough taper angles α, as can be seen from Equation 12.8. The situation is similar to a wedge that is driven into a crack, where the "splitting force" of the wedge increases with decreasing taper angle. This dependence on α could be exploited to access different ranges of applied stress by using capillaries with different taper angles.

Both the magnitude of the obtained moduli K and G and the ratio between them are in fair agreement both with macroscopic measurements of the bulk elastic properties of poly-acrylamide hydrogels [26] and with theoretical predictions from Flory theory [24]. However, more detailed studies are necessary to test the validity of the presented method for different materials. For instance, AFM measurements should be performed and compared with capillary micromechanics measurements on the same particles in order to obtain a comparison at the single-particle level.

The capillary micromechanics method allows us to characterize the full elastic behavior of soft particles in the framework of linear elasticity. Moreover, it should be possible to extend the method to quantify the viscoelastic response of the particles, including the liquid-like response of the particles during their deformation. So far, we have considered only the situation where particles have reached their equilibrium shape at each respective level of externally applied stress. Because microgel particles are filled with liquid, their deformation will require liquid to be displaced through the network, which leads to large viscous stresses that are essential, for instance, in understanding the swelling kinetics of microgel particles. Upon an instantaneous change of the applied pressure p, the shape of the particle will therefore not change instantaneously and the time to reach the equilibrium shape is a measure of the

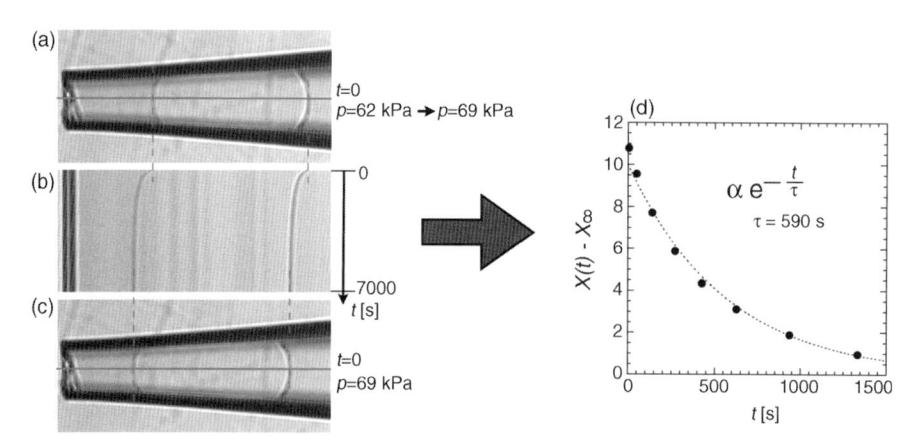

Figure 12.4 Transient deformation of a particle upon change in pressure (a) State of the particle at $t = 0$, before a change in pressure from 62 kPa to 69 kPa is applied. (b) The transient deformation is followed by tracking the centerline through the particle as a function of time. (c) State of the particle after $t = 7000$ s, in a equilibrium deformation state (d) The transient deformation can be tracked by plotting the edge position relative to the equilibrium position as a function of time, $x(t) - x_\infty$; this is well described by an exponential decay (dotted line) with a characteristic time scale of 590 s.

viscous response of the particles. This is illustrated for our poly-acrylamide particles in Figure 12.4, where we show the particle deformation immediately following a step in the applied pressure from $p = 62$ kPa to $p = 69$ kPa.

The shape of the particle at $t = 0$ and $t = 7000$ s is shown in Figure 12.4a and c, respectively, where $t = 0$ refers to the time of the pressure change. We follow the temporal shape change of the particle by video microscopy and track the position of the particle along the center of the capillary, marked as a solid line in Figure 12.4a and c. By taking only this central line from each of the frames of a movie and composing these lines into a two-dimensional image, shown in Figure 12.4b, we visualize the kinetics of deformation. This image represents the particle position in the horizontal direction, while t corresponds to the vertical direction.

We characterize the kinetics of the approach to the equilibrium position by plotting $x(t) - x_\infty$ in Figure 12.4d, where $x(t)$ is the position of the front of the particle in arbitrary units and x_∞ represents the equilibrium position, here taken as the position at $t = 7000$ s. The approach to equilibrium is well described by an exponential decay $[x(t) - x_\infty] \propto e^{-t/\tau}$ with a characteristic timescale $\tau \approx 590$ s, as shown by the dashed line.

A similar functional form of the deswelling kinetics has been observed for microgels exposed to a change in solvent quality, temperature, or pH [27]. In these cases, the kinetics of the swelling and deswelling of spherical gels is well accounted for by the theoretical description by Tanaka and Fillmore [28], which predicts a square dependence of the characteristic swelling time with the characteristic size of the gels: $\tau \propto a^2/D$, where a is the characteristic size and D is the network diffusion coefficient.

The geometry of the experiment shown in Figure 12.4 is similar to the case where transport of water through the polymer occurs in only one direction, such as in the swelling of a membrane. In this case, the characteristic size is the thickness L of the membrane or, in our case, corresponds to the length of the particle in the flow direction, $L \approx 125$ μm. The swelling time τ for this geometry can be expressed as [29]

$$\tau = \frac{L^2 f}{\pi^2 E} \tag{12.11}$$

where L is the gel thickness, $E = 9KG/(3K + G)$ is the Young's modulus of the material, and f is the polymer–water friction coefficient. Using our experimental values, $\tau \approx 600$ s, $E \approx 60$ kPa, and $L = 125$ μm, we obtain a friction coefficient $f = \pi^2 E \tau / L^2 \approx 2 \cdot 10^{16}$ Ns/m^4. This value is higher than typical values found in the literature for the friction coefficient of macroscopic poly-acrylamide gels; for instance, Suzuki et al. report a value of $f = 2 \cdot 10^{15}$ Ns/m^4 for a gel with $c_p = 8\%$ weight fraction of polymer. However, the relaxation test shown was carried out at the highest accessed pressures, where the gel is considerably compressed; therefore, the polymer concentration of the gel should be much higher than the 10 wt% of the original gels. Various experimental studies have shown that the dependence of the polymer–water friction coefficient as a function of the polymer concentration is well described by a power law, $f \propto c_p^n$, with an exponent in the range $n \approx 1.5, \ldots, 1.7$ [29, 30]. Therefore, in order to account for the value of f found in our experiment, we would have to assume a very high polymer concentration of $c_p \approx 30\%$ weight fraction. This is higher than the estimated polymer weight fraction, which does not exceed 20% by weight.

The reason for the remaining discrepancy could be the fact that the motion of fluid through the sample is additionally hindered by the glass capillary walls; moreover, the sliding friction between the glass wall and the polymer network could also affect the swelling kinetics in this geometry. Nevertheless, this simple test illustrates that reasonable values of the friction coefficient can also be obtained in capillary micromechanics; the transient compression behavior of these gels can be directly characterized by applying an instant step in the applied pressure and following the subsequent deformation as a function of time.

Moreover, this test illustrates that in order to obtain the equilibrium elastic behavior from capillary micromechanics we have to allow enough time for equilibration of the particle shape after a change in the applied pressure.

12.3
Discussion: Effects of Particle Softness on Suspension Rheology

The fascinating macroscopic behavior of microgel suspensions originates directly from the properties of the single microgel particles; however, the connection between macroscopic and microscopic behaviors is still poorly understood. The methods described above constitute an important starting point for experimentally studying this connection. It is essential to characterize and study microgels directly at the single-particle level, as the mechanical properties of microgel particles can differ significantly from those of macroscopic hydrogels of identical chemical composition. Moreover, the morphology of microgel spheres is in general not homogeneous; it can even exhibit a highly heterogeneous core–shell-like structure [31]. It is thus not sufficient to treat microgels as a homogeneous and isotropic bulk material, instead the properties should be directly characterized at the scale of a single particle.

The unique properties of microgels lead to significant differences in their suspension rheology compared to hard particles [7, 9, 14–16, 32–34]. The individual particles of a microgel suspension are soft and can readily change both their shape and their volume under an applied stress. Both aspects significantly affect the macroscopic rheological response of a microgel suspension. A simple illustration of this effect is seen if we compare the concentration-dependent viscosity for different materials. For suspensions of hard particles, a drastic increase in viscosity is observed as the concentration of particles approaches a volume fraction around $\phi \approx 60\%$, close to the volume fraction for random close packing ϕ_{RCP}, of spheres [1, 35]. Because they are not deformable or compressible, ϕ_{RCP} represents an upper limit for the concentration that can be achieved in these materials.

However, for soft microgel particles this limit no longer applies. Because the particles are highly compressible, they can shrink to accommodate more particles within the same volume. The viscosity of microgel suspensions thus depends much less sensitively on concentration than the viscosity of a suspension of hard particles. To achieve the same high levels of viscosity that are observed close to $\phi \approx 60\%$ in hard sphere suspensions, a microgel suspension of much higher concentration is required. The softer the particles are, the higher the concentration has to be increased in order to reach the same level of viscosity [7, 9, 14–16, 36].

12.4
Microgels as Model Glasses: Soft Particles Make Strong Glasses

Microgel suspensions have been identified as important model materials in the study of glass formation [7, 9, 14–16, 32–34]. Importantly, the role of the single-particle mechanical properties in controlling the glass transition behavior in microgel suspensions was recently demonstrated [16]. In molecular glass formers, the viscosity increases continuously as the material is cooled toward its glass transition temperature T_g; the temperature dependence of the viscosity varies significantly for different glass forming materials. This variety of observed behaviors is described by the concept of *fragility*, which quantifies the sensitivity of the viscosity or the relaxation time to a temperature change near its glass transition. *Fragile* glass formers are highly sensitive to temperature change; their relaxation time exhibits a dramatic increase as the glass transition is approached, reminiscent of critical behavior. *Strong* glass formers, on the other hand, exhibit a weaker exponential Arrhenius-type increase in their relaxation time.

For colloidal suspensions, a glass is formed by an increase in the colloid concentration, in analogy to decreasing the temperature in a molecular glass former. It was recently demonstrated [16] that the concentration-dependent viscosity, or the structural relaxation time, of suspensions of soft microgel particles is directly controlled by the mechanical properties of the single particles. By also defining fragility for colloids as sensitivity to a concentration change near the glass transition, it was found that particle softness directly controlled fragility. Fragility can thus be directly tuned by changing the mechanical properties of the single particles. *Hard particles behave like fragile glass formers* as their relaxation time and their viscosity increase dramatically within a small concentration range close to the glass transition concentration. However, for soft particles this same increase in viscosity and relaxation time is stretched out over a much broader range of concentrations, corresponding to a stronger dynamic behavior. Eventually, for the softest materials studied the relaxation follows an exponential increase in the relaxation time with concentration, in analogy to the Arrhenius behavior observed in molecular glass formers [16]. Thus, *soft particles behave like strong glass formers*. By changing the elastic properties of the single particles, it is therefore possible to achieve the same wide range of dynamic behaviors as observed in molecular glass formers in the colloidal model systems – from a highly fragile behavior for hard particles to a strong, Arrhenius-like behavior for soft, deformable particles. Moreover, it was demonstrated that a range of characteristic properties observed during dynamic arrest of molecular systems was mirrored in the dynamic arrest of colloidal suspensions [16]. Microgel suspensions thus constitute an important model system for studying glass formation.

12.5
Analogy to Emulsions and Foams

Another example of the importance of single-particle mechanics for understanding the macroscopic flow behavior is the behavior of suspensions of deformable but incompressible objects. In emulsion and foam systems, it is the surface tension between the two phases that leads to an elastic-like response of the single objects and in turn to an elastic-like response of the systems [37, 38]. Any deformation of a drop or bubble from the equilibrium spherical shape leads to an increase in surface area; thus, the surface tension provides a restoring force that is the origin of elasticity in these materials. The mechanical behavior of emulsions and foams has been extensively studied and is reasonably well understood [37–41].

The existing knowledge of emulsion mechanics should provide important insights into the role that the shear elastic modulus G of microgel particles plays in controlling suspension behavior. In addition, for microgels, the relatively small compressive elastic modulus K adds extra modes of relaxation that do not exist for emulsions or foams. Using our knowledge of the behavior of emulsions and foams and thus the role of G in controlling suspension dynamics should enable us to better isolate the importance of compressional deformations and thus K in the dynamics of microgel suspensions.

The mechanical behavior of microgel suspensions is at the heart of many industrial applications of microgels and is also increasingly studied in fundamental scientific research. Nevertheless, the link between local and macroscopic properties in these materials is still not well established. One major reason for this has been the lack of experiments that characterize the mechanics of single particles – which should be the basis for understanding the macroscopic mechanical behavior. The methods presented above are a good starting point for future studies aimed at a better understanding of the macroscopic response of microgel suspensions and their link to the mechanics at the single-particle level.

References

1 Mason, T.G. and Weitz, D.A. (1995) *Phys. Rev. Lett.*, **75**, 2770.
2 Larson, R.G. (1998) *The Structure and Rheology of Complex Fluids*, Topics in Chemical Engineering, 1st edn, Oxford University Press, Oxford, New York.
3 Segre, P.N., Meeker, S.P., Pusey, P.N., and Poon, W.C.K. (1995) *Phys. Rev. Lett.*, **75**, 958.
4 Pusey, P.N. and van Megen, W. (1987) *Phys. Rev. Lett.*, **59**, 2083.
5 Brady, J. and Morris, J. (1997) *Fluid Mech.*, **348**, 103.
6 Krieger, I. and Dougherty, T. (1959) *Soc. Rheol.*, **3**, 137.
7 Borrega, R., Cloitre, M., Betremieux, I., Ernst, B., and Leibler, L. (1999) *Europhys. Lett.*, **47**, 729.
8 Saunders, B. and Vincent, B. (1999) *Adv. Colloid Interface Sci.*, **80**, 1.
9 Senff, H. and Richtering, W. (1999) *Chem. Phys.*, **111**, 1705.
10 Cloitre, M., Borrega, R., and Leibler, L. (2000) *Phys. Rev. Lett.*, **85**, 4819.
11 Wu, J., Zhou, B., and Hu, Z. (2003) *Phys. Rev. Lett.*, **90**, 048304.

12 Fernández-Nieves, A., Fernández-Barbero, A., Vincent, B., and de las Nieves, F. (2003) *Chem. Phys.*, **119**, 10383.
13 Meeker, S., Bonnecaze, R., and Cloitre, M. (2004) *Rheology*, **48**, 1295.
14 Purnomo, E.H., van den Ende, D., Mellema, J., and Mugele, F. (2006) *Europhys. Lett.*, **76**, 74.
15 Le Grand, A. and Petekidis, G. (2008) *Rheol. Acta*, **47**, 579.
16 Mattsson, J., Wyss, H.M., Fernandez-Nieves, A., Miyazaki, K., Hu, Z., Reichman, D.R., and Weitz, D.A. (2009) *Nature*, **462**, 83.
17 Alsayed, A., Islam, M., Zhang, J., Collings, P., and Yodh, A. (2005) *Science*, **309**, 1207.
18 Tagit, O., Tomczak, N., and Vancso, G. (2008) *Small*, **4** (1), 119.
19 Hashmi, S.M. and Dufresne, E.R. (2009) *Soft Matter*, **5** (19), 3682.
20 Bastide, J., Candau, S., and Leibler, L. (1981) *Macromolecules*, **14**, 719.
21 Cohen, Y., Ramon, O., Kopelman, I., and Mizrahi, S. (1992) *Polym. Sci. Pol. Phys.*, **30**, 1055.
22 Bradley, M., Ramos, J., and Vincent, B. (2005) *Langmuir*, **21**, 1209.
23 Routh, A., Fernandez-Nieves, A., Bradley, M., and Vincent, B. (2006) *J. Phys. Chem. B*, **110**, 12721.
24 Flory, P.J. and Rehner, J. (1943) *Chem. Phys.*, **11**, 521.
25 Barrat, J., Joanny, J., and Pincus, P. (1992) *J. Phys. II France*, **2**, 1531.
26 Wyss, H.M., Franke, T., Mele, E., and Weitz, D.A. (2010), *Soft Matter*, **6** (18), 4550.
27 Suarez, I., Fernandez-Nieves, A., and Marquez, M. (2006) *J. Phys. Chem. B*, **110**, 25729.
28 Tanaka, T. and Fillmore, D. (1979) *Chem. Phy.*, **70**, 1214.
29 Suzuki, Y., Tokita, M., and Mukai, S. (2009) *Eur. Phys. E: Soft Matter Biolo. Phys.*, **29**, 415.
30 Tokita, M. and Tanaka, T. (1991) *Chem. Phys.*, **95**, 4613.
31 Fernandez-Barbero, A., Fernandez-Nieves, A., Grillo, I., and Lopez-Cabarcos, E. (2002) *Phys. Rev. E*, **66**, 051803.
32 Yunker, P., Zhang, Z., Aptowicz, K., and Yodh, A. (2009) *Phys. Rev. Lett.*, **103**, 115701.
33 McKenna, G., Narita, T., and Lequeux, F. (2009) *Rheology*, **53**, 489.
34 Zhang, Z., Xu, N., Chen, D., Yunker, P., Alsayed, A., Aptowicz, K., Habdas, P., Liu, A., Nagel, S., and Yodh, A. (2009) *Nature*, **459**, 230.
35 Eckert, T. and Bartsch, E. (2002) *Phys. Rev. Lett.*, **89**, 125701.
36 Senff, H. and Richtering, W. (2000) *Colloid Polym. Sci.*, **278**, 830.
37 Princen, H. (1983) *J. Colloid Interface Sci.*, **91**, 160.
38 Mason, T., Bibette, J., and Weitz, D. (1995) *Phys. Rev. Lett.*, **75**, 2051.
39 Mason, T., Bibette, J., and Weitz, D. (1996) *J. Colloid Interface Sci.*, **179**, 439.
40 Cohen-Addad, S., Hoballah, H., and Hohler, R. (1998) *Phys. Rev. E*, **57**, 6897.
41 Gopal, A. and Durian, D. (2003) *Phys. Rev. Lett.*, **91**, 188303.

13
Rheology of Industrially Relevant Microgels
Jason R. Stokes

13.1
Introduction

Microgels are currently used in a *broad range of industrial applications and consumer products*, including surface coatings, paints, inks, oil recovery, controlled drug delivery, cosmetics, personal care, home care, food, and pharmaceuticals. An essential feature in many of these applications is the rheology of the microgel suspension during processing, storage, and transport, as well as during the application process itself (i.e., at the "in-use" or "end-use" stage). Microgel suspensions have a long history of *use for rheological control* due to their ability to swell in a suitable solvent to give increased viscosity and/or a gel-like consistency. Starches, which are essentially cross-linked carbohydrate polymers, are probably the oldest example of a microgel; they were separated from grains by the ancient Greeks for a variety of uses including foods, adhesives, and even wound healing, where the latter was achieved by mixing with saliva to form a honey-like coating [1].

The attraction of microgel suspensions in industrial applications is that their rheology can be controlled to meet the requirements for specific applications by varying composition, cross-link density, particle size, shape, surface properties, and solvent quality. A useful feature of microgels is their responsive nature to their environment; they can swell or deswell by altering solvent quality (through pH, temperature, ions, etc.), thus altering their effective phase volume and consequently the rheological properties of the suspension. The deformable and swellable nature of microgels allows a higher phase volume to be obtained compared to hard spheres of equivalent size. Their deformability also affects their response in shear flow due to their ability to change shape in response to perturbation, particularly at high volume fractions. The composition of microgels can be adapted and customized according to the application, and the microgel additives used have been made out of a wide variety of polymeric materials including natural or modified biopolymer-based components (e.g., polysaccharides, proteins, and starches), and synthesized components such as polyacrylates, poly(styrene sulfonates), poly(vinyl pyridine), and polyurethanes.

Microgel Suspensions: Fundamentals and Applications
Edited by Alberto Fernandez-Nieves, Hans M. Wyss, Johan Mattsson, and David A. Weitz
Copyright © 2011 WILEY-VCH Verlag GmbH & Co. KGaA, Weinheim
ISBN: 978-3-527-32158-2

In this chapter, the rheological properties of microgel suspensions are discussed from *applications* point of view, with specific discussion on the influence of various factors from a formulation and materials design perspective. Applications where rheological properties of microgel suspensions are a key feature are also highlighted.

13.2
Flow Behavior

Microgels are particles consisting of cross-linked polymeric molecules, and can be either colloidal (less than a micrometer) or noncolloidal (greater than a micrometer) in size. In terms of their flow behavior in industrially relevant situations, colloidal microgels are considered to fall in between the behaviors of a polymer molecule and a colloidal particle [2], as depicted in Figure 13.1. Microgel suspensions will typically not possess the nonlinear elastic and extensional properties in flow that are exhibited by polymer solutions, while they do show strongly increased viscosities and shear thinning at relatively low solids content. These factors combine to significantly affect the flow behavior and the textural response displayed by microgel suspensions, frequently leading to favorable consumer responses to products formulated with microgels rather than polymer solutions. At high phase volumes where particles are in contact and their relaxation is sterically confined, the microgel suspension can behave as a soft glass with a response that depends on the particle properties, while at

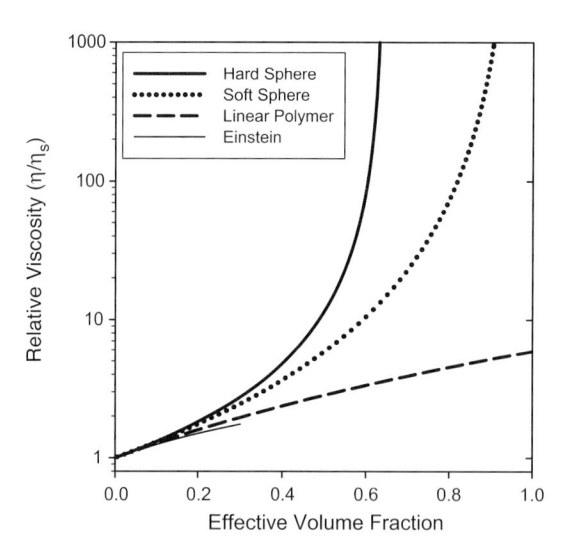

Figure 13.1 Qualitative comparison between the zero-shear viscosity of suspensions containing hard particles, microgels, and linear polymers. Also shown is the equation for the Einstein model (Equation 13.2). The line for hard spheres and soft spheres was drawn using the Kreiger–Dougherty model (Equation 13.5) with $\phi_c = 0.64$ and $\phi_c = 0.96$, respectively, and $[\eta] = 2.5$.

low concentrations they can flocculate to form soft gel structures with responses that depend more on the colloidal forces between particles [3]. In addition, microgels have the potential to incorporate solvent into their structure and they do not always have a well-defined boundary since the degree of cross-linking can vary through an individual particle. Microgel suspensions exhibit strong shear-thinning behavior, whether they are in the form of a soft glass or a soft gel; thus, they possess a viscosity at high shear rates that is typically lower than that found for a polymeric system with similar low-shear viscosity.

13.2.1
Influence of Phase Volume and Concentration

The rheological properties of dilute suspensions of polymers, microgels, and colloidal particles are dependent on the effective phase volume occupied by the dispersed component, as shown in Figure 13.1. While the phase volume occupied by a colloidal particle is usually known, this is not the case for microgels or polymers since they incorporate solvent into their structure. By measuring the viscosity of samples in the dilute solution regime, it is possible to obtain the effective hydrodynamic volume occupied by the polymer or microgel. This is achieved by determining the intrinsic viscosity ($[\eta]$), which is defined as the zero concentration limit of the reduced viscosity

$$[\eta] = \lim_{c \to 0} \frac{\eta_0 - \eta_s}{c \eta_s} \tag{13.1}$$

where η_0 is the zero-shear viscosity, η_s is the solvent viscosity, and c is the dry weight concentration. For polymer solutions, $[\eta]$ is related to the molecular weight and to the conformation of the molecule in response to the solvent quality. Similarly, $[\eta]$ will vary for microgels in terms of how they swell in relation to the solvent. The Einstein equation for the contribution of hard spheres to the viscosity is given by [4]

$$\eta_0 - \eta_s = 2.5 \eta_s \phi \tag{13.2}$$

where ϕ is the volume fraction of spheres. Substituting this into the equation for the intrinsic viscosity gives $[\eta] = 2.5\phi/c$. The effective phase volume (ϕ_{eff}) occupied by a swollen microgel in terms of concentration is thus [2] given by

$$\phi_{eff} = [\eta]c/2.5 \tag{13.3}$$

The effective phase volume can also be written in terms of a specific volume or swelling ratio (k ml/g) as $\phi_{eff} = kc$. This approach allows determination of the volume fraction at any particular weight concentration provided k is independent of concentration. k and $[\eta]$ can be used to evaluate the influence of parameters such as solvent quality (pH, temperature, salt, etc.) on the swellability of the microgels [5].

As the concentration is increased above the dilute regime, hydrodynamic interactions between hard spheres cause a deviation in the viscosity from Einstein's equation. These interactions were accounted for in a semiempirical treatment by Batchelor [6], and as for hard spheres, this approach has been found to be a reasonable prediction of the viscosity of microgel suspensions for effective volume fractions of up to about 10–20% [7]:

$$\eta_0 = \eta_s \left(1 + 2.5\phi_{\text{eff}} + 5.9\phi_{\text{eff}}^2\right) \tag{13.4}$$

At higher concentrations ($\phi_{\text{eff}} > 0.2$), the influence of multibody interactions and crowding effects must also be accounted for. As for hard spheres, the Krieger–Dougherty model [8] has been widely used to describe the rheology of microgels at high phase volumes. This is derived from Einstein's equation by considering that the viscosity increases incrementally ($d\eta$) with increases in phase volume of spheres relative to the space available, up to a critical (ϕ_c) packing fraction (i.e., $d\phi/(1-\phi/\phi_c)$). Integration leads to the following expression:

$$\eta_0 = \eta_s \left(1 - \phi_{\text{eff}}/\phi_c\right)^{-[\eta]\phi_{\max}} \tag{13.5}$$

$\phi_{\max} \sim 0.64$ for random close packed hard spheres and ~ 0.71 for hexagonal or face-center-cubic packed hard spheres. Quemada [9, 10] derived a similar, yet simpler, expression whereby $[\eta]\phi_c = 2$.

However, the viscosity can deviate from these rheological models at high phase volumes ($\phi_{\text{eff}} > 0.5$) if the specific volume is not constant with concentration. This can arise when microgels come into proximity such that interparticle forces become more prominent. Factors such as particle deformation/compression, solvent loss due to osmotic deswelling, interpenetration between microgels, and steric confinement where swelling is limited by the amount of available solvent can lead to the specific volume of the microgels decreasing with increasing concentration. This will in turn lead to a deviation of the rheological response from that predicted by the models discussed above, and alter the effective phase volume of the microgels, $\phi_{\text{eff}} = kc$.

Senff et al. [11] generated "master curves" using constant values of k (determined at low concentrations using the Batchelor equation and dynamic light scattering size analysis) for temperature sensitive poly(N-isopropylacrylamide) (PNIPAM) colloidal microgels [12, 13] and core–shell latex particles [11] in aqueous suspension; both systems were stable against flocculation. Deviation from the hard sphere behavior was found for effective phase volumes above about $\phi_{\text{eff}} \sim 0.5$, whereby a lower viscosity than for hard spheres is obtained at high volume fractions due to the soft interparticle potential and deformation/compression of the neutral PNIPAM microgels and shell layer, respectively. Similar observations on the influence of particle compression at high concentrations have been made for other colloidal microgels [2, 14] and core–shell or sterically stabilized particle [15] suspensions. For microgel systems with high central cross-link density but less dense outer core, the viscosity was found to be considerably lower than for hard spheres and interpenetration between microgels allowed the effective volume fraction to exceed unity (i.e., $\phi_{\text{eff}} > 1$). The polymer architecture of such systems closely resembles that of star polymers with low arm density; therefore, they are often referred to as "star-like" microgels.

Tan et al. [5, 7, 16, 17] introduced a semiempirical approach to investigate the change in k with concentration by assuming that $\phi_m = 0.64$ and varying k in order to mathematically fit the measured viscosity to the Krieger–Dougherty and Batchelor

($\phi_{eff} < 10\%$) models. The dependence of k on concentration was fitted according to Equation 13.6 for a range of conditions, including different ionic strength, pH, neutralization degree, cross-linked density, and molar ratio of acid groups, for a series of aqueous suspensions containing acrylate-based charged colloidal microgels (e.g., methacrylic acid-ethyl acrylate (MAA-EA) cross-linked with diallyl phthalate).

$$\frac{k - k_{min}}{k_0 - k_{min}} = \left(1 + \left(\frac{c}{c_0}\right)^2\right)^{-m} \tag{13.6}$$

It was postulated that the dependence of k on concentration, as shown in Figure 13.2, arises because the concentration of free counterions becomes sufficiently large at a critical particle concentration (c_0) to induce osmotic deswelling and

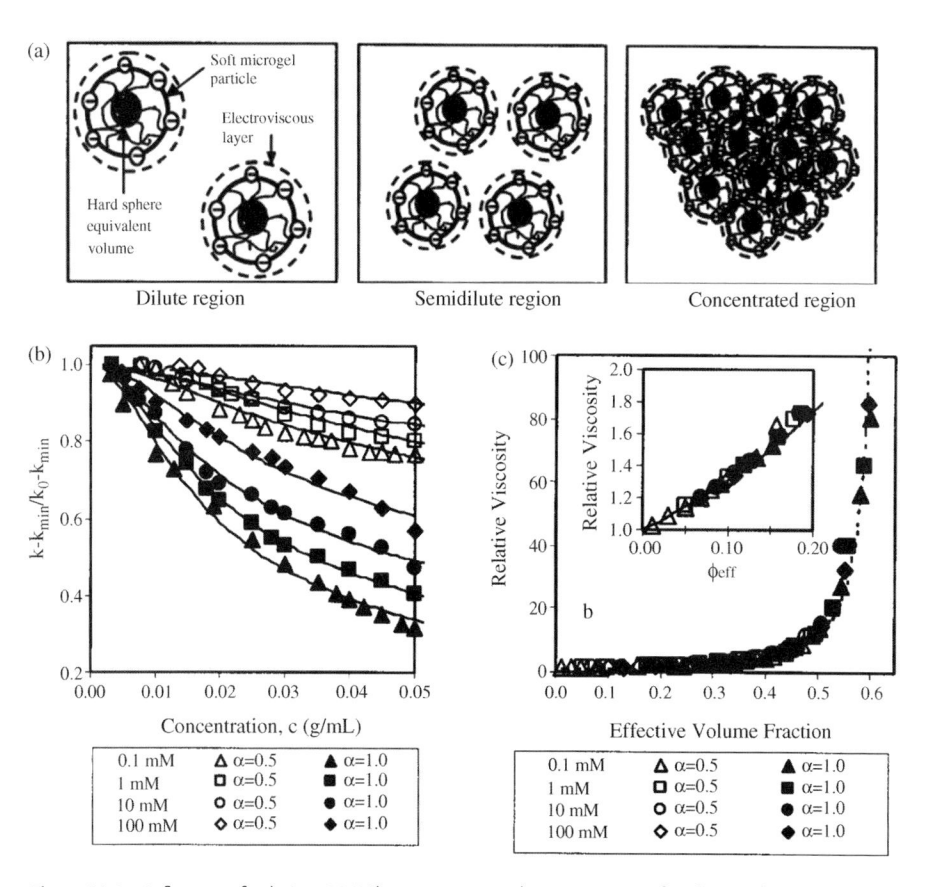

Figure 13.2 Influence of salt (~mM KCl) on the microstructure and rheology of cross-linked MAA-EA microgels [5]. (a) Pictorial representation of the change in specific volume (k) at three different microgel concentration regimes. (b) Specific volume as a function of microgel concentration, fitted according to Equation 13.6. (c) Master curve of relative viscosity using fitted values for k, where the solid line and dotted line denote hard sphere behavior according to Equations 13.4 and 13.5 respectively. Reproduced from Ref. [5].

subsequent shrinkage of the microgels. k_{min} describes the limiting condition when soft particles are compressed to the hard sphere equivalent volume at high concentrations and m is the rate of decrease of k with concentration. The influence of osmotic deswelling was calculated and quantified for microgel suspensions by Cloitre et al. [18], and following this approach Tan et al. [17] utilized an ion specific electrode to show that the free counterions increased with increasing microgel concentration. This resulted in stronger osmotic deswelling and confirmed that the specific volume (k) decreases with increasing concentration. In addition, osmotic pressure exerted by counterions in the solvent within these microgels helps to maintain particle conformation [16].

13.2.2
Shear Rheology of Concentrated Microgel Suspensions

Microgels are most widely used in the form of concentrated suspensions ($\phi_{eff} > 0.5$), where microgel particles are in intimate contact [12]. These microgel suspensions tend to respond strongly to an applied stress such that above a critical value they either flow or yield in what is often regarded as a solid-to-fluid transition, and below that value they behave in a solid-like fashion. These suspensions often display three distinct regions during flow, comprising a plateau in viscosity at low (termed the zero-shear viscosity, η_0) and high (termed infinite-shear viscosity, η_∞) shear rates, and a shear-thinning region in between where the viscosity decreases with increasing shear rate following a power law relationship. There are numerous empirical models used to describe shear-thinning fluids, and both the Carreau and Cross models have been widely used for microgel suspensions. The Cross model in terms of rate and stress, respectively, is given by

$$\eta = \eta_\infty + \frac{\eta_0 - \eta_\infty}{1 + \left(\frac{\dot{\gamma}}{\dot{\gamma}_c}\right)^m}, \quad \eta = \eta_\infty + \frac{\eta_0 - \eta_\infty}{1 + \left(\frac{\sigma}{\sigma_c}\right)^n} \qquad (13.7)$$

m and n are constants. The critical stress, σ_c, corresponds to the shear stress at the intermediate viscosity of $\eta = (\eta_0 - \eta_\infty)/2$ [2]. Senff and Richtering [13], as shown in Figure 13.3, and Rodriguez et al. [19] demonstrated that a master curve could be obtained, using aqueous suspensions of PNIPAM colloidal microgels of different cross-link density and temperatures and cross-linked polystyrene colloidal microgels in bromoform respectively, by plotting $(\eta - \eta_\infty)/(\eta_0 - \eta_\infty)$ against σ/σ_c. Such master curves are useful from an engineering point of view for designing rheological properties as a function of formulation variables as well as for highlighting the common physics responsible for observed properties. Wolfe and Scopazzi [2] utilized a similar equation to the Carreau model to empirically describe the stress dependence for the viscosity of colloidal polymethylmethacrylate (PMMA) microgels in better than θ solvent (butyl carbitol acetate):

$$\eta = \eta_\infty + \frac{\eta_0 - \eta_\infty}{(1 + (\delta\sigma)^2)^\beta} \qquad (13.8)$$

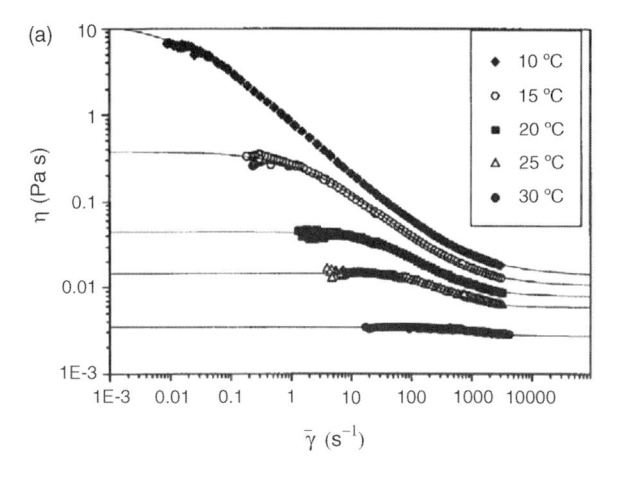

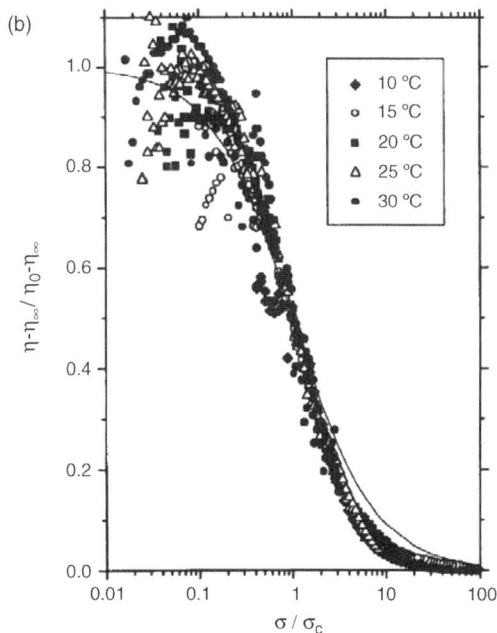

Figure 13.3 Viscosity of temperature-sensitive colloidal microgel suspension at constant weight concentration. (a) Shear rate dependent viscosity for different temperatures with the lines indicating fits to Equation 13.7. (b) "Master" curve showing $(\eta-\eta_\infty)/(\eta_0-\eta_\infty)$ as a function of σ/σ_c. Reproduced from Figures 4 and 5 from Ref. [13], with permission from Springer Science + Business Media.

δ and β are adjustable parameters. For suspensions that exhibit a distinct yield stress, which is a minimum stress required for flow to occur (i.e., no observable zero-shear viscosity plateau), the Hershel–Buckley model [20] is appropriate for describing the rheological properties:

$$\sigma = \sigma_y + K\dot\gamma^n \qquad (13.9)$$

where σ_y is the apparent yield stress. A variety of other empirical equations can also be used to fit the rheology of microgel suspensions as a function of stress or shear rate, as listed in common rheological text books [20].

In colloidal microgel systems, where the microgel size is of the order of a micrometer and below, the Peclet number is used to define the relative importance of shear compared with thermal diffusion and is defined as

$$Pe = \frac{6\pi\eta_s a^3 \dot\gamma}{k_b T} \qquad (13.10)$$

a is the particle radius, k_b is the Boltzmann constant, and T is the absolute temperature. Flow curves are examined as a function of Pe to determine the relative importance of thermal motion and microgel deformability on the observed rheological response; this way of plotting will generally lead to a master curve. Master curves are also formed by scaling the viscosity, shear rate or stress with the η_0, or the critical values, $\dot\gamma_c$ or σ_c, respectively. Interestingly, σ_c has been found to be concentration dependent, and for some colloidal microgel systems, it shows a peak at around $\phi_{eff} \sim 0.5$. However, it remains unclear why there is sometimes a maximum in σ_c and it does not appear to be universal. The volume fraction at which the maximum occurs (ϕ_{peak}) varies for different microgels due to a range of competing effects. For example, microgels can deswell at high phase volumes leading to an increase in ϕ_{peak}, while the presence of polymer tails at the microgel surface may cause a long-range interparticle interaction and thus a decrease in ϕ_{max} [2]. The values of σ_c and ϕ_{peak} have been found to decrease with decreasing cross-linker content and thus increasing particle softness [2, 13]. With further decreases in cross-linker content, this behavior would be expected to approach that observed for linear polymer solutions whereby σ_c is constant or zero in dilute solution and increases monotonically with polymer concentration above that required for chain overlap.

Microgels may undergo some degree of deformation during shear flow, particularly in concentrated suspensions, where the microgel particles are in close proximity. This has been used as an explanation for measurable η_∞ values at effective phase volumes exceeding those characteristic of arrested hard sphere suspensions, $\phi_m \sim 0.71$, and in some cases exceeding unity (e.g., $\phi_{eff} > 1$ for aqueous suspensions of colloidal PNIPAM microgel [12, 13] and agar microgels [21]). η_∞ is not sensitive to the interparticle potential since the equilibrium microstructure is distorted when shear forces dominate.

Adams et al. [21] examined the influence of particle modulus on the shear viscosity of noncolloidal agar microgels (10–40 µm diameter) as a function of shear stress. Agar microgels present a good model system in which to probe the influence of particle deformability since they have a uniform internal structure and well defined phase volume. In addition, they do not display osmotic deswelling at high concentrations, although crowding and shear effects can cause some deswelling if the particles are sufficiently soft. It is found that the particle modulus (ranging from 2.4 to 185 kPa) had a significant impact on the viscosity, particularly at high shear stresses,

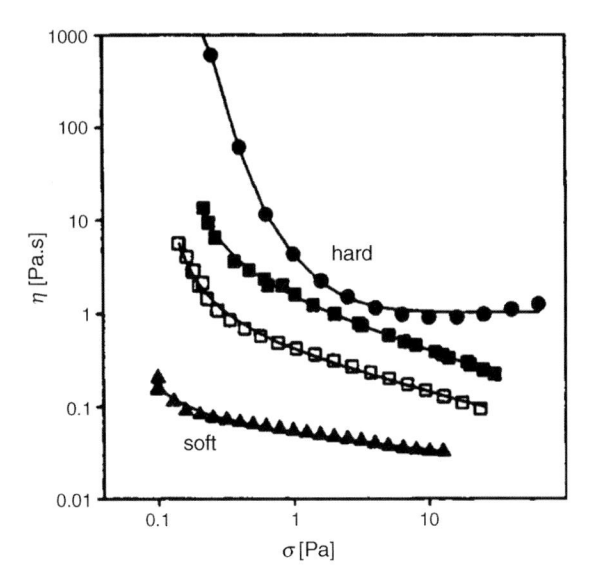

Figure 13.4 Stress dependent viscosity of noncolloidal agar microgel suspensions at an effective phase volume of $\phi_r = 0.96$, where $\phi_r = \phi_{eff}/\phi_c$, showing that the viscosity is strongly influenced by the elasticity of the microgel. Symbols correspond to the agar concentration in the microgel: ● 5%, ■ 1%, □ 0.75%, and ▲ 0.5%. Reproduced with permission from Ref. [21].

as shown in Figure 13.4. At high microgel concentrations, the viscosity is significantly lower for softer particles and hence controlling the particle modulus is a means in which to control the bulk rheological properties of the microgel suspension. The critical stress for shear thinning (or apparent yield stress) is also found to depend on the particle modulus, with a lower critical stress occurring for softer microgels at the same effective phase volume.

13.2.3
Linear Viscoelasticity of Concentrated Microgel Suspensions

When particles are in proximity ($\phi_{eff} > 0.5$), the linear viscoelastic response depends strongly on the individual particle modulus, the separation distance between neighboring particles, and on the microstructure when there is minimal interpenetration. Particle deformation gives rise to strong repulsive interactions [12, 23], and a general expression can be formulated for the high frequency plateau modulus of monodisperse spherical particles as a function of the interaction potential and pair correlation function using statistical mechanics [22, 23]:

$$G = Nk_bT + (2\pi N^2/15) \int_0^\infty g(r) \frac{d}{dR}\left[r^4\left(\frac{dV(r)}{dr}\right)\right] \qquad (13.11)$$

k_b is the Boltzmann constant, N is the particle number density, $g(r)$ is the radial distribution function and $V(r)$ is the pair interaction potential. $dV(r)/dr$ is the force that particles exert on one another as a function of their center-center separation, r. For an electrostatically stabilized concentrated polystyrene microgel system, Buscall et al. [24, 25] assumed a lattice-like microstructure so that $g(r)$ could be approximated as a delta function centered at the nearest neighbor, to yield a simple expression:

$$G \propto \frac{1}{r}\left(\frac{\partial^2 V}{\partial r^2}\right) \tag{13.12}$$

Assuming that the interparticle distance is related to the closest packed volume fraction ($r^3 = d^3(\phi_c/\phi_{\text{eff}})$) and the interparticle potential is of the form $V \propto 1/r^n$, then [26],

$$G \propto \phi_{\text{eff}}^m \tag{13.13}$$

The power law exponent is given by $m = (n/3) + 1$. Although this approach neglects hydrodynamic interactions and an osmotic term($\propto dV/dr$), it has been found to follow experimental data with exponents ranging from $m = 3$ to $m = 8$ for a range of different concentrated colloidal and noncolloidal microgel suspensions just above close packing [11, 13, 21, 23, 26]; Senff and Richtering [13] found that n and m increased with increasing cross-link density for aqueous suspensions of colloidal PNIPAM microgels.

For two elastic spheres in contact, the force as a function of separation is dependent on the particle elasticity according to the Hertz model [27], with the Hertzian pair potential given as a function of the particle shear elastic modulus (G_p) and Poisson ratio (ν) as follows [23]:

$$V(r) = (32/15)\, R_0^{1/2}(R_0 - r/2)^{5/2}\,\frac{G_p}{1-\nu} \tag{13.14}$$

Evans and Lips [23] assumed that noncentral forces arising from friction and adhesion, which are described by JKR theory [28], are negligible in the case of highly swollen microgels and that only nearest neighbors interact. Nk_bT in Equation 13.11 is also assumed to be negligible. By substituting $g(r)$ with a delta function, the shear plateau modulus of the suspension G can be expressed in terms of the shear modulus of the individual particles:

$$G = \frac{\phi_c z G_p}{5\pi(1-\nu)}\left[(\phi_r^{2/3} - \phi_r^{1/3})^{1/2} - \frac{3}{3}(\phi_r - \phi_r^{2/3})^{3/2}\right] \tag{13.15}$$

Here, z is the number of nearest neighbors and the reduced packing fraction is given by $\phi_r = \phi_{\text{eff}}/\phi_c$. This model was found to predict essential features of a number of microgel systems, including sephadex [23], with the result shown in Figure 13.5, as well as gelatinized starch [29] and agar microgels [21]. Such features include the prediction of a rapid increase in the shear modulus at a critical concentration followed by a leveling off at higher concentrations, as well as an

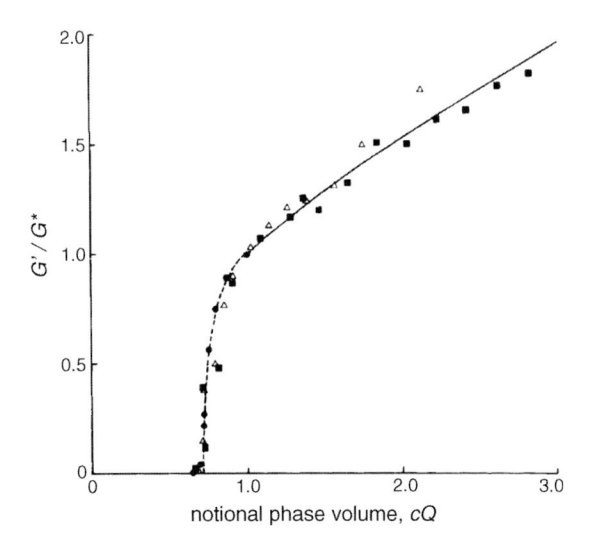

Figure 13.5 Normalized elasticity at a frequency of 0.25 Hz, where G^* is the value of G' at $cQ = 1$ (Q is the swelling ratio that is equivalent to the specific volume k), for noncolloidal deformable sephadex microgels at varying swelling ratios. The dotted line represents the theoretical fit to the data in the intermediate range, $0.7 < cQ < 1$ according to Equation 13.15. The solid line represents the best-fit power law (exponent 0.61) for $cQ > 1$. Reproduced from Ref. [23] with permission of The Royal Society of Chemistry.

overall dependence on particle modulus. However, this failed to quantitatively predict the dynamic response of concentrated agar microgel suspensions of Adams et al. [21] as a function of phase volume and particle modulus as shown in Figure 13.6. Given that the distance between neighboring particles decreases with increasing phase volume, $R \propto \phi^{-1/3}$, the following empirical equation was utilized to describe the modulus of the agar microgel suspension:

$$G' = aG'^b_p \left[1 - (\phi_{\text{eff}}/\phi_c)^{\frac{1}{3}}\right] \tag{13.16}$$

where a and b are constants that can be determined experimentally for a particular system and G'_p is the plateau storage modulus of individual microgels. ϕ_c is here the critical concentration at which G' is effectively negligible, determined empirically by extrapolation.

Seth et al. [30] use a micromechanical model to predict the modulus and osmotic pressure for a disordered array of compressed elastic spheres interacting through a pairwise Hertzian potential. Qualitative agreement was found with experimental results whereby the modulus increases dramatically by several orders of magnitude at ϕ_c. Above this concentration, due to packing constraints the particles alter their shape (forming flat facets at contact) and their volume; the modulus in this regime has been described to scale either as $\sim \phi^6$ or as $\sim (\phi - \phi_c)$. These scaling laws also fit the data for agar spherical microgels, as shown in Figure 13.6. The micromechanical model also

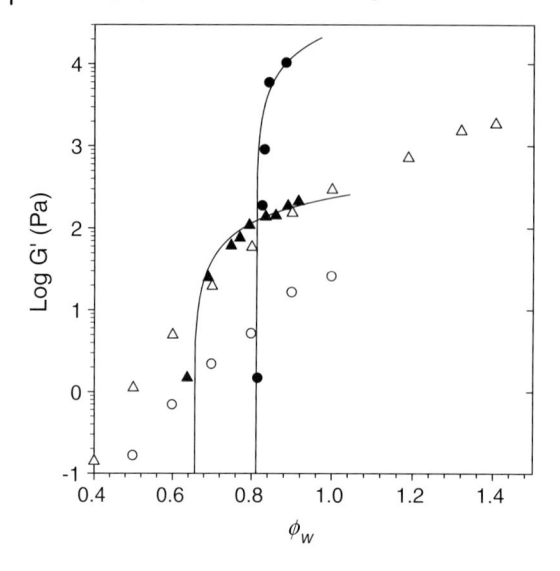

Figure 13.6 Rheology of agar microgels highlighting the influence of particle morphology. Comparison between spherical microgels comprising 5 (●) and 0.5 (▲) wt% agar compared with anisotropic shear gel microgels comprising 0.75 (o) and 1.75 (△) wt% agar. The solid black lines are predictions for spherical microgels using Equation 13.16; the dashed lines follow the percolation model ($G \sim \phi - \phi_c$) and the gray solid line follows $G \sim \phi^6$. Adapted from Ref. [3].

indicated that the individual microgel modulus effectively sets the absolute magnitude of the suspension modulus, but not the relative variation with phase volume, which is controlled by the overall microstructure.

At extremely high concentrations, where $\phi_{\text{eff}} > 1$, the shape of the microgels is likely to be fixed and further increases in concentration will only serve to decrease the swelling of the microgel. The suspension may then be expected to deform homogeneously so that the elasticity of the suspension scales with the elasticity of the intraparticulate material. Under such conditions, it is anticipated that the microgel suspension behaves like a macroscopic gel in terms of its elasticity where only a weak power law dependence on concentration is predicted and observed ($m \sim 1/3$ to $2/3$) [23, 29–32]. However, it is not necessarily a universal response since microgels do not behave as bulk macroscopic gels; they can retain their identity as single particles even at very high effective phase volumes.

13.3
Microgel Suspension Rheology in Applications

After discussing the generic rheological properties of microgel suspensions, this section gives more specific examples with regard to rheology for a number of important applications of microgels. This also includes practical concerns such as the interaction of microgels with other formulation components, the utilization and impact of their responsive properties, and the development of nonspherical microgels including gel fibers. While specific examples of industrially relevant microgels are given in their respective fields, the understandings from each of these areas have wide implications to other microgel systems and applications.

13.3.1
Coating Formulations

Microgel suspensions have been widely investigated in the coating and printing industry primarily for rheology control, although they are also helping to improve the quality of coatings. Their introduction into coatings arose due to environmental concerns with regard to hazardous components in paints. In particular, due to government regulations, it became necessary to reduce the volatile organic components (VOC) from paints. This requires formulations with high solids solvent-borne or aqueous-based compositions that are still capable of forming a smooth finish with optimal appearance and resistance to its environment (e.g., weather and scratch resistance). However, microgel suspensions have also created new possibilities due to their unique rheological properties and potential for chemical modification, and have contributed to improved quality and appearance of coatings and prints including reinforcing properties in cured paint films and optical effects.

The rheological properties of coatings to a large extent determine product performance. For example, in the automotive paint industry, application of paint typically involves tank storage, pipe flow, spraying, and film formation onto the substrate, each step requiring different rheological properties [33]. In order to reduce the amount of organic solvent used in coatings, the phase volume of solids was originally increased by lowering the molecular weight of the polymers used within the resin; while this lowered the viscosity at high shear rates so that it could be readily applied (good leveling and sprayability), it also resulted in coatings that tended to sag when baking/drying. Sagging here refers to the runs and drips that occur in paint coatings due to factors such as localized build up of paint around edges and holes in the substrate, increased surface tension due to solvent evaporation, and so on. On the other hand, high molecular weight polymers prevented sag due to their high viscosity at low stresses but these had poor leveling and atomization properties since they still had a high viscosity at high shear rates. The development of microgel-based systems has provided some control of the rheological properties at each step; their low viscosity at high shear rates aides atomization and sprayability, while their very high viscosity and/or yield stress at low shear provides resistance to sagging while retaining good leveling properties [34]. Microgel suspensions improve sagging properties considerably by quickly reestablishing a high viscosity after application, particularly when there is little solvent evaporation as for dispersions containing a high concentration of solids. In comparison, solvent evaporation controls the maximum film thickness obtainable without sagging for Newtonian paints, and only low film thicknesses are achieved in a single spray cycle when these are water-based due to a slow rate of evaporation that is strongly dependent on humidity [33].

The development of highly shear-thinning formulations based around microgel technologies has offered improved control over coating films.

Coatings typically contain multiple components, and formulators use a variety of additives as agents beneficial to the rheology, the film formation, the protection, or the appearance of the coatings. It is essential to achieve a good control of the rheological properties and particularly the low and high shear viscosity, the yield stress and the linear viscoelastic properties. To obtain a yield stress using microgel particles for coating formulations, it is necessary to either (i) flocculate the particles to form a network, a so-called soft gel structure or (ii) close-packed particles to form a soft glass structure. Both can be achieved at relatively low polymer weight fractions.

In the case of flocculated colloidal microgel dispersions, the rheological properties depend on the surface forces between microgels and their interaction with various components in the formulation. For example, Ishikura [35] showed that the rheology of polyester microgels in melamine-formaldehyde resin and xylene can vary widely depending on the state of aggregation of the colloidal microgel. A yield stress, which is considered necessary for a suitable coating formulation, is only apparent when the microgels aggregate to form a network structure and this depends strongly on the amount of melamine-formaldehyde in solution when low concentrations of microgels are used. Outside specific formulation concentration, the suspension was Newtonian and the particles are present as dispersed clusters rather than a network structure possessing a yield stress. Boggs et al. [36] found that colloidal microgels dispersed into acrylic resin had a yield stress at relatively low concentrations. The yield stress is only apparent above a critical concentration, which is considered to be a percolation threshold. Boggs et al. [36] and Bauer et al. [37] found that the yield stress for colloidal microgels dispersed in acrylic polymer resins followed the relationship

$$\sigma_y = \eta_{sol} A . c^{3.1} \tag{13.17}$$

where c is the weight concentration of microgel, η_{sol} is the viscosity of acrylic polymer resin, and A is a constant that depends on the polymer resin polarity and microgel structure. This matches similar power law scalings found for colloid particle suspensions where the yield stress depends both on the interactive forces between particles and on the particle size [38–40]. In the case of colloidal microgels dispersed in polymer resins, flocculation occurs primarily from either depletion or bridging mechanisms. Depletion flocculation occurs when the free polymer does not adsorb to the particle surfaces but is excluded from the interparticle space leading to an osmotic pressure that acts to push particles together; this is considered to be a reversible process. Bridging flocculation occurs when polymer adsorbs onto multiple particle surfaces and is considered to be irreversible.

For nonflocculating and/or noncolloidal microgel suspensions used in coating formulations, the highly shear-thinning rheology typically desired is obtained through a high effective phase volume of microgel particles. The effective phase volume can be controlled through factors such as cross-link density and solvent quality. On improving solvent quality, by use of different solvents, pH, salt content, and so on, microgels swell to take in more solvent, thereby increasing the effective phase volume occupied by the microgel as previously discussed. In addition, other components in coating formulations can also affect the rheology. A typical method in traditional coating formulations for increasing viscosity is to include additional linear polymers [41]. However, for microgel suspensions, introducing linear chains of polymer with similar molecular composition to the microgel can have two competing effects on the suspension rheology: an increase in viscosity of the matrix phase and a decrease in viscosity arising from osmotic deswelling of the particles. Osmotic deswelling arises because of limited diffusion of the polymer chains into the particles; the excluded polymer that is also swollen by the solvent exerts an osmotic pressure and the particles subsequently lose solvent accordingly. These two competing processes depend on concentration and cross-link density. For high concentrations of soft spheres, deswelling of the microgels dominates and a reduction in the modulus and viscosity is apparent, as demonstrated by Rayuois et al. [41] for styrene/acrylic microgels in toluene and shown in Figure 13.7. Similar results are also found by Wolfe [42] for PMMA in butyl carbitol acetate. For harder spheres, where swelling effects are limited, an increase in viscosity due to increased matrix viscosity is more pronounced. Wolfe [42] showed that for low concentrations of PMMA microgels in butyl carbitol acetate with added PMMA linear polymer of varying molecular weight, the viscosity increased for the suspension, as one would expect due to an increase in matrix viscosity. At high effective phase volumes ($\phi_{eff} > 0.6$), increasing linear polymer concentration and molecular weight caused the viscosity to increase at high shear stresses and decrease at low shear stresses.

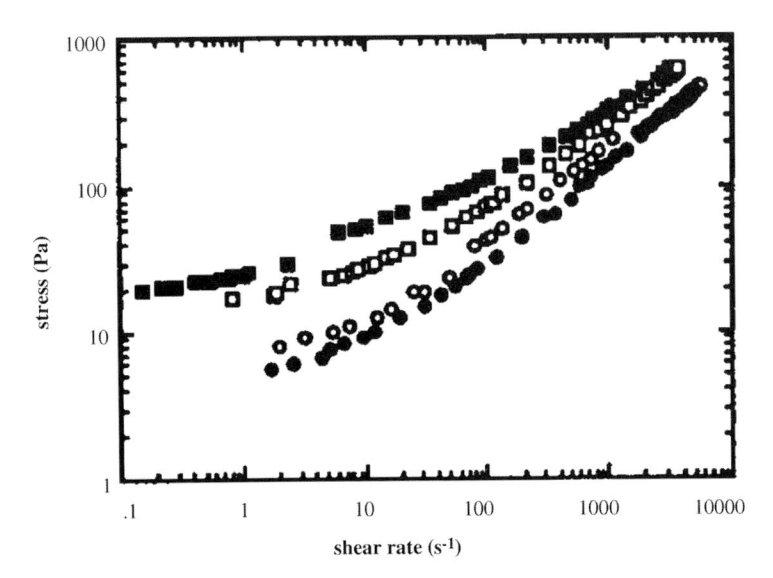

Figure 13.7 Rheology of 10 wt% spherical microgel (copolymer of styrene and acrylic) of ~40 nm diameter in toluene, showing the influence of the addition of linear polymer of the same composition as microgel (MW = 63 000 g/mol) at concentrations of 0% (■), 2% (□), 4% (o), and 5% (●). Reprinted from Ref. [41] with permission from Elsevier.

The viscoelastic properties of microgel suspensions are also important in coating applications. For example, the angle of orientation of metallic flakes on coated substrates has been found to depend on the viscoelastic properties of the coating when applied to the surface [43, 44]. The flakes create a visual effect whereby metallic coatings appear as different colors that depend on the angle of observation, which is measured as a so-called "flip–flop" value, which quantifies the variation of brightness with observation angle. Interestingly, the flip–flop value was found to be at maximum for a waterborne coating when the suspension of core–shell acrylic microgel suspension had a $G'/G'' \sim 1$; this illustrates the importance of viscoelasticity even on the optical properties of coatings. In addition, the elastic nature and high molecular weight of microgels have also provided reinforcing properties in coatings, such as improved elasticity and hardness in cured paint films; these properties for instance provide improved resistance to stone chips [33]. Bauer et al. [37] and Ishikura [35] have particularly discussed the use of polymeric microgels for automotive topcoats that have to have good transparency and durability.

13.3.2
Biomedical, Pharmaceutical, Personal Care, and Cosmetic Products

Microgels are highly attractive for use in biomedical, pharmaceutical, cosmetics, and personal care applications due to their stimulus-responsive nature, potential for surface fictionalization, as well as their ability to encapsulate and release functional species [45]. The use of microgels in biomedical applications typically revolves around either a fluid-to-gel or a gel-to-fluid transition of the microgel suspension at an area of interest within the body, and/or adhesion (mucoadhesion/bioadhesion) of the microgel to a specific site. For example, microgels can be designed to swell at the neutral pH and physiological temperature of the body, so that they become adhesive at the mucosal surfaces due to the increase in viscosity and viscoelasticity in areas lined with mucous, that is, *in situ* gelation [46]. Adhesion in the vicinity of target sites within the body provides the potential for slow release of medicinal aides by increasing the residence time of the formulation. Muco- or bioadhesion may also be achieved through a physical interaction and binding to the site of interest, for example, through the attachment or incorporation of receptor-specific molecules to the surface of the microgel, which subsequently swells or deswells to release drugs [45]. Several reviews are available on mucoadhesive polymers [47–49] and on the use of microgels in drug delivery applications [50]. First generation mucoadhesives relied upon hydrogen bonds from hydroxyl, carboxyl, and amine groups, which were activated in a moist environment to nonspecifically bind to many surfaces [49]. However, rapid hydration of hydrophilic polymers can also result in breakdown of the gel structure and ultimately adhesive failure [51, 52], while overhydration can cause a slippery mucilage to form [49]. To combat such effects and to have improved control over release rates, molecular adhesion promoters have been used, for example, poly(ethylene glycol) chains grafted onto polymethacrylic acid microgels have been used to improve control of the rate of release of insulin [53], while the inclusion of polyethylene glycols into polyacrylic acid microgels improved bioadhesive function by slowing the hydration rate of the microgel [54]. Biocompatible chitosan-based microgels have been utilized in order to provide electrostatic interaction between the microgels' positive charge and the typically negatively charged mucosal fluids and surfaces. In addition, chitosan-based microgels can swell at low pH in order to release encapsulated compounds within the digestive tract [55] (see Chapter 15).

One of the microgel systems most extensively used as a general structuring aid and thickener in a wide variety of consumer products including skin creams, lotions, pharmaceutical gels, and mucoadhesion is that of polyacrylic-based microgels. The most common commercially known product is Carbopol®, which consists primarily of cross-linked polyacrylic acid although there are numerous polyacrylate variations, some of which incorporate various copolymers and functional groups. Carbopol has been extensively used in consumer products for the last 50 years. They offer numerous advantages in pharmaceutical and dermocosmetic fields, including [56] (i) high viscosity and formation of a yield stress at low concentrations, (ii) compatibility with many active ingredients, (iii) bioadhesive properties, (iv) good thermal stability, (v) no significant thixotropy, (vi) good sensorial properties, (vii) patient/consumer acceptability, and (viii) it formulates into a relatively optically clear product. Under conditions where Carbopol microgels are swollen, the suspensions show a yield stress at relatively low polymer concentrations, leading to very low high-shear viscosities. This allows a low viscosity to be apparent during coating and rubbing-type processes involving film formation and spreading, while during storage and transport the yield stress prevents the sedimentation or creaming of particulates, air and oil droplets over long periods of time. To characterize Carbopol under conditions relevant to coating and rubbing-type processes, Davies and Stokes [57] measured the rheology of Carbopol solutions up to shear rates of 10^5 s^{-1} and also studied its gap dependency down to narrow gaps of about 10 μm; they summarize a range of complications that can arise during rheological characterization including "slip" that causes a lower viscosity to be measured than expected due to depletion of the microgels from the wall. The microgel nature of Carbopol plays a key role in its extensive use, and the rheology of Carbopol suspensions has been studied extensively as a function of solution conditions, as reviewed by Piau [58].

Carbopol is typically supplied in the form of a polyacrylic resin that consists of agglomerated polydisperse particles. On mixing in water, the suspension has an acidic pH and the particles slowly hydrate. Good hydration is necessary for reproducible and stable rheological properties; many industrial problems such as variations in product rheology during storage are a result of poor hydration. The rheology of Carbopol depends on the swelling of the individual microgels, which is controlled by the osmotic pressure of the ions within the particle, as mentioned in Section 13.2.1. When neutralized with a suitable base, such as organic amines (e.g., triethanolamine, TEA), the particles swell to their maximum extent due to the movement of ions into the microgel in order to balance the osmotic pressure. The addition of salt ions causes a decrease in particle swelling, due to a similar balancing of osmotic pressure that is also dependent on the type of ions. As for many microgel suspensions, the inclusion of other components in Carbopol-based formulations influences the bulk rheological properties. For example, the addition of polyelectrolyte polymer that is sterically excluded from the microgels draws solvent from the microgel due to osmotic forces; this effect is reduced by using lower molecular weight polymers [59].

In assessing the properties of bioadhesives and mucoadhesives, determination of the mechanism for interaction between the material and the biofluid or biosurface can be particularly difficult for ion sensitive and pH sensitive materials. For example, in assessing the interaction of Carbopol with mucin for ocular applications, it was found that Carbopol responded more strongly to the ions present in the mucin samples and simulated tear fluid than with the mucin itself, as shown in Figure 13.8 [60]. Hence, the interactions between a potential mucoadhesive and a biofluid/biosurface may result from changes in ions, pH, or temperature rather than associations through hydrogen bonding or electrostatic interactions commonly thought to be responsible for the observed mucoadhesive interactions. In addition, while many studies use model fluids such as mucin solutions to investigate mucoadhesive interactions, such fluids may not be good mimics of the real biofluid. In particular, it is questionable how good of a mimic mucin solutions are for human biofluids such as tear fluid, saliva and mucous; for example, model saliva samples that include mucin do not mimic or resemble neither the highly elastic yet low viscosity nature of human saliva nor its interfacial properties [61].

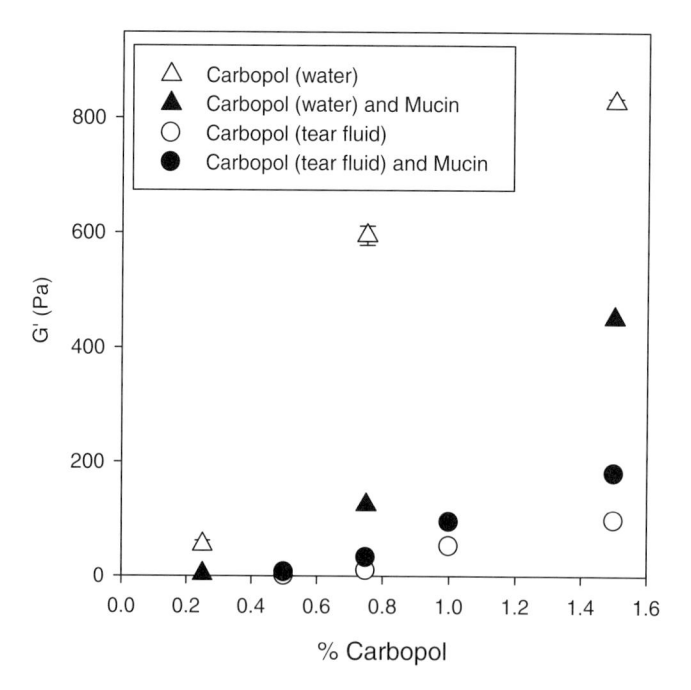

Figure 13.8 Elastic modulus of Carbopol 934 in water and simulated tear fluid (salt solution comprising Na^+, K^+, and Ca^{2+}), as well as when mixed with 4% bovine submaxillary mucin, neutralized to physiological pH (7.4) using 1 M NaOH. The modulus of Carbopol in water decreases in mucin and tear fluid due to the presence of ions that shields the carboxylic groups such that the polymer is less expanded. Under physiological conditions, there is a slight increase in modulus due to potential interaction with mucin.

13.3.3
Biopolymer Microgels for Food and Other Applications

Microgels made from natural ingredients such as starch, polysaccharides, and proteins provide a renewable source of structuring and control release agents [62] that have the potential to be used in many of the aforementioned applications of microgels. Particular interest in these systems has revolved around their biocompatibility, but they have also been used in oil related applications.

Microgels have long been used in food products, mainly in the form of starch, since they can produce an attractive soft solid texture [3, 62]. Starch granules swell upon heating in water, whereby they pass through a gelatinization stage and the granule becomes a swollen cross-linked particulate, the nature of which is retained upon cooling. Gelatinized starch particles are thus used extensively to thicken food, including sauces, dressings, and low-fat mayonnaise. They give the food an apparent yield stress and thick appearance, while also imparting a creamy mouthfeel, making them suitable as a replacement for fat in many food products. Moreover, starch is readily broken down by enzymes such as amylase in the mouth and in the digestive system. The breakdown process in the mouth enhances the mouthfeel of starch thickened systems, although the primary evolutionary purpose is for nutrition [63].

New microgel systems for use in foods have recently been introduced that consist of individual or mixed biopolymers such as polysaccharides and/or proteins. As for starches, they offer the potential to control the texture and "mouthfeel" of food. This can be important in applications aimed at fat replacement since they can potentially be designed to mimic some of the physical and sensory attributes of fat. Starches have also been receiving increasing attention for encapsulation and satiety control. Therefore, the rheology of starch and biopolymer microgel systems is particularly influential to product performance and stability. While the principal applications of starch and biopolymer microgels are in foods, they are also utilized across a diverse range of applications including cosmetics, coatings, material composites, and packaging. In addition, due to their unique rheological properties combined with low cost, both starch [64] and polysaccharide [65–68] microgels have been utilized in oil field applications (see also Chapter 16).

13.3.3.1 Starch Microgels
The desirable rheological properties of starch occur following a gelatinization step induced by heating above a critical temperature. Prior to gelatinization, the granules behave as a suspension of hard particles [69, 70], although it is apparent that they exhibit shear thickening at high-particle concentrations; for example, corn flour and custard powders are popularly used to demonstrate shear thickening and the "quicksand" or "walking on water" effect to students (i.e., at high stresses the suspension of nonswollen starch particles jam so that the suspension behaves as a solid and can be walked upon, but if the person stops, it behaves as a liquid because the particles slide past each other more easily due to lubrication provided by water phase and so the person sinks) [20]. Gelatinization causes the granules to lose their semicrystalline structure so that they are able to swell in water as well as release

polymer chains, namely, amylose (a linear polymer of glucose units) and/or amylopectin (a branched polymer of glucose units) [70], into the continuous phase. This allows the granules to adsorb large quantities of water, thus enhancing the viscosity relative to the nonheat-treated starch suspension [70, 71]. Provided that the starch has not been exposed to excessive shear or temperature, the suspension rheology is due to the presence of a tightly packed array of swollen, deformable granules [71] together with the contribution from the continuous phase that is influenced by the release of polymeric components from within the starch granule during gelatinization [72, 73]. Gelatinized starch suspensions are thus non-Newtonian, time dependent and viscoelastic [73], typically possessing a gel-like nature and a yield stress as well as a degree of thixotropy at high concentrations [70]. The rheology is highly dependent on processing, particularly by parameters such as shear and temperature, and this has proved to be a major hurdle when interpreting the rheological properties of starch suspensions [70, 73]. In addition, native and modified starches also vary in their degree of cross-linking, granule size, and composition, all of which influence the rheology. Starches can also be obtained that have been modified and pregelatinized so they are cold water swellable.

Investigations into the influence of starch concentration on the viscosity and elastic modulus [69–72, 74, 75] have shown that the rheology of starch suspensions largely follows that of other microgel systems, as discussed in Section 13.2, despite the fact that swollen granules typically show some degree of anisotropy. In dilute systems, the viscosity is proportional to the volume fraction of the swollen particles, which in turn depends on swelling capacity. In concentrated systems, on the other hand, the viscosity depends on particle rigidity [70]. Higher swelling starches are found to be more viscous in the dilute regime, while in the concentrated regime lower swelling starches are more viscous. The relationship between the elastic modulus and the concentration for native and modified starches is also similar to that of model microgel suspensions, although some differences at high phase volumes are observed, particularly between native and modified starches as shown in Figure 13.9. These differences were considered to be due to factors such as granule anisotropy, deformability, interactions between the granular amylopectin chains and the soluble amylase, microphase separation, and associated osmotic equilibria [71]. In addition, as opposed to model microgel systems, the rheology is complicated by the changing microstructure during the gelatinization process and the release of polymer into the continuous phase; it is also often found that the rheology of gelatinized starch suspensions can evolve over time as well as showing thixotropic (decrease in viscosity with time during shear) and antithixotropic [76] (increase in viscosity with time during shear) behavior due to the evolving microstructure [73].

Starch solutions can also show "stringiness," with the general terms of "long texture" and "short texture" being used to describe the formation of elongated filaments as often observed when cooking sauces and deserts based on starch. This behavior is likely due to the polymers released from the granule during gelatinization, which can impart extensional viscosity to the mixture. Of importance is also the interaction between the granules and amylose and amylopectin with added hydrocolloids (e.g., carrageenans and xanthan) as well as with naturally present and added

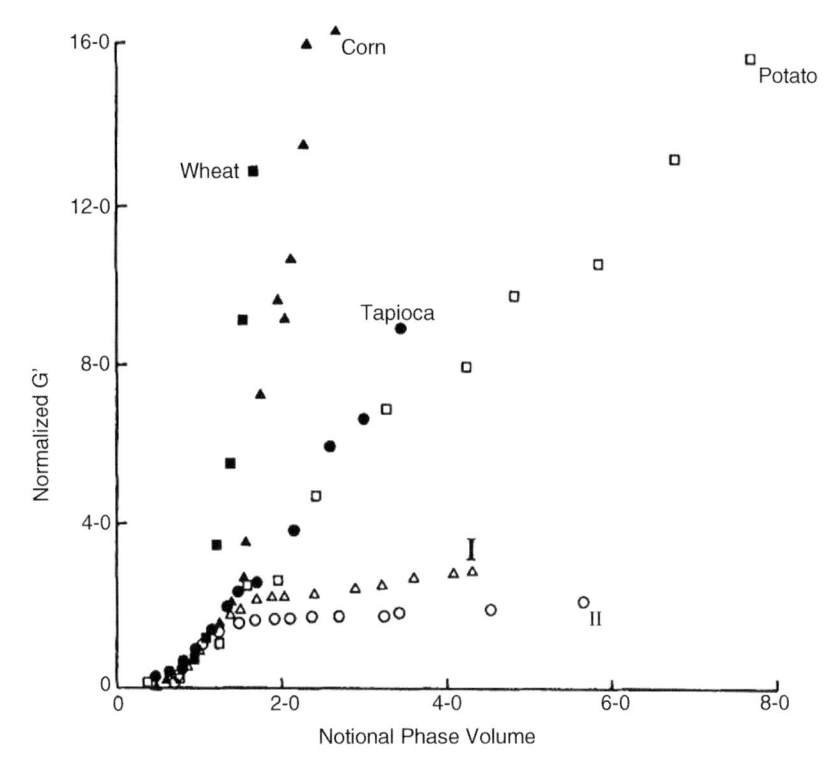

Figure 13.9 Normalized elastic modulus as a function of phase volume "master" plot of gelatinized starch suspensions. (I) At low notional phase volumes, all starches fall on the same curve and (II) above close packing, divergent behavior is observed between natural starches (wheat, corn, tapioca, and potato) and modified waxy maize starch (Types I and II). Reproduced from Ref. [71] with permission from Wiley-Blackwell.

lipids (e.g., oil and fat) [77]. Composite gel biopolymer–starch composites, where the biopolymer is a gelling polysaccharide and/or protein, are used to produce a wide range of rheologies for enhanced control over texture, flavor and mouthfeel attributes, as well as for shear stability and pumpability [76, 78, 79].

13.3.3.2 Biopolymer Microgels and Particle Anisotropy

The rheology of biopolymer microgels follows the same rules of behavior as for other microgel systems, but the desire to create soft materials with a range of microstructures for food applications has led to extensive studies into the influence of processing and shear conditions on the formation of microgels with a variety of morphologies [80]. Such studies may also provide useful insights for other microgel systems.

Spherical biopolymer microgels, which are typically above a micron in size, can be created in a number of ways [62]. The principal technique is to form a gel from a spherical droplet phase originating from a dispersion in oil (i.e., emulsion) or air (spray); for example, emulsion gelation [21] involves dispersing an aqueous biopolymer solution into oil using a homogenizer and then instigating gelation of the aqueous phase primarily through either a temperature change, introduction of ions, or addition of cross-linking agents; the choice depends on the particular biopolymer or biopolymers present and the mechanism for network formation that is necessary for gelation. Sieving can be utilized to obtain microgels of specific size with a relatively monodisperse size distribution [21]. Alternative homogenization techniques to produce microgels with more precise control over their size and monodispersity include membrane emulsification [81] and microfluidics [55, 82]; microfluidics also offers the potential for greater control over composition and facilitates the incorporation of active ingredients. Alternatively, spherical biopolymer microgel particles can be produced using a technique that involves spray drying a biopolymer solution and then controlling the rehydration process in a gelling medium [83].

Biopolymer microgels are particularly attractive because it is relatively easy to modulate the modulus of individual particles, which is typically a function of the biopolymer concentration or extent of cross-linking that depends on salt concentration or concentration of the cross-linking agent. By controlling only the particle modulus, a range of rheologies and textures can be created as shown in Figures 13.4 and 13.6; the modulus, yield stress, and high-shear viscosity are found to decrease with decreasing particle modulus for concentrated agar microgel suspensions [21].

Particle morphology is an additional factor that strongly influences the rheology of microgel suspensions. Two of the simplest methods for producing anisotropic biopolymer microgels are shearing a pregelled biopolymer network (often referred to as a "broken gel") [84] or shearing the biopolymer solution during the gelation process (often referred to as a "fluid gel" or "shear gel") [85–91]. These terms arise because without the application of shear, they would otherwise be a cross-linked gel. The fluid gel process produces microgel suspensions by causing localized phase separation during the gelation process, and produces microgels that are slightly elongated with a thin tail, an appearance akin to "tadpoles" as shown in Figure 13.10a. Such simple techniques can be used to make thickened fluids that are pourable, yet also able to suspend particulates (e.g., herbs) and oil, due to their low yet finite yield stress, without the negative mouthfeel sensations (e.g., sliminess) often associated with polymer thickeners such as xanthan gum. The dilution behavior of these anisotropic microgels is markedly different from that of spherical microgels; for example, the plateau modulus and apparent yield stress for spherical agar microgels largely disappears below the concentration corresponding to close packing, while a significant G' and yield stress is maintained at low concentrations of sheared gel suspensions [3, 80] as shown in Figure 13.6. The resulting yield stress is highly dependent on the microgel size and shear during formation; the larger the particle size the larger the yield stress [91, 92]. The amount of cross-linker present during the shear gelation process has also been shown to be a means in which to effectively control the final fluid gel rheology [92]. In addition, shear gels formed from thermosetting polymers can be reheated to their liquid state, and subsequently a

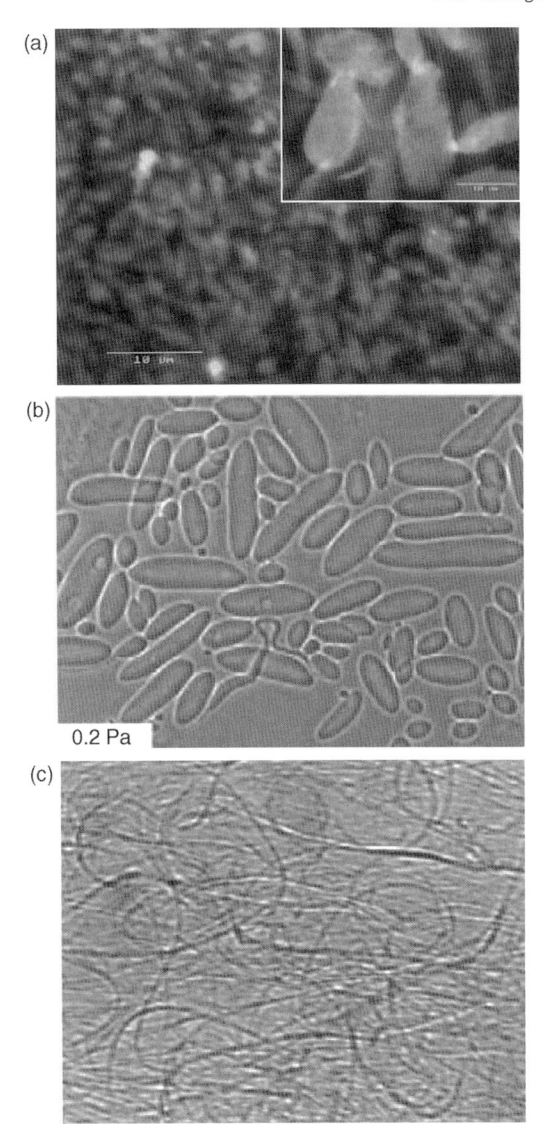

Figure 13.10 Anisotropic microgels comprising biopolymers. (a) Shear gel particles formed at high shear (image width 63 μm) and low shear (inset, image width 40 μm). Reprinted from Ref. [88] with permission from Elsevier. (b) Spheroids (image width 630 μm). Reprinted from Ref. [97] with permission from Elsevier. (c) Gel fiber (image width 630 μm). Reprinted from Ref. [98] with permission from Springer Science + Business Media.

solid gel can be formed through quiescent cooling. One practical application is to use a "fluid gel" in the food service industry for the quick formation of desserts (e.g., cheesecakes) without the necessary problems associated with hydration of biopolymer or starch powder granules.

In oil field applications, the shear gel process has also been particularly relevant to the use of cross-linked polymers in fracturing fluids [66–68]. Hydraulic fracturing involves fracturing low permeable rock by pumping at high rate and pressure a thickened fluid containing suspended solids (proppants) that prevent sealing of the fractures after well treatment. This increases the permeability for fluid flow to the wellbore. Cross-linked polymers are used as the thickener, which subsequently form microgels upon shear during the fracturing process. These have the advantage over noncross-linked polymer thickeners that absorb onto rock surfaces and decrease permeability, and because cross-link breakers can be used to reduce the viscosity of the fracturing fluid in order to optimize its recovery when returning the well to production [93] (see also Chapter 16).

A number of approaches have been taken to produce microgel particles with more control over their morphology. The emulsion route has been used to produce elongated spheroids by kinetically trapping a biopolymer droplet while it is in an elongated shape due to shearing or application of an extensional force. This can be achieved using a highly viscous oil phase, which slows down the retraction of the droplet during the gelation process [94]; however, it is difficult to process and source food oils of a suitably high viscosity. Alternatively, phase separating mixed biopolymers can be used to form water-in-water emulsions that have extremely low interfacial tension; these emulsions behave in the same manner as normal emulsions and polymer blends, but have interfacial tensions of order 10–100 μN/m. Since the relaxation time for the droplets is inversely proportional to the interfacial tension, the driving force for retraction of the droplet into a sphere is sufficiently low to enable the droplet to be kinetically trapped in an anisotropic shape by the gelation process [95]. This technique makes it possible to form a range of microstructures, ranging from spheroids (see Figure 13.10b) to flexible and rigid gel fibers (see Figure 13.10c) [96–99]. The rheology depends greatly on particle shape, with gel fibers forming a yield stress at much lower phase volumes than for spherical microgels. The rheology depends on the gel fiber stiffness and size, although these factors have not been studied in detail [99].

13.4 Outlook

Microgels have a long history of being used in everyday life, and controlling their rheological properties in suspension is often essential to their successful application. It is hoped that this chapter of their rheological complexity will help those utilizing engineering and designing microgel systems for practical applications. It is particularly exciting to see the growth in the design and the use of microgel systems based on natural polymers, which due to their biocompatibility are suitable for in-body use (food, biomedical, pharmaceutical) as well as being potentially more environmentally friendly in other applications. Developing routes by which to manipulate their design without compromising their biocompatibility is, however, still a challenge. While there is a general understanding of the rheology of microgel suspensions, both the diverse nature of microgel systems found in industrial applications and the complexity of many real formulations lead to a richness and complexity that is still difficult to model in a predictive way.

Acknowledgments

I would like to thank Dr. William Frith (Unilever, UK) for many helpful discussions and suggestions, as well as for reviewing the final manuscript. I would also like to thank the following scientists with whom I have interacted and discussed microgels with over the last 10 years at Unilever: Sarah Adams, Bettina Wolf, Julia Telford, Bronwyn Elliot, Dan Jarvis, Alex Lips, Ian Norton, Tim Foster, Gleb Yakubov, Georgina Davies, Juan de Vicente, Pip Rayment, Alan Clarke, and Mike Adams, as well as the following non-Unilever academic researchers: Dave Dunstan (University of Melbourne), Nara Altmann (University of Melbourne), Justin Cooper-White (The University of Queensland), Yan Yan (University of Birmingham), and Zhibing Zhang (University of Birmingham).

References

1 Mason, W.R. *100 Years of Food Starch Technology*, National Starch and Chemical Company, http://eu.foodinnovation.com/pdfs/100years.pdf
2 Wolfe, M.S. and Scopazzi, C. (1989) *J. Colloid Interface Sci.*, **133**, 265.
3 Stokes, J.R. and Frith, W.J. (2008) *Soft Matter*, **4**, 1133.
4 Larson, R.G. (1999) *The Structure and Rheology of Complex Fluids*, Oxford University Press, New York.
5 Tan, B.H., Tam, K.C., Lam, Y.C., and Tan, C.B. (2004) *J. Rheol.*, **48**, 915.
6 Batchelor, G.K. (1977) *J. Fluid Mech.*, **83**, 97.
7 Tan, B.H. and Tam, K.C. (2008) *Adv. Colloid Interface Sci.*, **136**, 25.
8 Krieger, I.M. and Dougherty, T.J. (1959) *Trans. Soc. Rheol.*, **3**, 137.
9 Quemada, D. (1978) *Rheol. Acta*, **17**, 632.
10 Quemada, D. (1978) *Rheol. Acta*, **17**, 643.
11 Senff, H., Richtering, W., Norhausen, C., Weiss, A., and Ballauff, M. (1999) *Langmuir*, **15**, 102.
12 Senff, H. and Richtering, W. (1999) *J. Chem. Phys.*, **111**, 1705.
13 Senff, H. and Richtering, W. (2000) *Colloid Polym. Sci.*, **278**, 830.
14 Nieuwenhuis, E.A., Pathmamanoharan, C., and Vrij, A. (1981) *J. Colloid Interface Sci.*, **81**, 196.
15 Mewis, J., Frith, W.J., Strivens, T.A., and Russel, W.B. (1989) *AICHE J.*, **35**, 415.
16 Tan, B.H., Tam, K.C., Lam, Y.C., and Tan, C.B. (2005) *Langmuir*, **21**, 4283.
17 Tan, B.H., Tam, K.C., Lam, Y.C., and Tan, C.B. (2004) *Polymer*, **45**, 5515.
18 Cloitre, M., Borrega, R., Monti, F., and Leibler, L. (2003) *CR Physique*, **4**, 221.
19 Rodriguez, B.E., Kaler, E.W., and Wolfe, M.S. (1992) *Langmuir*, **8**, 2382.
20 Steffe, J.F. (1996) *Rheological Methods in Food Process Engineering*, 2nd edn, Freeman Press.
21 Adams, S., Frith, W.J., and Stokes, J.R. (2004) *J. Rheol.*, **48**, 1195.
22 Zwanzig, R. and Mountain, R.D. (1965) *J. Chem. Phys.*, **43**, 4464.
23 Evans, I.D. and Lips, A. (1990) *J. Chem. Soc. Faraday Trans.*, **86**, 3413.
24 Buscall, R., Goodwin, J.W., Hawkins, M.W., and Ottewill, R.H. (1982) *J. Chem. Soc. Faraday Trans. 1*, **78**, 2873.
25 Buscall, R., Goodwin, J.W., Hawkins, M.W., and Ottewill, R.H. (1982) *J. Chem. Soc. Faraday Trans. 1*, **78**, 2889.
26 Paulin, S.E., Ackerson, B.J., and Wolfe, M.S. (1996) *J. Colloid Interface Sci.*, **178**, 251.
27 Hertz, H. (1895) *Gesammelte Werke*, Aufbau.

28. Johnson, K.L., Kendall, K., and Roberts, A.D. (1971) *Proc. Roy. Soc. Lond. A*, **324**, 301.
29. Evans, I.D. and Lips, A. (1992) *J. Texture Stud.*, **23**, 69.
30. Seth, J.R., Cloitre, M., and Bonnecaze, R.T. (2006) *J. Rheol.*, **50**, 353.
31. Treloar, L.R.G. (1975) *The Physics of Rubber Elasticity*, Clarendon Press, New York.
32. McEvoy, H., Ross-Murphy, S.B., and Clarke, A.H. (1985) *Polymer*, **25**, 1493.
33. Saatweber, D. and VogtBirnbrich, B. (1996) *Prog. Org. Coat.*, **28**, 33.
34. Bosma, M., Haldankar, G., DeGooyer, W., and Shalati, M. (2002) Proceedings of the 29th International Waterbourne, High Solids, and Powder Coating Symposium, New Orleans, LA, 408.
35. Ishikura, S. (1997) *Polymer News*, **22**, 344.
36. Boggs, L.J., Rivers, M., and Bike, S.G. (1996) *J. Coating Technol.*, **68**, 63.
37. Bauer, D.R., Briggs, L.M., and Dickie, R.A. (1982) *Ind. Eng. Chem. Prod. Res. Dev.*, **21**, 686.
38. Buscall, R. et al. (1987) *J. Nonnewton. Fluid Mech.*, **24**, 183.
39. Buscall, R., Mills, P.D.A., Goodwin, J.W., and Lawson, D.W. (1988) *J. Chem. Soc. Faraday Trans. 1*, **84**, 4249.
40. Johnson, S.B., Franks, G.V., Scales, P.J., Boger, D.V., and Healy, T.W. (2000) *Int. J. Miner. Process.*, **58**, 267.
41. Raquois, C., Tassin, J.F., Rezaiguia, S., and Gindre, A.V. (1995) *Prog. Org. Coat.*, **26**, 239.
42. Wolfe, M.S. (1992) *Prog. Org. Coat.*, **20**, 487.
43. Hong, S.M., Kim, H.S., Kim, Y.B., and Park, J.M. (1999) *Korea. Polym. J.*, **7**, 213.
44. Backhouse, A.J. (1982) *J. Coating Technol.*, **54**, 83.
45. Das, M., Zhang, H., and Kumacheva, E. (2006) *Ann. Rev. Mater. Res.*, **36**, 117.
46. Wang, Q., Zhao, Y.B., Yang, Y.J., Xu, H.B., and Yang, X.L. (2007) *Colloid Polym. Sci.*, **285**, 515.
47. Peppas, N.A., and Huang, Y.B. (2004) *Adv. Drug Deliv. Rev.*, **56**, 1675.
48. Smart, J.D. (2006) *J. Pharm. Pharmacol.*, **58**, A93.
49. Smart, J.D. (2005) *Adv. Drug Deliv. Rev.*, **57**, 1556.
50. Oh, J.K., Drumright, R., Siegwart, D.J., and Matyjaszewski, K. (2008) *Prog. Polym. Sci.*, **33**, 448.
51. Zaman, M.A., Martin, G.P., Rees, G.D., and Royall, P.G. (2004) *Thermochim. Acta*, **417**, 251.
52. Huang, Y.B., Leobandung, W., Foss, A., and Peppas, N.A. (2000) *J. Control. Release*, **65**, 63.
53. Peppas, N.A. (2004) *Int. J. Pharm.*, **277**, 11.
54. Zaman, M.A., Martin, G.P., and Rees, G.D. (2008) *J. Dent.*, **36**, 351.
55. Zhang, H., Mardyami, S., Chan, W.C.W., and Kumacheva, E. (2006) *Biomacromolecules*, **7**, 1568.
56. Islam, M.T., Rodriguez-Hornedo, N., Ciotti, S., and Ackermann, C. (2004) *Pharm. Res.*, **21**, 1192.
57. Davies, G.A. and Stokes, J.R. (2008) *J. Nonnewton. Fluid Mech.*, **148** (1–3), 73.
58. Piau, J.M. (2007) *J. Nonnewton. Fluid Mech.*, **144**, 1.
59. Kiefer, J., Naser, M., Kamel, A., and Carnali, J. (1993) *Colloid Polym. Sci.*, **271**, 253.
60. Hagerstrom, H., Paulsson, M., and Edsman, K. (2000) *Eur. J. Pharm. Sci.*, **9**, 301.
61. Stokes, J.R. and Davies, G.A. (2007) *Biorheology*, **44**, 141.
62. Burey, P., Bhandari, B.R., Howes, T., and Gidley, M.J. (2008) *Crit. Rev. Food Sci.*, **48**, 361.
63. Perry, G.H. et al. (2007) *Nat. Genet.*, **39**, 1256.
64. Zhang, L.M. (2001) *Starch-Starke*, **53**, 401.
65. Power, D., Larson, I., Hartley, P., Dunstan, D., and Boger, D.V. (1998) *Macromolecules*, **31**, 8744.
66. Power, D.J., Paterson, L., and Boger, D.V. (2001) *Spe Drill. Completion*, **16**, 239.
67. Chauveteau, G., Omari, A., Tabary, R., Renard, M., and Rose, J. (2001) *J. Petrol. Technol.*, **53**, 51.
68. Omari, A. et al. (2006) *J. Colloid Interface Sci.*, **302**, 537.
69. Rao, M.A., Okechukwu, P.E., Da Silva, P.M.S., and Oliveira, J.C. (1997) *Carbohydr. Polym.*, **33**, 273.
70. Steeneken, P.A.M. (1989) *Carbohydr. Polym.*, **11**, 23.

71 Evans, I.D. and Lips, A. (1992) *J. Texture Stud.*, **23**, 69.
72 Eliasson, A.C. (1986) *J. Texture Stud.*, **17**, 253.
73 Lagarrigue, S. and Alvarez, G. (2001) *J. Food Eng.*, **50**, 189.
74 Frith, W.J. and Lips, A. (1995) *Adv. Colloid Interface Sci.*, **61**, 161.
75 Jacquier, J.C., Kar, A., Lyng, J.G., Morgan, D.J., and McKenna, B.M. (2006) *Carbohydr. Polym.*, **66**, 425.
76 Tecante, A. and Doublier, J.L. (1999) *Carbohydr. Polym.*, **40**, 221.
77 Raphaelides, S.N. and Georgiadis, N. (2008) *Food Res. Int.*, **41**, 75.
78 Savary, G., Handschin, S., Conde-Petit, B., Cayot, N., and Doublier, J.L. (2008) *Food Hydrocolloids*, **22**, 520.
79 Carvalho, C.W.P., Onwulata, C.I., and Tornasula, P.M. (2007) *Food Sci. Technol. Int.*, **13**, 207.
80 Norton, I.T., Frith, W.J., and Ablett, S. (2006) *Food Hydrocolloids*, **20**, 229.
81 Yan, Y., Zhang, Z., Stokes, J.R., Zhou, Q., Ma, G., and Adams, M. (2009) *Powder Technol.*, **192**, 122.
82 Amici, E., Tetradis-Meris, G., de Torres, P., and Jousse, F. (2008) *Food Hydrocolloids*, **22**, 97.
83 Burey, P., Bhandari, B.R., Howes, T., and Gidley, M.J. (2009) *Carbohydr. Polym.*, **76**, 206.
84 Ellis, A. and Jacquier, J.C. (2009) *J. Food Eng.*, **90**, 141.
85 Brown, C.R.T. and Norton, I.T. (1991) European Patent EP 3355908.
86 Hedges, N.D. and Norton, I.T. (1991) European Patent EP 432835.
87 de Carvalho, W. and Djabourov, M. (1997) *Rheol. Acta*, **36**, 591.
88 Norton, I.T., Jarvis, D.A., and Foster, T.J. (1999) *Int. J. Biol. Macromol.*, **26**, 255.
89 Hilliou, L. and Goncalves, M.P. (2007) *Int. J. Food Sci. Tech.*, **42**, 678.
90 Omari, A., Chauveteau, G., and Tabary, R. (2003) *Colloids Surf. A*, **225**, 37.
91 Altmann, N., Cooper-White, J.J., Dunstan, D.E., and Stokes, J.R. (2004) *J. Nonnewton. Fluid Mech.*, **124**, 129.
92 Nakamoto, R., and Yasue, R. (2003) *Nihon Reoroji Gakk.*, **31**, 337.
93 Borchardt, J.K. (1987) in *Encylcopedia of Polymer Science and Engineering*, 2nd edn, vol. 10 (eds A. Klinsberg, R.M. Piccininni, A Salvatore, and E. Mannarino), John Wiley & Sons, p. 328.
94 Stokes J.R. and Frith, W.J. (2000) *XIIIth International Congress on Rheology*, Cambridge, UK.
95 Stokes, J.R., Wolf, B., and Frith, W.J. (2001) *J. Rheol.*, **45**, 1173.
96 Wolf, B., Frith, W.J., and Norton, I.T. (2001) *J. Rheol.*, **45**, 1141.
97 Wolf, B., Frith, W.J., Singleton, S., Tassieri, M., and Norton, I.T. (2001) *Rheol. Acta*, **40**, 238.
98 Wolf, B., Scirocco, R., Frith, W.J., and Norton, I.T. (2000) *Food Hydrocolloids*, **14**, 217.
99 Wolf, B., White, D., Melrose, J.R., and Frith, W.J. (2007) *Rheol. Acta*, **46**, 531.

Part Five
Applications

14
Exploiting the Optical Properties of Microgels and Hydrogels as Microlenses and Photonic Crystals in Sensing Applications

L. Andrew Lyon, Grant R. Hendrickson, Zhiyong Meng, and Ashlee N. St. John Iyer

14.1
Introduction

Polymeric materials that respond to external stimuli are becoming increasingly attractive in a wide range of applications, including tissue replacement [1–6], biological coating technologies [7–9], drug delivery [1, 5, 10–22], and biosensing [5, 10, 13, 15]. Hydrogels, in particular, are of interest for these applications due to the tunability of their mechanical and chemical properties, and therefore their versatility as device components. As discussed elsewhere in this book, the use of hydrogel particles (i.e., nanogels and microgels) offers particular advantages in terms of material architecture, response time, and versatility. Responses to different stimuli can be engineered into a microgel by the choice of monomers, comonomers, and/or cross-linkers. The most common responsive microgels incorporate either a thermoresponsive monomer that shows a phase change at a given temperature or a pH-responsive monomer that exhibits more or less ionic osmotic pressure as a function of the pH and salt content of the surrounding media. Temperature and pH are advantageous stimuli for some applications, but to design materials for specific biointerfacial applications, it is desirable to employ a generalizable construct that permits incorporation of a variety of response elements [5, 10, 13, 15, 23–26].

In this chapter, the use of hydrogels and microgels in sensing applications will be in focus. These types of gels have been shown to respond differently to analyte recognition, depending on their design and architecture. Typical examples of gel responses include an expansion or contraction of the polymer network, a change in fluorescence response of a fluorophore in the gel, a change in the diffracted wavelength in a colloidal assembly, or a change in the optical properties of the gel, such as in the case of microlenses. Select examples of each type of response will be discussed.

Microgel Suspensions: Fundamentals and Applications
Edited by Alberto Fernandez-Nieves, Hans M. Wyss, Johan Mattsson, and David A. Weitz
Copyright © 2011 WILEY-VCH Verlag GmbH & Co. KGaA, Weinheim
ISBN: 978-3-527-32158-2

14.2
Responsive Microgel and Hydrogel-Based Lenses

The most straightforward method for studying the swelling of a gel is through direct observation of its dimensional changes. At the nanometer or micrometer length scale, this is not trivial due to the limitations of quantitative size determination via optical microscopy. To circumvent these problems, alternative approaches to particle investigation have been developed. For example, our group has developed microgel-based microlenses that change their optical properties, such as their focal length, due to a change in refractive index and/or radius of curvature. The swelling response of such a system is conveniently monitored by observing the focusing power of the lens by projecting an image through the microlens. In response to a stimulus that changes the focal length of the lens, a focusing or defocusing of the projected image is obtained.

In our group, microlens arrays have been developed using well-defined, individual microgels as lenses. These microgels are synthesized via free radical polymerization of N-isopropylacrylamide (NIPAM) and acrylic acid (AAc) with N,N'-methylene (bisacrylamide) (BIS) as a typical covalent cross-linker. Adsorption of the microgels to a solid substrate (e.g., glass) results in deformation into hemispherical structures [27]. The resulting microgels behave as independent optical elements due to their curvature and refractive index contrast relative to the surrounding solvent. Due to the incorporation of temperature and pH-sensitive monomers into the microgels, the resulting microlenses are similarly responsive to changes in temperature [28] and pH [27]. The optical response, as shown in Figure 14.1, is due to a change in pH, which changes the degree of AAc protonation and therefore the osmotic pressure in the gel. Decreasing the pH of the surrounding media protonates the acid groups in

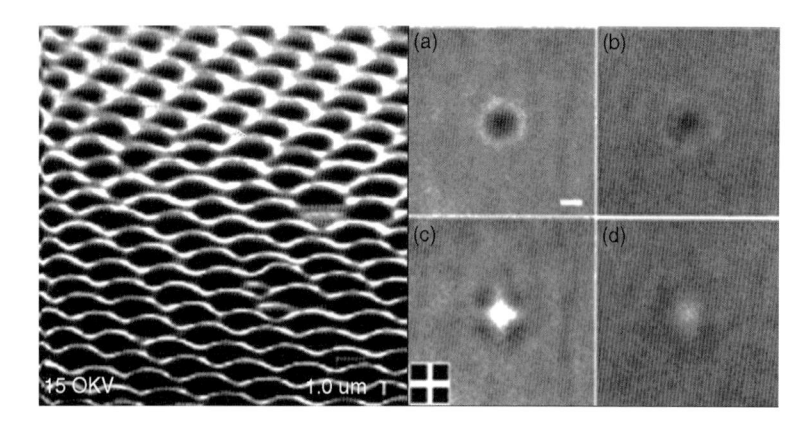

Figure 14.1 Microgel microlenses: SEM image (left panel) taken at a grazing angle of an array of microlenses (a)–(d) differential interference contrast (DIC) microscopy images depicting a microgel at pH 3.0 (a) and 6.5 (b), together with the corresponding projection images (c) and (d), respectively. Scale bar = 1 μm. Reprinted with permission from Ref. [27]. Copyright 2004 American Chemical Society.

the gel, which causes deswelling and thus an increase in the refractive index, in turn leading to a decrease of the focal length of the lens. In the thermoresponsive case, a temperature increase causes a contraction of the polymer network and a subsequent decrease in focal length. In order to investigate thermoresponsive lenses, a multilayer structure was created with a gold nanoparticle array located under a microlens array [28]. Laser light is focused onto the sample to irradiate the gold nanoparticles. The irradiation creates thermal energy, which heats the lenses, causes deswelling and thus results in an optical response. Since this study [27] was performed on microlenses containing not only a thermoresponsive monomer but also an acid functionality, the laser power required to give an optical response changed for different pH values and temperatures. It was shown that the swelling rate at different conditions could be obtained by observing the reswelling of the lenses after an irradiation pulse, as a function of laser pulse frequency. When the swelling rate is slower than the pulse frequency, the microlenses are not allowed to reswell and therefore look optically identical during and in between pulses. The responsivity of these lenses may allow their extension to devices for bioanalytical sensing applications.

To further investigate the utility of microgels in sensing applications, we have designed microlenses that display a change in refractive index in response to protein binding [8, 28, 29]. Specifically, microlenses have been designed for two different sensing pathways: a direct binding-induced response and a displacement-induced response (Figure 14.2). To illustrate each method, the small vitamin biotin was conjugated to the acrylic acid groups on the microgels. For the first approach, (Figure 14.2, Route a), avidin or anti-biotin (antibody) was added to the solution surrounding the microlens, resulting in multivalent binding of the protein to the microlens surface. Since both avidin (four binding sites) and anti-biotin (two binding sites) are able to bind multiple equivalents of biotin, the protein binding events increase the surface cross-linking of the microlens. This cross-linking induces a refractive index change and a visual signal is observed, as shown in Figure 14.3 [28]. In this example, the lens response in projection mode changes from a single square to a near double image projection. The projection mode is achieved simply by placing a square pattern in between the light source and the analyzer in an inverted

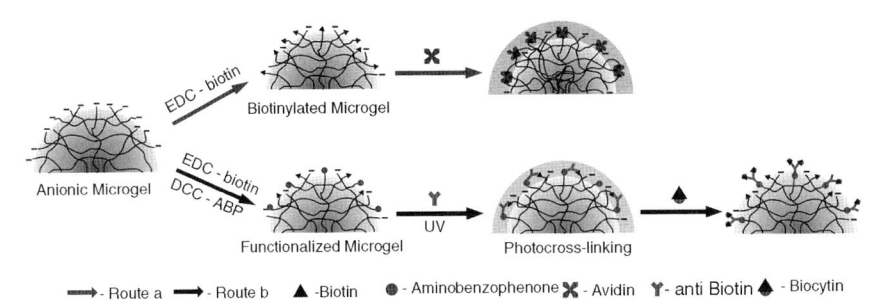

Figure 14.2 Scheme showing two different microlens sensing strategies: the binding-induced deswelling method (Route a) and the displacement-induced swelling method (Route b). Reprinted with permission from Ref. [8]. Copyright 2007 American Chemical Society.

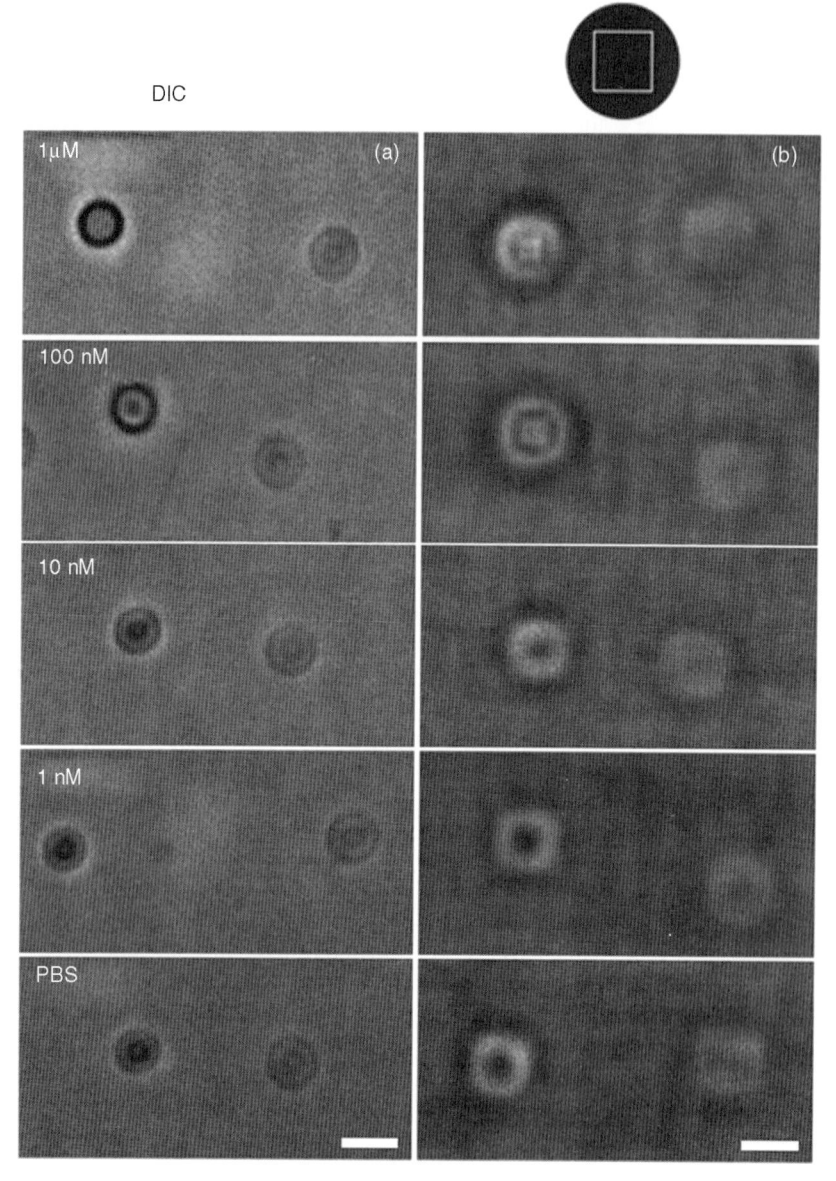

Figure 14.3 Microlens response to increasing amounts of avidin, showing the DIC (a) and projection images (b) for microlenses with (b) and without (a) biotin. Reprinted with permission from Ref. [28]. Copyright 2005 American Chemical Society.

microscope. This method is simple and could be applied to many different protein binding applications.

A displacement-induced method, (Figure 14.2, Route b), can alternatively be achieved by designing a reversible antibody–antigen cross-linking construct. In this case, a photoaffinity approach is used to couple a bound antibody to the antigen-laden microlens. When the free biotin disrupts the cross-links via displacement, the microlens swells and the focal length increases accordingly. A biotin-free buffer wash removes the free biotin, allowing for re-cross-linking of the gel and regeneration of the sensor [15, 30]. The reversibility of this system has been demonstrated, and the changes in the response rate due to different concentrations of photocross-linked antibody and free biotin have been studied [15, 30]. In this study, the microgel was initially cross-linked with the photobound antibody, free biotin was added, and the optical response of the system was monitored over time. The results showed that the response rate was directly proportional to the free biotin concentration and inversely proportional to antibody incorporation. As expected, with more free biotin in the system cross-links were disrupted faster, and the corresponding optical signal was detected more rapidly. Conversely, when the cross-linking concentration was increased, it took longer to disrupt enough cross-links to change the optical properties of the lens [8]. It has also been demonstrated via fluorescence labeling that these lenses are only sensitive to cross-link formation or disruption and are unaffected by nonspecific adsorption of proteins [8]. Despite these successes, in order for these microgels to be realized as a useful diagnostic tool, their application must be extended to more clinically relevant problems and could perhaps be employed in a microfluidic device format to produce highly parallelized sensor arrays. Also, recent structure–function relationship studies of these microgels [8, 27–29] have shown that it is nontrivial to control the optimal ligand density or microgel structure that give rise to the greatest optical response, using the lowest possible number of substrate binding events. However, using microgels as microlenses in sensing applications is attractive because it is possible to use postpolymerization, solution-based bioconjugation methods enabling their application to many different sensing targets without modifying the intrinsic sensor construction.

Moreover, a "bulk" hydrogel approach to responsive microlenses has been developed, using microlenses synthesized by polymerizing a hydrogel precursor solution onto a glass substrate in the form of a "microdome" [31]. In one example, the microdome was an acrylamide-based gel that incorporated both covalently attached calmodulin (CaM) and phenothiazine, which bind to each other creating a noncovalent cross-link in the gel. When the competing ligand chlorpromazine (CPZ) is present in the surrounding solution, it binds to CaM and displaces phenothiazine, thereby disrupting the cross-link, thus inducing gel swelling. The swelling response changes the curvature of the microdome, as well as the refractive index, resulting in a concomitant change in the microdome focal length, as shown in Figure 14.4.

Another example of a microlens construct based on a "bulk" hydrogel, used a hydrogel ring to manipulate the curvature of a water/oil interface [32]. This construct was made to have both pH and temperature responsivity by incorporating acid or amino groups and thermoresponsive monomers into the hydrogel ring, respectively. In the temperature responsive system, the hydrogel ring was formed largely from PNIPAM. Similarly, in the pH-responsive case the gel contained acrylic acid or 2-(dimethylamino)ethyl methacrylate (DMAEMA). These microlenses were fabricated by sandwiching a hydrogel ring between a solid glass surface and a

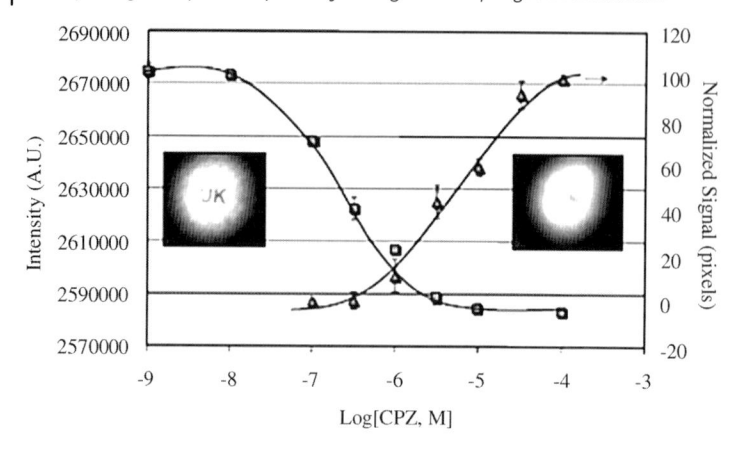

Figure 14.4 Calibration curve showing the swelling response of the microdomes as a function of chlorpromazine (CPZ) concentration. The left axis and squares show decreasing intensity of light focused on the photomultiplier tube by the microdome as it swells. The right axis and triangles show an increase in the number of bright pixels inside of the microdome as it swells. Optical micrographs are also included to show the visual response. Reprinted with permission from Ref [31].

surface containing a hole aligned with the hole in the ring. Water was incorporated into this region of the device, and oil was sandwiched between the top surface that contained the hole and another glass surface. This created a system that had a water/oil interface in the middle of the hydrogel ring. The curvature of this interface was then tuned by swelling or deswelling of the gel, resulting from either a pH change or a change in temperature.

14.3
Photonic Crystals

Concentrated suspensions of spherical colloids are known to order, due to repulsive interparticle interactions, into periodic arrays that mimic the structures of atomic crystals [33, 34]. Monodisperse suspensions of such particles most often take on face-centered cubic (FCC) and body-centered cubic (BCC) lattice structures, depending on the form and length-scale of the interaction potential [35, 36]. When the refractive index of the particles is different from that of the suspension medium, the colloidal crystal creates a periodic variation in the dielectric function that results in diffraction of photons from the assembly. This phenomenon is akin to the effect that the periodic potentials of atomic crystals have on electrons in semiconductor materials, as shown in Figure 14.5 [37]. In fact, there has been great interest in using colloidal crystals in the development of optoelectronic and all-optical circuit technologies since the idea of a photonic bandgap (PBG) material was suggested in the late 1980s in separate work by Yablonovitch [38] and John [39]. A full 3D PBG will inhibit a large range of wavelengths from propagating through the material in any direction, as represented

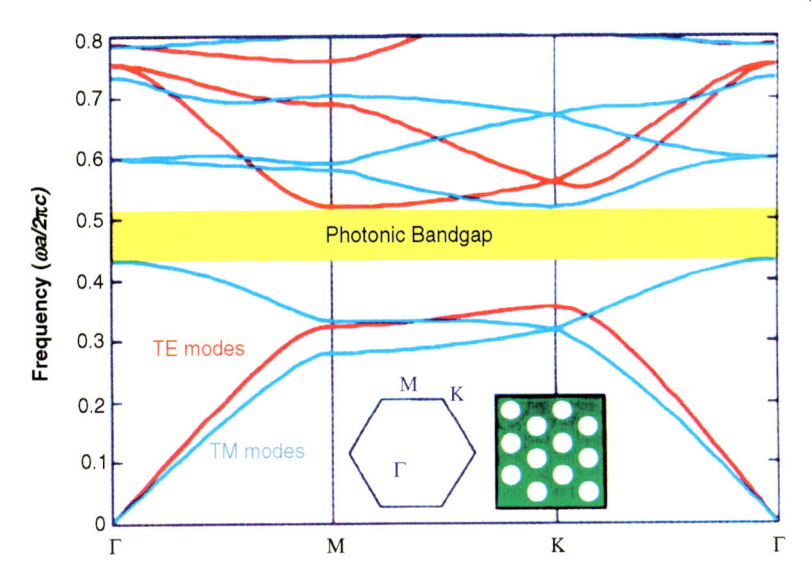

Figure 14.5 Photonic band structure for a triangular lattice of air cylinders with a radius $r \sim 0.48a$ in a dielectric (permittivity, $\varepsilon = 13$), where a is the lattice constant. The solid lines are from theory, wherein TM (transverse magnetic) modes are shown in blue and TE (transverse electric) modes in red. Note the presence of a complete photonic bandgap for both TE and TM polarizations, as shown by the solid yellow bar. Also the high dielectric constant material is indicated in green in the insets. Reprinted with permission from ref [37].

by the band diagram shown in Figure 14.5. Defect free 3D periodic materials with high refractive index contrast exhibit this behavior and are highly desired for applications such as low-loss optical waveguides [40–42] and optical integrated circuits [43–45].

Fabrication of such structures, with lattice constants that are commensurate with the optical and near-IR regions of the spectrum, is of particular interest for these applications. Whereas top–down approaches, such as lithography, can provide greater control over structure and defects, such approaches are limited in fabrication speed, and also in how small a lattice constant they can create, since optical fabrication is typically employed [45, 46]. For this reason, assembly from colloidal building blocks has become important to the development of optical PBG materials. A number of fabrication methods have been developed to create relatively defect free structures from colloidal particles on shorter time scales than traditional slow sedimentation or serial optical fabrication approaches. These methods include vertical deposition [47], spin coating [48], isothermal heating and evaporation [49], controlled dip coating [50], electric field induced assembly [51], and surface templated assembly [52]. Representative examples of such colloidal assemblies are shown in Figure 14.6 [44]. Colloidal crystals have also been used as sacrificial templates for the formation of inverse opal structures from high dielectric constant materials [53, 54]. These opal structures diffract specific wavelengths of light giving them strong iridescence resulting from a highly ordered microscopic crystalline structure.

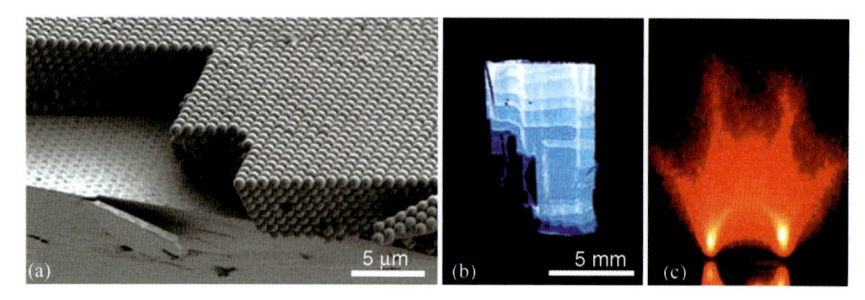

Figure 14.6 (a) Cross-sectional SEM image. (b) Large-scale optical photograph, looking down on the wafer. The opal is formed as the meniscus is swept from right to left. The horizontal lines represent monolayer steps in the crystal (lighter shades of blue represent single-layer increases in thickness). (c) Optical diffraction pattern obtained from the sample in (b). Reprinted with permission from Ref. [44].

Interest in the optical properties of colloidal crystalline materials has not been limited to the production of PBG materials and applications such as sensors [55, 56] and photonic inks [57, 58] have taken advantage of the dependence of the wavelength of diffraction on the lattice constant of the assembly. As mentioned above, the diffractive properties of simple colloidal crystals are similar to their atomic counterparts and can therefore be described using a modified Bragg's Law [27, 55],

$$m\lambda = 2nd \sin \theta \tag{14.1}$$

which relates the diffracted wavelength (λ) and the diffraction order ($m = 1, 2, 3...$) to the refractive index of the medium (n), the lattice spacing (d), and the angle of incidence relative to the lattice plane (θ). For colloidal particles of a known diameter, D, in cubic lattices, the lattice spacing can be calculated for any set of Miller indices, h, k, and l,

$$d_{hkl} = \frac{1.414 D}{(h^2 + k^2 + l^2)^{1/2}} \tag{14.2}$$

To a first approximation, this simple mathematical treatment is sufficient for systems where the refractive index of the assembly is accurately represented by a weighted sum of the refractive indices of the two phases. This approximation is applicable for the low index contrast microgel crystals that will be described below.

Microgels have been demonstrated to be excellent building blocks for the fabrication of photonic crystals via self-assembly [27, 59–63]. In particular, PNIPAM-based microgels have been intensively investigated due their thermoresponsivity [60, 63, 64]. Microgels can be centrifuged [60] or self-assembled [27] to form photonic crystals [61], for which crystallization and melting is a completely thermoreversible process [60, 63]. Figure 14.7 shows a 2D image of a 3D PNIPAM-AAc microgel crystal made by our group. The concentrated suspension (>1.0 wt%) of PNIPAM-AAc microgel particles was introduced, at room temperature, into rectangular capillaries using capillary forces and sealed with epoxy resin. After aging for over 1 month, the

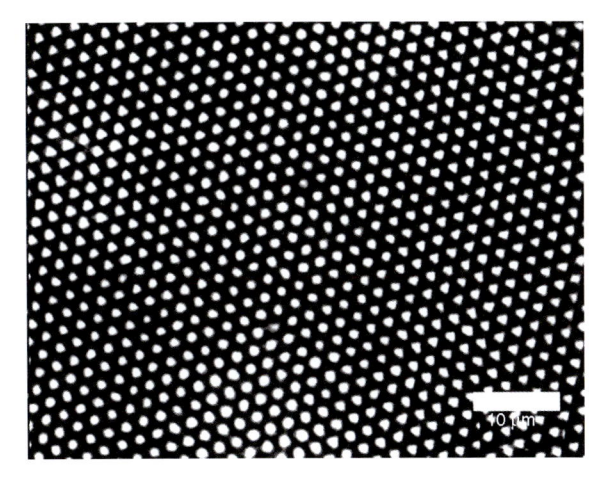

Figure 14.7 Optical microscopy images of colloidal crystals made by poly(N-isopropylacrylamide-coacrylic acid) (PNIPAM-AAc) copolymeric microgels with 2.5 wt% of polymer at pH 3.0.

sample self-assembled into a crystalline phase. Because of the intrinsic thermoresponsivity of the component microgels, the color of PNIPAM-based photonic crystals is tunable by manipulating temperature and concentration, as shown in Figure 14.8 [60, 65, 66]. Also, the PNIPAM nanogels could be self-assembled into crystalline colloidal arrays (CCA) in a polyacrylamide bulk gel for thermosensitive optical switching devices [62]. However, due to the small difference between the refractive index of the microgels and that of the solvent (water), the current focus of

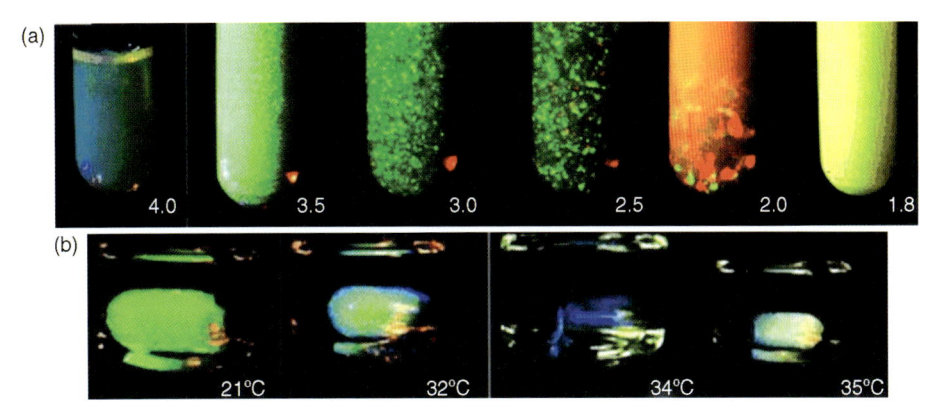

Figure 14.8 (a) Photographs of PNIPAM-coallylamine nanoparticle dispersions at various polymer concentrations at 23 °C. From left to right: 4.0, 3.5, 3.0, 2.5, 2.0, and 1.8 wt%. The Bragg diffraction peak shifts to shorter wavelength as the polymer concentration increases. (b) Photographs of PNIPAM-coallylamine crystal hydrogel changes its iridescent color with temperature. The diameter of the vial is 2.73 cm. From left to right: 21, 32, 34, and 35 °C. Reprinted with permission from Ref. [66].

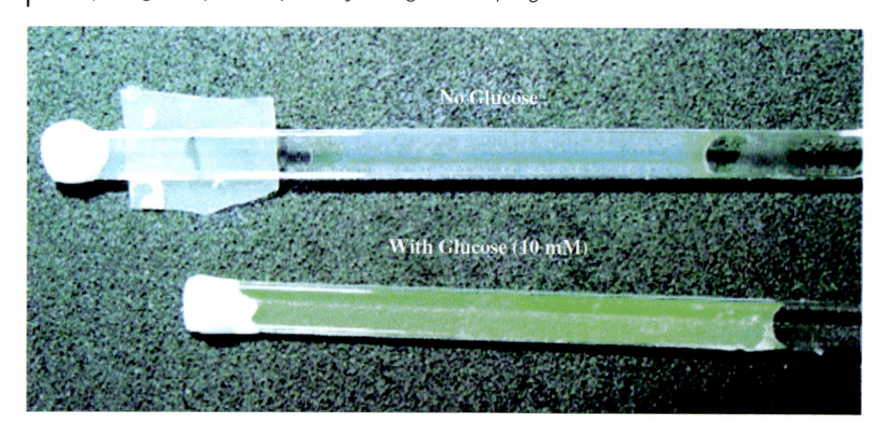

Figure 14.9 Glucose sensing microgel colloidal crystals with and without glucose. Reprinted with permission from Ref. [30]. Copyright 2006 American Chemical Society.

photonic crystals for true optical applications remains with rigid organic and inorganic materials [44, 67]. Such materials have also benefited from the numerous approaches to assembly, including nano- or microfabrication [68, 69], lithography [68], and self-assembly [44, 68] methods for three-dimensional [44, 68, 69] and two-dimensional [68, 70] photonic structures.

One of the few current illustrations of sensing, using a diffraction construct in which the colloidal crystal is directly formed from microgels, is the photonic glucose sensor [30]. In this approach, 3-acrylamidophenylboronic acid (APBA) is incorporated into PNIPAM microgels. The APBA is in equilibrium between neutral and hydrolyzed negatively charged species. Upon addition of glucose, the glucose binds to the charged form, yielding two water molecules and shifting the equilibrium towards the hydrolyzed products. This increases the number of locally charged species and the ionic osmotic pressure, thereby causing a swelling of the microgels. Since these microgels are assembled into a colloidal crystal, the expansion of the particles yields an increase in the interparticle distance and an increase in the Bragg diffraction wavelength and therefore a color change, as shown in Figure 14.9.

One key example of a photonic crystal sensor is given by the polymerized crystalline colloidal arrays (PCCA) studied by the Asher group at the University of Pittsburgh. These PCCAs are not strictly microgel based, but are typically polystyrene particle-based colloidal crystals embedded in a responsive hydrogel network, as shown in Figure 14.10. A number of different sensing applications have been achieved using PCCAs, including glucose sensing [71–74]. For this application, acrylamide-based hydrogels were modified to contain boronic acid units that bind to glucose. When glucose is present, multiple acid groups complex the glucose molecule, thereby cross-linking the gel. The deswelling caused by this induced cross-linking causes the interparticle spacing of the colloidal crystalline array to decrease, thereby blueshifting the Bragg diffraction [71–73]. The blueshifted response is found when the ionic strength of the medium is high enough to shield Coulombic repulsion between the negatively charged glucose–boronic acid

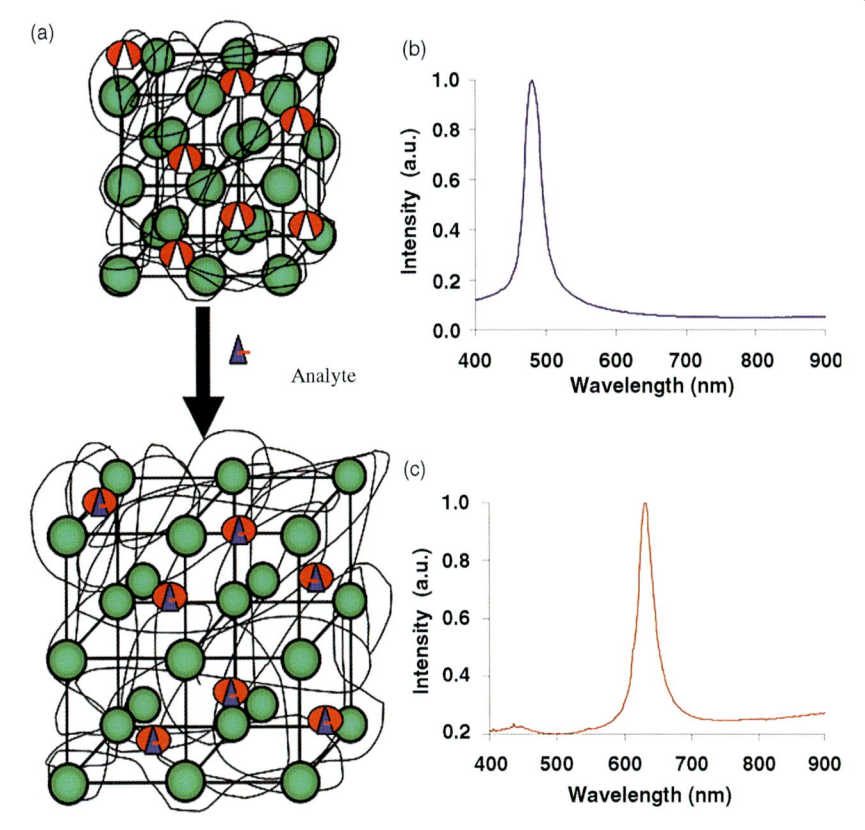

Figure 14.10 Scheme (a) and corresponding diffraction spectra of a PCCA in the absence (b) and presence (c) of analyte. Reprinted with permission from *Journal of the American Chemical Society*, 2003, **125**, 3322–3329. Copyright 2003 American Chemical Society.

complexes and diminish ionic osmotic pressure in the complexes. Conversely, when the ionic strength is lower, osmotic pressure in the gel causes a swelling response, increasing the interparticle spacing and inducing a redshift in the diffraction spectrum (Figure 14.10) [72].

PCCAs have also been employed for nerve agent sensing [75, 76]. Within this methodology, an enzyme is covalently attached to the hydrogel network in a manner that allows binding of the analyte (the nerve agent) to the enzyme. Upon analyte binding or enzyme conversion, the charge in the network is either increased or decreased and swelling or deswelling occurs. The analytes in two specific cases were organophosphorus (OP) compounds [75, 76]. In one case, the enzyme acetylcholinesterase was bound to the hydrogel network and upon OP binding, an anionic complex was formed, causing the hydrogel to swell, increasing the particle spacing, and shifting the Bragg diffraction peak [75]. In a similar case, organophosphorous hydrolase was attached to the hydrogel along with a phenol moiety [76]. When the analyte was introduced at a pH of 9.7, which is above the pK_a of the pendant phenol groups, these moieties became deprotonated and charged. When the OP was introduced and hydrolyzed by the enzyme, protons were released, thereby lowering the local pH below the phenol pK_a and protonating the phenols. As a result, the internal network charge, osmotic pressure, interparticle spacing, and diffraction wavelength were all decreased. The PCCA method has been widely applied to other sensing applications, including metal ion [55], ammonia [13], and creatinine sensing [25]. PCCAs set an excellent sensing example, which respond to many different types of analytes; however, it is sometimes overly sensitive to sample matrix components (e.g., pH or ionic strength), which in some cases may not be easily controlled.

14.4
Other Responsive Systems

A physical change in the size of the gel is the basis for all of the responsive hydrogel systems discussed below. This sensing modality is particularly useful, given the preponderance of physical observables that are modulated as a result of hydrogel swelling. These include, but are not limited to, size, porosity, density, refractive index, and modulus. Furthermore, the extreme porosity of hydrogels permits rapid analyte diffusion into the network, thereby taking advantage of the entire three-dimensional structure. As a result of these advantages, the presence of an analyte could be converted into a wide range of responses using hydrogel-based bioresponsive materials. Note, however, that many of the examples described below have not been specifically engineered into a sensing system or device. For many bioresponsive hydrogels, this remains a relatively untapped area of research, where far more effort has been spent on the development of new materials than on their applications as sensors. Bulk gel examples are included here, as they may be useful sensing methods to extend into microgel constructs.

An early bioresponsive hydrogel was developed by Miyata *et al.* [77] that utilized a poly(2-glucosyloxyethyl methacrylate) (poly(GEMA)) hydrogel, which contains glucose moieties. The multivalent lectin concancavalin A (Con A) was introduced to the polymer, which results in Con A binding to two to four glucose moieties, thereby further cross-linking the poly(GEMA) hydrogel. This noncovalent cross-linking was interrupted when free glucose was added to the solution, causing displacement of the glucose moieties and expansion of the hydrogel network. A compression apparatus was used to detect the swelling of the gel. The construct was slightly modified in another study by covalently attaching the Con A to the hydrogel network, thereby limiting the diffusion of Con A from the gel [78]. The covalently attached Con A could then recross-link the network after the gel was rinsed with buffer, freeing the glucose and binding the glucose moieties attached to the polymer network.

In a similar vein, Miyata *et al.* [33] have published two different examples of antigen–antibody responsive gels. In one example, an acrylamide-based hydrogel was synthesized using a comonomer displaying a covalently attached "antigen" (rabbit IgG). An anti-rabbit IgG antibody was added to the material, thereby forming cross-links by binding two polymer-bound antigen equivalents. Consequently, when free rabbit IgG was present in an analyte solution, it displaced the covalently attached antigen from the bound antibody, disrupting the cross-links, and causing swelling of the gel [33]. In the other example, both the antibody and the antigen were covalently linked to comonomers in an interpenetrating network (IPN) [33]. The IPN was naturally cross-linked by the antigen–antibody interactions and upon addition of free antigen these cross-links were again disrupted and caused swelling of the gel. The advantage of a covalently attached sensing element is that it allows for reversibility of the cross-linking and therefore possibility of regeneration of the sensor after washing out the free analyte.

A different method for introducing binding specificity into hydrogels is by molecular imprinting, which has been utilized in both bulk hydrogels [34] and microgels [79]. In the microgel example, the imprinted molecule was a therapeutic drug, theophylline and the steroid 17β-estradiol. These molecules were introduced into a prepolymer solution of methacrylic acid (MMA) and the cross-linker trimethylolpropane trimethacrylate (TRIM). After the photoinitiated synthesis and several rounds of washing to remove the imprinted molecules, affinity and competitive binding studies were performed showing good specificity and selectivity for the target molecules. Molecular imprinting is an interesting way of improving specificity although it might not be easily applied to multiple types of analytes because of difficulties in the introduction of multiple affinity sites as well as problems in removing the template molecules.

Biospecificity can be added to hydrogels by incorporating enzymes into the system [15]. In this chapter, we will focus on two specific examples of enzyme-incorporated hydrogels in which two different types of transitions are observed: Sol-to-gel/gel-to-sol and swelling/deswelling transitions. A sol is simply a polymer solution, whereas gel is cross-linked polymer swollen in good solvent. One example of a gel-to-sol transition was demonstrated using an acrylamide-based hydrogel synthesized with a tetrapeptide cross-linker [80], wherein an enzyme-cleavable cross-linker was used to control a gel-to-sol transition. In this study, two different peptide cross-linkers were used in different hydrogel synthesis: a tyrosine–lysine linkage cleavable by α-chemotrypsin and a serine–lysine linkage that could not be cleaved by the enzyme. The hydrogels were polymerized in a circular disk in a microfluidic channel, and in the presence of α-chemotrypsin a gel-to-sol transition of the hydrogel containing the enzyme-cleavable peptide was observed by optical microscopy. Figure 14.11 shows the hydrogel containing the enzyme-degradable cross-linker and the hydrogel containing the nondegradable cross-linker before and after exposure to the enzyme. Upon cross-linker proteolysis, the peptide cross-linked gel dissolves due to a destruction of the network. This primitive example illustrates that this method is a promising one for generating enzyme responsive materials that are dynamically reconfigurable in response to changes in enzyme activity. Such materials would be particularly useful as mimics for extracellular matrices in regenerative medicine applications.

An example of a swelling response to enzymatic activity involved the cleavage of a peptide linker that contained both anionic and cationic residues [19]. In this case, a

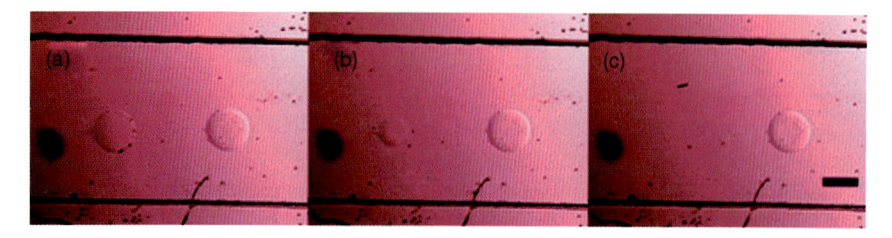

Figure 14.11 Optical micrograph of an enzyme-responsive gel, containing a cleavable peptide sequence (left) and a noncleavable sequence (right) 0 (a), 5 (b), and 20 min (c) after enzyme addition. Scale bar = 500 µm. Reprinted with permission from Ref. [80]. Copyright 2005 American Chemical Society.

poly(ethylene glycol) (PEG) bead was synthesized with a pendent zwitterionic peptide sequence. Two different hydrogels were made with different peptide sequences, where the sequences contained the following critical components: a positively charged arginine directly attached to the hydrogel, a dialanine or diglycine linkage, a pendent aspartic acid with an 9H-fluoren-9-ylmethoxycarbonyl (FMOC) protected amine and an acid group contributing the anionic part of the polyelectrolyte. The

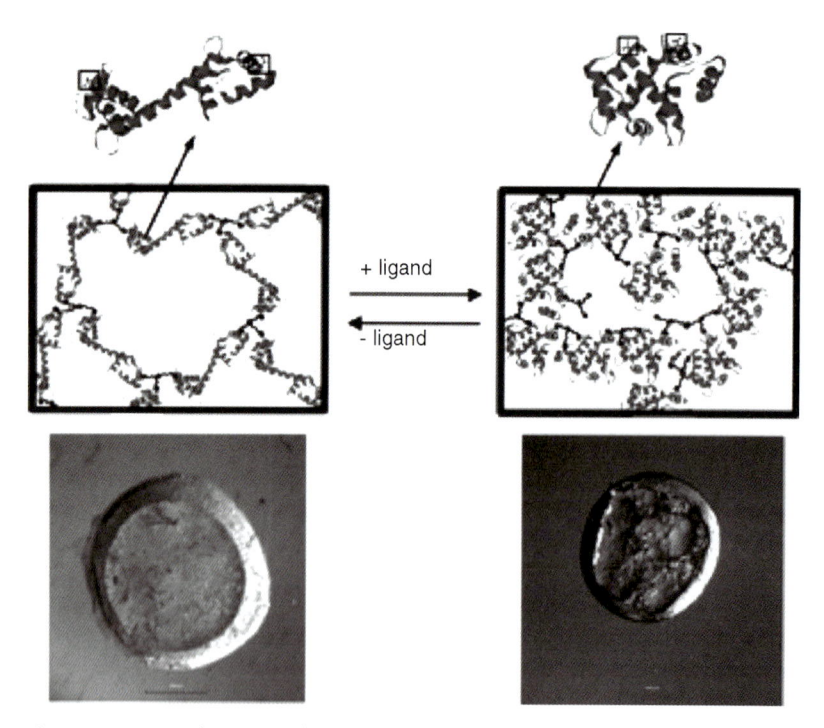

Figure 14.12 Crystal Structure, diagram, and optical micrograph of the CaM containing gel before (left) and after (right) ligand binding. Reprinted with permission from Ref [81].

diglycine or dialanine linkages were cleaved with various enzymes, the most effective for both being thermolysin. As these linkages were cleaved, the negative part of the polyelectrolyte was lost into solution, and the positive fragment was left attached to the gel increasing the osmotic pressure in the hydrogel and causing swelling. This change could conceivably be used for enzyme responsive drug delivery, or in the screening of enzyme inhibitors, as many proteases are involved in specific disease states, including many cancers.

Another sensing modality couples physical changes in hydrogels that include proteins, with protein conformational changes that occur during a ligand binding event. In one example, a PEG hydrogel was synthesized incorporating the protein calmodulin (CaM) as a hydrogel cross-linker [74, 81]. CaM has an extended conformation in the presence of calcium ions and upon ligand binding the protein collapses into a more compact structure. Due to this dramatic change in conformation, the CaM-cross-linked hydrogel collapses significantly upon binding of the ligand trifluoperazine (TFP). This response is shown in Figure 14.12. Moreover, a hydrogel glucose sensing construct was developed in which fluorescence inside the gel increased due to the presence of glucose [82]. In this case, a PEG-based hydrogel was synthesized with a

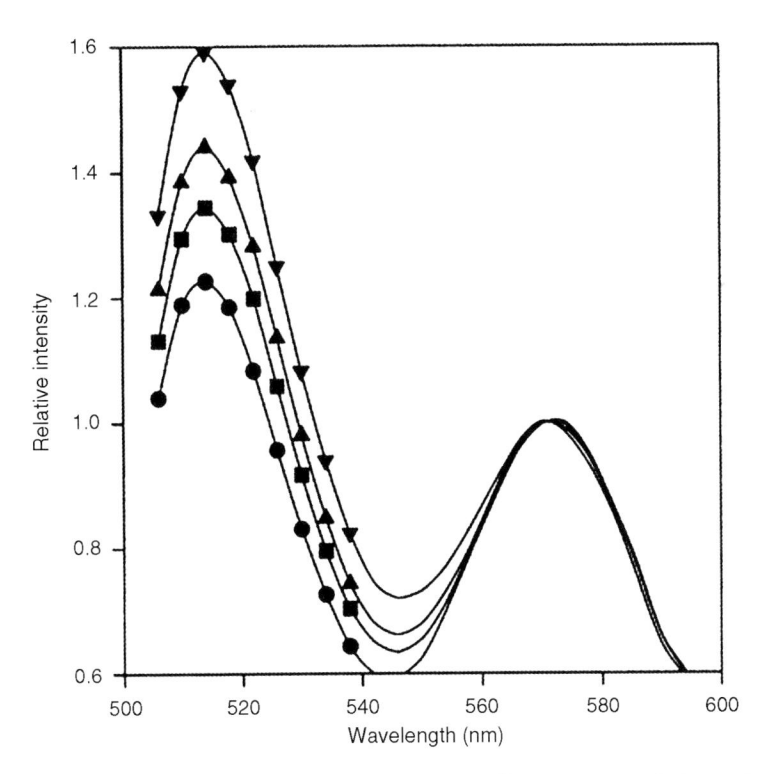

Figure 14.13 Fluorescence spectra showing the *fluorescein isothiocyanate* (FITC)-dextran signal (~515 nm) and FRET signal (~575 nm) with 0 (●), 200 (■), 400 (▲), and 1000 (▼) mg/dl concentrations of glucose. Reprinted with permission from Ref. [82]. Copyright 1999 American Chemical Society.

covalently attached, fluorescently-labeled Con A, and physically trapped fluorescein isothiocyanate (FITC) labeled dextran. The fluorophore attached to Con A was tetramethylrhodamine isothiocyanate (TRITC), which undergoes fluorescence resonance energy transfer (FRET) with fluorescein at short distances due to their strong spectral overlap. Therefore, when the dextran binds Con A, the two chromophores undergo FRET and a decreased amount of fluorescein fluorescence is observed. However, introduction of glucose displaces the dextran from the Con A and increases the fluorescein fluorescent signal, as shown in Figure 14.13.

14.5
Summary

We have discussed different examples of microgel-based and bulk hydrogel-based sensing constructs, with an emphasis on constructs based on optical lenses and photonic crystals. Constructing lenses using hydrogels and specifically microgels is interesting due to the unambiguous optical response that can be achieved in response to specific changes in their environment, as well as the ability to tailor the fundamental optical properties by changing the structure and composition of the polymer network. Photonic crystals will continue to be studied due to their relatively simple colorimetric response to changes in the crystal lattice structure. Although they are in various stages of development, we have discussed different examples of responsive hydrogels and microgels that due to their network flexibility could possibly be extended to a wide variety of sensing applications. In some cases, as with the well-developed PCCA approach, the pathway to true sensing systems is well established. However, other methods are still in the materials discovery and development stage; developing these approaches into commercially applicable sensor constructs remains as a logical next step in device development.

References

1 Deng, Y.H., Wang, C.C., Shen, X.Z., Yang, W.L., An, L., Gao, H., and Fu, S.K. (2005) *Chem. Eur. J.*, **11**, 6006–6013.
2 Moschou, E.A., Madou, M.J., Bachas, L.G., and Daunert, S. (2006) *Sens. Actuators, B*, **115**, 379–383.
3 Moschou, E.A., Peteu, S.F., Bachas, L.G., Madou, M.J., and Daunert, S. (2004) *Chem. Mater.*, **16**, 2499–2502.
4 Seliktar, D., Zisch, A.H., Lutolf, M.P., Wrana, J.L., and Hubbell, J.A. (2004) *J. Biomed. Mater Res. A*, **68**, 704–716.
5 Tae, G., Kim, Y.J., Choi, W.I., Kim, M., Stayton, P.S., and Hoffman, A.S. (2007) *Biomacromolecules*, **8**, 1979–1986.
6 Yamaguchi, N., Zhang, L., Chae, B.S., Palla, C.S., Furst, E.M., and Kiick., K.L. (2007) *J. Am. Chem. Soc.*, **129**, 3040.
7 Gan, D.J. and Lyon, L.A. (2002) *Macromolecules*, **35**, 9634–9639.
8 Kim, J., Singh, N., and Lyon, L.A. (2007) *Biomacromolecules*, **8**, 1157–1161.
9 Nolan, C.M., Reyes, C.D., Debord, J.D., Garcia, A.J., and Lyon, L.A. (2005) *Biomacromolecules*, **6**, 2032–2039.
10 Alarcon, C.D.H., Pennadam, S., and Alexander, C. (2005) *Chem. Soc. Rev.*, **34**, 276–285.
11 De Geest, B.G., Dejugnat, C., Sukhorukov, G.B., Braeckmans, K., De Smedt, S.C.,

and Demeester, J. (2005) *Adv. Mater.*, **17**, 2357.

12 Ehrick, J.D., Deo, S.K., Browning, T.W., Bachas, L.G., Madou, M.J., and Daunert, S. (2005) *Nat. Mater.*, **4**, 298–302.

13 Galaev, I.Y. and Mattiasson, B. (1999) *Trends Biotechnol.*, **17**, 335–340.

14 Gu, J.X., Xia, F., Wu, Y., Qu, X.Z., Yang, Z.Z., and Jiang, L. (2007) *J. Control. Release*, **117**, 396–402.

15 Mart, R.J., Osborne, R.D., Stevens, M.M., and Ulijn, R.V. (2006) *Soft Matter*, **2**, 822–835.

16 Rao, K., Naidu, B.V.K., Subha, M.C.S., Sairam, M., and Aminabhavi, T.M. (2006) *Carbohydr. Polym.*, **66**, 333–344.

17 Soppimath, K.S., Kulkarni, A.R., and Aminabhavi, T.M. (2001) *J. Control. Release*, **75**, 331–345.

18 Tae, G., Scatena, M., Stayton, P.S., and Hoffman., A.S. (2006) *J. Biomater. Sci. Polym. Ed.*, **17**, 187–197.

19 Ulijn, R.V., Bibi, N., Jayawarna, V., Thornton, P.D., Todd, S.J., Mart, R.J., Smith, A.M., and Gough, J.E. (2007) *Materials Today*, **10**, 40–48.

20 Wheeldon, I.R., Barton, S.C., and Banta, S. (2007) *Biomacromolecules*, **8**, 2990–2994.

21 Wood, K.C., Chuang, H.F., Batten, R.D., Lynn, D.M., and Hammond, P.T. (2006) *Proc. Natl. Acad. Sci. USA*, **103**, 10207–10212.

22 Zhou, J., Wang, G.N., Zou, L., Tang, L.P., Marquez, M., and Hu, Z.B. (2008) *Biomacromolecules*, **9**, 142–148.

23 Deo, S., Moschou, E., Peteu, S., Eisenhardt, P., Bachas, L., Madou, M., and Daunert, S. (2003) *Anal. Chem.*, **75**, 207A–213.

24 Kikuchi, A. and Okano, T. (2002) *Adv. Drug Deliv. Rev.*, **54**, 53–77.

25 Miyata, T., Uragami, T., and Nakamae, K. (2002) *Adv. Drug Deliv. Rev.*, **54**, 79–98.

26 Thornton, P.D., Mart, R.J., and Ulijn, R.V. (2007) *Adv. Mater.*, **19**, 1252.

27 Kim, J., Serpe, M.J., and Lyon., L.A. (2004) *J. Am. Chem. Soc.*, **127**, 9512–9513.

28 Kim, J., Nayak, S., and Lyon, L.A. (2005) *J. Am. Chem. Soc.*, **127**, 9588–9592.

29 Kim, J.S., Singh, N., and Lyon, L.A. (2006) *Angew. Chem. Int. Ed.*, **45**, 1446–1449.

30 Lapeyre, V., Gosse, I., Chevreux, S., and Ravaine, V. (2006) *Biomacromolecules*, **7**, 3356–3363.

31 Ehrick, J.D., Stokes, S., Bachas-Daunert, S., Moschou, E.A., Deo, S.K., Bachas, L.G., and Daunert, S. (2007) *Adv. Mater.*, **19**, 4024.

32 Dong, L., Agarwal, A.K., Beebe, D.J., and Jiang, H.R. (2006) *Nature*, **442**, 551–554.

33 Miyata, T., Asami, N., and Uragami, T. (1999) *Macromolecules*, **32**, 2082–2084.

34 Miyata, T., Jige, M., Nakaminami, T., and Uragami, T. (2006) *Proc. Natl. Acad Sci. USA*, **103**, 1190–1193.

35 Pusey, P.N. and Vanmegen, W. (1986) *Nature*, **320**, 340–342.

36 Robbins, M.O., Kremer, K., and Grest., G.S. (1988) *J. Chem. Phys.*, **88**, 3286–3312.

37 Joannopoulos, J.D., Villeneuve, P.R., and Fan, S.H. (1997) *Nature*, **386**, 143–149.

38 Yablonovitch, E. (1987) *Phys. Rev. Lett.*, **58**, 2059.

39 John, S. (1987) *Phys. Rev. Lett.*, **58**, 2486–2489.

40 Johnson, S.G., Villeneuve, P.R., Fan, S.H., and Joannopoulos, J.D. (2000) *Phys. Rev. B*, **62**, 8212–8222.

41 Mekis, A., Chen, J.C., Kurland, I., Fan, S.H., Villeneuve, P.R., and Joannopoulos., J.D. (1996) *Phys. Rev. Lett.*, **77**, 3787–3790.

42 Vlasov, Y.A., O'Boyle, M., Hamann, H.F., and McNab, S.J. (2005) *Nature*, **438**, 65–69.

43 Almeida, V.R., Barrios, C.A., Panepucci, R.R., and Lipson, M. (2004) *Nature*, **431**, 1081–1084.

44 Norris, D.J. and Vlasov, Y.A. (2001) *Adv. Mater.*, **13**, 371–376.

45 Thylen, L., Qiu, M., and Anand, S. (2004) *ChemPhysChem*, **5**, 1268–1283.

46 Xia, Y.N., Gates, B., Yin, Y.D., and Lu, Y. (2000) *Adv. Mater.*, **12**, 693–713.

47 Jiang, P., Bertone, J.F., Hwang, K.S., and Colvin, V.L. (1999) *Chem. Mater.*, **11**, 2132–2140.

48 Jiang, P. and McFarland., M.J. (2004) *J. Am. Chem. Soc.*, **126**, 13778–13786.

49 Wong, S., Kitaev, V., and Ozin, G.A. (2003) *J. Am. Chem. Soc.*, **125**, 15589–15598.

50 Gu, Z.Z., Fujishima, A., and Sato, O. (2002) *Chem. Mater.*, **14**, 760–765.

51 Trau, M., Saville, D.A., and Aksay, I.A. (1996) *Science*, **272**, 706–709.

52 vanBlaaderen, A., Ruel, R., and Wiltzius, P. (1997) *Nature*, **385**, 321–324.

53 Schroden, R.C., Al-Daous, M., Blanford, C.F., and Stein, A. (2002) *Chem. Mater*, **14**, 3305–3315.
54 Stein, A. and Schroden., R.C. (2001) *Curr. Opin. Solid State Mater. Sci.*, **5**, 553–564.
55 Holtz, J.H. and Asher, S.A. (1997) *Nature*, **389**, 829–832.
56 Shipway, A.N., Katz, E., and Willner, I. (2000) *ChemPhysChem*, **1**, 18–52.
57 McGrath, J.G., Bock, R.D., Cathcart, J.M., and Lyon, L.A. (2007) *Chem. Mater.*, **19**, 1584–1591.
58 Park, J., Moon, J., Shin, H., Wang, D., and Park., M. (2006) *J. Colloid Interface Sci.*, **298**, 713–719.
59 Das, M., Zhang, H., and Kumacheva, E. (2006) *Annu. Rev. Mater. Res.*, **36**, 117–142.
60 Debord, J.D. and Lyon., L.A. (2000) *J. Phys. Chem. B*, **104**, 6327–6331.
61 Hellweg, T., Dewhurst, C.D., Bruckner, E., Kratz, K., and Eimer, W. (2000) *Colloid Polym. Sci.*, **278**, 972–978.
62 Reese, C.E., Mikhonin, A.V., Kamenjicki, M., Tikhonov, A., and Asher, S.A. (2004) *J. Am. Chem. Soc.*, **126**, 1493–1496.
63 Senff, H. and Richtering., W. (1999) *J. Chem. Phys.*, **111**, 1705–1711.
64 Wu, J.Z., Zhou, B., and Hu, Z.B. (2003) *Phys. Rev. Lett.*, **90**, 4.
65 Debord, J.D., Eustis, S., Debord, S.B., Lofye, M.T., and Lyon, L.A. (2002) *Adv. Mater.*, **14**, 658–662.
66 Hu, Z.B. and Huang, G. (2003) *Angew. Chem. Int. Ed.*, **42**, 4799–4802.
67 Norris, D.J., Arlinghaus, E.G., Meng, L.L., Heiny, R., and Scriven, L.E. (2004) *Adv. Mater.*, **16**, 1393–1399.
68 Prather, D.W., Shi, S.Y., Murakowski, J., Schneider, G.J., Sharkawy, A., Chen, C.H., and Miao, B.L. (2006) *IEEE J. Sel. Top. Quantum Electron.*, **12**, 1416–1437.
69 Yablonovitch, E., Gmitter, T.J., and Leung., K.M. (1991) *Phys. Rev. Lett.*, **67**, 2295–2298.
70 Song, B.S., Asano, T., and Noda, S. (2007) *NANO*, **2**, 1–13.
71 Alexeev, V.L., Das, S., Finegold, D.N., and Asher, S.A. (2004) *Clin. Chem.*, **50**, 2353–2360.
72 Alexeev, V.L., Sharma, A.C., Goponenko, A.V., Das, S., Lednev, I.K., Wilcox, C.S., Finegold, D.N., and Asher, S.A. (2003) *Anal. Chem.*, **75**, 2316–2323.
73 Ben-Moshe, M., Alexeev, V.L., and Asher, S.A. (2006) *Anal. Chem.*, **78**, 5149–5157.
74 Murphy, W.L., Dillmore, W.S., Modica, J., and Mrksich, M. (2007) *Angew. Chem. Int. Ed.*, **46**, 3066–3069.
75 Walker, J.P. and Asher, S.A. (2005) *Anal. Chem.*, **77**, 1596–1600.
76 Walker, J.P., Kimble, K.W., and Asher, S.A. (2007) *Anal. Bioanal. Chem.*, **389**, 2115–2124.
77 Miyata, T., Jikihara, A., Nakamae, K., and Hoffman., A.S. (1996) *Macromol. Chem. Phys.*, **197**, 1135–1146.
78 Miyata, T., Jikihara, A., Nakamae, K., and Hoffman, A.S. (2004) *J. Biomater. Sci. Polym. Ed.*, **15**, 1085–1098.
79 Ye, L., Cormack, P.A.G., and Mosbach., K. (2001) *Anal. Chim. Acta*, **435**, 187–196.
80 Plunkett, K.N., Berkowski, K.L., and Moore, J.S. (2005) *Biomacromolecules*, **6**, 632–637.
81 Sui, Z.J., King, W.J., and Murphy, W.L. (2007) *Adv. Mater.*, **19**, 3377.
82 Russell, R.J., Pishko, M.V., Gefrides, C.C., McShane, M.J., and Cote, G.L. (1999) *Anal. Chem.*, **71**, 3126–3132.

15
Microgels in Drug Delivery
Martin Malmsten

15.1
Introduction

As a result of new and increasing demands from candidate drugs, advanced delivery systems are receiving increasing attention. For example, many low molecular weight drug candidates display low, or very low, aqueous solubility, and delivery systems are therefore required to provide a sufficient drug bioavailability and/or to facilitate clinical or even preclinical research and development work. On the other end of the spectrum, a large and strongly increasing fraction of candidate drugs are proteins and peptides, requiring delivery systems for providing increased bioavailability. For such drugs, the delivery system is critical to maintain the native conformation and to control aggregation. The latter is essential for avoiding problems related to loss or alteration of biological effect of such drugs. Advanced delivery systems may offer not only solutions in both these contexts but also a range of other advantages, including protection from drug hydrolysis and other types of chemical and enzymatic degradation, reduction of toxicity, controlled drug release rate, and improvement of drug bioavailability. Many advanced delivery systems may also be triggered to go from one structure to another by parameters such as temperature, ionic strength, pH, water content, or presence of specific metabolites. This facilitates storage and administration of the formulation in one form, for example, as a low viscous solution, followed by transition to another, for example, a highly viscous gel, after administration.

Out of the many different advanced delivery systems attracting attention, this brief review focuses on polymer microgel and nanogel particles (ranging from about 100 nm to about 10 μm in diameter), as well as closely related polyelectrolyte microcapsules and nanocapsules. Rather than providing a complete inventory of the research field, an attempt is made to provide some illustrative examples on the basis and application of microgels and related systems. More complete discussions on other types of soft drug delivery systems may be found elsewhere [1–4].

15.2
Polymer Gels

Before turning to microgels and polyelectrolyte microcapsules as delivery systems, it is instructive to first have a look at the use of macroscopic gel systems in drug delivery as much more work has been done in this field and as many of the physicochemical aspects are similar for the macroscopic and the microscopic systems. Whether macroscopic or microscopic, polymer solutions and gels responding to external stimuli offer interesting opportunities in drug delivery [1–10]. For both chemical and physical gel systems, where cross-links are formed by chemical bonds and transient intermolecular interactions, respectively, swelling transitions allow triggered exposure of a drug encapsulated in the gel (particle) to the surrounding aqueous solution and resulting drug release. Polyacids are particularly interesting in this context since they are uncharged at low pH (e.g., in the stomach), resulting in a network collapse, a low drug release rate, and protection against acid-catalyzed hydrolysis. At higher pH (e.g., in the small intestine), on the other hand, the polymer swells as a result of electrostatic interactions, thereby facilitating drug release in a region where it is absorbed more effectively and where it is more stable against hydrolytic degradation [1].

Considerable interest has also been placed on macroscopic polymer gels displaying reversed temperature solubility since such systems can be loaded at a high degree of swelling at low temperature and then achieve a sustained release after administration due to a temperature-induced collapse of the polymer network (Figure 15.1) [11, 12]. As with many other types of delivery systems, the properties of the drug influence the behavior of such responsive gel delivery systems. Figure 15.1 illustrates this for a thermally responsive hydrogel prepared from N-isopropylacrylamide-containing hydrophobic comonomers. Thus, the relatively hydrophobic drug ibuprofen results in a gel collapse at all temperatures, whereas the hydrophilic drug ephedrine has essentially no influence on the deswelling behavior in this system [12]. Since both drug hydrolysis and release depend on the network swelling in aqueous polymer-based drug delivery systems, this type of drug–carrier interactions must be considered in the design of responding polymer gels for drug delivery.

Many different types of polymer-based systems are interesting in the context of drug delivery, including self-assembled liquid crystalline phases formed by block copolymers, polymer-surfactant gels, polysaccharide-based gels, and chemically cross-linked gels [1]. Although "polymer gels" based on block copolymer self-assembly are interesting as delivery systems and are receiving considerable attention, our main focus in this chapter will be on the chemically and physically cross-linked gel systems not based on surfactant-like self-assembly. Due to the general lack of substantial hydrophobic domains of such nonself-assembly gels, at least in comparison to, for example, surfactant- and lipid-based alternatives, such systems are not very efficient in solubilizing uncharged hydrophobic drugs and can solubilize only either fully soluble (hydrophilic) drugs or dispersed drug (-containing) colloids. Nevertheless, a considerable drug loading in particulate form can be reached also for

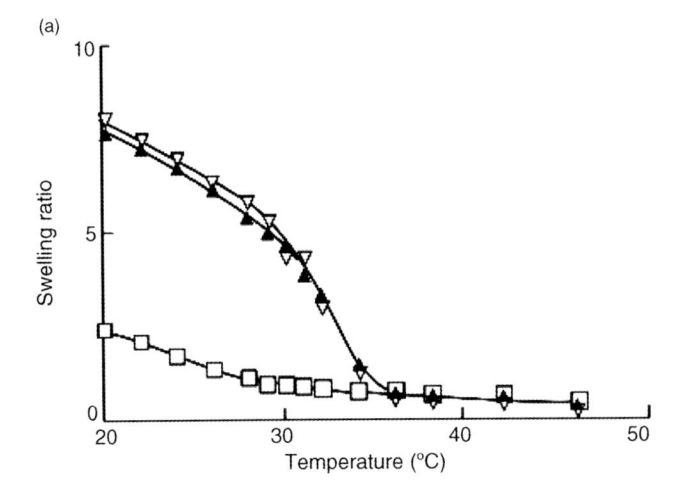

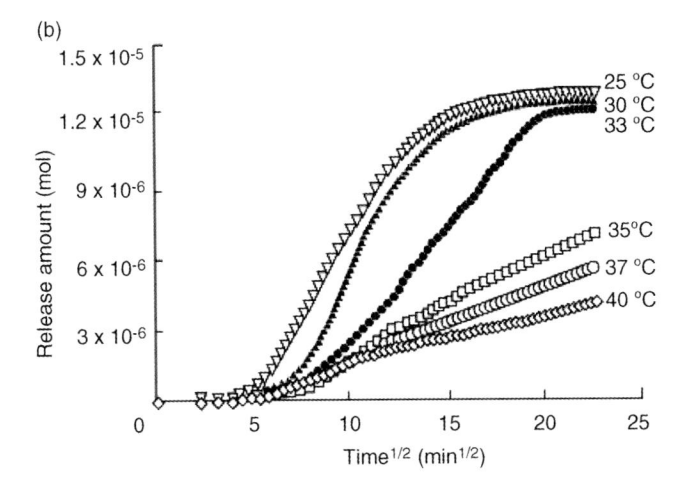

Figure 15.1 (a) Effect of ibuprofen (□) and ephedrin (△) on the temperature-induced deswelling of poly-NIPAAM gels. Also shown are results obtained in the absence of drugs (▲). (Redrawn from Ref. [12].) (b) Release of sodium benzoate from poly-NIPAAM gels as a function of temperature. The lower consolute temperature for this polymer system is 33 °C. Note that the gel collapses at $T > 33$ °C, resulting in a drastic decrease in the drug release rate. (Redrawn from Ref. [11].)

gel systems lacking larger hydrophobic domains by utilizing poor solvency conditions for the drug.

Of the different macroscopic gel systems, polysaccharide gels have received considerable interest in drug delivery and may be designed to respond to various external stimuli. Examples of this include alginate and gellan gum, which form gels in the presence of Ca^{2+} and other divalent cations. By varying the Ca^{2+} concentration, the effective "cross-linking" density of the gels, as well as drug

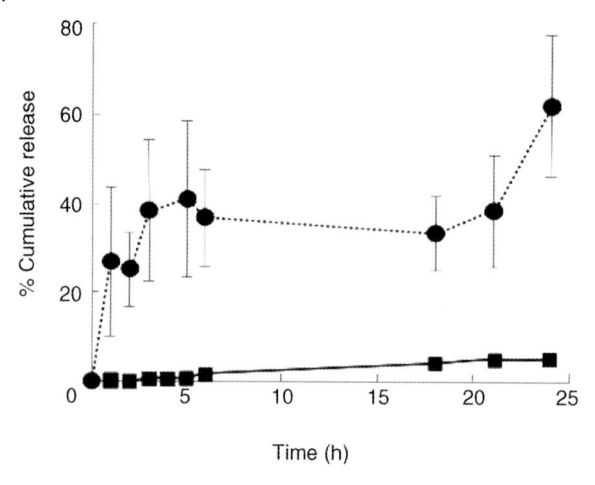

Figure 15.2 Cumulative release of indomethacin from a calcium pectate gel in citrate buffer in the presence (●) and absence (■) of pectinolytic enzyme. (Redrawn from Ref. [13].)

diffusion rate, may be tailored [1]. An area where related polysaccharide gels are of interest is within colon drug delivery [13]. One of the reasons for this is that polysaccharide degradation by microbial enzymes allows localized drug release in the colon and the large intestine, interesting, for example, for local therapies against colon cancer or Crohn's disease. Administration to the colon is also interesting for systemic absorption of peptide and protein drugs, which are extensively degraded in the gastrointestinal tract. Figure 15.2 illustrates this enzyme-triggered release for the case of indomethacin formulated in a calcium pectate gel in the absence and the presence of pectinolytic enzymes, known to be present in the colonic region. As can be seen, a significant release is observed in the presence of pectinolytic enzymes, while essentially no release is observed in the absence of digestive enzymes. The release to the colon may therefore be controlled by the stability of the matrix to enzymatic degradation, and a specific targeting obtained. As will be discussed below, related localized drug release has also been applied to microgel systems.

15.3
Polymer Microgels

As with macroscopic polymer gels, microgels may be designed to respond to a number of stimuli, including temperature, ionic strength, pH, the presence of divalent ions, specific metabolites, and external fields. Given the small size of these gel particles, they have potential also in areas where macroscopic polymer gels have not found use, for example, in parenteral administration (injectables). They are also expected to provide advantageous effects in other administration routes, including oral and nasal administration, due to their small size. As with macroscopic gels, microgels have particular potential as delivery systems for protein and other biomacromolecular drugs since they are generally hydrophilic and contain a lot of water, which allows proteins to be incorporated into the microgels with only moderate conformational changes and with limited aggregation, thus facilitating maintained biological effect of the protein drug. This effect is demonstrated in Figure 15.3, where CD spectra of hemoglobin in buffer are compared with that released to buffer after incorporation in a microgel system [14].

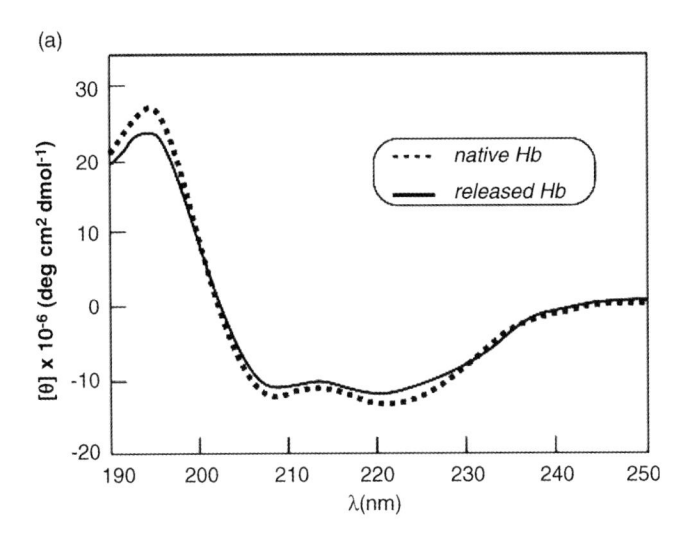

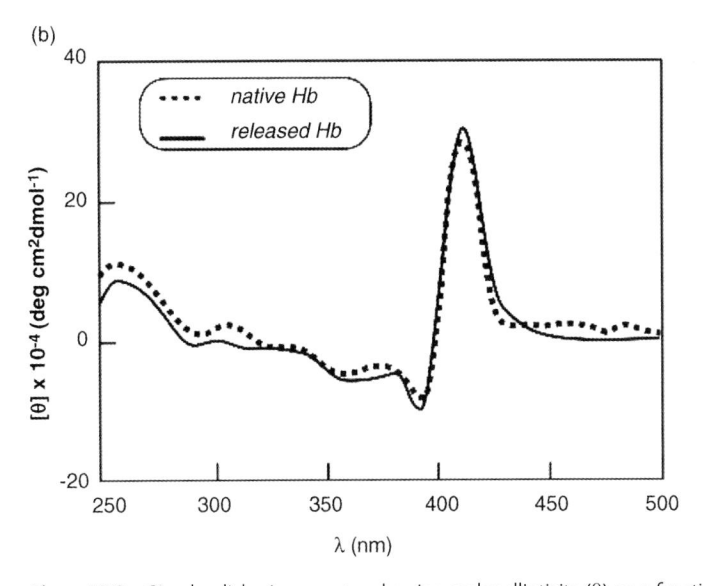

Figure 15.3 Circular dichroism spectra showing molar ellipticity (θ) as a function of wavelength around the far-UV bands (a) and the Soret bands (b) of native hemoglobin and of hemoglobin released from lactic acid/Pluronic F127 microgels. (Redrawn from Ref. [14].)

15.3.1
Temperature Triggering of Microgels

Of the different response triggers for microgels, temperature is probably the most extensively investigated one. Several different types of polymers exhibit temperature-dependent swelling–deswelling transitions, including systems based on polyethylene oxide derivatives, those based on cellulose ethers, and those including variants of poly(N-isopropylacrylamide) (PNIPAM) [8, 14–30]. In common for all these systems is a reduced solvency with increasing temperature, resulting in a dramatic deswelling with increasing temperature. In some (rare) cases, this may allow a temperature-induced "squashing release." Much more frequently, however, drug release decreases with gel deswelling. Thus, gel deswelling may be used both for encapsulation of drugs (Figure 15.1) and for protection of encapsulated drugs from enzymatic degradation after administration. For example, Saunders, Vincent, and others have in a number of investigations studied various aspects of PNIPAM microgels, which undergo temperature-induced deswelling on increasing temperature, and the transition temperature of which may be controlled, for example, by inclusion of charged/titratable groups [15, 22, 24–26]. This also facilitates an additional pH and electrolyte concentration control (i.e., dual or multiple responses) by addition of electrolytes or other cosolutes.

As far as drug delivery applications of such systems are concerned, Nayak et al. reported folate-mediated cell targeting with PNIPAM microgels that displayed temperature-dependent toxicity, attributed to particle aggregation in the cell interior at elevated temperatures [31]. Using the same type of microgels, Nolan et al. investigated temperature-induced insulin release and noted that insulin release could be enhanced by increasing the temperature. As the microgels undergo a drastic deswelling during the same temperature increase, the microgels essentially behave as "sponges," releasing insulin when "squashed" (Figure 15.4) [32]. Of course, there are also limitations to the use of microgels as drug delivery vehicles. Again working with PNIPAM-based microgels, Lopez et al. investigated the use of such systems for transdermal delivery of ibuprofen and salicylamide [33, 34]. Although such systems may have advantages in terms of an increased drug stability and reduced peak concentrations after administration, these come at a cost. In the case of the transdermal microgel-based delivery systems investigated by Lopez et al., the transdermal diffusion rate of both ibuprofen and salicylamide were strongly reduced compared to the corresponding saturated aqueous solutions.

A perhaps somewhat more exotic application of PNIPAM and related microgels in drug delivery is given by composite microgels also containing superparamagnetic γ-Fe_2O_3 or Fe_3O_4 particles [19, 20]. Although the temperature response of the PNIPAM microgels is strongly affected by the presence of the magnetic particles, a significant amount of particles may nevertheless be incorporated at largely maintained deswelling behavior. The resulting composite particles are therefore magnetic and can be localized by external magnetic fields to parts of the body in much the same way as other magnetic nanoparticles [35, 36], thereby facilitating localized delivery. As an alternative, such particles may be combined with ultrasound or light fields to

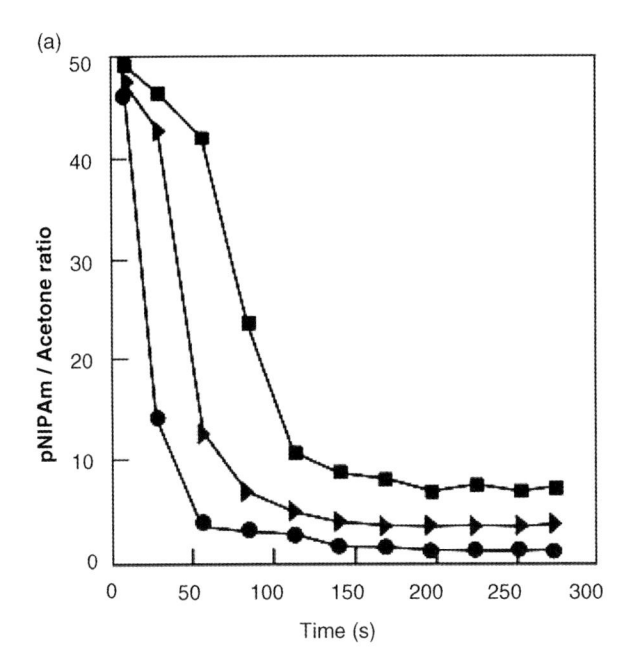

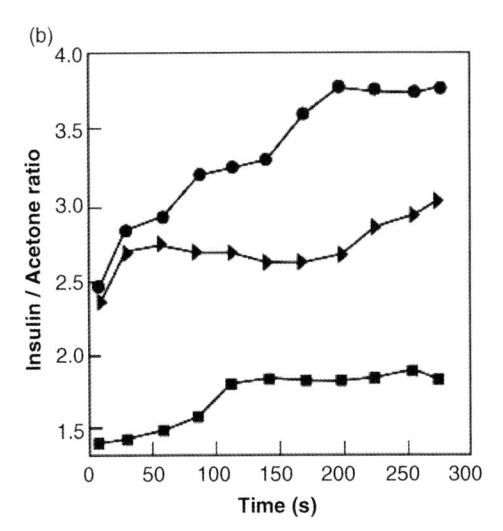

Figure 15.4 Microgel deswelling (a) and insulin release (b) from PNIPAM microgels subjected to a temperature jump from 25 to 34 °C (■), to 37 °C (▲), and to 40 °C (●). (Redrawn from Ref. [32].)

increase temperature locally, and thus reach delivery of encapsulated drug at the desired site of action [20, 35–41].

It should also be noted that swelling/deswelling in microgel systems may also affect the activity of incorporated (protein) drugs. For example, by incorporating an enzyme in a responsive gel particle and varying the temperature of such a system between one temperature just above, and one just below, the lower consolute temperature of a thermally responsive microgel was found to result in a reversible variation in the probed enzymatic activity [42].

15.3.2
Electrostatic Triggering of Microgels

Apart from thermally triggered microgel systems, there is also extensive interest in electrostatically triggered ones [27, 43–52]. Among the latter, particular attention has been directed to pH as a triggering mechanism, which is interesting particularly for polyacids. Thus, at low pH (such as in the stomach) such systems are collapsed, protecting encapsulated drug from acid-catalyzed hydrolysis. With increasing pH (such as in the small intestine), ionization of the polyacid network occurs and results in a swelling of the network and in drug release. In the context of oral delivery, not only is the drug chemically protected from the harsh environment of the stomach but also does the drug release occur in a region where drug uptake is most efficient. Examples of studies on pH-induced swelling/deswelling of microgel systems include the work by Eichenbaum *et al.* [44–46], who prepared poly(methacrylic) microgels by precipitation polymerization and investigated their response to pH and ionic strength. As expected, the gels expand as a result of the dissociation of the carboxyl groups in microgels, an effect that can be screened in a normal fashion by electrolyte (Figure 15.5) [46], and can also be described with basic electrostatic modeling. These microgels, and variants thereof, were subsequently investigated as drug carriers for a series of drugs (doxorubicin, dibucaine, and benzylamine). By controlled variations on the functional group density of the microgels, a direct correlation could be established between the microgel loading capacity for protons and the amount of drug loaded, although the extent of drug loading into the microgel was still concluded to be a complex issue. Similarly, Tan and Tam studied pH-induced swelling of a methacrylic-ethyl acrylate microgel system, as well as its consequences for procaine hydrochloride release. As can be seen in Figure 15.5, increasing pH in this systems leads to an increased gel swelling due to an increased microgel charge build-up, which in turn causes an increased drug release [47]. A similar result was obtained also by Babu *et al.* [48] for interpenetrating network microgels of sodium alginate-acrylic acid, as well as by Zhang *et al.* [50] for chitosan-based microgels containing methotrexate.

The triggered drug release, in turn, is of course coupled to biological effects. For example, LaVan *et al.* [53] reported pH-triggered drug release from microgels, which allowed triggered uptake in macrophages. Furthermore, Das *et al.* studied pH-responsive microgels consisting of PNIPAM and poly(N-isopropylacrylamide acrylic acid) in the context of cancer targeting [49]. These microgels were loaded with doxorubicin and conjugated with transferrin to target the cancer cells. Using this approach, these authors demonstrated a specific doxorubicin delivery to HeLa cells based on the viability of the latter following administration. Since doxorubicin is known to result in toxic side effects, including cardiotoxicity and myelosuppression, such targeted delivery with the drug encapsulated is interesting. Focusing on

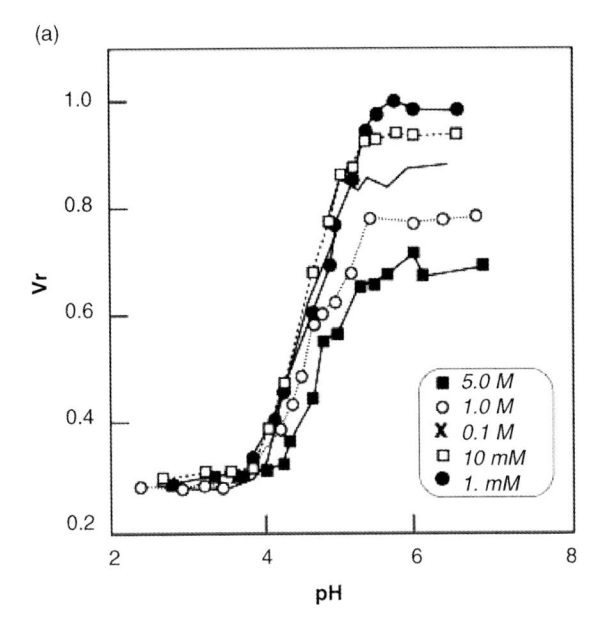

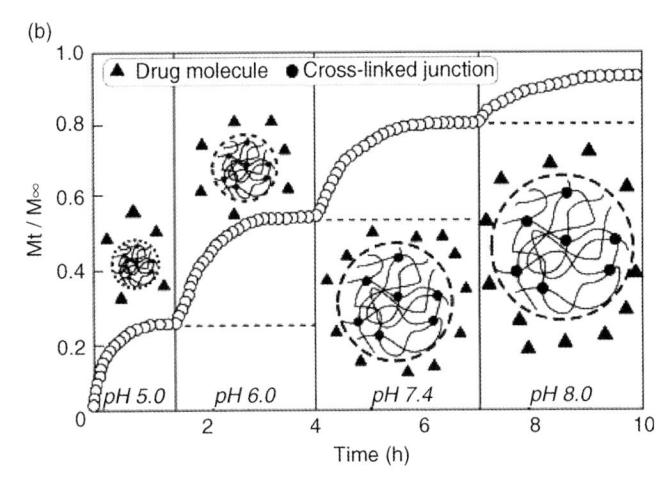

Figure 15.5 (a) Microgel equilibrium volume ratio (V_r) for poly(methacrylic acid) microgels versus excess electrolyte concentration and pH. (Redrawn from Ref. [46].). (b) *In vitro* release profile of procaine hydrochloride from an anionic methacrylic acid/ethyl acrylate microgel as a function of pH, where M_t and M_∞ refer to fraction released at time t and limiting release, respectively. (Redrawn from Ref. [47].)

doxorubicin delivery with microgel systems, Oishi *et al.* also investigated doxorubicin encapsulated in a pH-sensitive PEG-modified system based on diethylamino ethyl methacrylate and found that doxorubicin could be loaded into such microgels and that these did not display any burst (initial peak) release, the latter an advantageous

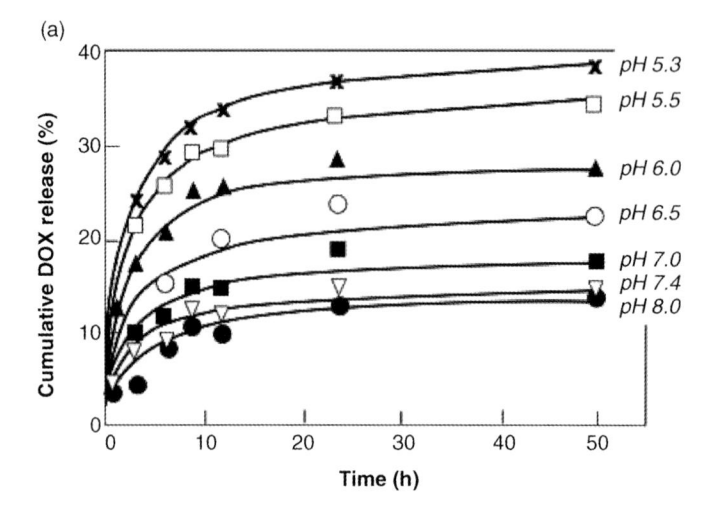

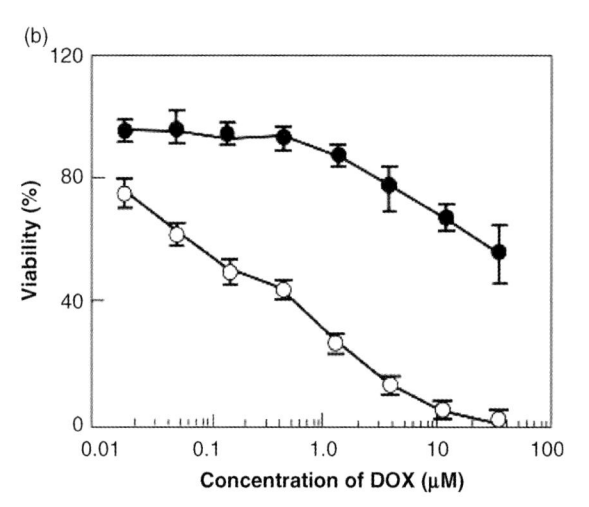

Figure 15.6 (a) pH dependence of doxorubicin release from PEG-PEAMA microgels. (b) Antitumor activity (expressed as cell viability) of free doxorubicin (●) and doxorubicin incorporated in the PEG-PEAMA microgels (○). (Redrawn from Ref. [51].)

effect from a toxicity perspective as discussed above [51]. Furthermore, doxorubicin release strongly depended on pH (Figure 15.6), which is interesting for endosomal release of the drug during acidification. The antitumor activity of the doxorubicin-loaded, pH-sensitive, microgel system against the MCF-7 human breast cancer cell line, frequently drug resistant, was found to be much higher than that of both free doxorubicin and doxorubicin loaded into a PEG-modified non-pH-sensitive microgel. As also shown by fluorescence microscopy, doxorubicin is initially contained in the pH-responding microgels, but is released in the endosomes in the cell interior as a result of the acidification process there, and then diffuses through the cytoplasm and finally exerts its effect in the cell nuclei.

15.3.3
Triggering of Microgels by Specific Metabolites

Polymer (micro)gels may also be designed to display drug release in response to the concentration of a specific metabolite. Insulin has received particular attention in this context, and a number of different systems have been designed to yield insulin release in response to an increasing glucose concentration [8–10, 54–56]. For example, concanavalin A (ConA) is a lectin with ability to bind carbohydrates. When mixed with dextran derivatives, the four-valent ConA acts as a cross-linker, and by choosing a sufficiently high ConA concentration, the effective cross-linking density is sufficiently high, and the mesh size sufficiently low, for insulin incorporated into the gels to experience restricted diffusion. When such a composite insulin-containing gel is exposed to an increased glucose concentration, there is competition between dextran and free glucose for the ConA binding sites. This results in the rupture of the gel cross-links and hence also in insulin release (Figure 15.7). Along a similar line of thinking based on linkages removable by competition from the aqueous solution surrounding the microgels, RNA/DNA base [57] and antigen–antibody [58, 59] pairs have also been used in a much similar fashion, for example, with semi-interpenetrating networks prepared by grafting antigen and antibody to the different networks and forming the gel particles by mixing these together. In the presence of free antibody/antigen, competition will occur, resulting in swelling and eventually dissolution of the network (Figure 15.8). Yet another related principle for specific drug release was reported by Tanihara *et al.* [60, 61], in which thrombin-like proteolytic activity displayed by *Pseudomonas aeruginosa* and other bacteria strains toward specific peptide sequences was employed. By using such peptide linkers in gel cross-linking, infection-specific drug release is reached.

As for the frequently investigated glucose-depending release systems, other mechanisms for triggering glucose-specific insulin release have also been investigated. For example, Ito *et al.* investigated PDEAEM gel systems in which glucose oxidase has been conjugated. Upon exposure to glucose, this enzyme generates gluconic acid, thereby resulting in a reduced pH that facilitates swelling of the cationic gel network and hence also insulin release [56].

15.3.4
Microgel Triggering by External Fields

Triggering by external fields such as ultrasound, light, and magnetic fields constitutes yet another triggering mechanism for microgel delivery systems. For example, Patnaik *et al.* investigated photoregulation of drug release in azo-dextran microgels [40]. By using trans-cis isomerization of an azobenzene moiety present in the cross-linker of the microgel, drug release could be regulated by light. More specifically, the release of rhodamine and aspirin from such gels was found to be slower

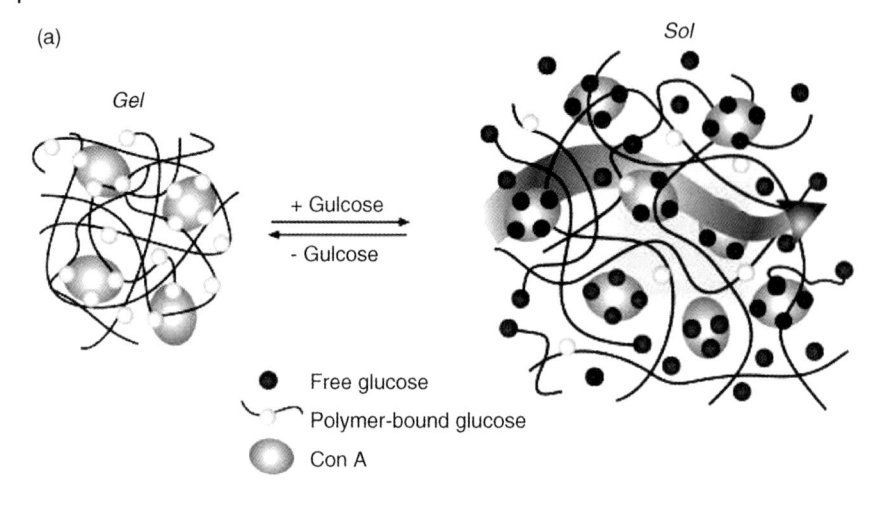

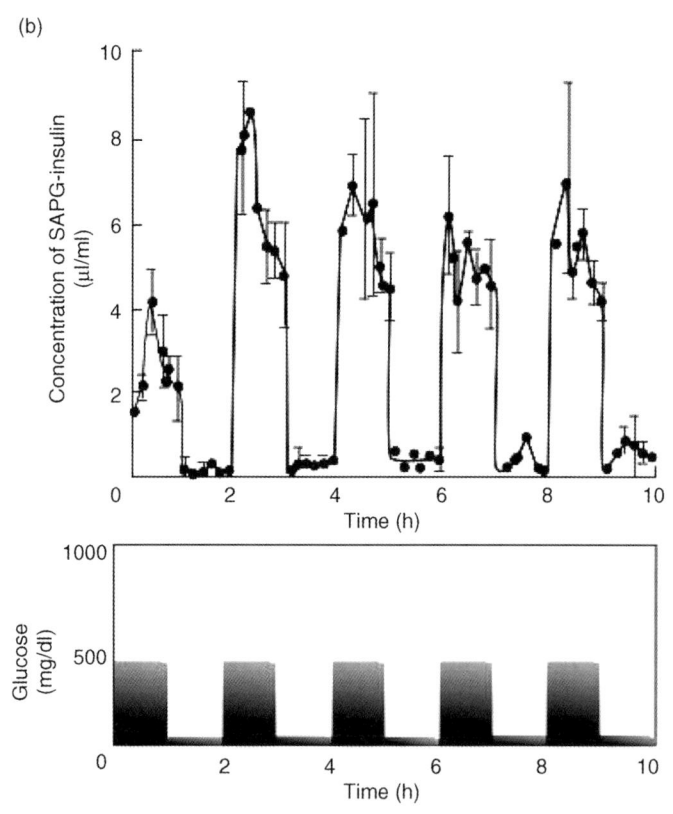

Figure 15.7 (a) Schematic illustration of a glucose-responding gel system based on concanavalin A and a dextran derivative. (b) Glucose-initiated insulin release from such a system during repeated variations in the external glucose concentration. (Redrawn from Ref. [55].)

(a)

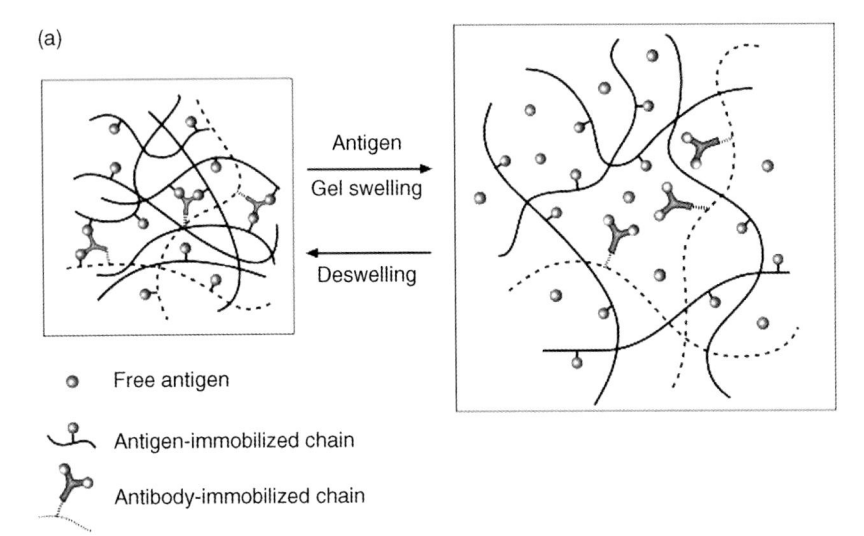

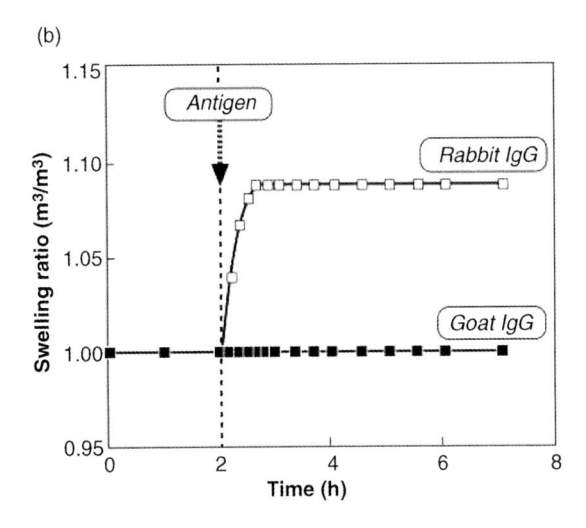

- ○ Free antigen
- Antigen-immobilized chain
- Antibody-immobilized chain

(b)

Figure 15.8 (a) Schematic illustration of the build-up of an antigen-specific microgel. Antibodies and antigens are covalently conjugated to separate polymer chains. By mixing the two polymer components, interpenetrating networks are formed together with antibody–antigen "cross-links." On addition of either antigen or antibody to the solution surrounding the microgel particles, these compete with the corresponding component conjugated to the polymer chain. With increasing concentration of free antigen/antibody, an increasing fraction of the antigen–antibody "cross-links" is dissolved, as is eventually the entire microgel particle. (Redrawn from Ref. [8].) (b) Swelling ratio changes of antibody/antigen microgel particles on addition of the specific antibody (rabbit IgG), as well as on an irrelevant antibody (goat IgG). (Redrawn from Ref. [59].)

with the azo moiety in E-configuration and faster in Z-configuration. Analogous effects have been found for polymer microcapsules, for example, by Angelatos et al. [39], by Radt et al. [37], by Skirtach et al. [38], and by Gorelikov et al. (Figure 15.9) [62], although the light triggering in those cases was achieved by

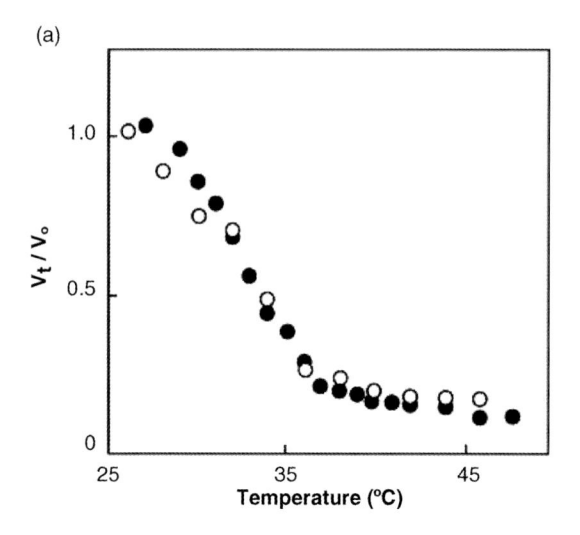

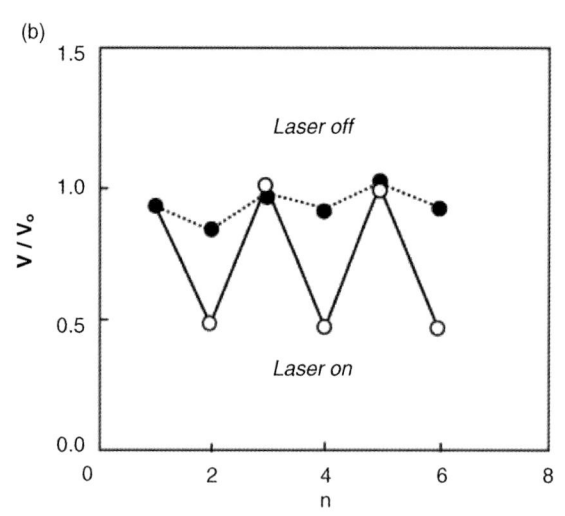

Figure 15.9 (a) Deswelling ratio for poly(NIPAM-AA) microgels at pH 4 as a function of temperature. Results are shown both for pure poly(NIPAM-AA) microgels (●) and for hybrid microgels containing gold nanorods (○). (b) Light-triggered and thermally induced deswelling of the same hybrid microgel system (○), as well as for the reference system not responding to light-induced thermal heating (●). (Redrawn from Ref. [62].)

incorporation of metal nanoparticles, which generate local heat by light illumination, thus causing a change in temperature-sensitive polymer network permeability. Furthermore, De Geest et al. investigated ultrasound-triggered release, albeit again from multilayered microcapsules rather than microgels [41]. Again, the external field, in this case ultrasound, is used to generate thermal waves, which trigger an increased permeability. In this investigation, FITC-labeled dextran was encapsulated in the microcapsule interior by coprecipitation with $CaCO_3$, an elegant method to include macromolecular drugs into such systems. Upon application of ultrasound, mechanical disruption of the capsule wall occurs, an effect that increases with sonication time and power. Such light- and ultrasound-triggered drug release is of interest in localized delivery, for example, in photodynamic therapy and analogous localized cancer and other therapies.

15.3.5
Microgel Triggering by Degradation

Apart from triggered response of microgel delivery systems, there is a substantial interest in microgel systems not requiring any triggers [63–71]. Such "self-exploding" systems are of interest in pulsatile release systems, for example, in vaccines and hormonal therapies. In general, such systems involve degradable microgels, sometimes surrounded by a shell impermeable to the drug. In one type of such systems, gel core degradation results in an increased swelling pressure, eventually causing gel rupture and drug release. By combining such systems with different rupture times, achieved by different cross-linking density, different charge density, or different mechanical strength of a shell surrounding the microgel particles and opposing their swelling, a pulsatile release could in principle be reached for the ensemble system. As an example of shells investigated in this context could be mentioned those formed by polyelectrolyte multilayers, the mechanical strength and permeability of which may be controlled by the number of layers, as well as a range of external parameters such as ionic strength and pH. For example, De Geest et al. investigated dextran microgels coated by a number of different multilayer systems in this context [67]. However, as investigated by Kraft et al. [68], by Kiser et al. [69], and by De Geest et al. [70], lipid-coated microgels are also interesting in this context, as their rupture may be triggered, for example, by membrane-active peptides and enzymes, and the strength and permeability of which may be varied by pH, ionic strength, presence of divalent cations, peptide length, polar head group charge, presence/absence of cholesterol, degree of lipid saturation, temperature, and copresence of surfactants. The concept of microgel particles coated with lipid layers was particularly elegantly demonstrated by Kiser et al. [69]. Using micromanipulator-assisted light microscopy, these authors could demonstrate that such coated microgels release incorporated doxorubicin as a result of surfactant-induced disruption of the lipid layer and resulting microgel swelling, and that a similar effect can be obtained by electroporation (Figure 15.10). As expected, the leakage induction in the lipid-coated microgels and the corresponding "empty" liposomes (i.e., liposomes not containing a microgel core) is comparable.

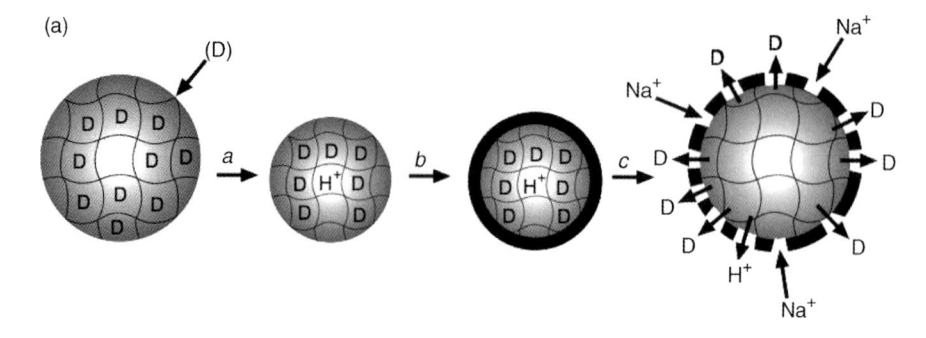

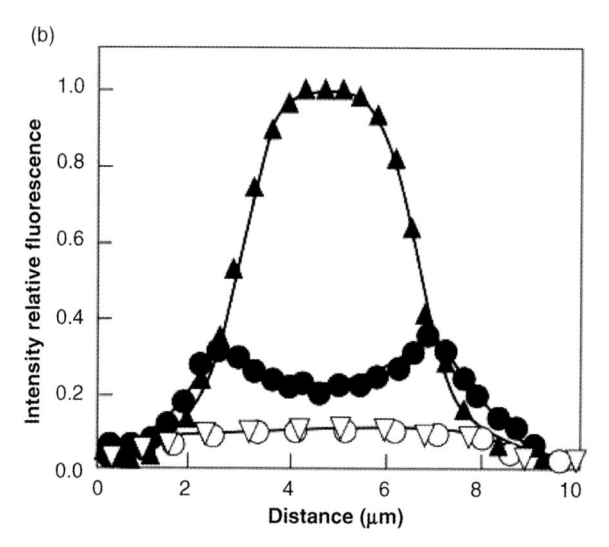

Figure 15.10 (a) Schematic illustration of lipid-coated microgels. The network is loaded with drug in its swollen state and condensed after drug loading by a reduction of pH. The condensed gel particles are then coated by a lipid bilayer (e.g., by vesicle fusion). Upon creation of defects in the lipid bilayer, the microgel undergoes swelling and causes rupture of the coated particles. (b) Fluorescence intensity ("concentration") profiles of a single microgel particle coated by a lipid bilayer (●, ▲), where circles and triangles refer to lipid and doxorubicin, respectively. Upon addition of the surfactant SDS, the lipid bilayer is disintegrated and initially incorporated doxorubicin released (○, △). (c) Release triggering can be achieved also by electroporation above a critical bias. While no volume increase was observed at 4.50 kV/cm, substantial defect formation and network swelling was observed at 4.75 kV/cm, demonstrating the critical bias to be in the range of 4.50–4.75 kV/cm. (d) The critical bias varies between different lipid systems, but is comparable to that of vesicle bilayers in the absence of incorporated microgel particles, and is in agreement with predictions from electrocompression theory. (Redrawn from Ref. [69].)

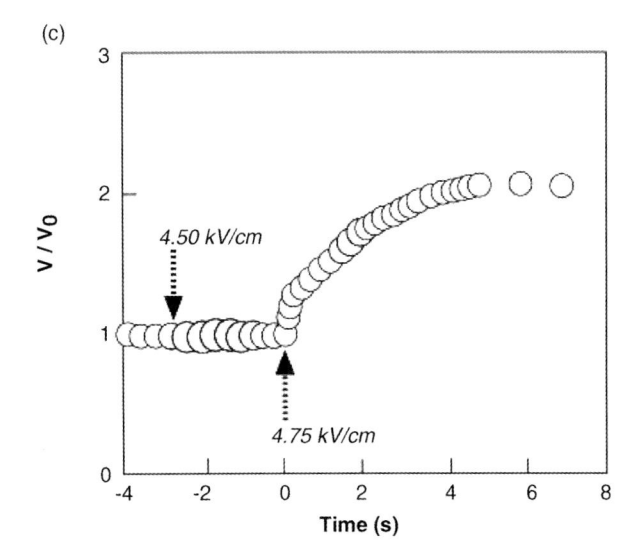

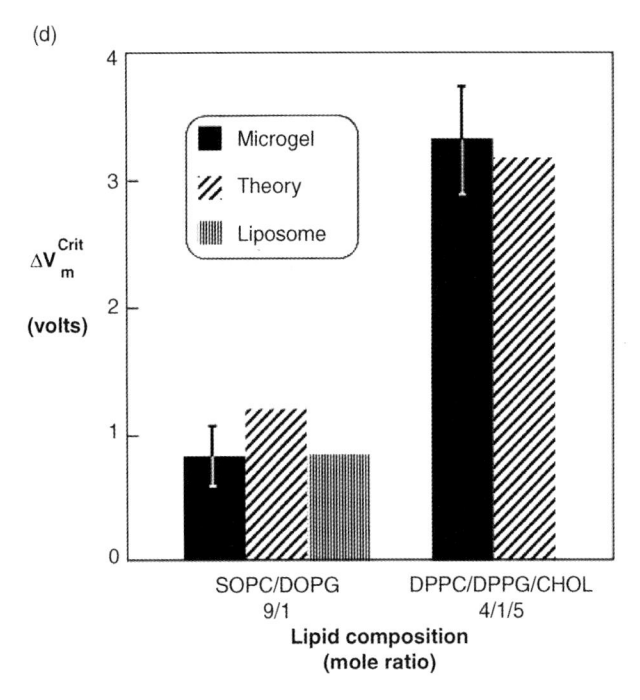

Figure 15.10 (*Continued*)

Triggerable biodegradation as a controlled drug release mechanism for microgels has been investigated also more generally. For example, Murthy et al. prepared microgels from acrylamide/bisacrylamide using acetal linkers, thereby obtaining microgels displaying biodegradation that is triggered by low pH as a result of acid-catalyzed acetal hydrolysis [65, 66]. Specifically, hydrolysis half-life at pH 7.4 and 5.0 is 24 h and 5.5 min, respectively. Using such systems, these authors could demonstrate pH-triggered release induction of FITC-albumin. The same authors could also demonstrate pH-triggered degradation of microgels as a tool not only for protein release but also for pH-triggered potency of protein-based vaccines (Figure 15.11) [66]. Similarly, Bromberg et al. investigated microgels composed of poly(acrylic acid) also containing poly(ethylene oxide)-poly(propylene oxide)-poly(ethylene oxide) copolymers, and cross-linked with disulfide groups, and could demonstrate increased swelling triggered by the reduction of the disulfide bonds [43]. Also, Oh et al. obtained largely analogous results for the disulfide-functionalized dimethacrylate cross-linker used [64]. Of course, these chemically degradable microgels are rather similar to physically cross-linked microgels using specific interactions (discussed above), such as antigen–antibody, biotin–streptavidin, DNA/RNA base, or polysaccharide–concanavalin A pairs. In all these cases, as well as in cyclodextrin-based microgels [71], "biodegradation" can also be caused by rupturing of physical "cross-links" within the microgel particles, for example, by competition with specific solutes, or as a response to physical parameters.

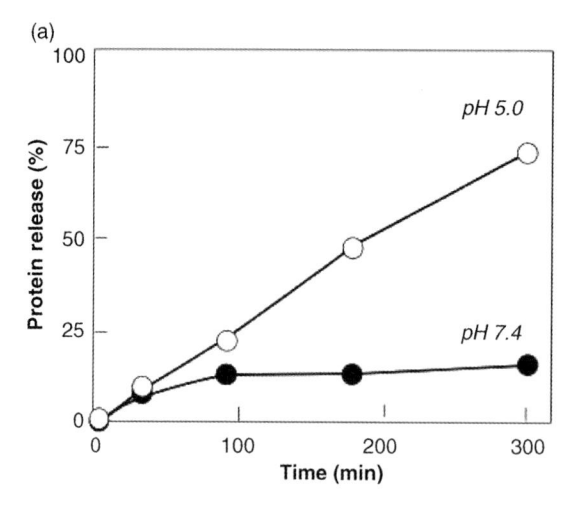

Figure 15.11 (a) pH-dependent protein release from acid-degradable microgels. (b) Class I antigen presentation of phagocytosed ovalbumin-loaded acid-degradable microgels compared to that of free ovalbumin. As can be seen, improved antigen presentation ("vaccination") is achieved when ovalbumin is incorporated in the microgels. (c) Dose-dependent toxicity of ovalbumin-loaded acid-degradable microgels to RAW 309.1CR cells, demonstrating limited toxicity of the ovalbumin-loaded microgels up to concentrations as high as 5 mg/ml. (Redrawn from Ref. [66].)

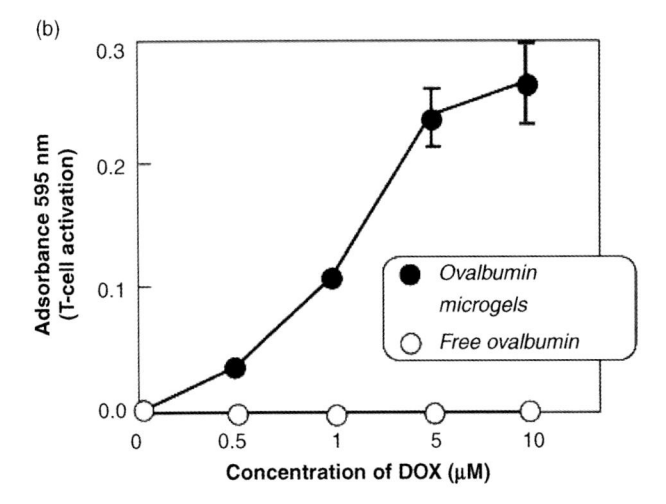

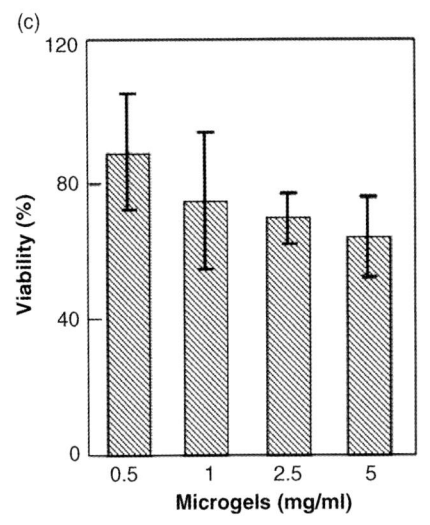

Figure 15.11 (Continued)

15.4
Polymer Microcapsules

Although fairly complex in their structure and preparation, polyelectrolyte multilayer capsules are receiving increasing interest as potential drug delivery systems [37–39, 41, 72–93]. Such capsules consist of a shell composed of alternating anionic and cationic polyelectrolytes, surrounding a core containing the drug either in aqueous solution or in its solid state. In some special cases, the drug may also be incorporated in the shell multilayer structure. The main feature making such multilayer capsules interesting in drug delivery is the barrier function generated by the multilayer, which may be controlled by a range of factors, including the number of layers, pH, ionic strength, temperature, polarity of the surrounding solution, and even specific metabolites. For example, Liu et al. investigated a layer-by-layer assembly of oppositely charged polyelectrolytes (PSS and PAH/PDADMAC) onto melamine formaldehyde colloidal particles, followed by removal of the core by degradation at low pH [83]. Using such systems, efficient loading of daunorubicin could be achieved. Moreover, loading and release could be controlled by seeding concentrations, temperature, pH, and salt concentration. Furthermore, Zhang et al. investigated hepatic targeting microcapsules prepared by conjugation of galactose branches to PGEDMS polymers, and subsequent generation of multilayers through layer-by-layer deposition of this polymer and PSS on acyclovir (ACV) microparticles. As expected, the drug release decreases with increasing number of layers in the capsule wall (Figure 15.12) [81]. The targeting of these carriers was further improved by incorporation of PNA lectin. Such and related surface modifications of microcapsules can be performed using essentially the same methods as for other types of colloidal drug carriers, for example, for active targeting and also for reducing serum protein adsorption and resulting opsonization in parenteral administration [94].

Of course, low molecular weight drugs are not only the ones that may be delivered using multilayer microcapsules. As shown by Kreft et al. [89] and Dejugnat et al. [88], such structures are potentially attractive delivery systems also for biomacromolecular drugs such as DNA/RNA and proteins, allowing fine control of the amount of drug incorporated into each microcapsule. In addition, an interesting application of polyelectrolyte multilayer capsules surrounding (charged) macromolecules is that of induced precipitation of poorly soluble drugs within such capsules. As the encapsulated macromolecules cannot penetrate the polyelectrolyte multilayers while solutes and water can, this facilitates the establishment of a polarity gradient across the capsule wall, which may be used to precipitate poorly water-soluble drugs into the capsule interior [86].

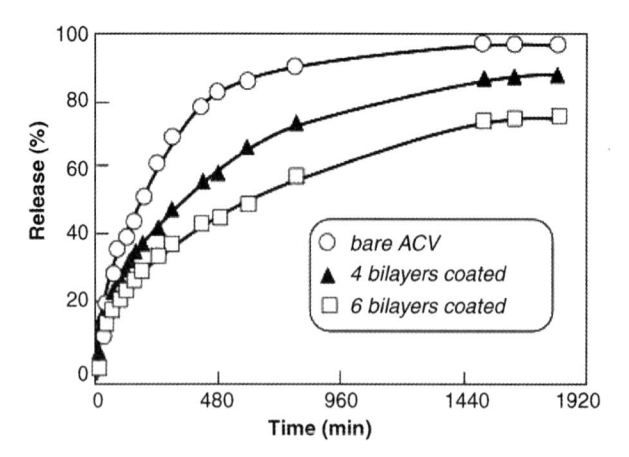

Figure 15.12 Release of acyclovir from ACV particles, as well as such particles coated by different number of oppositely charged polyelectrolyte layers. (Redrawn from Ref. [81].)

15.4.1
Microcapsule Triggering

In analogy to microgels, drug release from microcapsules may be triggered by a range of factors, including temperature, ionic strength, pH, and specific metabolites. For example, Quinn and Caruso [87], as well as Nolan et al. [92], investigated layer-by-layer structures formed by poly(acrylic acid) and poly(N-isopropylacrylamide). Through the temperature dependence of systems like these, such structures allow incorporated drugs to be released by increasing temperature (Figure 15.13) [92].

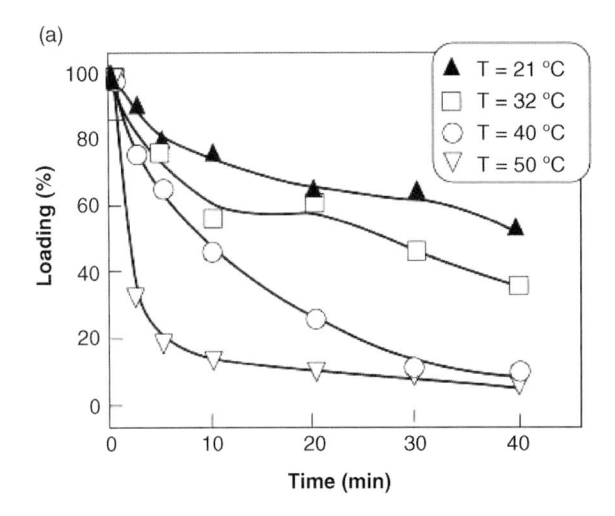

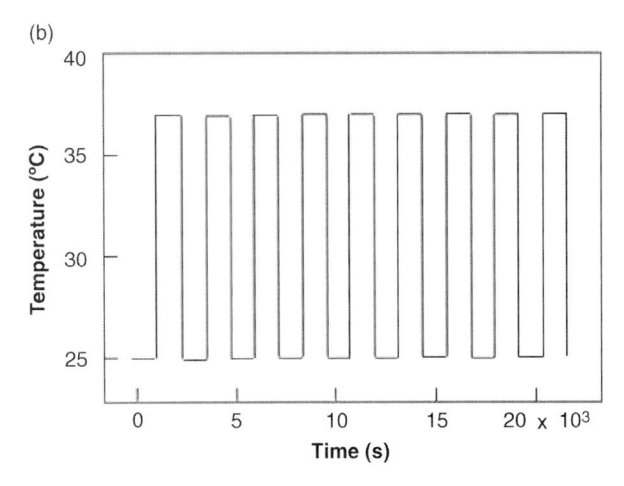

Figure 15.13 (a) Temperature-dependent release of rhodamine B from (PAA/PNIPAAm)$_{10}$ multilayers. (Redrawn from Ref. [87].) Shown also is thermally induced pulsatile release of FITC-insulin from a 30-layer microcapsule film, where (b) and (c) denote temperature program and FITC-insulin release, respectively. (Redrawn from Ref. [92].)

(c)

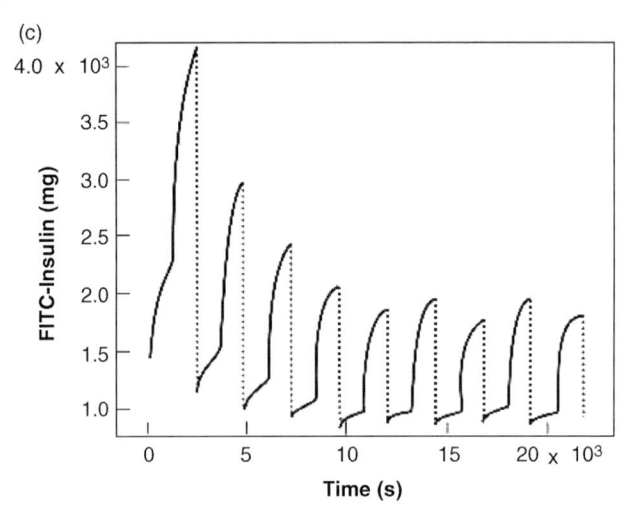

[Graph: FITC-Insulin (mg) vs Time (s), showing repeated sawtooth-like release profiles. Y-axis from 1.0 to 4.0 × 10³, X-axis from 0 to 20 × 10³.]

Figure 15.13 (Continued)

As for microgels, the permeability of polyelectrolyte multilayer capsules may be controlled through pH and electrolyte concentration. For example, Antipov et al. investigated polyelectrolyte capsules formed by poly(styrene sulfonate) and poly (allylamine hydrochloride), and found fluorescein permeability to increase with ionic strength and decrease with increasing pH (Figure 15.14), both results from an increased number of noncomplexed charges in the polyelectrolyte multilayer at higher ionic strength and at lower charge contrast, respectively [75]. Somewhat related to this pH response is the charge-controlled selectivity displayed by some microcapsule systems. For example, Tong et al. investigated polyelectrolyte multilayer capsules formed by poly(styrene sulfonate) and poly(allylamine), also containing "free" polystyrene sulfonate localized in the capsule interior [90]. Through a delicate interplay between this polystyrene sulfonate and the capsule wall, a high degree of permeation selectivity can be reached, such that negatively charged probes are completely rejected by the capsule interior while positively charged ones are attracted to the capsule core. Combined with the pH-dependent ionization displayed by most drugs, such systems may therefore be used for reaching pH-dependence in both drug loading and release. In a subsequent investigation, these authors extended their study to other polyelectrolytes forming the microcapsule and to cross-linked capsules [91].

As with microgels, polyelectrolyte multilayer microcapsules may be designed to also respond to specific metabolites. One example of this are capsules containing phenylboronic acids, which form covalent complexes with polyol compounds such as glucose. By including such phenylboronic acid compounds in polyelectrolyte multilayer capsules, a glucose-triggered permeability increase of the capsule can be achieved, which makes such capsules interesting for glucose-responsive insulin

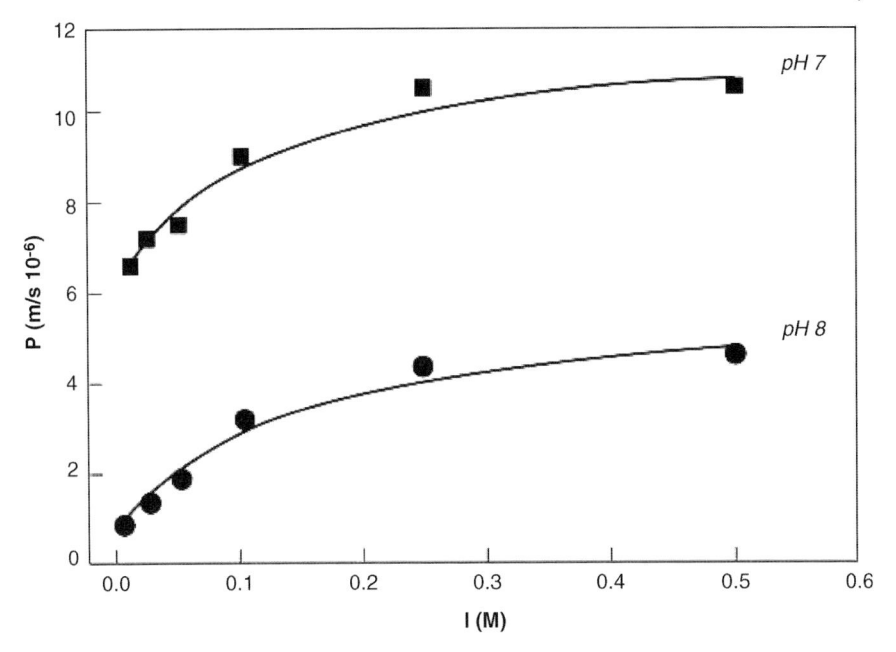

Figure 15.14 Permeability of poly(styrene sulfonate)/poly(allylamine hydrochloride) multilayer capsules as a function of ionic strength and pH. (Redrawn from Ref. [75].)

release [72]. Furthermore, in analogy to cross-linking discussed in relation to glucose-responding microgels above, glucose-sensitive layer-by-layer structures may be formed by alternating cancanavalin A and glycogen [93]. As for the corresponding microgel systems, free glucose competes with glycogen for concanavalin A, resulting in a reduction in the number of attachment points in the multilayer, in the swelling of the latter, and in an increased drug release.

Also in analogy to microgel particles, attention has been placed on self-degrading microcapsules displaying increased drug release as a function of capsule degradation, which is of interest, for example, in pulsatile drug release. For example, Borodina et al. investigated degradable multilayer microcapsules containing proteases in the capsule interior, combined with capsules formed by poly(L-arginine) and poly(L-aspartic acid) [80]. As the degradation of the capsule wall progresses, the release from the capsule interior increases (Figure 15.15). By varying the amount of protease present in the capsule interior, capsule degradation time and drug release rate can be varied from seconds to days. Furthermore, and in analogy to similar approaches taken for microgel systems, Zelikin et al. investigated multilayer capsules containing disulfide "cross-links," the stability of which may be controlled by the reducing environment, which also provides a self-degradation route for microcapsules [82].

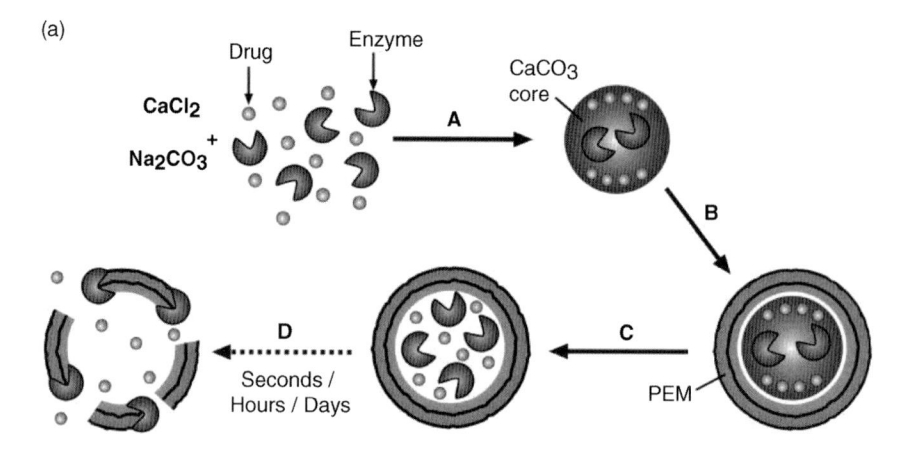

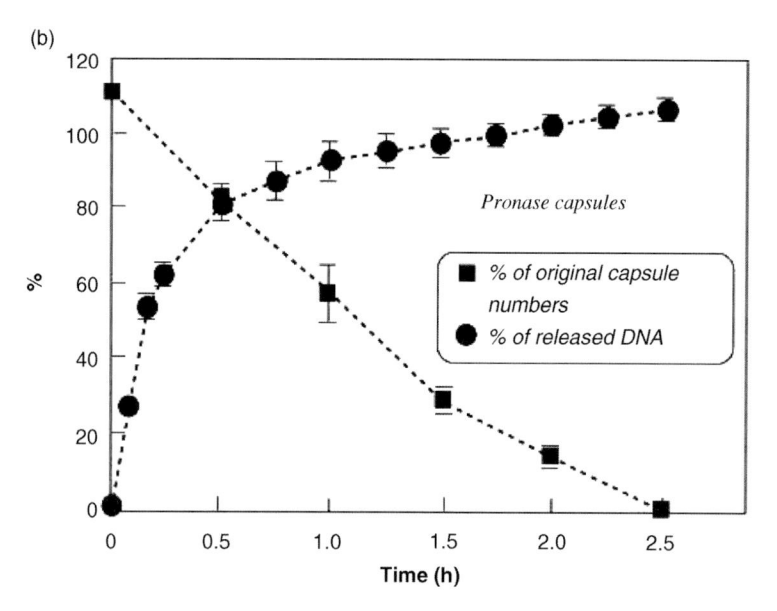

Figure 15.15 (a) Schematic illustration of enzyme-containing self-degrading multilayer capsules. A proteolytic enzyme is coprecipitated with the drug to be encapsulated together with $CaCl_2$ and Na_2CO_3. Polyelectrolyte multilayers based on homopolypeptides are formed around the particles generated, whereafter the particle core is dissolved by Ca^{2+} extraction through EDTA addition. Through proteolytic action, the encapsulated enzyme starts degrading the homopolypeptides forming the capsule wall, resulting in its rupture and in drug release. (b) Degradation rate of, and DNA release from, enzyme-containing ("Pronase") polyelectrolyte microcapsules formed by homopolypeptides. Shown also in (c) are results for the corresponding control system not containing any proteolytic enzyme. (Redrawn from Ref. [80].).

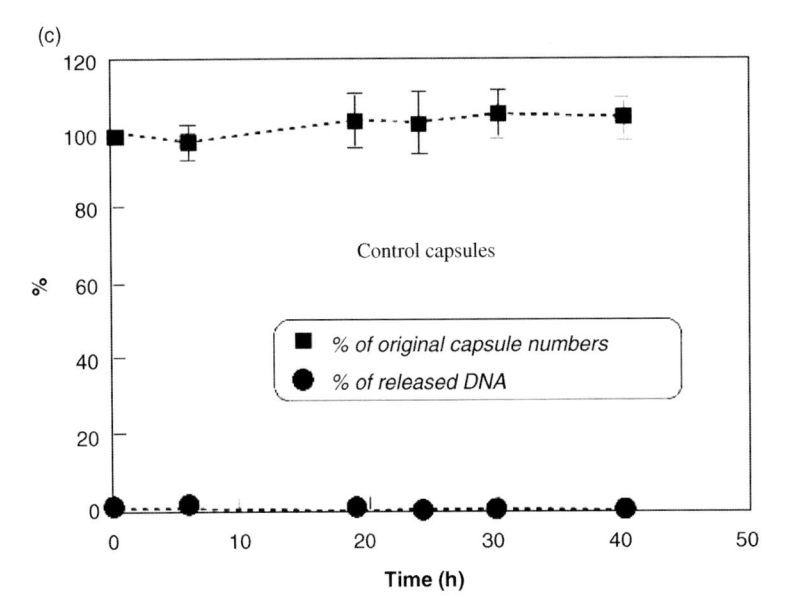

(c) [Graph showing % vs Time (h) for Control capsules, with % of original capsule numbers remaining near 100% and % of released DNA near 0% over 0–50 h.]

Figure 15.15 (Continued)

15.5
Swelling, Loading, and Release Kinetics

Apart from the degree of swelling, critical factors in relation to loading and release kinetics of microgels and related systems include the particle size, cross-linking density, and network homogeneity [95–103]. One of the limiting factors in the use of microgel systems in drug delivery is their relatively slow response, and much effort has thus been dedicated to obtaining a faster response. Given the above, this has involved work with smaller microgel or nanogel particles, with particles of lower cross-linking density, and with particles of more/less uniform cross-linking distribution. As predicted in early theoretical work by Tanaka and Fillmore [95], the gel (de) swelling time scales with the diameter-squared. Although this early work was developed for macroscopic gels, this relation seems to hold also for microgels. To give just one example on this, Dupin et al. investigated the swelling behavior of poly (2-vinylpyridine) microgels and found swelling times in good agreement with predictions from the Tanaka equation [96].

Furthermore, the higher the microgel cross-linking density, the smaller the average mesh size and the maximum swelling, the slower the response, and the smaller the maximum drug size possible to incorporate in the gel network. For example, Bromberg et al. investigated microgels formed by poly(acrylic acid) onto which polyether chains (Pluronic F127) had been grafted [97]. With increasing cross-linking density the swelling rate decreases as expected. As also expected, increasing the effective length of the subchain between cross-links results in an increasing uptake of host molecules such as doxorubicin. In analogy, decreasing the effective mesh size of the network by increasing temperature for this system also results in a reduced doxorubicin uptake. Similar findings have also been reported for microgels containing degradable cross-links. For example, both Plunkett et al. [98] and Bromberg et al. [43, 97] investigated microgels cross-linked by disulfide bonds. Using microgels formed by 2-hydroxyethyl methacrylate, Plunkett et al. found that both the network swelling and the swelling rate increase when the number of cross-links is reduced by adding a reducing agent [98]. Similarly, Bromberg et al. studied poly(acrylic acid) microgels containing reversible disulfide or biodegradable azoaromatic cross-links [43, 97]. Again, degradation-limited swelling kinetics was observed. Combined with localized degradation, as discussed above in the context of macroscopic polysaccharide gels, such degradation-mediated swelling and drug release offer opportunities in localized drug delivery. This was also realized by Bromberg et al., who could demonstrate microgel swelling triggered by cleavage of the azoaromatic cross-links by azoreductases from the rat intestinal cecum [97]. Such systems are therefore of interest, for example, in colon-specific drug delivery.

Also, the nature of the cross-links and of the network naturally affects the swelling kinetics. In particular, the more hydrophobic the microgel, the slower the swelling kinetics [99–101]. For example, Loxley and Vincent studied cationic microgels formed by copolymerization of 2-vinylpyridine and styrene, and found that higher styrene contents reduce not only the extent of swelling in this system but also the swelling kinetics [100]. A similar observation was made by Amalvy et al. who found microgels formed by 2-(diethylamino)ethyl methacrylate to swell faster than those formed by the more hydrophobic 2-(diisopropylamine)ethyl methacrylate [99]. Along the same line, Gan and Lyon investigated core–shell microgels formed by poly(N-isopropylacrylamide) and found that inclusion of a hydrophobic monomer, butyl methacrylate, into the particle shell strongly reduces the kinetics of the temperature-induced collapse of these microgels [101]. An interesting aspect of Gan and Lyon's work is that the (de)swelling rate thus can be straightforward reduced for essentially any microgel system by convenient hydrophobic postmodification of the microparticle surface region.

Interesting in the context of loading and release kinetics is also the cross-linking homogeneity of microgels. Thus, larger microgels formed by emulsion polymerization are likely to display a cross-linking density decreasing from the particle centers toward the periphery [22]. In contrast, smaller microgels may be characterized by more homogeneous cross-link distribution. For example, Saunders investigated deswelling of poly(NIPAM/xBA) microgel particles at various degrees of swelling and found an essentially linear relationship between microgel particle and average mesh size [29]. In the context of microgel homogeneity, it should also be noted that microgel parameters may vary both between individual particles and within a single microgel particle. As an example of the former, Johansson et al. investigated lysozyme-induced shell formation in poly(acrylic acid) microgels and observed that, under some conditions, such shell formation occurred in some of the microgel particles within a given microgel batch, but not in others, clearly indicating particle heterogeneities [104]. Furthermore, Hoare and Pelton investigated PNIPAM-based microgels and could demonstrate a "core–shell" structure with primarily surface functionalization [18]. In a later investigation, the same authors could also show that such surface charge may result in shell formation due to electrostatic collapse of the neutralized outer network on interaction with an oppositely charged drug, clearly demonstrating the practical significance of such heterogeneities [105]. Interesting in the context of microgel uniformity is the effect of the cross-linking distribution within the microgels on the (de)swelling kinetics. Perhaps somewhat counterintuitively, Chu et al. found, when investigating PNIPAM microgels containing voids, that microgel particles containing such voids indeed respond faster to temperature-induced changes than those which are void free and that the phase transition kinetics can be fine-tuned by controlling the number and size of the voids [102]. Although much remains to be done to clarify the role of microgel heterogeneity in swelling transitions, this still offers some seemingly interesting opportunities both in drug delivery and in biosensor applications.

As a result of the possible importance of size polydispersity and other types of heterogeneity on microgel performance in drug delivery, attention has been paid to the preparation of monodisperse microparticles. One of the numerous techniques that have emerged as interesting in this context is PRINT (particle replication in nonwetting templates) [16]. The technique is based on perfluoropolyester molds, which are liquid at room temperature and can be photochemically cross-linked into elastic solids to enable high-resolution imprint lithography. The technique does allow rather precise control not only of the microgel size in the range from 20 nm to >100 μm but also of the shape, content, modulus, and so on.

A key factor influencing not only loading and release kinetics but also secondary parameters such as the amount of drugs encapsulated and drug susceptibility to chemical and proteolytic degradation is the drug distribution within the microgels. While extensive work has been done on protein and peptide absorption to, and distribution within, chromatography beads, limited corresponding work has been done on lightly cross-linked microgels displaying the pronounced swelling/deswelling characteristics central to the performance of microgels in drug delivery. Of the investigations performed to date on this subject, a series of papers by Malmsten et al. could be mentioned [104, 106–109], in which the incorporation of both homopolypeptides and spherical hard proteins (e.g., lysozyme and cytochrome C) into poly(acrylic acid) microgels were investigated. Since charge contrast was found to be a requirement for peptide/protein incorporation into the microgels for this systems, focus was placed on positively charged proteins and polypeptides. On the basis of a method combination of micromanipulator-assisted light microscopy and confocal microscopy, it was found that both peptide/protein distribution and gel deswelling kinetics are strongly influenced by the peptide/protein size, originating partly from limited entry of large molecules into the gel particle core. Also, pH was shown to significantly influence deswelling, incorporation kinetics, and protein/peptide distribution. These effects are determined by a complex interplay between the pH-dependence of both peptide and the gel network, also influencing volume transitions of the latter. Finally, salt concentration was shown to have a significant effect on both gel deswelling rate and peptide transport, with an increased electrolyte concentration, resulting not only in a decreased deswelling rate but also in an increased peptide transport rate within the microgel particles.

The importance of such "shell formation" for low molecular drugs in microgel systems was also demonstrated in a recent publication by Hoare and Pelton, who showed that when cationic drugs bind to carboxylic acid groups located at the surface of poly(N-isopropylacrylamide) microgels, a locally collapsed surface layer is formed, preventing additional drug uptake into the core and resulting in decreased deswelling [105]. Thus, although the formation of such surface layers primarily depends on the size of the incorporated compound and the mesh size of the gel network, other factors such as pH and ionic strength also play a key role.

Finally, it should be noted that while microgels and microcapsules have been investigated mainly as dispersed systems, their applicability in drug delivery extends beyond this. Thus, microgels may also act as a component in composite drug delivery systems (so-called "plum pudding" formulations, where microgels are dispersed in a gel or polymer solution matrix [110, 111]) for injectable gel formulations with prolonged action. In order to strengthen the potential of microgels and microcapsules as delivery systems, their incorporation in dry solid matrices is also relevant, hence further work on the fate of such systems in lyophilization processes such as spray drying and freeze drying needs to be further investigated in much the same way as previously done, for example, for liposomal drug delivery systems [1].

15.6
Outlook

The importance of soft drug delivery systems is likely to increase in the coming years. Partly, this is related to an increasing fraction of drug candidates very sparingly soluble in water emerging from present drug discovery work. Thus, there are increasing demands to get the drug into aqueous solution, not only for administration of drug but also for facilitating drug development itself, for example, in clinical trials, in animal experiments, or even in early-stage toxicity screening. The likely increased importance of soft drug delivery systems also stems from an increasing fraction of biopharmaceutical drugs, notably recombinant proteins and antibodies, emerging on the market and in the drug discovery work. Here, advanced delivery systems such as those formed by microgels and microcapsules offer real opportunities for providing "biological stability" through preservation of protein secondary and tertiary structure, avoidance of aggregation, and elimination of chemical and enzymatic degradation. Furthermore, the range of triggering factors available to such systems make them very versatile tools in a range of drug delivery contexts. Delivery systems are also likely to be used earlier in the drug development chain as a consequence of the merger between the drug discovery and the drug delivery processes. Which of the different types of soft delivery systems that will be most successful in this context is difficult to foresee, but clearly stability, ease of preparation, and versatility will be important concerns, so will be the biological response to the different carrier systems.

Acknowledgments

This work was financed by the Swedish Foundation for Strategic Research. Expert assistance with illustrations by Ms. Maud Norberg is gratefully acknowledged.

References

1. Malmsten, M. (2002) *Surfactants and Polymers in Drug Delivery*, Marcel Dekker, New York.
2. Malmsten, M. (2000) Chapter 14, in *Amphiphilic Block Copolymers: Self-Assembly and Applications* (eds P. Alexandridis and B. Lindman), Elsevier, Amsterdam.
3. Malmsten, M. (2006) *Soft Matter*, **2**, 760.
4. Malmsten, M. (2007) *J. Dispers. Sci. Technol.*, **28**, 63.
5. Scherlund, M., Malmsten, M., and Brodin, A. (1998) *Int. J. Pharm.*, **173**, 103.
6. Scherlund, M., Malmsten, M., Holmqvist, P., and Brodin, A. (2000) *Int. J. Pharm.*, **194**, 103.
7. Scherlund, M., Welin-Berger, K., Brodin, A., and Malmsten, M. (2001) *Eur. J. Pharm. Sci.*, **14**, 53.
8. Qui, Y. and Park, K. (2001) *Adv. Drug Deliv. Rev.*, **53**, 321.
9. Miyata, T., Uragami, T., and Nakamae, K. (2002) *Adv. Drug Deliv. Rev.*, **54**, 79.
10. Vinogradov, S.V. (2006) *Curr. Pharm. Des.*, **12**, 4703.
11. Makino, K., Hiyoshi, J., and Oshima, H. (2001) *Colloids Surf. B*, **20**, 341.
12. Lowe, T.L., Virtanen, J., and Tenhu, H. (1999) *Polymer*, **40**, 2595.
13. Rubinstein, A. and Sintov, A. (1992) Chapter 5, in *Oral Colon-Specific Drug Delivery* (ed. D.R. Friend), CRC Press, Boca Raton.
14. Zhang, F.Y., Zhu, W., Wang, B., and Ding, J. (2005) *J. Control. Release*, **105**, 260.
15. Murray, M.J. and Snowden, M.J. (1995) *Adv. Colloid Interface Sci.*, **54**, 73.
16. Napier, M.E. and Desimone, J.M. (2007) *Polymer Rev.*, **47**, 321.
17. Goldberg, M., Langer, R., and Jia, X. (2007) *J. Biomater. Sci. Polym. Ed.*, **18**, 241.
18. Hoare, T. and Pelton, R. (2004) *Langmuir*, **20**, 2123.
19. Rubio-Retama, J., Zafeiropoulos, N.E., Serafinelli, C., Rojas-Reyna, R., Voit, B., Cabarcos, E.L., and Stamm, M. (2007) *Langmuir*, **23**, 10280.
20. Bhattacharya, S., Eckert, F., Boyko, V., and Pich, A. (2007) *Small*, **3**, 650.
21. Das, M., Zhang, H., and Kumacheva, E. (2006) *Annu. Rev. Mater. Res.*, **36**, 117.
22. Saunders, B. and Vincent, B. (1999) *Adv. Colloid Interface Sci.*, **80**, 1.
23. Shimoboji, T., Larenas, E., Fowler, T., Hoffman, A.S., and Stayton, P.S. (2003) *Bioconjug. Chem.*, **14**, 517.
24. Crowther, H.M., Saunders, B.R., Mears, S.J., Cosgrove, T., Vincent, B., King, S.M., and Yu, G.E. (1999) *Colloids Surf. A*, **152**, 327.
25. Dowding, P.J., Vincent, B., and Williams, E. (2000) *J. Colloid Interface Sci.*, **221**, 268.
26. Saunders, B.R., Crowther, H., Morris, G.E., Mears, S.J., Cosgrove, T., and Vincent, B. (1999) *Colloids Surf. A*, **149**, 57.
27. Dong, L., Yan, Q., and Hoffman, A.S. (1992) *J. Control. Release*, **19**, 171.
28. Sakai, T. and Yoshida, R. (2004) *Langmuir*, **20**, 1036.
29. Saunders, B.R. (2004) *Langmuir*, **20**, 3925.
30. Pelton, R. (2000). *Adv. Colloid Interface Sci.*, **85** 1.
31. Nayak, S., Lee, H., Chielewski, J., and Lyon, L.A. (2004) *J. Am. Chem. Soc.*, **126**, 10258.
32. Nolan, C.M., Gelbaum, L.T., and Lyon, L.A. (2006) *Biomacromolecules*, **7**, 2918.
33. Lopez, V.C., Raghavan, S.L., and Snowden, M.J. (2004) *React. Funct. Polym.*, **58**, 175.
34. Lopez, V.C., Hadgraft, J., and Snowden, M.J. (2005) *Int. J. Pharm.*, **292**, 137.
35. Arruebo, M., Fernandez-Pacheco, R., Ibarra, M.R., and Santamaria, J. (2007) *NanoToday*, **2**, 22.

36 Alexiou, C., Arnold, W., Klein, R.J., Parak, F.G., Hulin, P., Bergemann, C., Erhardt, W., Wagenpfeil, S., and Lubbe, A.S. (2000) *Cancer Res.*, **60**, 6641.

37 Radt, B., Smith, T.A., and Caruso, F. (2004) *Adv. Mater.*, **16**, 2184.

38 Skirtach, A.G., Dejugnat, C., Braun, D., Susha, A.S., Rogach, A.L., Parak, W.J., Möhwald, H., and Sukhorukov, G.B. (2005) *Nano Lett.*, **5**, 1371.

39 Angelatos, A.S., Radt, B., and Caruso, F. (2005) *J. Phys. Chem. B*, **109**, 3071.

40 Patnaik, S., Sharma, A.K., Garg, B.S., Gandhi, R.P., and Gupta, K.C. (2007) *Int. J. Pharm.*, **342**, 184.

41 De Geest, B.G., Skirtach, A.G., Mamedov, A.A., Antipov, A.A., Kotov, N.A., De Smedt, S.C., and Sukhorukov, G.B. (2007) *Small*, **3**, 804.

42 Hoffman, A.S. (1987) *J. Control. Release*, **6**, 297.

43 Bromberg, L., Temchenko, M., Alakhov, V., and Hatton, T.A. (2005) *Langmuir*, **21**, 1590.

44 Eichenbaum, G.M., Kiser, P.F., Shah, D., Simon, S.A., and Needham, D. (1999) *Macromolecules*, **32**, 8996.

45 Eichenbaum, G.M., Kiser, P.F., Dobrynin, A.V., Simon, S.A., and Needham, D. (1999) *Macromolecules*, **32**, 4876.

46 Eichenbaum, G.M., Kiser, P.F., Simon, S.A., and Needham, D. (1998) *Macromolecules*, **31**, 5084.

47 Tan, J.P.K. and Tam, K.C. (2007) *J. Control. Release*, **118**, 87.

48 Babu, V.R., Rao, K.S.V.K., Sairam, M., Naidu, B.V.K., Hosamani, K.M., and Aminabhavi, T.M. (2006) *J. Appl. Polym. Sci.*, **99**, 2671.

49 Das, M., Mardyani, S., Chan, W.C.W., and Kumacheva, E. (2006) *Adv. Mater.*, **18**, 80.

50 Zhang, H., Mardyani, S., Chan, W.C.W., and Kumacheva, E. (2006) *Biomacromolecules*, **7**, 1568.

51 Oishi, M., Hayashi, H., Iijima, M., and Nagasaki, Y. (2007) *J. Mater. Chem.*, **17**, 3720.

52 Neyret, S. and Vincent, B. (1997) *Polymer*, **38**, 6129.

53 LaVan, D.A., Lynn, D.M., and Langer, R. (2002) *Nat. Rev. Drug Discov.*, **1**, 77.

54 Miyata, T., Jikihara, A., Nakamae, K., and Hoffmann, A.S. (2004) *J. Biomater. Sci. Polym. Ed.*, **15**, 1085.

55 Kim, J.J. and Park, K. (2001) *J. Control. Release*, **77**, 39.

56 Ito, Y., Casolaro, M., Kono, K., and Yukio, I. (1989) *J. Control. Release*, **10**, 195.

57 Aoki, T., Nakamura, K., Sanui, K., Kikuchi, A., Okano, T., Sakurai, Y., and Ogata, N. (1999) *Polym. J.*, **31**, 1185.

58 Miyata, T., Asami, N., and Uragami, T. (1999) *Macromolecules*, **32**, 2082.

59 Miyata, M., Asami, T., and Uragami, T. (1999) *Nature*, **399**, 766.

60 Suzuki, Y., Tanihara, M., Nishimura, Y., Suzuki, K., Kakimaru, Y., and Shimizu, Y. (1998) *J. Biomed. Mater. Res.*, **42**, 112.

61 Tanihara, M., Suzuki, Y., Nishimura, Y., Suzuki, K., Kakimaru, Y., and Fukunishi, Y. (1999) *J. Pharm. Sci.*, **88**, 510.

62 Gorelikov, I., Field, L.M., and Kumacheva, E. (2004) *J. Am. Chem. Soc.*, **126**, 15938.

63 Van Thienen, T.G., Raemdonck, K., Demeester, J., and De Smedt, S.C. (2007) *Langmuir*, **23**, 9794.

64 Oh, J.K., Siegwart, D.J., and Matyjaszewski, K. (2007) *Biomacromolecules*, **8**, 3326.

65 Murthy, N., Thng, Y.X., Schuck, S., Xu, M.C., and Frechet, J.M. (2002) *J. Am. Chem. Soc.*, **124**, 12398.

66 Murthy, N., Xu, M., Schuck, S., Kunisawa, J., Shastri, N., and Frechet, J.M.J. (2003) *Proc. Natnl. Acad. Sci. USA*, **100**, 4995.

67 De Geest, B.G., Dejugnat, C., Prevot, M., Sukhorukov, G.B., Demeester, J., and De Smedt, S.C. (2007) *Adv. Funct. Mater.*, **17**, 531.

68 Kraft, M.L. and Moore, J.S. (2004) *Langmuir*, **20**, 1111.

69 Kiser, P.F., Wilson, G., and Needham, D. (2000) *J. Control. Release*, **68**, 9.

70 De Geest, B.G., Stubbe, B.G., Jonas, A.M., Van Thienen, T., Hinrichs, W.L.J., Demeester, J., and De Smedt, S.C. (2006) *Biomacromolecules*, **7**, 373.

71 Liu, Y.-Y., Fan, X.-D., Kang, T., and Sun, L. (2004) *Macromol. Rapid Commun.*, **25**, 1912.

72 De Geest, B.G., Jonas, A.M., Demeester, J., and De Smedt, S.C. (2006) *Langmuir*, **22**, 5070.

73. Pargaonkar, N., Lvov, Y.M., Steenekamp, J.H., and de Villiers, M.M. (2005) *Pharm. Res.*, **22**, 826.
74. Tong, W., Gao, C., and Möhwald, H. (2006) *Macromolecules*, **39**, 335.
75. Antipov, A.A., Sukhorukov, G.B., and Möhwald, H. (2003) *Langmuir*, **19**, 2444.
76. Shi, X., Wang, S., Chen, X., Meshinchi, S., and Baker, J.R., Jr. (2006) *Mol. Pharm.*, **3**, 144.
77. Shi, X. and Caruso, F. (2001) *Langmuir*, **17**, 2036.
78. Ai, H., Jones, S.A., de Villiers, M.M., and Lvov, Y.M. (2003) *J. Control. Release*, **86**, 59.
79. Wang, K., He, Q., Yan, X., Cui, Y., Duan, L., and Li, J. (2007) *J. Mater. Chem.*, **17**, 4018.
80. Borodina, T., Markvicheva, E., Kunizhev, S., Möhwald, H., Sukhorukov, G.B., and Kreft, O. (2007) *Macromol. Rapid Commun.*, **28**, 1894.
81. Zhang, F., Wu, Q., Chen, Z.-C., Zhang, M., and Lin, X.-F. (2008) *J. Colloid Interface Sci.*, **317**, 477.
82. Zelikin, A.N., Li, Q., and Caruso, F. (2006) *Angew. Chem., Int. Ed.*, **45**, 7743.
83. Liu, X., Gao, C., Shen, J., and Möhwald, H. (2005) *Macromol. Biosci.*, **5**, 1209.
84. De Geest, B.G., Dejugnat, C., Verhoeven, E., Sukhorukov, G.B., Jonas, A.M., Plain, J., Demesteer, J., and De Smedt, S.C. (2006) *J. Control. Release*, **116**, 159.
85. Khopade, A.J. and Caruso, F. (2003) *Langmuir*, **19**, 6219.
86. Radtchenko, I.L., Sukhorukov, G.B., and Möhwald, H. (2002) *H. Int. J. Pharm.*, **242**, 219.
87. Quinn, J.F. and Caruso, F. (2004) *Langmuir*, **20**, 20.
88. Dejugnat, C., Halozan, D., and Sukhorukov, G.B. (2005) *Macromol. Rapid Commun.*, **26**, 961.
89. Kreft, O., Georgieva, R., Bäumler, H., Steup, M., Müller-Röber, B., Sukhorukov, G.B., and Möhwald, H. (2006) *Macromol. Rapid Commun.*, **27**, 435.
90. Tong, W., Gao, C., and Möhwald, H. (2006) *Macromolecules*, **39**, 335.
91. Tong, W., Dong, W., Gao, C., and Möhwald, H. (2005) *J. Phys. Chem. B*, **109**, 13159.
92. Nolan, C.M., Serpe, M.J., and Lyon, L.A. (2005) *Macromol. Symp.*, **227**, 285.
93. Sato, K., Imoto, Y., Sugama, J., Seki, S., Inoue, H., Odagiri, T., Hoshi, T., and Anzai, J. (2005) *Langmuir*, **21**, 797.
94. Malmsten, M. (2003) Chapter 26, in *Biopolymers at Interfaces* (ed. M. Malmsten), Marcel Dekker, New York.
95. Tanaka T. and Fillmore, D.J. (1979) *J. Chem. Phys.*, **70**, 1214.
96. Dupin, D., Fujii, S., Armes, S.P., Reeve, P., and Baxter, S.M. (2006) *Langmuir*, **22**, 3381.
97. Bromberg, L., Temchenko, M., and Hatton, T.A. (2002) *Langmuir*, **18**, 4944.
98. Plunkett, K.N., Kraf, M.L., Yu, Q., and Moore, J.S. (2003) *Macromolecules*, **36**, 3960.
99. Amalvy, J.I., Wanless, E.J., Michailidou, V., and Armes, S.P. (2004) *Langmuir*, **20**, 8992.
100. Loxley, A. and Vincent, B. (2004) *Colloid Polym. Sci.*, **275**, 1108.
101. Gan, D. and Lyon, L.A. (2001) *J. Am. Chem. Soc.*, **123**, 7511.
102. Chu, L.-Y., Kim, J.-W., Shah, R.K., and Weitz, D.A. (2007) *Adv. Funct. Mater.*, **17**, 3499.
103. Nieuwenhuis, E.A. and Vrij, A. (1979) *J. Colloid Interface Sci.*, **72**, 321.
104. Johansson, C., Hansson, P., and Malmsten, M. (2007) *J. Colloid Interface Sci.*, **316**, 350.
105. Hoare, T and Pelton, R. (2008) *Langmuir*, **24**, 1005.
106. Bysell, H. and Malmsten, M. (2006) *Langmuir*, **22**, 5476.
107. Bysell, H., Hansson, P., and Malmsten, M. (2008) *J. Colloid Interface Sci.*, **323**, 60.
108. Bysell, H. and Malmsten, M. (2009) *Langmuir*, **25**, 522.
109. author>Johansson, C., Hansson, P., and Malmsten, M., *J. Phys. Chem. B*, **113**, 6183.
110. McGillicuddy, F.C., Lynch, I., Rochev, Y.A., Burke, M., Dawson, K.A., Gallagher, W.M., and Keenan, A.K. (2006) *J. Biomed. Mater. Res.*, **79**, 923.
111. Salvati, A., Söderman, O., and Lynch, I. (2007) *J. Phys. Chem. B*, **111**, 7367.

16
Microgels for Oil Recovery
Yuxing Ben, Ian Robb, Peng Tonmukayakul, and Qiang Wang

16.1
Introduction

The recovery of oil and gas from porous rock formations is one of the most technically complex procedures in the chemical industry. This is largely because the hydrocarbons are commonly located a few miles below the earth's surface at high temperatures and pressures, and the drilling is often done from rigs floating on the ocean, high above the seabed. The pressure from the earth's crust upon the oil-bearing formation has to be taken into account when treating the well and care must be taken to avoid uncontrolled release of oil or gas by proper design of the fluids used in the treatment. Oil and gas are not located in large caverns or hollows of a rock formation, but rather within tiny pores in the rock formation. The objective of most well treatments is therefore to drive passages from the main wellbore to as many of these pores as possible. After the initial drilling has released, a small percentage of the hydrocarbon content output declines and it is usually necessary to create fractures extending radially from the wellbore into the formation. This is achieved by applying very high pressures to the rock formation via an aqueous fluid. In order to prevent that the pressure due to the earth's crust closes the fractures after completion of this treatment, sand is included in the aqueous fluid, and this sand keeps the fracture open. A range of fluids, mostly aqueous, is used at various stages in the treatment of wells.

One of the major costs of hydrocarbon recovery comes from the capital required for the hardware such as drilling rigs, supply vessels and pumping equipment. It is thus vital to use this hardware as economically as possible and this is where microgels are of great use. During both the initial drilling and the subsequent treatments of the wellbore, it is important to control the loss to the formation of the many different, mainly aqueous, fluids that are used. The loss of fluid to the formation means that additional pumping capacity is required, which can raise the cost of oil extraction significantly. Fluid loss to the formation will be discussed in more detail in the sections below together with the microgel systems appropriate for each stage of the oil recovery process. Microgel particles, being deformable, are capable of entering

Microgel Suspensions: Fundamentals and Applications
Edited by Alberto Fernandez-Nieves, Hans M. Wyss, Johan Mattsson, and David A. Weitz
Copyright © 2011 WILEY-VCH Verlag GmbH & Co. KGaA, Weinheim
ISBN: 978-3-527-32158-2

and blocking the pores in rocks thus minimizing the loss of fluid to the rock formation. When recovery of the hydrocarbons is to start, these blocked pores need, however, to be cleared, which can be achieved by using microgels that are self-destructing over time in a manner similar to internal sutures used in medical procedures. Such microgels with a delayed reaction can save time for a rig that would otherwise have to be cleaned by more time consuming oxidative treatments.

The different stages [1] involved in creating and operating a well can be briefly summarized as follows:

- **Drilling**. A hole is drilled from the earth's surface down to the oil-bearing rocks, commonly 1–3 miles deep. The process requires that the drill cuttings (or rock fragments) are removed continually and this is achieved by suspending them in either water-based or oil-based fluids known as muds. These muds are sufficiently dense such that the hydrostatic pressure pointing the well, arising purely from the weight of the fluid, is sufficient to oppose the pressure of the earth's crust on muds trapped within the rock. The density of the muds is adjusted by the addition of heavy colloidal particles (such as barium sulfate) or by the use of high salt concentrations (sometimes near the solubility limit of a salt). Clearly, it is in the interests of both economy and the environment to lose as little drilling fluid as possible, thus preventing it from flowing into the porous rock formation; for this purpose, fluid loss agents are incorporated into the muds. These fluid loss agents are often polymer gels or microgel particles that are filtered from the drilling fluid and deposit near the rock surface as a so-called "filter cake". This filter cake is essentially a film formed at the rock surface by the accumulation of microgel particles; it reduces the loss of fluid to the formation and thereby greatly reduces the power required to fracture the rock (see below). After the initial drilling, this filter cake is usually left undisturbed; however, it is sometimes removed by subsequent treatment with acids that dissolve some components of the drilling fluid.
- **Casing and Cementing**. After completing the drilling, a steel pipe or casing is run down the hole leaving an annulus between the outer surface of the casing and the rock. This annulus needs to be filled otherwise hydrocarbons would escape from the rock strata directly to the atmosphere resulting in a probable explosion or blowout. This annulus is therefore filled by pumping cement down the inside of the casing and forcing it up between the outer casing surface and the rock. This ensures good filling of the annulus. The rate of setting of the cement from a slurry to a solid needs to be designed so that setting does not occur until the cement is in place. However, once in place the cement should set quickly so that gas migration through the cement is minimized. The rate of setting is dependent on both temperature and pressure and thus often a complex mixture of retarding and accelerating agents is used to control it. In addition, the rheology of the cement must be controlled to achieve a complete filling of the annulus while preventing separation of any of the components of the cement mixture.
- **Perforating**. Direct contact is achieved between the inside of the casing and the rock formation by perforating the casing and cement layers using specially

designed and located explosive charges. The pressure of the earth's crust usually provides an initial burst of hydrocarbon flow that may be sufficient to produce 5–10% of the oil in the reservoir. However, after some time, the output due to the earth's pressure declines and the well requires "stimulation" to maintain production.

- **Stimulation**. There are several forms of stimulation such as (i) matrix (rock formation) acidizing, where the oil-bearing formation near to the wellbore is partially dissolved by acid; (ii) water flooding, where water is injected in one well to push oil out of a nearby well; and (iii) fracturing, which commonly involves the application of very high pressures ($\sim 10\,000$ psi) to the oil-bearing regions of the rock to open one or more cracks away from the initial wellbore. These cracks or fractures then act as more extensive ducts for the flow of hydrocarbons from the formation to the wellbore. In order to keep these fractures open after releasing the high pressures, it is common practice to transport special sand or ceramic beads, known as proppants, down the wellbore and into the fracture, thus holding or "propping" it open. The width of a fracture is often 1–2 cm and the length of a fracture may be up to 300 m long, depending mainly on the nature of the rock. Transport of the proppant requires a suspending agent that can withstand the high temperatures, pressures and salt concentrations without significant degradation over the time required to place the proppant within the fracture; this time is generally up to a few hours. Cross-linked polymers such as guar are used to suspend the sand while it is being pumped down the drill hole and whenever pumping stops. The rheology of these suspending fluids is the subject of intense research in the oil industry. In addition, it is important to design the suspending fluid such that it degrades shortly after the proppant has been placed in the fracture; otherwise, the presence of the suspending agent is likely to retard the flow of hydrocarbons to the wellbore. To achieve this, oxidizers or enzymes are usually included in the fracturing fluid that carries the proppant.
- **Fluid Loss Control**. The various stimulation and completion treatments usually require that the pressure in the rock pores, which is essentially the hydrostatic pressure in the earth's crust, is opposed and exceeded by the pressure of the fluids in the wellbore or fracture. This applies to most wellbore treatments but is especially important for the processes of fracturing and drilling, where a variety of fluids are used. There is generally a net pressure acting on these fluids that tends to force fluid into the rock. Clearly, the greater the permeability of the rock or the higher the net pressure, the greater is the volume of fluid likely to enter the formation. Loss of fluid into the formation has two potentially detrimental effects: firstly, the energy (and hence cost) to pump this fluid becomes too high and secondly, the surface of the rock can be damaged, producing a "skin" that may retard the return flow of hydrocarbons. Thus, it is highly important to control the loss of fluid to the rock formation in most well treatments. Microgels can play a vital role in the control of fluid loss. Soft deformable microgel particles that can penetrate the pores give a better control of fluid loss than more rigid microgels that tend to stay on the rock face without forming a coherent film or "skin." Thus, controlling the microgel cross-linking density is critically important.

16.2
Microgels Used in Oil Recovery

Microgels are formed from polymers by a variety of cross-linking mechanisms, the nature of which has a strong effect on the rheological properties of the gel. The types of polymer used in oil recovery nearly all contain hydroxyl, amine, or amide groups, as these are the only ones that are thermodynamically stable under conditions of high temperature and pressure. Polymers that contain carbonyl (e.g., poly(vinyl pyrrolidone)), ether (e.g., poly(ethylene oxide)), or ester groups generally phase separate at high temperatures, pressures, or salt concentrations due to their unfavorable entropy of hydration [2]. Instead, the most commonly used polymers are polyacrylamides, polysaccharides, or well-designed polyelectrolytes.

16.2.1
Guar

The workhorse of fracturing is guar. Guar is the polysaccharide obtained from the endosperm of guar seed *Cyamopsis tetragnonolobus*. It is a galactomannan and consists of a main chain of $(1 \rightarrow 4)$ linked β-D-mannopyranosyl units with single α-D-galactopyranosyl units connected by a $(1 \rightarrow 6)$ linkage (Figure 16.1). The mannose to galactose (M/G) ratio is $1.6 \sim 1.8$ for guar. McCleary [3] showed that the solution viscosity of guar was mainly dependent on the length of the mannan backbone with the galactose side chains playing a very important role in determining the ease with which guar can be dissolved and retained in solution. Galactose and mannose both contain a *cis*-diol group (Figure 16.1) that can complex with the borate ion. Polymannose is insoluble in water as are galactomannans [3, 4] with $M/G > 5$.

At the oil field guar powder is mixed with water immediately prior to pumping the fracturing fluid down the wellbore, making hydration rate very important. Hydration rates and water binding properties of guar are dependent on its processing history and its particle size [5, 6]. Kesavan *et al.* [6] showed that a guar having a D_{50} (a standard measure of particle size giving the average particle diameter at 50 mass% of polymer powder) particle size of less than 40 μm reached at least 70% hydration within 60 s at about 70 °F. The production of finer powder sizes, which requires a higher milling

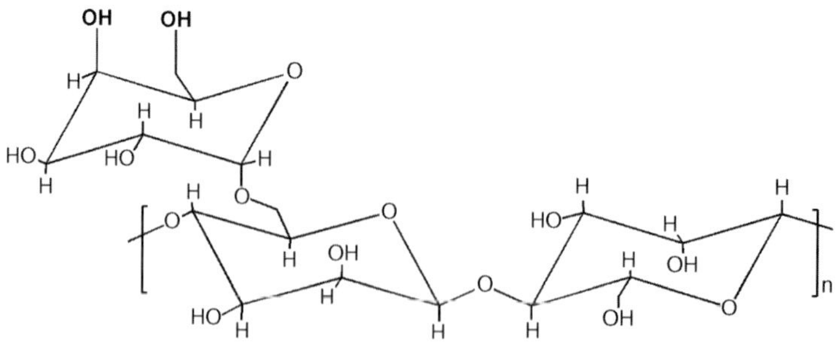

Figure 16.1 Molecular structure of guar.

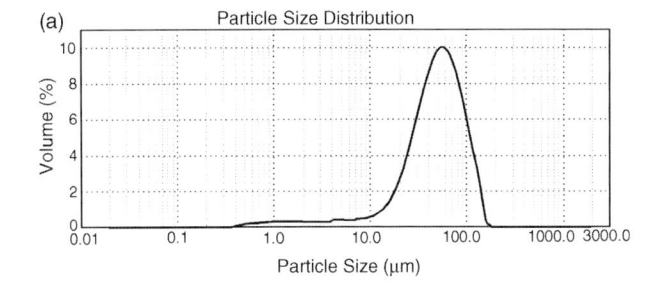

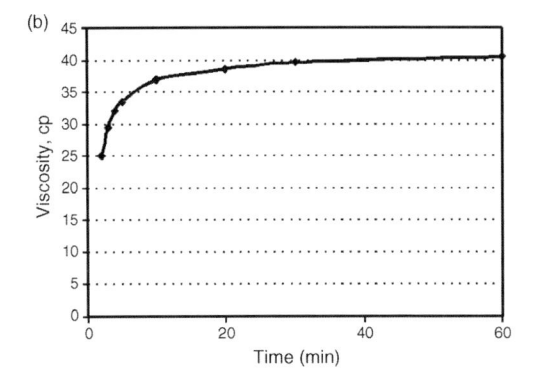

Figure 16.2 (a) Typical particle size distribution of guar powder. (b) Hydration rate for a 0.48% guar dispersion.

power, probably resulted in degradation of the polymer. The typical particle size distribution and hydration rate of guar, as reflected in the viscosity, are shown in Figure 16.2.

The maximum viscosities of guar gum dispersions are observed [6] at temperatures of about 20–40 °C. In oil field operations, the mixing energy of the blender [7] and the pH also affect the hydration rate. In Figure 16.3, we plot the logarithm of the radius of gyration R_g versus the logarithm of the molecular weight M_w, as obtained by GPC and multiangle light scattering measurements. The data are well described by a power law behavior [8] with an exponent of 0.55, which is in agreement with a random coil conformation of the chains.

The random coil conformation is also confirmed [9] by the Mark–Houwink relationship, relating the intrinsic viscosity of a polymer, $[\eta]$, to its average molecular weight, M_w, $[\eta] = K' M_w^\alpha$. For guar, α has been reported to be around 0.7–0.75, which falls within the range of 0.7–0.8, as expected for a random coil conformation. The viscosity of guar increases dramatically if the concentration is above the overlap concentration C^*, 0.05 ~ 0.1 wt%.

16.2.1.1 Gel Formation

Gels of guar or its derivatives, such as hydroxypropyl guar (HPG) can be cross-linked by various means. The behavior and properties of the gel must allow good mixing or complete dissolution on the surface and provide support for the proppant but not be

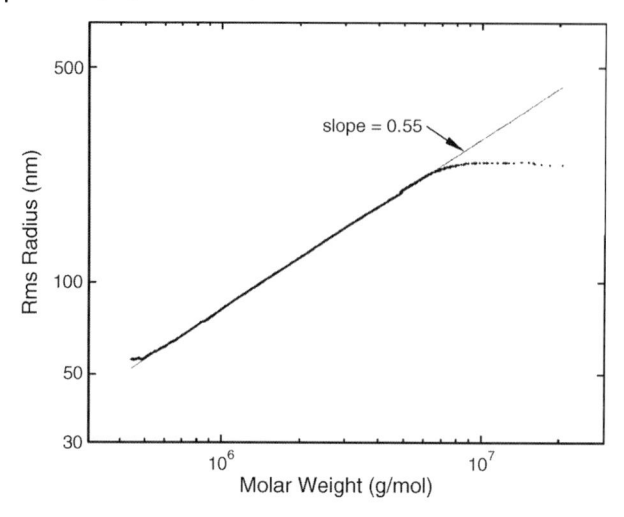

Figure 16.3 Radius of gyration versus molecular mass of guar.

so rigid that pumping requires high power. The cross-linking agents are designed to start cross-linking the gel as the fluid passes down the wellbore so that the forming gel can provide support for the proppant within the fracture. The most common cross-linking agent is sodium borate, which can be dissolved rapidly and thus cross-link the system quickly or after a short delay. Borate cross-links have short lifetimes (millisecond) that allow the gel to move under shear and the cross-links weaken with increasing temperature and decreasing pH. The depth of the rock formation mainly determines the ultimate temperature the fluid must withstand. At higher temperatures, where borate becomes less effective, metal oxide nanoparticles are often used, as discussed below.

An example of a temperature and viscosity profile [10] for a typical fracturing fluid is shown in Figure 16.4. Point (a) represents the viscosity of the fluid before cross-linking. As the temperature increases to about 40 °C (100 °F), the borate dissolves and the cross-linking reaction starts. The viscosity increases (b) with cross-linking until a maximum is reached at (c), which is followed by thermal thinning (d) as temperature increases. Reliable results are difficult to obtain [11] due to the reactive nature of the cross-link and the presence of large normal forces that can cause the fluid to crawl out of the concentric cylinders.

Fracturing requires that a "skin" is produced on the fracturing rock surface to reduce the loss of fluid to the formation, without which more hydraulic power would be needed to complete the fracture. With guar-based fluids, this skin or filter cake is supplied by the insoluble components from the cell wall of the guar seed that form a gel consisting of a small percentage of guar mixed with hemicelluloses. This filter cake and the cross-linked guar fluid need to be removed after the fracturing procedure to allow for the unimpeded back flow of hydrocarbons. One of the main problems [12] in removing the filter cake is its high viscosity, which retards reaction rates and access of breakers to the filter cake. The gel breakers that are used at this

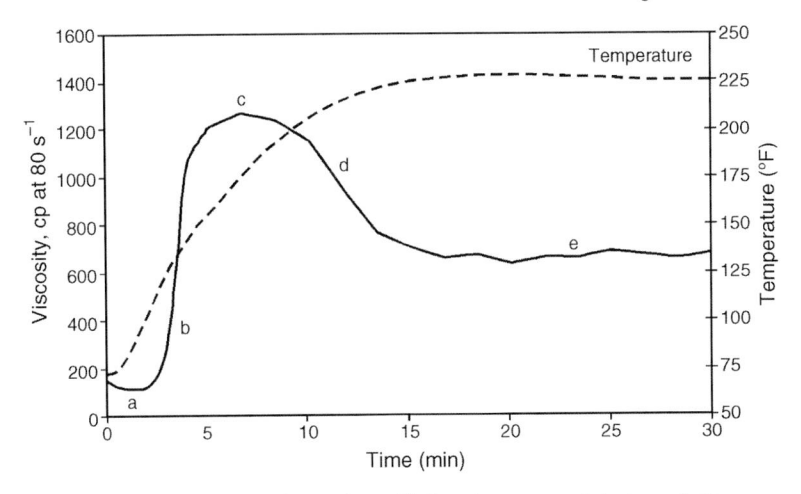

Figure 16.4 Temperature and viscosity profile for a borate cross-linked guar fluid measured on a Fann 50 Couette viscometer.

stage mainly consist of enzymes or oxidizers. Enzymes are effective at lower temperatures although they denature at elevated temperatures. Oxidizers are typically salts of persulfate, perborate, chlorite, or hydroperoxides. They may be coated or used with various catalysts to control their rate of decomposition according to the conditions in the fracture and the time to complete the placement of the proppant.

At high temperatures, where borate cross-links are not effective, metal oxide nanoparticles (consisting of titanium or zirconium) cross-link the polymer, which is usually carboxymethyl hydroxypropyl guar (CMHPG). The nanoparticles are formed from simple zirconium or titanium salts that are stabilized by ligands. These exist as dimers and tetramers at high concentrations as shown by the mass spectroscopy data in Figure 16.5 and are around 1.4 nm in size.

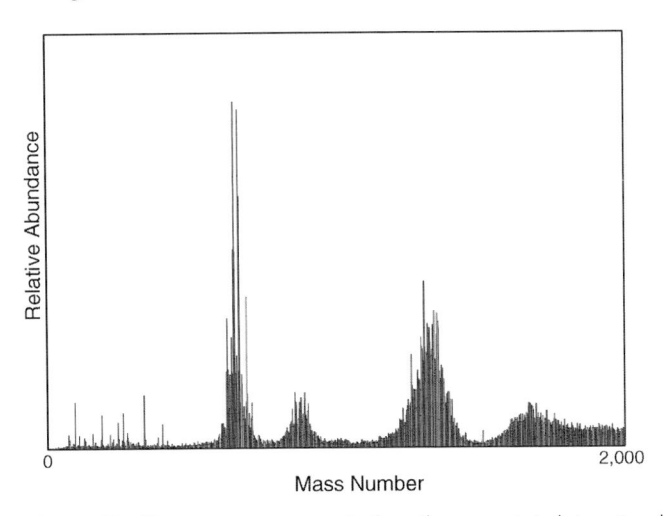

Figure 16.5 Mass spectroscopy results from the concentrated zirconium ligand solution.

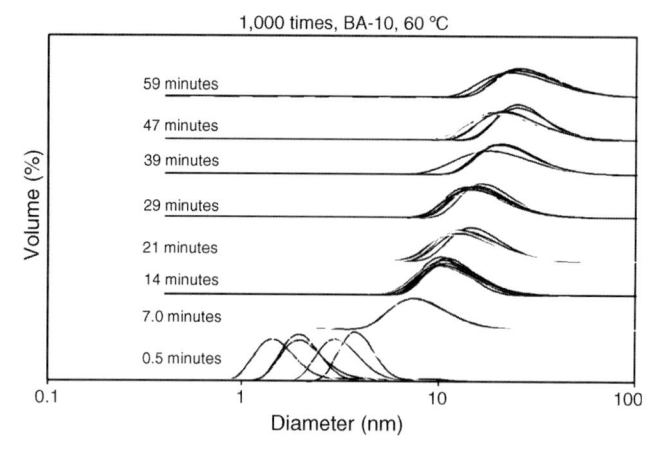

Figure 16.6 Size evolution of zirconium nanoparticles at 60 °C after 1 : 1000 dilution, as measured by dynamic light scattering.

When this cross-linking solution is diluted over 1000 times with water at 60 °C, hydrolysis of the zirconium ions occurs and nanoparticles are formed, which can be attributed to a nucleation and growth process (Figure 16.6).

When dilution of the concentrated zirconium solution occurs in the presence of CMHPG the polymer adsorbs onto these growing zirconium nanoparticles thus forming the cross-link for the gel. However, it is shown in Figure 16.7 that as the zirconium grows into larger but fewer particles, the storage modulus of the cross-linked gel decreases, which is probably caused by the reduced cross-linking density in the gel.

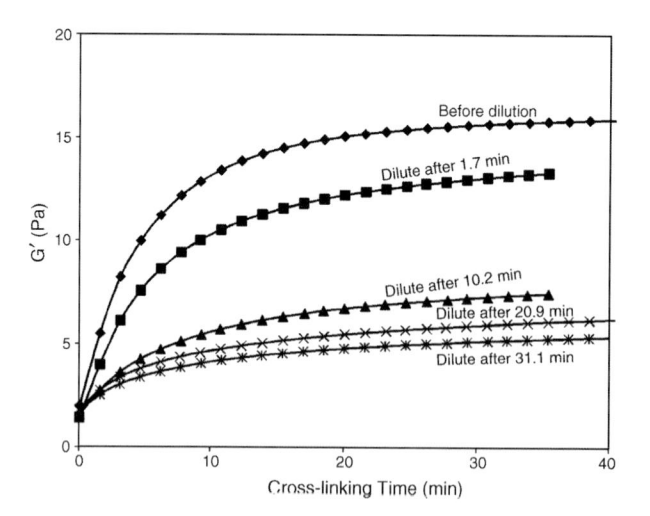

Figure 16.7 Storage modulus of a zirconium cross-linked CMHPG solution (0.48%) where the polymer has been added after the delay times shown on the graph.

The fact that the polymer is cross-linked by adsorption to the surface of small solid particles giving several bonds per cross-link means that the energy per cross-link is greater than for the borate ion. This makes these metal cross-linking systems appropriate for higher temperatures.

16.2.2
Rheology of Guar Gels and its Relation to Proppant Transport

Fracturing begins by pumping a mixture of the polymeric fluid and proppant particles at high shear rate ($\dot{\gamma}_{ave} \geq 500 \, s^{-1}$) into the reservoir via the wellbore. The time in the wellbore is relatively short (10–15 min), while the slurry spends longer (up to a few hours) at low a shear rate ($10 \geq \dot{\gamma}_{ave} \geq 100 \, s^{-1}$) in the fracture. Fracturing slurries are typically pseudoplastic (shear thinning) and exhibit power law behavior of the viscosity within the range of shear rates (wall shear rate) expected in the fracture. Fracture models assume that the fluid velocity is zero at the fracture wall, a maximum velocity at the middle of the fracture with no fluid leak-off inside the fracture path [13]. A typical average shear rate along the fracture length is generally expected to be larger than $20 \, s^{-1}$.

It is known that the shear rate distribution inside the fracture length depends on (i) the fracture width, (ii) the position of fluid along the fracture length, (iii) fluid loss or leak-off, and (iv) fluid rheology. In most fracture designs, the shear rate is assumed to be constant over the fracture length but depends on the fracture width. Fluid loss has a large effect on the shear rate distribution within the fracture, as differences in the fluid loss at different distances into the fracture must directly affect the flow rate through the fracture. Surprisingly, the fluid loss or leak-off is typically lowest directly adjacent to the wellbore [13], as here the filter cake has had most time to form. The highest leak-off occurs near the growing tip of the fracture that is furthest from the wellbore since this is where the filter cake has had least time to form. Such effects of fluid leak-off on the shear rate distribution have not yet been incorporated into current fracture designs. It is, however, well known that the shear rate distribution depends strongly on the fluid's rheological properties especially when a highly viscoelastic fluid is used [14, 15]. Current models used by the oil and gas industry typically assume a power law behavior for the shear rate-dependent viscosity of the fluids.

The temperature of the fracturing fluid as it passes down the fracture increases [16] with distance from the wellbore. The typical effect of temperature on the viscosity of fracturing gels is given in Figure 16.8 [11]. The sample is a 0.48 wt% of HPG solution, cross-linked with titanium dioxide nanoparticles and sheared at a constant shear rate of $37.7 \, s^{-1}$ using a viscometer equipped with a narrow gap Couette geometry (Fann Model 50). The sample was heated from 24 to 90 °C and subsequently cooled back to 24 °C while the viscosity of the sample was measured as a function of temperature.

The data show that the initial viscosity at 24 °C is lower than that obtained after the sample has been subjected to a heating and cooling cycle. This suggests that temperature enhances the cross-linking process (as described above) and leads to higher viscosity of the sample after cooling. Thermal thinning of viscosity was also observed during the heating cycle.

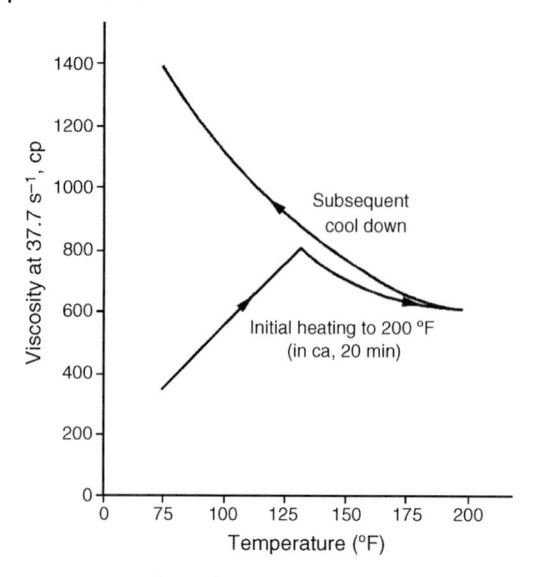

Figure 16.8 Effects of temperature history on the viscosity of 0.48 wt% of HPG solution cross-linked with titanium oxide nanoparticles [11].

16.2.2.1 Proppant Transport

The effect of proppant size on proppant settling is estimated by using the Stokes' law relationship for Newtonian and non-Newtonian fluids, assuming spherical particles. Several attempts have been made to use Stokes' law to describe particle-settling rates in cross-linked gels. Novotny [13] and Hannah and Harrington [17] measured particle settling in cross-linked water-based gels by using a rotating concentric cylinder and found the settling velocity decreased gradually as particle diameter, solid volume fraction and viscosity increased. Roodhart [18] used a parallel plate system at different gap separations to measure proppant-settling velocity and suggested incorporating zero shear viscosity to calculate the settling velocity. However, Roodhart's relationship is applicable for calculating settling velocity at shear rates only below $25\,s^{-1}$. At higher shear rates, Roodhart recommended using a three-parameter model such as the Ellis model to describe fluid behavior. Dunand and Soucemarianadin [19] reported that the rate of settling increased with increasing proppant concentration. Figure 16.9 shows experimental and calculated terminal-settling velocities [19] as a function of fluid viscosity and particle size.

16.2.3
Xanthan

Xanthan is regularly used in various aspects of oil recovery. Xanthan is most often used in solutions without cross-linking additives, but sometimes multivalent ions are used to create cross-links. In its solid state, xanthan exists as a fivefold helix with a pitch of about 4.7 nm. Its backbone is composed of β-D-glucan units, that is,

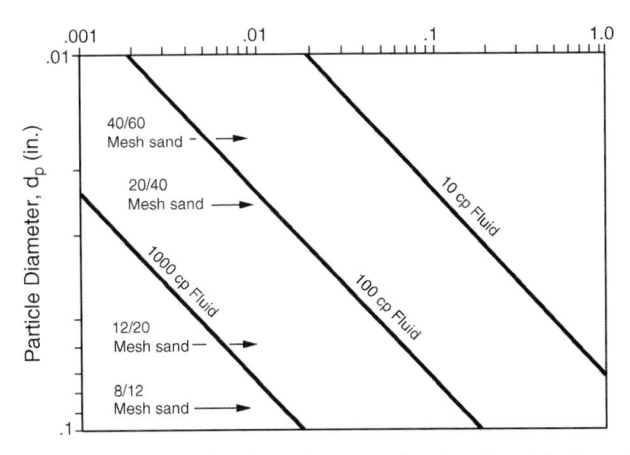

Figure 16.9 Terminal-settling velocity as a function of particle size and fluid viscosity.

essentially a cellulose chain having anionic trisaccharide units branching from alternate glucan rings. In solution, xanthan forms dimers although it is not clear whether the dimer is a coaxial double helix or a simple side-by-side association of single helices. These dimers, in turn, associate to give a network that weakens with increasing temperature. Addition of most salts raises the temperature at which this weakening occurs as the increased ionic strength reduces the electrostatic repulsion between the charged dimers, allowing them to associate more easily. Addition of Ca^{2+} to xanthan [20] results in a maximum in the elastic modulus of the system (Figure 16.10), attributed to association of Ca^{2+} between carboxyl groups on different chains, thus forming a cross-link.

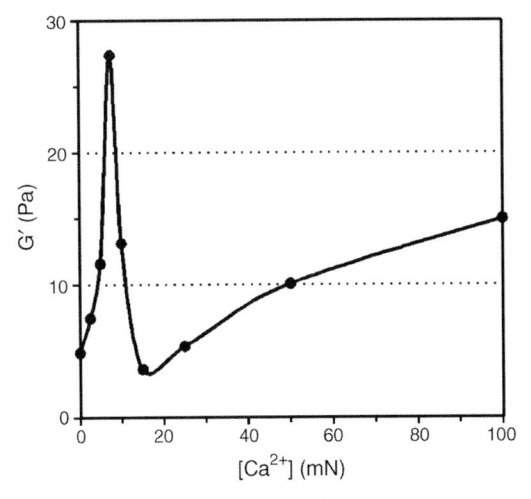

Figure 16.10 The effect of added Ca^{2+} on the elastic modulus of 0.5 wt% xanthan in 10 mN NaCl at 5 °C [20].

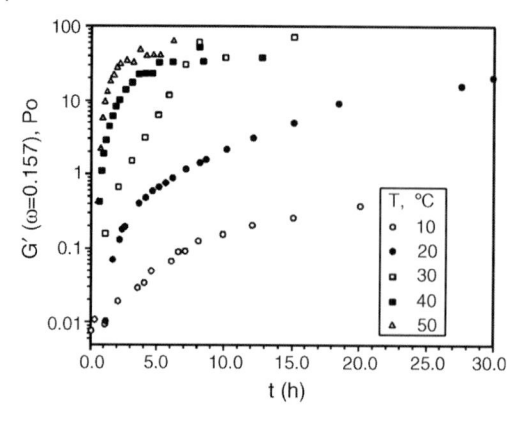

Figure 16.11 Storage modulus G' as a function of t_G – the reaction time after mixing for 0.125 wt% xanthan with 5 mM Cr^{3+}.

Trivalent ions are more effective than the divalent calcium, with Fe^{3+}, Cr^{3+}, and Al^{3+} all cross-linking xanthan [21]. The cross-linking reaction is strongly pH dependent, showing a maximum in the elastic modulus for wt% xanthan gels with 50 mM of the cations at pH: ~3 (Fe^{3+}), ~4 (Al^{3+}), or between 5 and 6 (Cr^{3+}). This pH influence on the gel strength suggests [22] that the cross-linking species is probably a dimer of the hydrated ion. The rate of gelation of xanthan with Cr^{3+} was found to increase with increasing xanthan concentration. Extrapolation of the gelling times indicated that a minimum xanthan concentration of ~0.035% was needed to form a gel. This critical concentration for gel formation is close to the overlap concentration C^* for xanthan.

The rate of gelation of xanthan increases with increasing temperature [21] for both Cr^{3+} and Al^{3+} as shown in Figure 16.11. This influence of temperature and of the xanthan and metal ion concentrations must be taken into account in designing a system for transporting proppant into fractures. The concentrations of xanthan and trivalent metal ions have to be chosen to fit the conditions of time of travel down the wellbore, temperature of the well and the porosity of the rock formation.

16.2.4
Gels for Gravel Packing

A further application of gelled systems occurs in gravel packing. This procedure is applied to wells in softer, poorly consolidated formations of higher permeability sandstones where the unwanted production of sand is a common problem. To avoid sand being removed from the formation and blocking the wellbore, it is common practice to place a slotted screen liner (roughly a perforated tube wrapped with wire to form a screen) within the wellbore and place sized gravel in the annulus between the liner and the outer wellbore surface. The gravel size (approximately five to six times the diameter of the formation sand) is chosen to be small enough to prevent the movement of sand from the formation to the wellbore, yet large enough to allow hydrocarbons to move from the formation to the wellbore. Gravel packing can be performed in vertical or horizontal wells and the gravel is usually carried by simple brines, the densities of which are designed to oppose the pore pressure of the formation. The gravel is carried down the annulus between the rock and the slotted liner until it reaches the end of the wellbore when the brine either leaks off to the formation or returns to the surface via the inside of the slotted liner. As horizontal wells become longer (up to 300–400 m) polymer gel systems (usually based on xanthan) are being used to carry the gravel the complete distance without it prematurely blocking the annulus. The slots are then opened and the carrying fluid passes through the slots back to the surface, leaving the gravel packed in the annular region. It is important for gravel packing that the transporting fluid is subsequently removed entirely from the residual gravel pack; otherwise, it will retard the flow of hydrocarbons when the well is put on production. So called "breakers" are thus added to the system to degrade the polymer gel after the gravel has been placed.

16.2.5
Gels for Fluid Loss Control

The earliest gels used for fluid loss control were starches that are still commonly used in drilling fluids. When drilling through the oil-bearing formation, special care is taken to minimize any damage to the freshly exposed surface. This damage can arise from fines or polymer gels becoming embedded in the surface of the formation and retarding the return flow of the hydrocarbons. For fluids used during drilling, an expanded form of starch granules provide a means of blocking small pores. Starch granules are generally 10–50 µm in diameter and are chemically treated to improve their stability with respect to high temperatures. The inside of a starch granule can be considered to have a gel structure; thus, a fluid containing starch granules is really an aqueous dispersion of microgel particles. The continuous fluid phase usually contains various salts used to raise its density so that the fluid pressure at the hydrocarbon-bearing rock layer exceeds the pressure from the earth's crust. The microgel particles reduce fluid loss of the continuous aqueous phase to the rock formation either by entering the pores of the rock or by forming a film on the rock surface, a filter cake, that retards fluid ingress to the rock.

Fluid loss control is improved if fine solid particles are included in the drilling fluid. However, these solids need to be of a material that can be removed later. Microgel suspensions containing small-sized solids probably hinder fluid loss since the microgel particles penetrate and block the space between the solids that aggregate on the rock surface. The fine solids used in conjunction with starch are generally calcium carbonate, which can be removed by acid treatment in preparation for hydrocarbon production.

An alternative approach is to make the microgels with degradable cross-links so that they dissolve over time. One example [23] is microgels made of polyacrylamide and cross- linked by molecules containing ester groups, which degrade at rates that are determined by the pH and/or temperature. Microgel particles made from about 5 wt% polyacrylamide and cross-linked using poly(ethylene glycol) diacrylate, with about one cross-link per 20 acrylamide units can make a satisfactory microgel that will self destruct under acid or alkaline conditions within a reasonable time.

A related system [24] uses microgel particles cross-linked by two different cross-linking agents, one stable and the other hydrolysable at elevated temperatures and/or increased pH. This system has been used to restrict the flow of water through so called "thief" zones that can occur in injector wells, which are wells where water is injected into a formation to push oil toward a nearby wellbore. Thief zones are channels of high permeability that can transport much of the water, preventing the "pushing" effect required to remove oil from the lower permeability zones. Using microgels with cross-links degrading over time, the microgel particles are initially small, allowing them to penetrate pores, especially in the thief zones. However, with time the labile cross-links degrade allowing the particles to swell, thus blocking the thief zone; this diverts more fluid through the remaining oil-bearing formation, improving the yield of hydrocarbons from the well.

16.2.6
Gels for Pills

Another use of gels for fluid loss control occurs when some temporary control of the well is required. An example of this is when a change of drilling or completion hardware is needed and small volumes of a gel, so called "pills," are used to fill the wellbore around a perforation. The gels may need to last several hours, which means that fluid loss from the gel to the formation must be minimized. In addition, the gel needs to be able to heal so that drill strings or other hardware can be moved through without permanently destroying the gel.

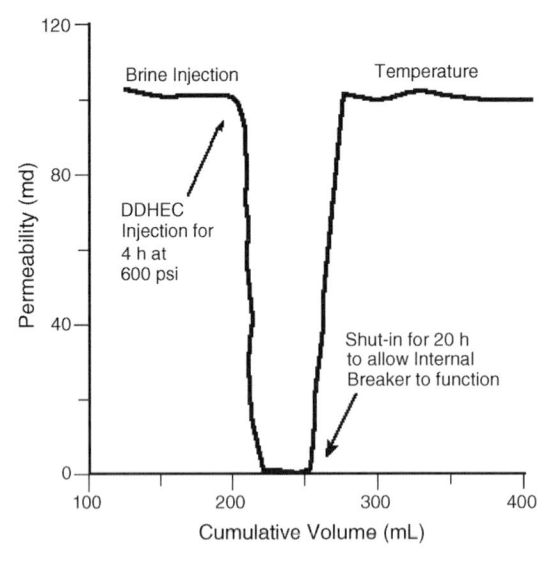

Figure 16.12 Change of permeability of a Berea core as a result of treatment with a cross-linked HEC derivative.

A gel system that fulfils these requirements has been achieved using a vinyl phosphonic acid derivative of hydroxyethyl cellulose – a double derived material double derived hydroxyethyl cellulose (DDHEC). This polymer is initially hydrated in water at low pH, after which magnesium oxide powder is added. The oxide dissolves in the acid, raising the pH of the system, until magnesium hydroxide begins to precipitate. At high magnesium concentrations, this process probably starts by homogeneous nucleation, resulting in many small hydroxide particles. These particles act as cross-linking agents for the DDHEC, which adsorbs via the phosphonate groups to the surface of the metal hydroxide particles. The resulting gel is stable at alkaline pH, has low fluid loss and can be removed using acid treatment. Figure 16.12 shows the effect [25] on the permeability of a Berea core treated with this DDHEC. Initially, when brine is injected through the core, the permeability of the core is high. The permeability, however, decreases when the DDHEC and magnesium oxide are injected and the hydroxide acts as a cross-linker for the polymer. After treatment with acid, the gel is dissolved and the permeability of the core recovers.

16.3
Concluding Remarks

The process of recovering oil from rocks is one of the most complex technical processes undertaken by any industry on daily basis. This is because of the great variety of temperatures, pressures and types of rock formations that are encountered. Microgels play a vital role in this process, where the main means of controlling their properties is the selection and design of the cross-linking agents. The kinetics of cross-linking need to take account of the time involved in drilling from the surface until the final stage of the oil-bearing formation is reached, as well as the associated changes in temperature and pressure. In many instances, the microgels have to contain the seeds of their own destruction, to allow maximum return flow of hydrocarbons.

Due to their rich range of behaviors and their response to external stimuli, microgels are already tremendously important in oil recovery. Future developments in the technology of oil recovery will undoubtedly further exploit the remarkably versatile behaviors of these materials.

References

1 Economides., M.J., Walters, L.T., and Dunn-Norman, S. (1988) *Petroleum Well Construction*, John Wiley & Sons.
2 Dormidontova, E. (2002) *Macromolecules*, **35**, 987.
3 McCleary, B.V. (1981) *Carbohydr. Res.*, **92**, 269.
4 Wientjes, R.H.W., Duits, M.H.G., Bakker, J.W.P., Jongschaap, R.J.J., and Mellema, J. (2001) *Macromolecules*, **34**, 6014.
5 Wang, Q., Ellis, P.R., and Ross-Murphy, S. (2006) *Carbohydr. Polym.*, **64**, 239; Wang, Q., Ellis, P.R., and Ross-Murphy, S., (2003) **53**, 75.

6 Kesavan, S., Neyraval, P., and Boukhelifa, A. US Patent US2006/0073988 A1.
7 Stromberg, J.L., Brown, D., and Curtice, R.J. SPE 21857. Paper presented at the Rocky Mountain Regional Meeting, 1991, Denver, CO.
8 Teraoka, I. (2002) *Polymer Solutions: An Introduction to Physical Properties*, John Wiley & Sons, Inc., New York.
9 Picout, D.R. and Ross-Murphy, S.B. (2007) *Carbohydr. Polym.*, **70**, 145.
10 Harris, P.C. Halliburton Technology Center, Duncan, OK, USA. Private communication.
11 Cameron, J.R. and Prud'homme, R.K. (1989) Rheology of Fracturing Fluids, in *Recent Advances in Hydraulic Fracturing* (eds J.L. Gidley, S.A. Holditch, D.E. Nierode, and R.W. Veatch Jr.), SPE Monograph, vol. **12**, Chapter 9, SPE, Richardson, TX.
12 Cheng, Y. and Prud'homme, R.K. (2000) *Biomacromolecules*, **1**, 782.
13 Novotny, E.J. (1977) Proppant transport. 52nd Annual Fall Technical Conference and Exhibition of the Society of Professional Engineers of AIME, Denver, CO, p. 12.
14 Patankar, N.A. and Hu, H.H. (2001) *J. Nonnewtonian Fluid Mech.*, **96**, 427.
15 Metzner, A.B., Cohen, Y., and Rangel-Nafaile, C. (1979) *Nonnewtonian Fluid Mech.*, **5**, 13.
16 Biot, M.A., Mass'e, L., and Medlin, W.L. (1987) *J. Petroleum Engineering*, **39**, 1389.
17 Hannah, R.R. and Harrington, L.J. (1981) *J. Petrol. Technol.*, **33**, 909.
18 Roodhart, L.P. (1985) SPE/DOE 13905 10.
19 Dunand, A. and Soucemarianadin, A. (1985) SPE 14259.
20 Mohammed, Z.H., Haque, A., Richardson, R.K., and Morris, E.R. (2007) *Carbohydr. Polym.*, **70**, 38.
21 Nolte, H., John, S., Smidsrod, O., and Stokke, B. (1992) *Carbohydr. Polym.*, **18**, 243.
22 Mesmer, R.E. and Bates, C.F. (1971) *J. Inorg. Chem.*, **10**, 2290.
23 Robb, I.D., Saini, R.K., Sarkar, D., Todd, B.L., and Griffin, J.M. (2007) US Patent 7,306,040.
24 Chang, K.Y., Frampton, H., and Morgan, J.C. (2007) US Patent 7,300,973.
25 Clay Cole, R., Ali, S., and Foley, K. (1995) SPE 29525.

17
Applications of Biopolymer Microgels
Eugene Pashkovski

17.1
Introduction

As a result of their structural versatility, rich dynamic behavior, biocompatibility, and sustainability, biopolymer microgels are employed in a wide range of industrial applications. Most biopolymers are linear or branched macromolecules forming microgel particles due to intermolecular binding. As was discussed in previous chapters, an essential property of microgel particles is their ability to swell, thus occupying a large volume in solution at very low-weight fractions of polymer. In the case of synthetic microgels, swelling is due to osmotic effects and the equilibrium swelling ratio is given by a balance between the entropic elasticity of the macromolecules, the polymer–solvent interactions [1–3], and, for flexible polyelectrolytes, the ionic strength and/or solution pH. By contrast, microgels of biological origin are composed of relatively rigid macromolecules; it is this intrinsic rigidity that largely determines the swelling of these biopolymer microgels.

Above a critical overlap, concentrated suspensions of microgel particles behave as "jammed materials" [4, 5], or more precisely, as "soft glassy materials" [6], which possess both a high elastic modulus and a yield stress. In the linear viscoelastic regime, the stress relaxation is very slow because the stress relaxes via collective rearrangements of the microgel particles. This makes microgels ideal for applications in consumer and food industries, where they are used to stabilize emulsions and suspensions and to control the rheological characteristics of the products. Microgel particles formed by biopolymers are preferred in pharmaceutical, food, and oral care industries due to their natural origin and biocompatibility [7–9]. In addition, microgels of biological origin have found applications in oil recovery because of their high solubility in water, low sensitivity to temperature, and high thickening potential. Biogels also find applications in drug delivery, particularly for oral [10] and ocular [11] delivery.

There are other important reasons for using microgels of biological origin. First, high prices of petrochemicals make synthetic microgels, such as polyacrylates or polyamides, less affordable. Second, growing interest in sustainable resources has become an important factor for the growing industrial consumption and applications of microgels. A preference for products made from natural ingredients is a consumer-driven trend for many product categories, such as shampoos, body wash liquids, lotions, creams, toothpastes, and mouthwashes. This is also true for foods, where biopolymers are preferred due to their ability to prevent crystallization and form glasses in low moisture products and frozen foods [12]. Most biopolymers have rigid chains; this reduces the mobility of polymer segments and promotes vitrification. Glassy biopolymer components affect the physical properties of most food products since they influence the physical and chemical stability and the crispness of the foodstuffs. In the food industry, biopolymer microgels are used in many different ways, depending on their function. In ice-cream, for instance, xanthan is used as an additive, as it prevents ice formation in the water-rich continuous phase [13]. The sensory perception of many microgel-containing food products depends on the behavior of the system under high shear rates [14]. In addition to influencing the rheology and texture, microgels of biological origin can also affect the formation of complexes with proteins [15] and the release of flavor in food products [16, 17].

It is not possible to cover all aspects related to microgels of biological origin in a single chapter; therefore, we will mainly give some specific examples of microgel applications in the stabilization of particle suspensions. Microgels based on polysaccharides produced by bacteria enjoy a unique position on the market. This is because bacteria-produced polysaccharides belong to the class of recoverable materials, as opposed to synthetic microgels or polysaccharides manufactured from plants and wood. Among these, microgels based on polysaccharides extracted from seaweeds, such as carrageenans [18] and agar (a mixture of agaropectin and agarose), are the most commonly produced and used. These polysaccharides can form microgels when the polymer solution is sheared during a sol–gel transition [19, 20]. The molecules of these polysaccharides are quite rigid and form double helices. Some sections of the chains form bundles, which play the role of cross-linkers between the chains [21]. In Europe, the use of carrageenan started more than 600 years ago [18] in the village of Carraghen on the South Irish coast where pies were made by cooking red seaweed species in milk. The world production of carrageenan is today based on farming seaweeds and exceeds 26 000 tons annually. In comparison, the production of the microbial polysaccharide xanthan gum via biotechnology exceeds 30 000 tons annually [22]. Both carrageenan and xanthan are used in many applications, including foods, oral care products, cosmetics, and drug delivery systems [7, 23–26]. Due to the abundance of carbohydrate sources and the well-established biotechnological routes for its production, xanthan gum is of great commercial significance, as demonstrated by the data in Table 17.1 [27–30]. Comprehensive information on production, molecular structure, and applications of xanthan gum is provided in Ref. [8]. Here, we focus on summarizing relevant information both on its molecular structure and on its microstructure.

Table 17.1 Main industrial applications of xanthan gum.

Industrial applications	Food and pharmaceutical applications
Abrasives (viscosity control)	Beer (foam stabilizer)
Ceramic glazes, polishes, thixotropic paints (stabilization, pseudoplasticity)	Cheese (syneresis inhibitor)
Explosives (gelling agent)	Ice-cream (stabilizer, crystallization control)
Fire-fighting fluids (foam stabilizer)	Salad dressing, bakery fillings (emulsifying agent)
Hydraulic fracturing (viscosity/cross-linking)	Jams, sauces (thickening agent)
Oil-drilling muds (viscosity control, shear thinning)	Icings and glazes (adhesive)
Water clarification (flocculants)	Powdered flavors (encapsulation)
Textile dyeing (pseudoplasticity)	Syrups (pseudoplasticity)

17.2
Origin, Production, and Molecular Properties of Xanthan Gum

Xanthan is an extracellular polysaccharide produced by the *Xanthomonas campestris* in special reactors with carefully adjusted concentration of nutrients (glucose, sucrose) and small amount of glutamate. The structure of the main chain of xanthan is identical to that of cellulose. The two pendant trisaccharide side chains (Figure 17.1)

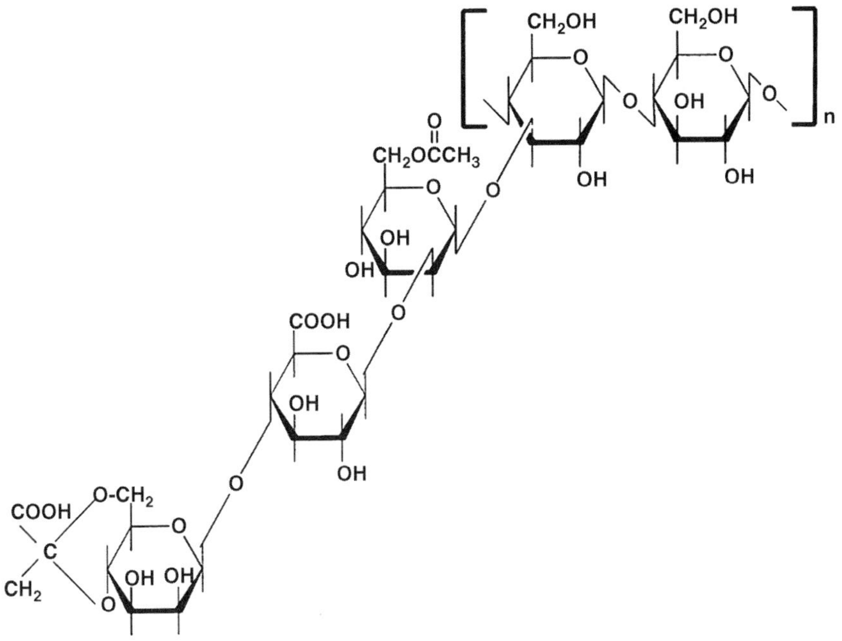

Figure 17.1 Molecular structure of xanthan.

are aligned closely to the main chain, making it extremely stiff as suggested by flow birefringence studies [31]. Isolated macromolecules of xanthan in water and salt solutions have exceptionally high persistence length values, $l_p = 120$ nm, which is consistent with its double helix structure [32]. This persistence length is roughly twice the persistence length of a DNA double helix and it is one order of magnitude higher than that of cellulose ($l_p = 13$ nm [32]). The trisaccharide side chains bear acidic groups, which ensure the solubility in water. Although the hydrodynamic radius of xanthan chains decreases with ionic strength [31], the optical birefringence remains unchanged, indicating that salts do not affect the chain rigidity. Unlike synthetic microgels based on cross-linking of flexible polyelectrolytes, the rheological behavior of xanthan is largely unaffected by the presence of salts; this makes them attractive for various applications, including dental creams, where fluoride salts are added for therapeutic reasons. In addition, xanthan is a nontoxic polymer that is approved by the Food and Drug Administration (FDA) without any specific quantity limitations [34]. The unique utility of xanthan microgels in various applications is due to the strong inter- and intramolecular aggregation. As shown by atomic force microscopy (AFM) [35–38], the entangled network of individual fibrous molecules is formed when a clear solution of xanthan is dried on a mica surface, as shown in Figure 17.2a, which is taken from Ref. [38]. The fibers in this image are individual xanthan helices and the white spots indicate locations were the profile is twice as high as an individual single chain; these are locations where one chain crosses another chain and correspond to cross-link points. Xanthan microgels thus form a network of long semirigid associative macromolecules, as shown in Figure 17.2b, which is also taken from Ref. [38]; the image corresponds to an area of $1.4 \times 1.4\,\mu m^2$ and was obtained from a dilute solution containing microgel particles.

Due to the intrinsic rigidity of xanthan macromolecules [31], the microgel particles also possess significant rigidity; from Figure 17.2a, which corresponds to an area of $1.2 \times 1.2\,\mu m^2$, the distance between cross-linking points is estimated to be ~300 nm. This distance is only ~2–3 times longer than the persistence length of xanthan [31]; this explains why the volume occupied by a single microgel particle is unaffected by ionic strength, temperature, or pH. This is in sharp contrast to the behavior of microgels based on flexible macromolecules [39]. Therefore, suspensions of xanthan microgel particles have high thickening potential over a wide range of pH, temperatures, and salt concentrations as well as in the presence of polyols (e.g., glycerol, sorbitol, etc.), which are used in oral care products either to reduce water activity and prevent bacterial growth or as moisturizers in skin care.

Typical measurements of the viscoelastic moduli of xanthan microgel suspensions in a 1:1 mixture of glycerol and water are shown in Figure 17.3a [40]. The values of the elastic modulus, G', are higher than the values of the loss modulus, G'', for xanthan concentrations, c, above a threshold, c_0, which defines the onset transition from a fluid to a jammed system. The slopes of the elastic moduli on the double log scale are $(d \log G'/d \log \omega) \approx 0.3$ in the entire frequency range, which indicates that the microgel particles are densely packed in this concentration range. Note that the value of $d \log G'/d \log \omega$ is associated with the balance between elastic energy storage and viscous losses, being nearly zero for absolutely elastic materials and close to 0.5

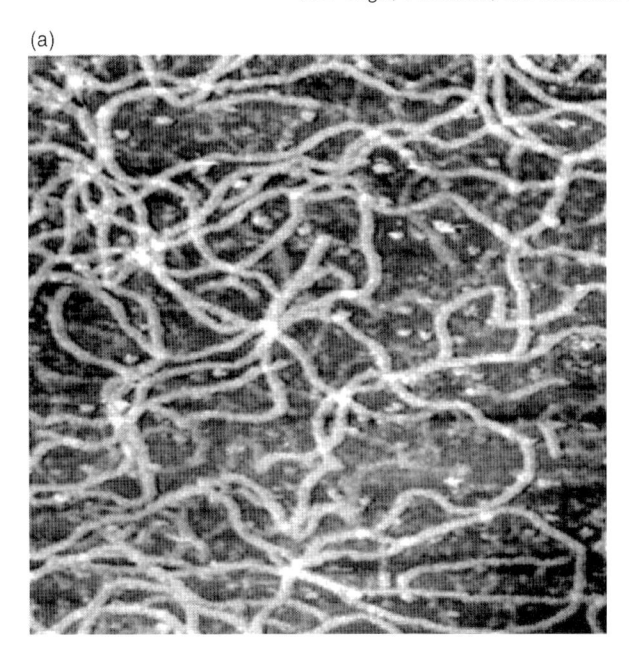

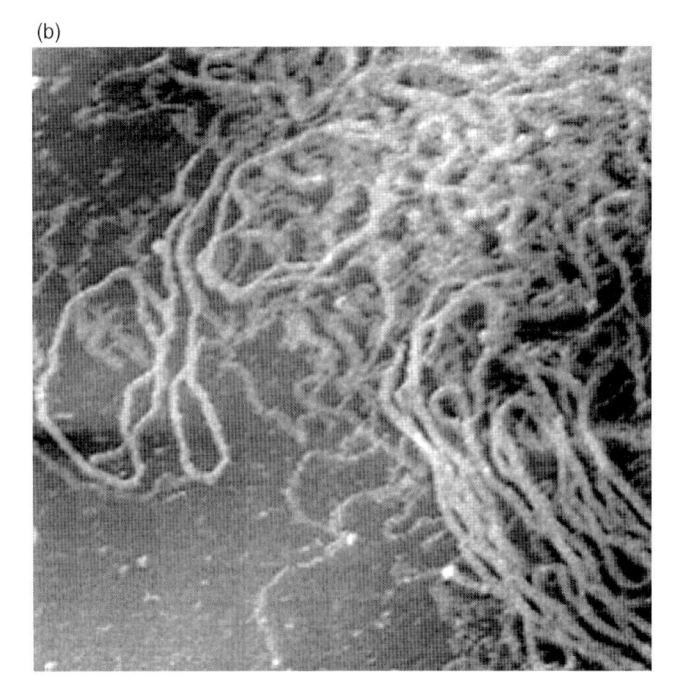

Figure 17.2 AFM images of aqueous xanthan samples. (a) Xanthan solution, scan size 1.2 × 1.2 μm. Note the bright spots on the molecules indicating doubling of the height as one chain crosses over another. (b) Xanthan microgel, scan size 1.4 × 1.4 μm. All images obtained by dc contact mode in butanol. Reproduced from Ref. [38], courtesy Elsevier.

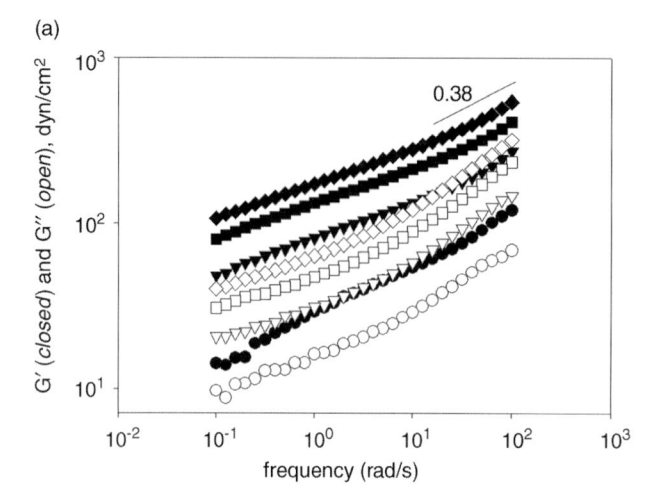

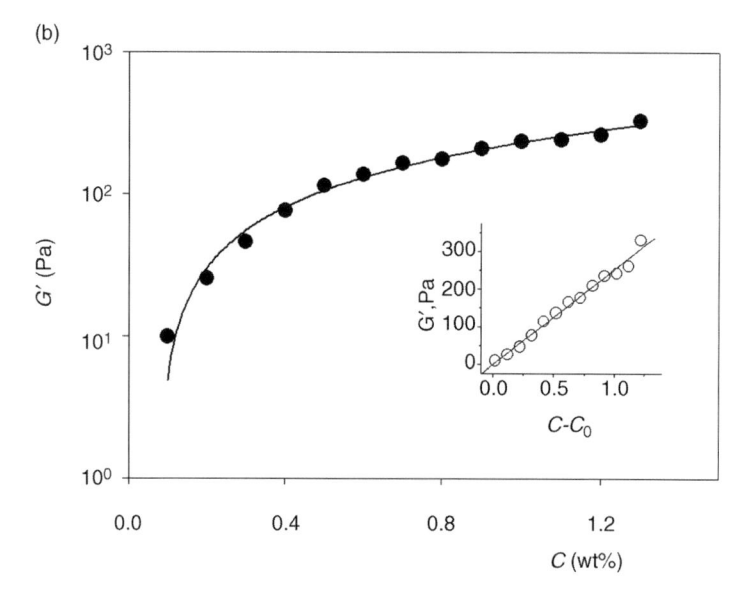

Figure 17.3 (a) The values of G' and G'' (filled and open symbols, respectively) measured as a function of frequency for xanthan solutions; $c = 0.3$ (○), 0.5 (△), 0.8 (□), and 1.1 wt% (◇). (b) Dependence of G' versus concentration (○, ●); solid line represents a fit to the equation $G' \propto (c - c_0)$ with $c_0 = 0.08 \pm 0.025\%$. Reproduced from Ref. [40], courtesy American Chemical Society.

for complex fluids near the gelation point. The elastic modulus depends linearly on $c - c_0$, where $c_0 = 0.08$ is the overlap concentration, as shown in Figure 17.3b. This behavior is very close to that of compressed emulsions [41], thus indicating that xanthan microgels behave as elastic repulsive particles. The stiffness of individual xanthan microgel particles is controlled by the high bending rigidity of the xanthan chains forming the microgels.

In Section 17.3, we will discuss the characterization of xanthan and carboximethyl cellulose (CMC) microgels using diffusing wave spectroscopy (DWS) and rheology. In particular, we will compare the rheology of xanthan microgels with that of linear CMC, which has as many industrial applications as xanthan; these include foods and oral care products, water-based paints, and detergents. However, the effect of CMC on the rheology of industrial products is quite different from that of xanthan because CMC is based on less rigid macromolecules. In Section 17.4, we compare the effects of CMC and xanthan on the rheology of suspensions of solid particles [40, 42, 43]. We emphasize that the methods for characterizing xanthan and CMC microgel suspensions can be applied to other strongly associating polysaccharides, such as carrageenans or gellan gum [44].

17.3
Characterization of Xanthan and CMC Microgels

To study xanthan microgels, as well as to illustrate the effect of these microgels have on the dynamic arrest of colloidal particles, we use DWS [40, 45, 46]. In contrast to classical dynamic light scattering, which probes the mobility of colloidal particles in dilute suspension in the single scattering limit, DWS allows to probe the mobility of particles in turbid systems (e.g., in the multiple scattering limit) provided that the dynamics is temporally and spatially homogeneous.

This technique is especially suitable for systems where the translational diffusion of particles is strongly suppressed, as DWS allows to measure very small displacements; this is especially relevant when colloidal particles are "caged" between microgels. The mean square displacement of the dispersed particles $\langle \Delta r^2(t) \rangle$ can be related to the field correlation function as

$$g_1(t) = \int_0^\infty P(S) \exp\left(-\frac{1}{3} k_0^2 \langle \Delta r^2(t) \rangle \frac{S}{l^*}\right) dS \qquad (17.1)$$

where $P(S)$ represents a distribution function for the light path length s through the sample, l^* is the mean free path of the photons, and $k_0 = 2\pi/\lambda$ is the wave number. DWS can be used in several geometries, for example, in transmission and backscattering geometries; the form of the function $P(S)$ in Equation 17.1 depends on the experimental geometry. By solving this equation numerically, one can determine $\langle \Delta r^2(t) \rangle$ using the experimentally measured $g_1(t)$, provided l^* is known. To determine l^*, a reference sample of known $\langle \Delta r^2(t) \rangle$ is required. As reference sample, one typically uses a relatively dilute suspension of colloidal particles, which is dilute enough for the particle motion to be diffusive but concentrated enough to be in a scattering regime fully dominated by multiple scattering. For such a sample, $\langle \Delta r^2(t) \rangle = 6Dt$, where D is the diffusion coefficient. Since the particle size is known, D can be determined from the Stokes–Einstein relation, which means that $\langle \Delta r^2(t) \rangle$ is also a known quantity. Thus, l^*_{REF} can be obtained from the measured field correlation function according to Equation 17.1. To obtain the mean free path of the photons in the sample of interest, l^*_{SAM}, one makes use of the fact that the transmitted intensity is approximately proportional to l^*. As a result, $l^*_{SAM}/l^*_{REF} = I_{SAM}/I_{REF}$, which allows the determination of l^*_{SAM}. In our experiments, we used three latex dispersions based on particles with sizes $d = 0.45$, 1.1, and 1.5 µm, which all corresponded to an $l^*_{REF} = 100–140$ µm. This allows the determination of l^* for all our samples of interest. Thus, from the measured $g_1(t)$, we can obtain $\langle \Delta r^2(t) \rangle$, which we use to calculate the macroscopic creep compliance, $J(t)$, from the generalized Stokes–Einstein relation (GSER) [47]:

$$\langle \Delta r^2(t) \rangle = J(t) \frac{k_B T}{\pi a} \tag{17.2}$$

where a is the particle radius and k_B is the Boltzmann's constant. The limitation of DWS for using GSER to determine $J(t)$ depends on several factors. Equation 17.2 assumes that the local viscoelasticity near the spheres is equivalent to the macroscopic viscoelasticity [47]. This assumption is valid if the particles are much larger than the typical length of the spatial inhomogeneities in the sample. This may not be true for heterogeneous strongly aggregated biopolymer solutions. Another source of GSER violation is spatially and temporally heterogeneous dynamics intrinsic to soft glassy materials. In order to analyze the limitations of GSER, it is necessary to compare $J(t)$ as obtained by DWS and rheology. In rheological experiments, the macroscopic creep compliance, $J(t)$, is defined as a response function to a suddenly applied stress, σ_0: $J(t) = \gamma(t)/\sigma_0$, and it characterizes viscoelastic materials. For small applied stresses with values below the yield stress, the response is purely elastic in the short time limit, $t \ll \tau_s$, where τ_s is the structural relaxation time of the system. However, for $t \gg \tau_s$, the creep compliance increases with a rate that is inversely proportional to the viscosity, $J(t) \propto t/\eta$.

The characteristic timescale in DWS is defined by the decay time of the intensity correlation function in Equation 17.1. This time is a function of the mean free path l^* and the particle mobility. In the semidilute regime, the mobility of dispersed tracers in xanthan and CMC solutions is still relatively high, and the samples remain ergodic and exhibit a decay time of several seconds. For further experimental details, the reader may consult Ref. [40]; here, we show only the main results of this work. To emphasize the difference between xanthan microgel suspensions and solutions of CMC, we show data for the mean square displacements in the presence of both systems.

In the presence of a 0.4 wt% CMC solution, microparticles of different sizes have values of $\langle \Delta r^2(t) \rangle \propto t^\alpha$, with α close to 1, as shown in Figure 17.4a. Surprisingly, the logarithmic slope $\alpha(t) \equiv d \ln \langle \Delta r^2(t) \rangle / d \ln t$ increases with particle size. For $t > 10$ ms, the slopes are found to be $\alpha \approx 0.8, 0.9$, and 1.05 for $d = 0.45, 1.1$, and 1.5 µm, respectively. This size dependence of α indicates the existence of structural heterogeneities of the CMC solutions at the length scale given by the size of the probe particles. In addition, the slightly subdiffusive motion of the 0.45 and 1.1 µm particles for $t \leq 10^{-1}$ s indicates that some particles could be trapped within aggregates formed

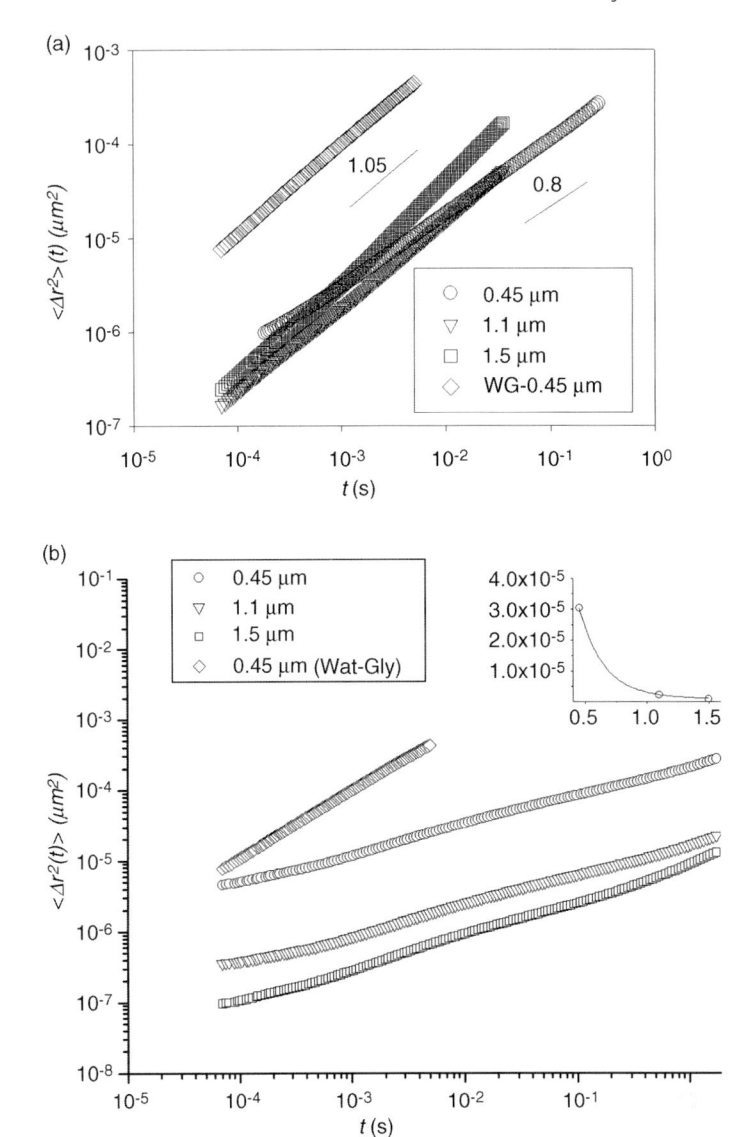

Figure 17.4 Characterization of the mobility of latex particles in CMC and xanthan solutions using DWS: the mean square displacements $\langle \Delta r^2(t) \rangle$ of carboxylated latex particles dispersed in 0.4% CMC solution (a) and 0.4% xanthan microgel paste (b), and the creep compliance for CMC (c) and xanthan (d) systems. The values of creep compliance were calculated from $\langle \Delta r^2(t) \rangle$ using Equation 17.2. The notations for particles of different size are $d = 0.45\,\mu m$ (○), $1.1\,\mu m$ (△), and $1.5\,\mu m$ (□). For comparison, the value of $\langle \Delta r^2(t) \rangle$ for $0.45\,\mu m$ particles in 50/50 water/glycerol mixture is presented (◇). Reproduced from Ref. [40], courtesy American Chemical Society.

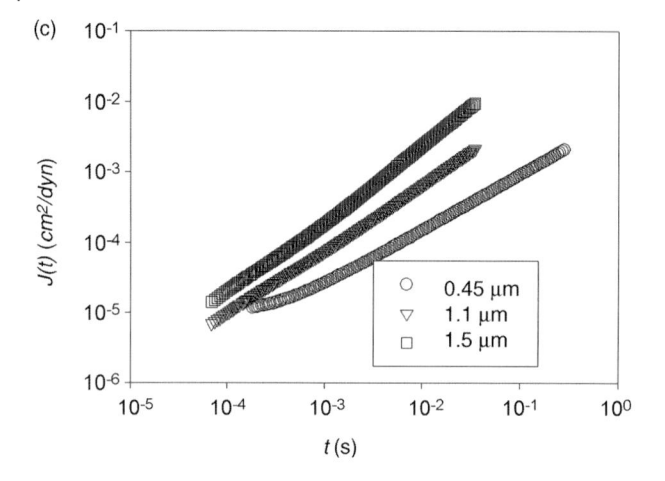

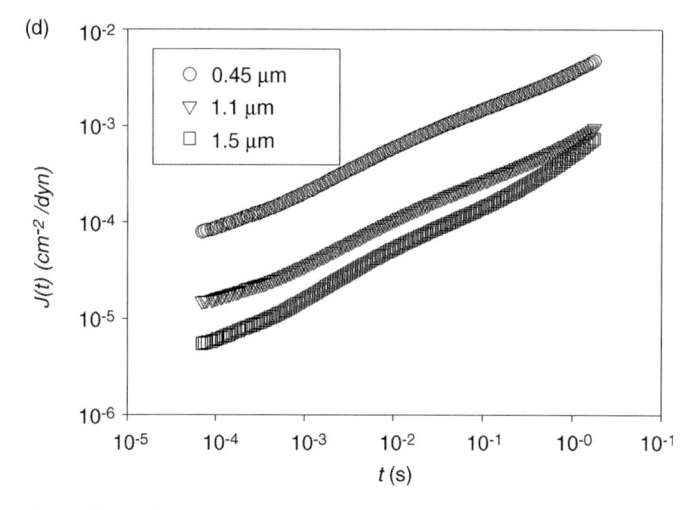

Figure 17.4 (Continued)

by CMC molecules. The slightly superdiffusive motion of the 1.5 μm particles shows that the dimensionality of the particle trajectories $d_t = 2/\alpha(t) \approx 1.9$ are lower than that for a Brownian trajectory, $d_t = 2$ [40]. For comparison, the dependence $\langle \Delta r^2(t) \rangle \propto t$, corresponding to particles experiencing Brownian diffusion, as obtained for a 1:1 water–glycerin mixture is also shown in Figure 17.4a. As a result of the structural inhomogeneities of the CMC solutions, the creep compliance estimated from Equation 17.2 is size dependent, as shown in Figure 17.4c. Particles with smaller sizes result in a smaller compliance; this indicates that in CMC solutions, small particles probe a more elastic environment than big particles, consistent with them being trapped within macromolecular aggregates of CMC.

By contrast to the behavior observed for CMC solutions, xanthan microgel suspensions are highly elastic even at relatively low concentrations. For a 0.4 wt%

xanthan suspension, the values of the mean square displacement for particles with $d = 0.45\,\mu m$ is two orders of magnitude smaller, even at short times (Figure 17.4b), than the values of the mean square displacements for particles with $d = 1.5\,\mu m$. The mobility of particles in xanthan suspensions decreases dramatically with particle size. In a homogeneous elastic medium, the value of $\langle \Delta r^2(t) \rangle$ would change with particle size as $\langle \Delta r^2(t) \rangle \propto 1/d$, according to Stokes law. By contrast, we find that the displacements decrease exponentially with increasing particle size, $\langle \Delta r^2(t) \rangle \propto e^{-d/\varsigma}$, with $\varsigma = 0.24\,\mu m$ for $c = 0.4\%$ (see insert in Figure 17.4b). The parameter ς sets the length scale for the spatial heterogeneities in xanthan suspensions. This result indicates that the probe particles are caged between xanthan microgels, creating local distortions and thus locally raising the stress; this explains the decrease in the creep compliance with particle size, as shown in Figure 17.4d. Thus, the mobility of colloidal (probe) particles dispersed in a xanthan paste is controlled by the rigidity of the microgels, consistent with the structure of xanthan microgel particles presented in Figure 17.2b. This also explains the low values of the logarithmic slopes of $\langle \Delta r^2(t) \rangle$ versus time for xanthan suspensions ($\alpha \sim 0.4$–0.5), indicating strongly subdiffusive motion of the probe, as shown in Figure 17.4c. Such low values of α, in combination with the small displacements observed, $\langle \Delta r^2(t) \rangle \approx 1\,nm$, indicate a very slow collective mobility of the xanthan paste. In the time window of our experiment, the system displays rubber-like behavior, whereas the behavior of the CMC solution is liquid-like. In both cases, the values of $J(t)$ calculated from Equation 17.2 depend on the probe particle size; however, the trends for CMC solutions and xanthan suspensions are totally different. In CMC solutions, larger particles result in higher values of $J(t)$ whereas in xanthan suspensions, smaller particles result in larger values of $J(t)$. Therefore, we conclude that xanthan suspension and CMC solutions have qualitatively different structures.

The linear scaling of G' with concentration observed for a xanthan paste differs from that of entangled polymer solutions in a good solvent, $G' \propto c^{9/4}$ [48]. At polymer concentrations significantly exceeding the value of critical overlap concentration ($c \gg c_0$), the elastic microgel particles of xanthan become compressed, and the storage and loss moduli are almost independent of frequency. At high frequencies, the slopes for G' and G'' are similar; for instance, $G'(\omega) \propto \omega^{0.38}$ and $G''(\omega) \propto \omega^{0.41}$ for $c = 0.3\%$, as shown in Figure 17.3a, and no high frequency crossover point is observed, at least for $\omega < 100\,rad/s$. The high-frequency behavior of xanthan pastes can be studied using the time–temperature superposition (TTS) principle since this system displays a thermorheologically simple behavior within the temperature range $T = 15$–$45\,°C$. Within this range, the viscoelastic moduli can be shifted using horizontal $a(T)$ and vertical $b(T)$ scaling factors. Such scaling yields a master curve that spans over a wider frequency range than the experimentally accessible one.

Figure 17.5 shows the master curves obtained for xanthan microgel solutions at $c = 0.6\,wt\%$ and $c = 1\,wt\%$; the use of TTS makes it possible to expand the upper frequency limit to $\omega\, a(T) = 10^3\,rad/s$. Even at $10^2 < \omega < 10^3$, the slopes of storage and loss moduli are quite small showing that the high-frequency modes associated with the bending of flexible polymer chains are not activated in this system, consistent with the high rigidity of xanthan chains [31]. In this case, the distance between the

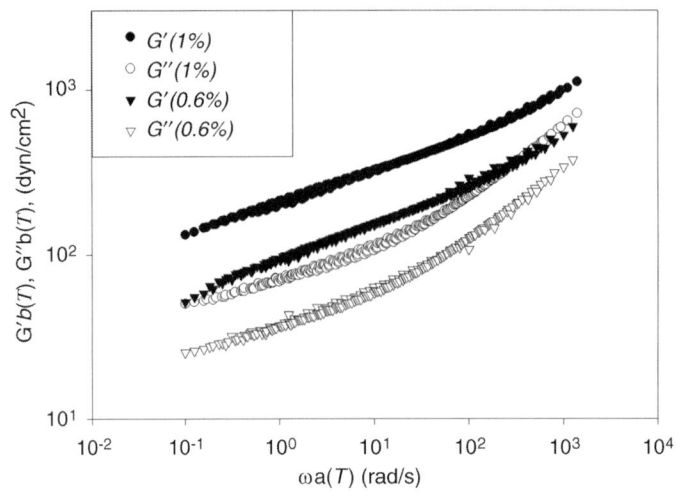

Figure 17.5 The master curves obtained using time–temperature superposition for xanthan solutions; $c = 0.6$ and 1% ($\bigcirc$, $\triangledown$ and $\bullet$, $\blacktriangledown$, respectively). The measurements of G' ($\bigcirc$, $\bullet$) and G'' ($\triangledown$, $\blacktriangledown$) were performed at $T = 15$–$45\,°C$. Reproduced from Ref. [40], courtesy American Chemical Society.

physical cross-links of the particles are significantly smaller than the persistence length of xanthan, which effectively suppresses the otherwise observable bending modes of the chains.

A microrheological approach can be used to reach a frequency range not accessible by mechanical rheology, even when TTS is taken advantage of. Figure 17.6a shows the values of G' and G'' obtained from DWS data for a $c = 0.27\,\text{wt\%}$ xanthan solution using 0.45 μm latex particles as probes. A good agreement between mechanical and DWS data for this particular system indicates that the 0.45 μm particles do not change the local elasticity of the paste and thus accurately provide the local elastic properties of the moderately concentrated paste. Larger particles would create excessive local stresses causing the viscoelastic moduli to be overestimated. In the range of $10^2 < \omega < 10^4\,\text{rad/s}$, the moduli scale as $G' \approx G'' \propto \omega^{0.45}$, which is consistent with the high-frequency behavior of rubber-like materials as predicted by the Mooney–Rivlin model [49]. In contrast to the xanthan microgel paste, the solution of CMC shows very different behavior even at $c = 0.4\,\text{wt\%}$, as shown in Figure 17.6b. There is a crossover at $\omega = 10\,\text{rad/s}$ where the loss modulus becomes larger than the storage modulus. Also, at very high frequencies, $3 \cdot 10^3 < \omega < 10 \cdot 10^3\,\text{rad/s}$, $G'(\omega) \propto \omega^{0.85}$, thus showing a larger power law exponent of ∼0.85 compared to the ∼0.45 obtained for xanthan; this supports the more flexible character of the CMC chains compared to the xanthan chains. In fact, the behavior for CMC microgels is consistent with the predictions of Morse [50], $G'(\omega) \propto \omega^{0.75}$, in terms of chain bending modes.

Despite the richness of information obtained using DWS, the microrheological analysis of xanthan pastes is limited to relatively low concentrations compared to c_0. For higher concentrations, the system becomes strongly nonergodic and dynamically

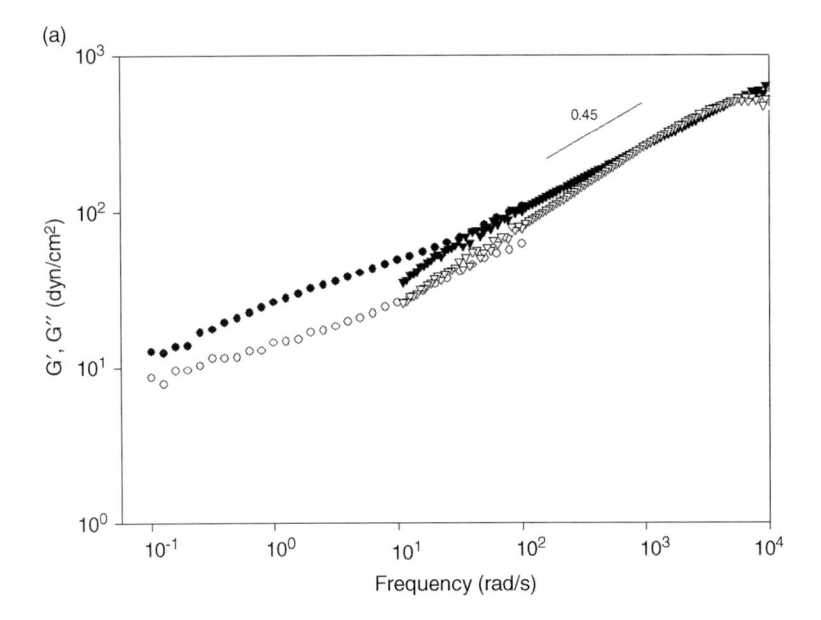

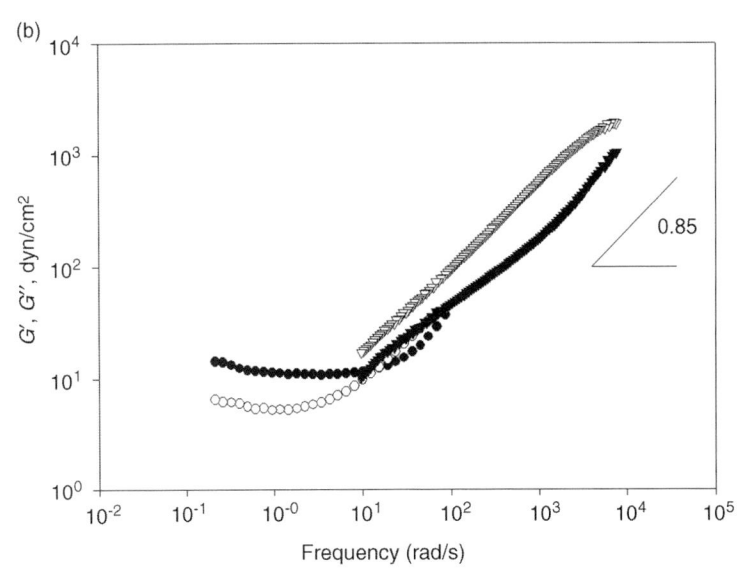

Figure 17.6 The storage moduli (●, ▼) and the loss moduli (○, ▽) for 0.27% xanthan (a) and 0.4% CMC (b) solutions. The circles and triangles correspond to the rheological and DWS measurements, respectively. At high frequency, the viscoelastic moduli of xanthan scale as $G' \simeq G'' \propto \omega^{0.45}$ and for CMC, $G' \propto \omega^{0.85}$ at $3 \cdot 10^3 < \omega < 10^4$ rad/s. Reproduced from Ref. [40], courtesy American Chemical Society.

heterogeneous [6, 51]. Moreover, the rheology of these systems becomes age dependent similar to that of polymeric glasses [52], and the dynamics becomes both temporally intermittent and spatially heterogeneous [51, 53–56]. These are general features of soft glassy materials, which have been described both theoretically [57–61] and experimentally [53, 62].

17.4
Rheology of Silica Suspensions in Xanthan Microgel Pastes

The striking difference in behavior for the CMC solutions and xanthan microgel pastes has dramatic implications in the rheological behavior of particle suspensions such as dental creams and other products with high-volume fractions of dispersed particles. Solid particles tend to aggregate and form gel networks with an elastic modulus that increases with the volume fraction of the dispersed particles, ϕ [63]. For particulate gels formed in Newtonian fluids, the elastic modulus scales as $G' \propto \phi^x$, with x around 4.0–5.0 [64–66]. The values of this exponent can be derived from theories that model the gel as a network of interconnected fractal clusters [63] and therefore, it reflects the type of colloidal aggregation (i.e., diffusion-limited or reaction-limited fractal aggregation) that takes place. For particles dispersed in concentrated biopolymer solutions, the rheology of the resulting particle gels depends on the contributions of both the viscoelastic continuous phase and the network formed by the particles. The problem of separating the contributions of the continuous phase and the particulate network is not a trivial one, but it may be important for complex industrial systems. The viscoelastic scaling algorithm developed by Trappe and Weitz [67] for particulate gels formed in Newtonian background fluids can serve to analyze the rheology of concentrated suspensions of silica particles in CMC solutions and xanthan microgels, although this scaling suggests that the background fluid contributes exclusively to the viscous (dissipative) component of the shear modulus while the particulate network is purely elastic.

The scaling implies that at some (high) frequency, the viscous contribution $G'' \propto \omega$ equals the elastic contribution $G' = G''(\omega_c) = \omega_c \eta$, which is almost frequency independent. The crossover frequency shifts to higher frequencies as the elastic modulus increases with ϕ. Therefore, by shifting the frequency dependences of the moduli, shown in Figure 17.7a, along frequency and moduli axes, a master curve results, as shown in Figure 17.7b. Despite some differences between the master curves for particulate gels formed in Newtonian fluids (see Figure 17.2, Ref. [67]) and those formed in CMC solutions, the linear relationship between the scaling factors still holds for both systems, as shown in Figure 17.7c. One relevant difference between the scalings pertains to the high-frequency slope of the loss modulus, $G'' \propto \omega^{0.56}$ for silica suspensions in CMC and $G'' \propto \omega$ for a Newtonian liquid. This difference reflects the high-frequency viscoelasticity of the CMC solution. Recently, a similar master curve was obtained for TiO_2 nanoparticles dispersed in a viscoelastic polymer melt [68]. In this case, similar to the CMC–silica system, the loss modulus was found to scale as $G'' \propto \omega^{0.6}$ at high frequencies. Thus, the difference between the

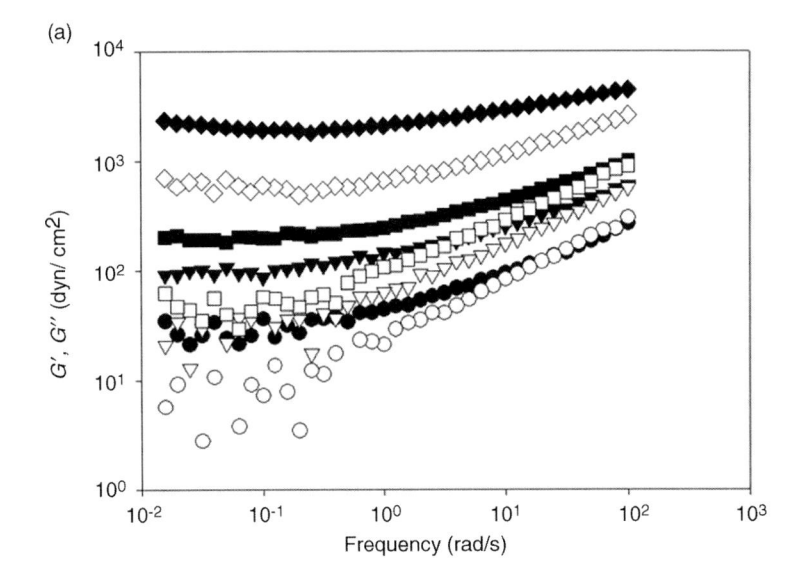

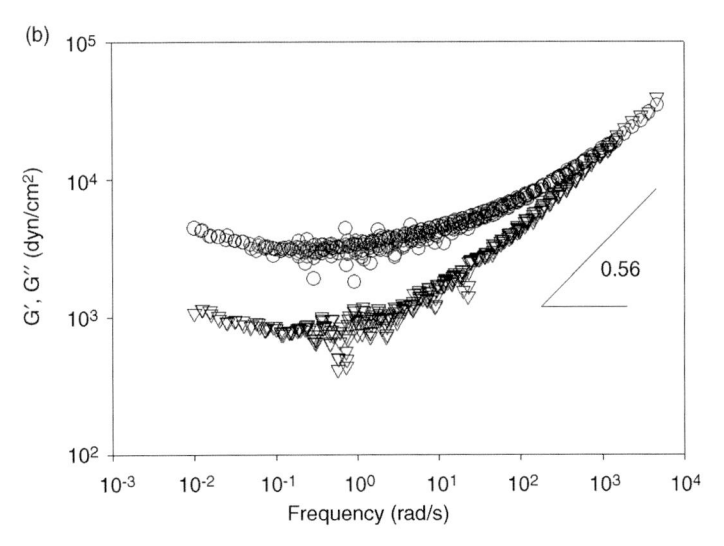

Figure 17.7 (a) The values of G' and G'' (●, ◆, ■, ▼ and ○, ◇, □, ▽, respectively) measured as a function of frequency for silica suspensions in 1% CMC solution with volume fractions $\phi = 0.101$ (●, ○), 0.127 (▼, ▽), 0.160 (■, □), and 0.195 (◆, ◇). (b) The master curve obtained by shifting the data along the moduli and frequency axes using two-dimensional minimization. Note that at high frequencies, $G'' \propto \omega^{0.56}$; this indicates the non-Newtonian character of losses for the polymeric background fluid. (c) The relationship between the vertical (b) and the horizontal (a) shift factors. Reproduced from Ref. [40], courtesy American Chemical Society.

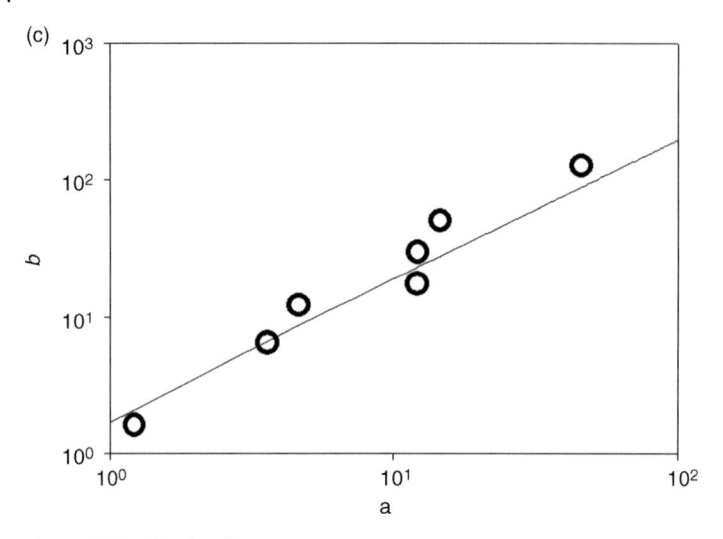

Figure 17.7 (Continued)

G''-scaling at high frequencies for colloidal gels formed in Newtonian [67] and polymeric fluids [40, 68] is due to the high-frequency rheological response of the continuous phase.

This further supports the approach by Trappe and Weitz based on the decoupling of viscoelastic responses of particulate gels and background fluids at high frequencies. Therefore, the scaling shown in Figure 17.7b is consistent with the dynamic decoupling of the continuous phase (e.g., CMC solution) and the colloidal network formed by the silica particles. Because of this decoupling, the colloidal network can consolidate [69] and eventually collapse under gravity [42, 70]. Interestingly, the particulate network collapses to a final volume fraction, ϕ_f, which depends, in a complex way, on the initial volume fraction of silica ϕ and the concentration of CMC solution, c_p. The final colloid volume fraction of the collapsed state for starting volume fractions $\phi = 0.11$ and $\phi = 0.055$, decreases as c_p is increased. The final state is denser for a starting volume fraction $\phi = 0.11$ than for $\phi = 0.055$, with ϕ_f approaching 0.20 at high polymer concentrations in both cases, independent of the initial volume fraction [42]. This suggests that samples with initial volume fractions above ϕ_f would become unstable at very long times. Such high loading of particles, however, may be not desirable due to cost limitations or because of the fact that the value of ϕ_f may be affected by other formula ingredients, such as salts and surfactants.

The model system used in Ref. [42] consists of 0.5 μm silica particles dispersed in a CMC solution ($M_w = 40$ kDa); the gravitational collapse of the gel network was observed for polymer concentrations $0.1 < c_p < 1.1$ wt%. Within this concentration range the polymer solution was essentially a viscous fluid exhibiting the typical viscosity dependence with concentration of polymer solutions, that is, $\eta \propto c_p^\alpha$, where α depends on the polymer–solvent interactions and on the interactions between the macromolecules. It is worth noting that the timescale associated with the dramatic collapse of the particle gel, τ_d, depends on polymer concentration, and this dependence exhibits certain similarities to the concentration dependence of the viscosity (see Figure 17.7, Ref. [42]). By contrast, the sedimentation velocity decreased with polymer concentration. However, when the velocity was normalized by the viscosity of the background fluid, it *increased* with polymer concentration reflecting the effect of interparticle attractions induced by the polymer. One possibility is that the attractive interaction due to depletion leads to growth of larger clusters that settle more rapidly.

The densification and collapse of the particle network lead to phase separation of the suspensions, as confirmed by direct confocal microscopy observations [42]. This occurs irrespective of the weak attractions between the silica particles, which still retain a relatively high mobility within the colloidal network formed in the presence of CMC solutions. As a result, this accelerated sedimentation can be explained by the densification of the particulate network during aging, which leads to the loss of connectivity between the clusters of colloidal particles. When the network is partially disrupted, the gel suddenly collapses because it can no longer support its own weight.

By contrast, the consolidation or compression of colloidal gels is a very slow process that depends on many parameters; therefore, it is poorly predictable. The interplay between the attractive forces induced by the linear biopolymers and the timescale of the onset of phase separation is rather complex and depends on the chemical nature of the continuous phase, that is, the presence of polyols (e.g., glycerol, sorbitol, etc.), the ionic strength, the size, and the density of particles and the surface chemistry of the particles. More recently, this qualitative picture was found to be consistent with the concept of gravitational compression [70], which sets a timescale for the sedimentation that increases with the viscosity of the suspending fluid and decreases with the permeability of the colloidal network.

The phenomenon of accelerated sedimentation and gel compression imposes significant limitations on applications of linear biopolymers in dental creams and other industrial suspensions. The standard solution to this problem is to increase the volume fraction of the dispersed particles to induce a kinetic arrest and thus increase the product shelf life. However, this route is not desirable due to both the increased viscosity and the cost of raw materials. As a result, instead of using linear biopolymers, xanthan microgels is used as it provides stability to the suspensions even at small concentrations of the dispersed particles. The microgels shown in Figure 17.2b have a highly porous rigid structure, which further stabilizes colloidal suspensions and gels. The obvious advantage of microgels for stabilizing particle suspensions can be demonstrated by comparing the master curves for silica particles dispersed in a 1 wt% CMC solution and a 1 wt% xanthan microgel paste. The particulate gels formed in the 1 wt% xanthan paste show no high-frequency crossover since the "background fluid" is highly elastic and the slopes of $G'(\omega)$ and $G''(\omega)$ are almost the same within the frequency window probed, 10^{-2}–10^2 rad/s, as shown in see Figure 17.8a. Therefore, no horizontal shifting is needed to scale the frequency dependences of the storage and loss moduli, and a vertical shift suffices, as shown in

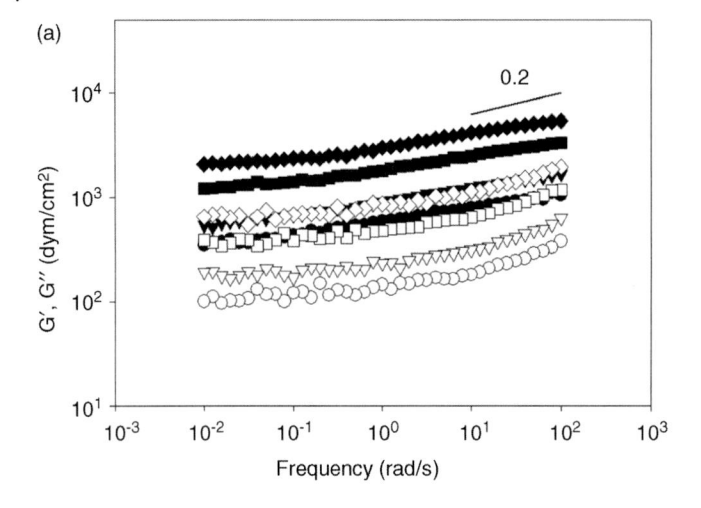

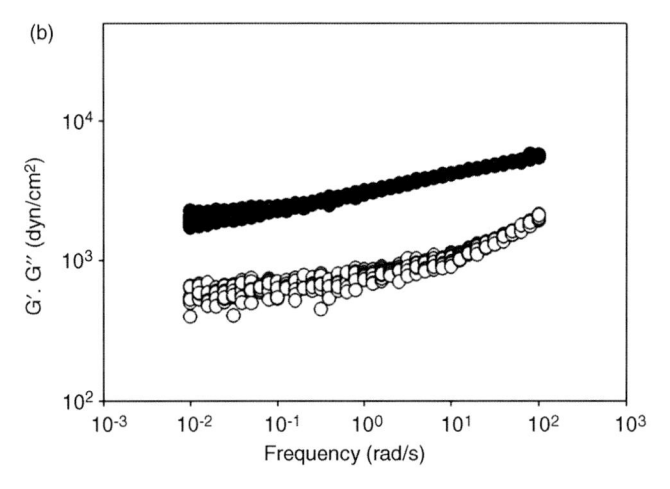

Figure 17.8 (a) The values of G' and G'' (●, ◆, ■, ▼ and ○, ◇, □, ▽, respectively) measured as a function of frequency for silica suspensions in 1% xanthan solution; $\phi = 0.114$ (●, ○), 0.146 (▼, ▽), 0.181 (■, □), and 0.209 (◆, ◇). (b) The master curve obtained by vertically shifting the data along the moduli axis. Note that G'' is almost frequency independent indicating very low level of viscous losses in the system. Reproduced from Ref. [40], courtesy American Chemical Society.

Figure 17.8b. This result can be explained by the fact that $G'(\omega) \propto \omega^{0.2}$ and $G''(\omega) \propto \omega^{0.21}$ for the 1 wt% xanthan paste without the dispersed particles (Figure 17.5), which remain essentially unaltered by the addition of particles (Figure 17.8a).

By looking at the vertical shift factors, we find that $b(\phi) \propto \phi^x$, with $x = 6.71 \pm 0.2$ for the CMC solution and $x = 3.15 \pm 0.2$ for the xanthan solution, as shown in

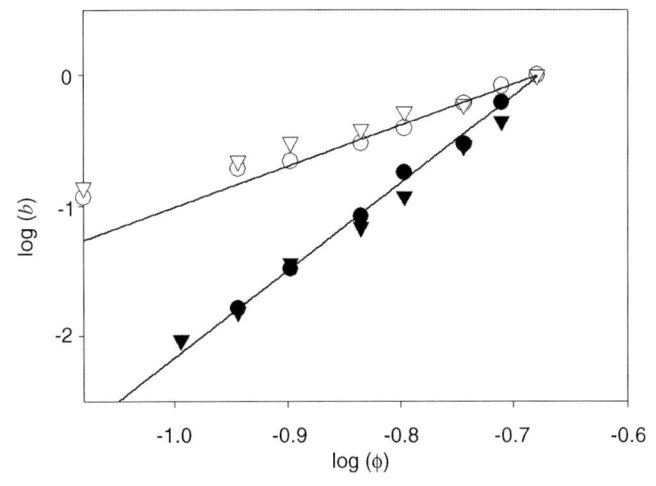

Figure 17.9 The vertical shift factors as a function of silica volume fraction for CMC (●, ▼) and xanthan (○, ▽) systems. Circles and triangles correspond to measurements before and after preshear, respectively. Reproduced from Ref. [40], courtesy American Chemical Society.

Figure 17.9. This power law dependence suggests the self-similarity of the particle networks. For xanthan-based systems, this behavior occurs for $\phi > 0.135$ suggesting that for $\phi < 0.135$ the particles do not form a fractal network. The exponent x can be associated with the fractal dimension of the gel, d_f, as $x = (2 + d_b)/(3 - d_f)$, where d_b is the fractal dimension of the backbone that supports the stress [66]. The backbone of the cluster is a fractal object for which the fractal dimension can be identified as $1 < d_b < d_f$. Based on the difference between the exponents, silica particles form denser clusters in CMC solutions than they do in xanthan pastes. However, the comparison based on this simplistic model of fractal colloidal aggregation may not be applicable to silica suspensions in xanthan gels, as the microgel particles are quite rigid and can themselves transmit forces and be compressed by the silica particles present in the suspension; this is in line with the stronger than $\sim 1/d$ size dependence found in DWS experiments, Figure 17.4b. Thus, the dispersions of solid particles in microgel pastes are expected to have much higher storage stability because the dynamics of the elastic particulate network and the microgels is strongly coupled. In addition, since the rigidity of xanthan molecules and the degree of chain aggregation is insensitive to temperature, pH, and ionic strength within a wide range of these parameters, commercial products with xanthan as suspending agent have excellent storage stability even at relatively low concentrations of xanthan. It is also worth mentioning that the rheology of xanthan changes with time after preparation of suspensions or emulsions. In general, the products thicken with aging. We now present some data related to this aging as studied using DWS [43, 71], small-angle dynamic light scattering [53], and interrupted creep measurements [43, 72]; conventional

17.5
Aging of Concentrated Xanthan Suspensions

The rheological properties of soft glassy materials, including microgel pastes, reflect their metastable structure and structural disorder [61]. Slow structural relaxation of such materials causes aging of the rheological response and mobility of the constituent particles. To study aging experimentally, one can measure the mobility of probe particles as a function of the time, t_w, elapsed since the sample was prepared or rejuvenated by shear. In Ref. [43], a $c = 1.45\%$ xanthan paste in a 1 : 1 water–glycerol solution was used to study the effect of t_w on the relaxation time, τ_S, estimated from the intensity correlation function, $g_2(t', t_w)$, measured using multispeckle DWS for different aging times, which we show in Figure 17.10a. In this case, we introduce a new time variable, $t' = t - t_w$, and fit the correlation function using the function $g_2(t', t_w) = g_2(0, t_w) \cdot \exp\{-[t'/\tau_s]^p\}$ [53]; the fits

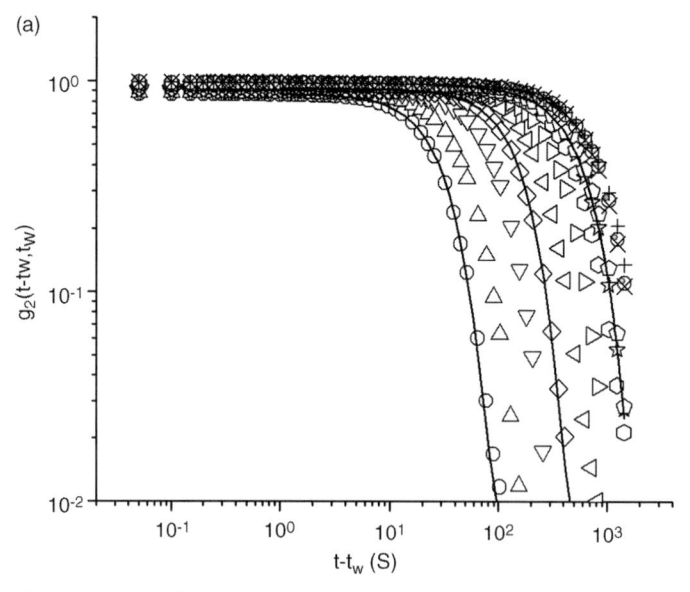

Figure 17.10 Xanthan paste (1.5 wt.% in 1 : 1 water–glycerin mixture): (a) the intensity correlation functions recorded at different aging times t_w; (b) the intensity correlation functions scaled using the reduced time $(t-t_w)/\tau$; and (c) the dependence of relaxation time versus aging time. Relaxation times were obtained by fitting the intensity correlation curves to the "compressed" exponential function (solid lines in Figure 17.10a). Reproduced from Ref. [43], courtesy American Chemical Society.

(b)

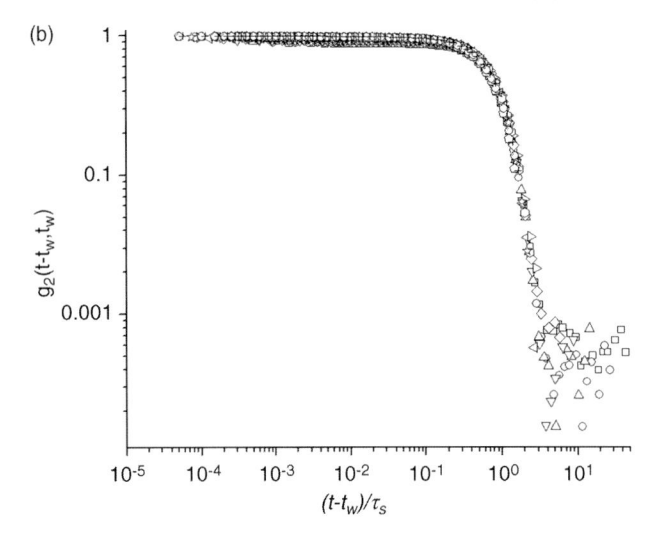

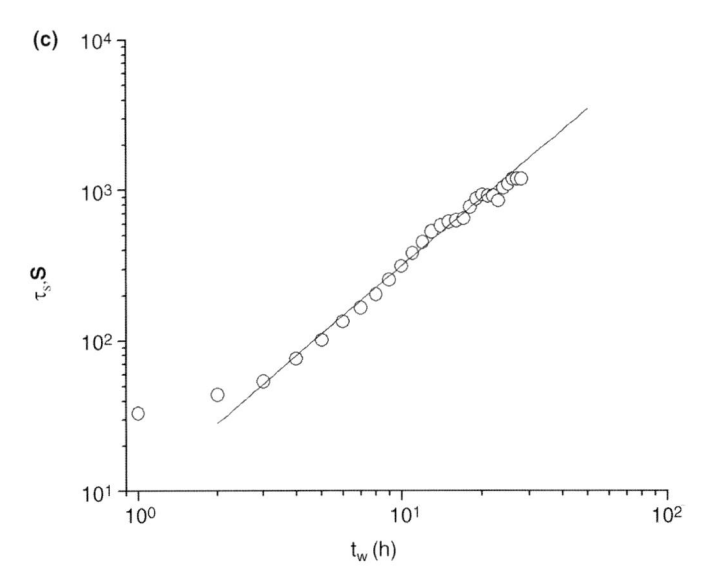

(c)

Figure 17.10 (Continued)

describe the data with "compressing" exponents $1.45 < p < 1.65$. Note that this is in contrast to what is usually found for colloidal glasses, which exhibit a "stretched" rather than a "compressed" exponential behavior. For the rigid xanthan microgels under consideration, however, the stress exerted on the particle by the presence of the neighbors causes collective rearrangements that result in exponents $p > 1$. Similar results were recently obtained for a soft glass made of a lamellar onion phase [73]. Theoretically, it was found that an exponent $p = 1.5$ can result from the presence of slow rearrangement events, which occur randomly in space, and can create long-ranged elastic strains in the system [74]; these strains create local dipolar forces that must be taken into account for deriving the equation of motion of the strain field. As a result, the dynamic structure factor associated with the rearrangement events involving the particles becomes $f(q,t) \propto \exp[-A(qt)^{1.5}]$, which is a compressed exponential. In addition, within this theoretical picture, the relaxation time scales inversely proportional to the scattering vector, $\tau_s \propto q^{-1}$ [53], by contrast to the usual diffusive scaling, $\tau_s \propto q^{-2}$; this emphasizes the ballistic character of these particle rearrangements.

As aging proceeds, xanthan microgel suspensions display a significant decrease in mobility [43], similar to compressed emulsions and onion phases [53], polymeric [52] and magnetic [75] glasses. By using the reduced timescale $(t-t_w)/\tau_s$, the curves for the dynamic structure factor shown in Figure 17.10a, all collapse onto a single master curve, as shown in Figure 17.10b. Interestingly, we find that for xanthan microgel suspensions, $\tau_S \propto t_w^\mu$, with $\mu = 1.5$, as shown in Figure 17.10c. This is consistent with what is found for Laponite [76] and for dense colloidal gels [77], but it is in contrast to the linear τ_s versus t_w dependence found for most common materials exhibiting glass-like aging phenomenon [52] and for some colloidal systems [78], which can also exhibit a behavior characterized by $\mu < 1$. Possibly, the particular value of μ depends, among other factors, on the sample preparation protocol or sample history and on the method used for determining the structural relaxation time. For instance, for a xanthan microgel paste, the values of the aging exponent obtained by DWS and rheology do not agree with each other [43].

The effect of aging on the rheological properties of xanthan microgel pastes can also be analyzed using the experimental protocol first developed by Cloitre et al. [72]. In this protocol, an initial preshear sets the sample age, followed by creep measurements using a very small constant value of the shear stress, $\sigma_0 \sim 1\,\mathrm{dyn\,cm^{-2}}$, compared to the yield stress, σ_Y, of the microgel suspension. As shown in Ref. [72] for a carbopol microgel paste, the aging exponent decreases from $\mu = 1$ (full aging), when $\sigma_0 \gg \sigma_Y$, to $\mu = 0$ (no aging), for $\sigma_0 \geq \sigma_Y$. For a $c = 1.5\,\mathrm{wt\%}$ xanthan paste [43], the yield stress was found to be around $\sim 15\,\mathrm{dyn/cm^2}$; thus, for an applied stress of $\sigma_0 \sim 1\,\mathrm{dyn\,cm^{-2}}$, the system remains solid-like and exhibits aging with time. This is clearly manifested in the creep response curves, as shown in Figure 17.11a. Interestingly, these curves can be scaled onto a single master curve by introducing horizontal shift factors that depend on sample age, as shown in Figure 17.11b. This scaling can be described by $J(t', t_w) \propto t'/t_w^{0.86}$, which quantifies the effects of aging over the relaxation time of the system. Alternatively, the relaxation time can be determined for each creep experiment separately by fitting the curves to the single-element Voight model, $J(t) = J_0 + \mathrm{const} \cdot [1-\exp(-t/\tau_s)]$. In this case, we obtain that the relaxation time changes almost linearly with aging time, $\tau_s \propto t_w^{0.95 \pm 0.1}$, as shown in Figure 17.11c. Note that these dependences are different from those presented above, despite the xanthan concentration being almost the same; these discrepancies most likely result from differences in sample history and emphasize the degree of care required in order to obtain reproducible results.

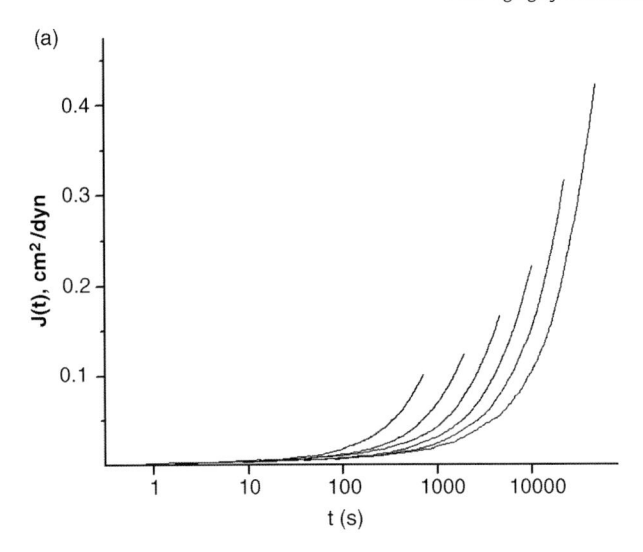

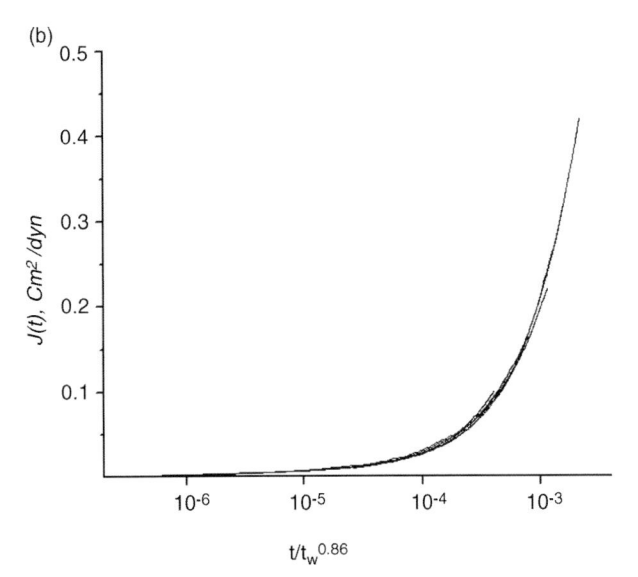

17.5 Aging of Concentrated Xanthan Suspensions

Figure 17.11 (a) Creep compliance curves recorded at different t_w using a small stress (1 dyn/cm^2); (b) the scaled creep compliance curves; and (c) the dependence of creep relaxation time versus aging time. Reproduced from Ref. [43], courtesy American Chemical Society.

The study of aging effects in microgel suspensions can be extended to the characterization of more complex systems containing microgels, for instance, shampoos, body wash formulations, lotions, dental creams, foodstuff, and so on. This may be useful for understanding the structural evolution of products containing microgels during the initial period of the product shelf life.

(c)

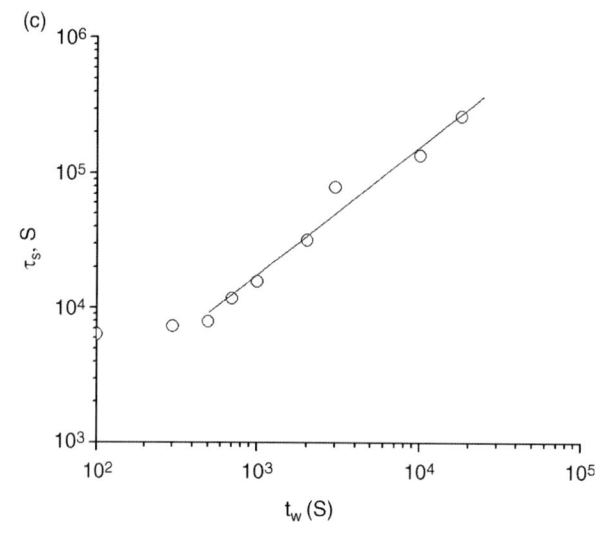

[Graph: τ_s, s vs t_w (S), log-log plot showing data points rising from ~10^4 at t_w=10^2 to ~3×10^5 at t_w=10^5]

Figure 17.11 (Continued)

17.6
Conclusions

Microgels are convenient systems for controlling stability and rheology of industrial suspensions and emulsions. Microgels of biological origin are particularly interesting materials as they represent a class of renewable materials [79] used in consumer and food industries in large volumes. Among extracellular polysaccharides, xanthan is the most commercially produced industrial gum obtained by fermentation. In addition, since it can be manufactured from many hydrocarbon sources [22], it provides excellent opportunities for reducing environmental pollution and partially replace petroleum-based microgels. Xanthan gum is particularly interesting because its thickening properties are not affected by temperature, pH, or ionic strength, as a result of the extreme rigidity of xanthan macromolecules. Therefore, xanthan gum is a very good suspending agent, as the dispersed particles have very low mobility.

The measurement of particle mobility in a xanthan paste using DWS suggests strongly subdiffusive behavior in contract to the nearly diffusive regime of the particles in CMC solutions.

The rheological characterization of concentrated suspensions of silica in xanthan pastes and CMC solutions can be performed in the spirit of the viscoelastic scaling developed for colloidal gel networks formed in Newtonian fluids [67]. For colloidal networks formed in polymeric suspending fluids such as CMC solutions and xanthan pastes, this scaling suggests that, in contrast to CMC-based colloidal gels where the dynamics of the continuous phase and the colloidal network are decoupled, the dynamics of particles and xanthan microgels is strongly coupled. This leads to a dramatic difference in the stability of colloidal gels formed in xanthan pastes and CMC solutions. In the latter case, the gravitational collapse of gels is a direct consequence of gel consolidation and high mobility of silica particles. In contrast to the case for CMC, xanthan microgel pastes are preferred for improved stability of suspensions and gels. Aging of xanthan microgel pastes can be followed using rheology and DWS techniques. The rheological creep response of xanthan pastes depends on the aging time, as for carbopol microgels [72]; the creep curves can be scaled together if one uses the age-dependent relaxation time as a scaling factor. The relaxation time increases with age as $\tau_s \propto t_w^{0.95 \pm 0.1}$. The multispeckle camera-based DWS provides a noninvasive method to study aging of xanthan pastes. In this case, the intensity correlation function is used to probe the mobility upon aging. This function has a compressed exponential form, which has also been observed for other soft glassy materials.

References

1 Flory, P.J. (1990) *Principles of Polymer Chemistry*, Cornell University, Ithaca, New York.
2 Sperling, L.H. (2006) *Introduction to Physical Polymer Science*, 4th edn, Wiley Interscience.
3 Strobl, G.R. (2007) *The Physics of Polymers: Concepts for Understanding their Structures and Behavior*, 3rd rev. and expanded edn, Springer-Verlag, Berlin
4 Liu, A.J. and Nagel, S.R.(eds) (2001) *Jamming and Rheology: Constrained Dynamics on Microscopic and Macroscopic Scales*, Taylor & Francis, Inc.
5 Miguel, M.C. and Rubí, J.M. (2006) *Jamming, Yielding, and Irreversible Deformation in Condensed Matter, Lecture Notes in Physics*, Springer-Verlag, Berlin.
6 Sollich, P. (1998) Rheological constitutive equation for a model of soft glassy materials. *Phys. Rev. E*, **58**, 738.
7 Becker, A., Katzen, F., Puhler, A., and Ielpi, L. (1998) Xanthan gum biosynthesis and application: a biochemical/genetic perspective. *Appl. Microbiol. Biotechnol.*, **50**, 145.
8 Garcia-Ochoa, F., Santos, V.E., Casas, J.A., and Gomez, E. (2000) Xanthan gum: production, recovery, and properties. *Biotechnol. Adv.*, **18**, 549.
9 Katzbauer, B. (1998) Properties and applications of xanthan gum. *Polym. Degrad. Stabil.*, **59**, 81.
10 Needleman, I.G., Smales, F.C., and Martin, G.P. (1997) An investigation of bioadhesion for periodontal and oral mucosal drug delivery. *J. Clin. Periodontol.*, **24**, 394.
11 Burgalassi, S., Chetoni, P., and Saettone, M.F. (1996) Hydrogels for ocular delivery of pilocarpine: preliminary evaluation in rabbits of the influence of viscosity and of drug solubility. *Eur. J. Pharm. Biopharm.*, **42**, 385.
12 Tolstoguzov, V.B. (2000) The importance of glassy biopolymer components in food. *Nahrung: Food*, **44**, 76.
13 Goff, H.D. (1997) Colloidal aspects of ice cream: a review. *Int. Dairy J.*, **7**, 363.
14 de Vicente, J., Stokes, J.R., and Spikes, H.A. (2006) Soft lubrication of model hydrocolloids. *Food Hydrocolloid.*, **20**, 483.
15 Burova, T.V., Grinberg, N.V., Grinberg, V.Y., Usov, A.I., Tolstoguzov, V.B., and de Kruif, C.G. (2007) Conformational changes in iota- and kappa-carrageenans induced by complex formation with bovine beta-casein. *Biomacromolecules*, **8**, 368.
16 Malone, M.E., Appelqvist, I.A.M., and Norton, I.T. (2003) Oral behaviour of food hydrocolloids and emulsions. Part 2. Taste and aroma release. *Food Hydrocolloid.*, **17**, 775.
17 Malone, M.E. and Appelqvist, I.A.M. (2003) Gelled emulsion particles for the controlled release of lipophilic volatiles during eating. *J. Control Release*, **90**, 227.
18 van de Velde, I.F. and de Ruiter, G.A. (2005) Carrageenan, in *Polysaccharides and*

Polyamides in the Food Industry (eds A. Steinbüchel and S.K. Rhee), Wiley-VCH Verlag GmbH, Weinheim, p. 85.

19 Altmann, F.N., Cooper-White, J.J., Dunstan, D.E., and Stokes, J.R. (2004) Strong through to weak 'sheared' gels. *J. Nonnewton. Fluid Mech.*, **124**, 129.

20 Stokes, J.R. and Frith, W.J. (2008) Rheology of gelling and yielding soft matter systems. *Soft Matter*, **4**, 1133.

21 Burchard, W. (2007) Macrogels, microgels and reversible gels: what is the difference? *Polym. Bull.*, **58**, 3.

22 Silva, M.F., Fornari, R.C.G., Mazutti, M.A., de Oliveira, D., Padilha, F.F., Cichoski, A.J., Cansian, R.L., Di Luccio, M., and Treichel, H. (2009) Production and characterization of xantham gum by *Xanthomonas campestris* using cheese whey as sole carbon source. *J. Food Eng.*, **90**, 119.

23 Sandford, P.A., Cottrell, I.W., and Pettitt, D.J. (1984) Microbial polysaccharides: new products and their commercial applications. *Pure Appl. Chem.*, **56**, 879.

24 Plank, J. (2004) Applications of biopolymers and other biotechnological products in building materials. *Appl. Microbiol. Biotechnol.*, **66**, 1.

25 Giavasis, I., Harvey, L.M., and McNeil, B. (2000) Gellan gum. *Crit. Rev. Biotechnol.*, **20**, 177.

26 Mott, C.L., Hettiarachchy, N.S., and Qi, M. (1999) Effect of xanthan gum on enhancing the foaming properties of whey protein isolate. *J. Am. Oil Chem. Soc.*, **76**, 1383.

27 Druzian, J.I. and Pagliarini, A.P. (2007) Xanthan gum production by fermentation from residue of apple juice. *Cienc. Tecnol. Alim.*, **27**, 26.

28 Rosalam, S. and England, R. (2006) Review of xanthan gum production from unmodified starches by *Xanthomonas comprestris* sp. *Enzyme Microb. Technol.*, **39**, 197.

29 Woiciechowski, A.L., Soccol, C.R., Rocha, S., and Pandey, A. (2004) Xanthan gum production from cassava bagasse hydrolysate with *Xanthomonas campestris* using alternative sources of nitrogen. *Appl. Biochem. Biotechnol.*, **118**, 305.

30 Sanderson, G.R. (1981) Applications of xanthan gum. *Brit. Polym. J.*, **13**, 71.

31 Yevlampieva, N.P., Pavlov, G.M., and Rjumtsev, E.I. (1999) Flow birefringence of xanthan and other polysaccharide solutions. *Int. J. Biol. Macromol.*, **26**, 295.

32 Sato, T., Norisuye, T., and Fujita, H. (1984) Double-stranded helix of xanthan: dimensional and hydrodynamic properties in 0.1-M aqueous sodium-chloride. *Macromolecules*, **17**, 2696;(1985) Double-stranded helix of xanthan: dissociation behavior in mixtures of water and cadoxen. *Polym. J.*, **17**, 729.

33 Hoogendam, C.W., de Keizer, A., Stuart, M.A.C., Bijsterbosch, B.H., Smit, J.A.M., van Dijk, J.A.P.P., van der Horst, P.M., and Batelaan, J.C. (1998) Persistence length of carboxymethyl cellulose as evaluated from size exclusion chromatography and potentiometric titrations. *Macromolecules*, **31**, 6297.

34 Kennedy, J.F. and Bradshaw, I.J. (1984) Production, properties and applications of xanthan. *Prog. Ind. Microbiol.*, **19**, 319.

35 Morris, V.J. (1994) Biological applications of scanning probe microscopies. *Prog. Biophys. Mol. Biol.*, **61**, 131.

36 Morris, V.J. and Wilde, P.J. (1997) Interactions of food biopolymers. *Curr. Opin. Colloid Interface Sci.*, **2**, 567.

37 Morris, V.J., Kirby, A.R., and Gunning, A.P. (1999) Using atomic force microscopy to probe food biopolymer functionality. *Scanning*, **21**, 287.

38 Morris, V.J., Mackie, A.R., Wilde, P.J., Kirby, A.R., Mills, E.C.N., and Gunning, A.P. (2001) Atomic force microscopy as a tool for interpreting the rheology of food biopolymers at the molecular level. *Lebensm. Wiss. Technol.*, **34**, 3.

39 Das, M. and Kumacheva, E. (2006) From polyelectrolyte to polyampholyte microgels: comparison of swelling properties. *Colloid Polym. Sci.*, **284**, 1073.

40 Pashkovski, E.E., Masters, J.G., and Mehreteab, A. (2003) Viscoelastic scaling of colloidal gels in polymer solutions. *Langmuir*, **19**, 3589.

41 Mason, T.G., Bibette, J., and Weitz, D.A. (1995) Elasticity of compressed emulsions. *Phys. Rev. Lett.*, **75**, 2051.

42 Kilfoil, M.L., Pashkovski, E.E., Masters, J.A., and Weitz, D.A. (2003) Dynamics of

43 Pashkovski, E.E., Cipelletti, L., Manley, S., and Weitz, D. (2005) Aging of soft glassy materials probed by rheology and light scattering, in *Mesoscale Phenomena in Fluid Systems*, vol. 861 (eds F. Case and P. Alexandridis), American Chemical Society.

weakly aggregated colloidal particles. *Philos. Trans. Roy. Soc. A*, **361**, 753.

44 Caggioni F M., Spicer, P.T., Blair, D.L., Lindberg, S.E., and Weitz, D.A. (2007) Rheology and microrheology of a microstructured fluid: the gellan gum case. *J. Rheol.*, **51**, 851.

45 Pine, D.J., Weitz, D.A., Chaikin, P.M., and Herbolzheimer, E. (1988) Diffusing-wave spectroscopy. *Phys. Rev. Lett.*, **60**, 1134.

46 Weitz, D.A., Zhu, J.X., Durian, D.J., Gang, H., and Pine, D.J. (1993) Diffusing-wave spectroscopy: the technique and some applications. *Phys. Scripta*, **T49**, 610.

47 Mason, T.G., Gisler, T., Kroy, K., Frey, E., and Weitz, D.A. (2000) Rheology of F-actin solutions determined from thermally driven tracer motion. *J. Rheol.*, **44**, 917.

48 Rubinstein, M. (1997) Theoretical challenges in polymer dynamics, in *Theoretical Challenges in the Dynamics of Complex Fluids*, Kluwer Academic Publishers, Dordrecht.

49 Ferry, J.D. (1980) *Viscoelastic Properties of Polymers*, 3rd edn, John Wiley & Sons, Inc., New York.

50 Morse, D.C. (1998) Viscoelasticity of concentrated isotropic solutions of semiflexible polymers. 2. Linear response. *Macromolecules*, **31**, 7044.

51 Cipelletti, L. and Ramos, L. (2005) Slow dynamics in glassy soft matter. *J. Phys. Condens. Matter*, **17**, R253.

52 Struik, L.C.E. (1978) *Physical Aging in Amorphous Polymers and Other Materials*, Elsevier Science Ltd.

53 Cipelletti, L., Ramos, L., Manley, S., Pitard, E., Weitz, D.A., Pashkovski, E.E., and Johansson, M. (2003) Universal non-diffusive slow dynamics in aging soft matter. *Faraday Discuss.*, **123**, 237.

54 Duri, A., Bissig, H., Trappe, V., and Cipelletti, L. (2005) Time-resolved-correlation measurements of temporally heterogeneous dynamics. *Phys. Rev. E*, **72**, 051401.

55 Duri, A. and Cipelletti, L. (2006) Length scale dependence of dynamical heterogeneity in a colloidal fractal gel. *Europhys. Lett.*, **76**, 972.

56 El Masri, D., Pierno, M., Berthier, L., and Cipelletti, L. (2005) Ageing and ultra-slow equilibration in concentrated colloidal hard spheres. *J. Phys. Condens. Matter*, **17**, S3543.

57 Cates, M.E. (2003) Arrest and flow of colloidal glasses. *Ann. Henri Poincare*, **4**, S647.

58 Cates, M.E., Fuchs, M., Kroy, K., Poon, W.C.K., and Puertas, A.M. (2004) Theory and simulation of gelation, arrest and yielding in attracting colloids. *J. Phys. Condens. Matter*, **16**, S4861.

59 Evans, R.M.L., Cates, M.E., and Sollich, P. (1999) Diffusion and rheology in a model of glassy materials. *Eur. Phys. J. B*, **10**, 705.

60 Fielding, S.M., Sollich, P., and Cates, M.E. (2000) Aging and rheology in soft materials. *J. Rheol.*, **44**, 323.

61 Sollich, P., Lequeux, F., Hebraud, P., and Cates, M.E. (1997) Rheology of soft glassy materials. *Phys. Rev. Lett.*, **78**, 2020.

62 Berthier, L., Biroli, G., Bouchaud, J.P., Cipelletti, L., El Masri, D., L'Hote, D., Ladieu, F., and Pierno, M. (2005) Direct experimental evidence of a growing length scale accompanying the glass transition. *Science*, **310**, 1797.

63 Larson, R.G. (1999) *The Structure and Rheology of Complex Fluids*, Oxford University Press.

64 Buscall, R., Mcgowan, I.J., Mills, P.D.A., Stewart, R.F., Sutton, D., White, L.R., and Yates, G.E. (1987) The rheology of strongly-flocculated suspensions. *J. Nonnewton. Fluid Mech.*, **24**, 183.

65 Rueb, C.J. and Zukoski, C.F. (1997) Viscoelastic properties of colloidal gels. *J. Rheol.*, **41**, 197.

66 Shih, W.H., Shih, W.Y., Kim, S.I., Liu, J., and Aksay, I.A. (1990) Scaling behavior of the elastic properties of colloidal gels. *Phys. Rev. A*, **42**, 4772.

67 Trappe, V. and Weitz, D.A. (2000) Scaling of the viscoelasticity of weakly attractive particles. *Phys. Rev. Lett.*, **85**, 449.

68 Romeo, G., Filippone, G., Fernández-Nieves, A., Russo, P., and Acierno, D. (2008) Elasticity and dynamics of particle gels in non-Newtonian melts. *Rheol. Acta*, **47**, 989.

69 Potanin, A.A. and Russel, W.B. (1996) Fractal model of consolidation of weakly aggregated colloidal dispersions. *Phys. Rev. E*, **53**, 3702.

70 Manley, S., Skotheim, J.M., Mahadevan, L., and Weitz, D.A. (2005) Gravitational collapse of colloidal gels. *Phys. Rev. Lett.*, **94**, 218302.

71 Knaebel, A., Bellour, M., Munch, J.P., Viasnoff, V., Lequeux, F., and Harden, J.L. (2000) Aging behavior of Laponite clay particle suspensions. *Europhys. Lett.*, **52**, 73.

72 Cloitre, M., Borrega, R., and Leibler, L. (2000) Rheological aging and rejuvenation in microgel pastes. *Phys. Rev. Lett.*, **85**, 4819.

73 Mazoyer, S., Cipelletti, L., and Ramos, L. (2009) Direct-space investigation of the ultraslow ballistic dynamics of a soft glass. *Phys. Rev. E.*, **79**, 011405.

74 Bouchaud, J.-P. (2008) From stretched to compressed exponentials: anomalous relaxation in complex systems, in *Anomalous Transport Foundations and Applications* (eds R. Klages, G. Radons, and I.M. Sokolov), Wiley-VCH Verlag GmbH, Weinheim.

75 Vincent, F.E., Hammann, J., Ocio, M., Bouchaud, J.-P., and Cugliandolo, L.F. (1997) *Slow Dynamics and Aging in Spin Glasses* (ed M. Rubí), Springer-Verlag, p. 184.

76 Bandyopadhyay, F.R., Liang, D., Yardimci, H., Sessoms, D.A., Borthwick, M.A., Mochrie, S.G.J., Harden, J.L., and Leheny, R.L. (2004) Evolution of particle-scale dynamics in an aging clay suspension. *Phys. Rev. Lett.*, **93**, 228302.

77 Bissig, H., Romer, S., Cipelletti, L., Trappe, V., and Schurtenberger, P. (2003) Intermittent dynamics and hyper-aging in dense colloidal gels. *Physchem. Comm.*, **6**, 21.

78 Ramos, L. and Cipelletti, L. (2001) Ultraslow dynamics and stress relaxation in the aging of a soft glassy system. *Phys. Rev. Lett.*, **87**, 245503.

79 Gavrilescu, M. and Chisti, Y. (2005) Biotechnology: a sustainable alternative for chemical industry. *Biotechnol. Adv.*, **23**, 471.

Index

a
accelerator 54, 59
acyclovir drug release 394
aggregation. *see also* flocculation
– colloidal 436
– of copolymer microgel particles 148 f
– depletion-induced 151 f
– electrolyte-induced 141 ff
– of ionic microgel particles 153
– of PNIPAM microgel particles 146 f
– SANS analysis of 150
– temperature-induced 141 ff
aging effect 301 ff
– of concentrated xanthan suspensions 442 f
– DWS-based study of 442 f
– in microgel suspensions 220 f, 445
alginate 377
alkali swellable microgel 7, 20, 292
ammonium persulfate (APS) 55, 58
angular fluctuation 251–253
annealing effect in microgel dispersion 220 f
antiferromagnetic order 270
antigen-antibody responsive hydrogel 368 f, 385
application of microgels 17, 33 f, 53, 65, 117, 208, 327 ff, 357
– in biomedical products 342 f
– in cancer therapy 378, 382 ff
– in coating formulations 338
– in cosmetic products 342 f
– in food 345 ff
– in oil recovery 350, 407 ff
– in personal care products 343
– in pharmaceutical products 342 f
– in pH-induced drug release 382
– sensing 359 ff
APS. *see* ammonium persulfate

aqueous "pregel" solution 15
attractive gel formation 158

b
bacterial contamination 22
bioadhesive microgel 342 ff
biofluid 344
biopolymer microgel 345 f, 423 ff
– application of 423 f
– control of particle morphology of 347 f
– formation of anisotropic 348
– limits of application in suspensions 439
bioresponsive hydrogel 358 ff, 368 ff
BIS. *see* N,N′-methylenebisacrylamide
block copolymer micelle 4
bond length fluctuation 261
Boson peak 195
Boyle temperature 93, 245
Bragg diffraction 208, 211 ff, 222, 244, 364
– effect of PCCA swelling response on 366 f
bright field microscopy 240, 245, 249
bulk modulus 102, 108 f, 168, 186, 188, 196, 202 ff. *see also* compressive modulus
BVU. *see* 1,3-divinylimidazolid-2-one

c
capillary microfluidic device 57 ff, 64, 66, 317
capillary micromechanics 315 ff
Carbopol (r) 343 f
carboxymethyl cellulose (CMC) 16, 429
carboxymethyl hydroxypropyl guar (CMHPG) 413
Careau model of shear-thinning fluids 332
carrageenan 424
CFT. *see* critical flocculation temperature
chemical coupling to microgels 19
classical nucleation theory (CNT) 222

CMC. *see* carboxymethyl cellulose
CMC microgel suspension   429 ff
coating brightness   342
colloid immobilization in microgels   65
colloidal "antiferromagnet"   266 f
colloidal behaviour   285
colloidal crystal   207
– diffractive properties of   362 ff
colloidal dispersion   3 ff, 33, 140, 303
colloidal gas monolayer   236, 240
colloidal gelation   207
colloidal monolayer   267 ff
colloidal polyelectrolyte complex   8
colloidal stability   4, 140 f, 439. *see also* dispersion stability
colloidal suspension   117, 207, 225, 229, 257, 291, 300
colloidosome   225
colon drug delivery   378
composite gel biopolymer-starch composite   347
compressive modulus   108, 312 ff, 315 f, 318 f, 324
ConA. *see* concanavalin A
concanavalin A (ConA)   368, 385 f, 397
conductometric titration   23
cononsolvency   104 f
contact deformation operator   199 f
contact force field   202 ff
contact modulus   296
core-shell assembly of nanoparticle-filled microgel   21
core-shell microgel   14 f, 63 f, 129 f, 400
creep compliance   432 f, 445
critical flocculation temperature (CFT)   141 ff, 147
Cross model of shear-thinning fluids   332
cross-linked polystyrene   151
cross-linked poly(4-vinylpyridine)-silica nanocomposite microgel   34
cross-linker concentration
– effect on osmotic pressure   89 ff
cross-linking agent   37
– in fracturing fluids   412 f
– influence on microgel structure   41
– metal oxide nanoparticles as   413 f
– in microchannel emulsification   55
– sodium borate as   412 f
cross-linking density   19, 377
– effect on loading/release kinetics   399 f
– effect on scattering experiments   119 f
– influence on swelling behaviour   97, 124
cross-linking monomer   20
crystal lattice distortion   271

crystal structure
– genetic algorithm for   178 f
crystallization of microgel spheres   207 ff
– kinetics of   222 ff
– uniaxial   224 f

d

daunorubicin drug release   394
deaggregation   159
Debye-Hückel theory   84 ff
decross-linking of microgels   19 f
defect
– in three-dimensional microgel crystals   244, 251
– in two-dimensional microgel crystals   255
defect concentration   261 f
degree of freedom   167 ff, 172, 175, 190
Delaunay triangulation   252
depletion interaction   134, 139 f, 151 f
dialysis   22
differential interference contrast (DIC) microscopy   358
diffuse wave spectroscopy (DWS)   288, 429
diffusion-limited transport of water   62
direct Coulombic interaction   74
disclination   261, 263
dislocation in two-dimensional colloidal crystals   261 ff
– free   261
– non-free dislocation pair   263
dispersion polymerization   232
– kinetics of   37 ff, 45
dispersion stability   140
displacement mode   197
1,3-divinylimidazolid-2-one (BVU)   10, 20
DLS. *see* dynamic light scattering
DLVO theory   140
Donnan equilibrium   82
double emulsion   63, 65, 66
doxorubicin drug release   382 ff, 389, 400
drop coalescence   57
drop formation in emulsification   56, 67
drop size in capillary microfluidic devices   61
drug delivery system   53, 65, 342, 375 ff
– based on block copolymer self-assembly   376
– composite   402
– containing magnetic nanoparticles   380
– drug distribution within the microgel   401 f
– polyacid-based gel   376, 382
– polyelectrolyte multilayer capsule   393 f
– polymer microgel   378 f
– polysaccharide gel   377
– pulsatile   389 ff

– requirements on   376, 399
drug distribution   401
drug encapsulation   380
drug release   377 ff
– antigen-antibody initiated   387
– biodegradation-induced   389 ff
– glucose-initiated   385 f
– from lipid-coated microgel   389 ff
– pH-induced   382, 392
– photoregulated   385, 388 f
– temperature-induced   380, 388
DWS. see diffuse wave spectroscopy
dynamic light scattering (DLS)   22, 117, 120, 209, 217, 314
dynamic structure factor   444
dynamics of frustrated colloidal systems   271 f

e
edge fracture   287
edge melting   259
effective charge of microgels   172 f
effective Hamiltonian of microgels   168 ff
effective interaction potential   184. see also effective interparticle potential
effective interparticle potential   167 ff
– charge dependence of   174
– density dependence of   172
Einstein model of zero-shear viscosity   328 ff
elastic deformation   79, 157, 321
elastic deformation force   167
elastic modulus   75, 108 f, 157, 195 ff, 202 f, 225, 288, 294, 296 f, 344 ff, 417 f, 428
– of Carbopol (r)   344
elasticity of microgel pastes   294
elastohydrodynamic lubrication   306 f
electric field autocorrelation function   239
electroneutrality condition   81 f, 86, 100, 168, 172, 178
electrophoretic mobility   17, 24 f
electroporation   389 f
electrostatic energy of charged microgels   83
electrostatic interaction   74, 137 f, 142
electrosterically stabilization   16 f
emulsification   8, 15 f
emulsion crystallization   223
emulsion gelation   348
emulsion polymerization (EP)   11 ff, 45
encapsulation by monodisperse microgels   65 f, 380
energy expansion   197 ff, 200
enzyme responsive hydrogel   370
EP. see emulsion polymerization
external stimuli   73, 208

f
field correlation function   429 f
filter cake   408, 412, 415
"first" melting of microgels   243 f
– near dislocations   250 f
– near grain boundaries   248 f
flip-flop value   342
flocculation   330
– depletion-induced   135, 151 f, 340
– effect of polymer addition on   151 f
– free energy of   140
Flory solubility parameter. see Flory solvency parameter
Flory solvency parameter   89, 91, 105
– temperature dependence of   94 f
Flory theory   314
flow behaviour of microgel suspensions   7, 328
flow curve   286, 299, 303 f
flow rate ratio   57, 61
fluctuation compressibility   175 f
fluid gel process   348
fluid loss   407 f
fluid loss agent   408, 419 ff
fluid loss control   409, 419 ff
fluid microgel phase   174 ff
– free energy of   177
– Hansen-Verlet criterion of freezing   186
– Lindemann criterion of melting   186, 251
– nucleation inside bulk solids   243 f
fluid-to-gel transition   342
focal length   359
force balance   199 f
form factor   122, 129 f
formation of monodisperse triple emulsion   67
fracture size   409
fracturing fluid   410, 412 f
– rheology of   415
fracturing of oilbearing formations   409 f, 415
fracturing shear rate   415
free radical polymerization   19, 36 f
– influence of functional monomers   43
– influence of initiators   39
friction coefficient   321

g
GA. see genetic algorithm
gel breaker   413
gel collapse   439
gel compression   439
gelatinized starch   345 f
– properties of   346
gellan gum   377

gel-to-fluid transition 342
gel-to-sol transition 369
GEMA. *see* 2-glucosyloxyethyl methacrylate
generalized Stokes-Einstein relation (GSER) 430
genetic alogorithm (GA) 178 f
– fitness value of 180
– structural unit of 179 f
geometric frustration 266
geometrically frustrated colloidal sample 268 ff
glass formation 207, 220, 323
glass fragility 323
glass transition 195, 211
2-glucosyloxyethyl methacrylate (GEMA) 368
grain boundary (GB) 243, 248
grain-boundary melting 248, 259
gravel packing 418 f
GSER. *see* generalized Stokes-Einstein relation
guar 409
– gel formation 411 f
– molecular structure of 410
– properties of guar dispersions 411
– rheology of guar gels 415 f

h

Hamaker constant 136, 145
Hansen-Verlet rule 186
HCN approximation. *see* hypernetted chain approximation
Herschel-Buckley equation 299, 303
heteroaggregation 153 f
hexagonal lattice 183
hexatic phase 254 ff, 264 ff
hexatic-liquid phase transition 265
HNC closure. *see* hypernetted chain closure
Hofmeister series. *see* lyotropic series
hollow colloidal particle 18
homoaggregation 153 f
homogeneous gelation 15 f
homogeneous nucleation 8, 10 f, 14
hormone therapy 389
hydraulic fracturing 350
hydrodynamic radius 118, 124, 142, 209
– temperature dependence of 239 f
hydrodynamic size 38 f, 136
hydrodynamic volume 34
hydrogel 35. *see also* microgel
– enzyme-binding to 367 ff
– fluorescently-labeled 371 f
– introduction of binding specifity into 368 f
– microsphere 53 (*see also* microgel)
– optical properties of 358 ff
– particle size 38

hydrogen bonding in polymer gels 74
hydrolysis of microgels 20
hydrophobic interaction in polymer gels 74
hydroxyethyl cellulose-base drilling fluid 421
hydroxypropyl guar (HPG) 411
hypernetted chain (HNC) closure 176, 184 ff, 242

i

ibuprofen drug delivery 376, 380
insuline drug release 381, 385, 395
intermolecular force
– in microgels 73 f
interparticle potential 219 ff, 226, 295, 330
– effective 167 ff
– video microscopy measurement of 240
interpenetrating network (IPN) microgel 18
inverse Debye screening length 170, 172
inverse emulsion polymerization 15, 18, 149
ionic microgel 74 f, 81 ff, 98, 166
– compressive modulus of 314 f
– counterion degrees of freedom in 167 ff, 172, 175, 190
– fluid phase of 174 f
– GA for the crystal structure of 181 f
– in microlenses 359
– phase behaviour of 182 ff
– phase diagram of 184
– salt concentration in 82
– swelling behaviour of 100 f
IPN microgel. *see* interpenetrating network microgel

j

jamming threshold 195, 198
jamming transition 195
Janus microgel particle 18

k

kerosene solution 55, 58
kinetics
– of crystallization of microgel dispersions 222 ff
– of microgel deswelling 321
– of microgel formation 45 ff
– of microgel loading/release 399 ff
– of microgel particle deformation 321
– of microgel swelling/shrinking 62
– of NIPAM polymerization 41, 43
– of radical polymerization 37 ff, 45 ff
– volume-phase transition 62
Kosterlitz-Thouless-Halperin-Nelson-Young (KTHNY) theory 254 f, 259 ff
Krieger-Dougherty relation 292, 328, 330

l

laser tweezer   229, 274
latex particle   3, 11, 14
latex-to-microgel transition   4
lattice fluid hydrogen bound (LFHB) theory   106
layer-by-layer assembly
– of charged polyelectrolytes on drugs   394
– of nanoparticle-filled microgel   21
LCST. see lower critical solution temperature
Lindemann parameter   246 f, 251, 261, 264. see also positional fluctuation
linear rheology   291 f
linear storage modulus   286, 289
linear viscoelasticity   286, 293 f, 298, 335 f
local dynamics of microgel suspensions   286
loss modulus   156, 219, 286, 289 f, 293 f, 297 f, 301, 426, 433 ff
lower critical solution temperature (LCST)   37, 53
lyophilization   22
lyotropic series   144
lyotropic suspension   229

m

magnetic nanoparticle   380
Mark-Houwink relationship   411
master curve   330, 333 f, 434
matrix acidizing   409
maximum freezing diameter   187
MBA. see N,N'-methylenebisacrylamide
MCT. see mode coupling theory
mean square displacement (MSD)   246 f, 288 ff, 429, 433
– angular MSD of bonds   253
– translational   253
mechanical stability   196 f
melting in three-dimensional colloidal crystals   242 ff
– bulk melting   247 f
– sample preparation for imaging   244 f
melting in two-dimensional colloidal crystals   254 ff
– analysis of experimental data   259 f
– experimental approach to analyzing   257 f
membrane emulsification   35, 53, 348
membrane osmometer   312
microcapsule   66, 393 ff
microcapsule triggering   395 ff
microchannel emulsification   53, 55, 65 f
microdome   361 f
microelectrophoresis   24
microencapsulation   393 ff
– of water   68

microfiltration   22
microfluidic chip   54 ff
microgel. see also generic microgel
– analogy to atomic fluids   93 f
– anisotropic   349
– antibody-binding to   361
– characteristics of   3 f
– characterization of   5 ff
– colloidal stability of   4 f, 140 f
– comparison with macrogel properties   7
– definition of   3
– flow properties of industrial relevant   328 ff
– for fluid loss control   419 ff
– formation from dilute polymer solution   11
– functionalization of   42, 64, 359, 369
– gelation of   348
– glass formation   155
– gold-filled   21
– for gravel packing   418 f
– ground-state configuration   182
– homopolymer   141
– hydration rate of   343
– with internal voids   62 f, 400
– lens based on   358
– lipid-coated   389
– magnetic   21
– matrix incorporation of   402
– mechanical behaviour of   311 ff
– molecular weight of   6
– monodisperse   35, 348, 401
– mucoadhesive   342 ff
– nanoparticle loading of   21
– nonaqueous   7
– nonionic precursor of   20
– optical response of   358 ff
– particle size of   3
– phase diagram of   188, 190
– pH-sensitive   357, 361, 383
– for pills   420 f
– polyampholyte-based   149
– polysaccharide-base   424
– polyvinylpyridine   102 f
– preparation strategies for   8
– protein-binding to   359 f
– response types to external stimuli   357
– as stabilizing agent   424
– starch-based   165, 345 ff
– structural effects of functional monomers   42 f
– structure of   5
– temperature-sensitive   34, 117 ff, 124, 129 ff, 229 ff, 330, 333, 357, 389
– used in oil recovery   410 ff
– water content of   6

microgel additive 327
microgel biodegradation 389 f
microgel bulk modulus 102, 108
microgel characterization technique 22 ff
microgel charge content 23
microgel deswelling 106
– cross-linking induced 366
– osmotically induced 315
microgel dispersion
– comparision to colloidal suspensions 291
– near-equilibrium property 291 f
– polymer-concentration dependence of phase behaviour 213 ff
– temperature dependence of phase behaviour 213 f
– thermodynamic model of phase behaviour of 215 f
– UV-VIS spectroscopic phase characterization 212 f
microgel dispersion rheology 285
– macroscopic shear rheology 286 f
– near confining surfaces 303 f
microgel network fluctuation 123
microgel particle
– anisotropy of 347 ff
– aspect ratio of 187
– in coating formulations 339 ff
– derivatization of 19 ff
– full elastic behaviour of a single 315 f
– osmotic compression of 312 ff
– viscoelastic response of 320
microgel particle interaction 133
– Coulomb potentials of 168 f
– dynamic rheological experiments for probing 154 ff
– elastic behaviour of 195 ff
– induced 170
– pair potential 216, 241
– steric contribution 189
– structure-independent 170
microgel particle number concentration 6, 154
microgel particle size
– control of, in preparation by capillary microfluidic device 60 f
– determination using SANS experiments 128
– effect of salt concentration on 101
– effect on loading/release kinetics 399 ff
– influence on microgel phase behaviour 187
– measurement of 22, 128
– temperature dependence of 94 f
– time dependence of 109

microgel paste 293 f
– aging of 301 ff
– disordered microstructure 296
– elastic properties of 294 f
– flow of 299
microgel phase separation 92
microgel purification 22 f
microgel radial distribution function 176
microgel storage 22 f
microgel strength 156
microgel structure factor 177
microgel suspension 285 ff, 311
– analogy to emulsions and foams 324
– colloidal 332 ff
– linear viscoelasticity of concentrated 335
– as model glasses 285, 323
microgel suspension rheology 322 ff, 327
– in coating formulations 339 ff
– rubber-like versus liquid-like behaviour 433
microgel swelling 6, 94 ff, 166, 423
– effect on drug activity 382
– effect on refractive index 108
– measurement of 358
microgel synthesis 8 ff
– by homogeneous nucleation 10 f
– from linear polymer 14
– from macrogels 8
– from monomer 8
– from polymer 8
– strategies for 8
– structures of vinyl monomers for 9 f
microgel triggering
– by degradation 389 f
– electrostatic 382
– by irradiation 385, 388 f
– by membran-active peptides 389 f
– by metabolites 385 f
– thermal 380 f
– by ultrasound 385, 389
microgel-based coating formulation 339 ff
– rheological properties of 339
microgel-based drug delivery 34, 342
microlens 358 ff
– response to avidin 360
– synthesis via "bulk" hydrogel approach 361
microlens sensing pathway
– direct binding-induced deswelling method 359 f
– displacement-induced swelling method 359 f
microrheology 288. see also microgel suspension rheology, microgel dispersion rheology
– of CMC microgel solutions 430 ff

– concentration dependence of 329 f
– DWS-based 288 f, 293, 429 ff
– influence of phase volume on 329 f
– real space particle-tracking 290 f
– salt effects on 331
– of silica suspensions in CMC solutions 436 ff
– of silica suspensions in xanthan microgel pastes 436 ff
– of xanthan microgel pastes 431 ff
microscopical tracking of particle motion 235 ff
miniemulsion polymerization 15. *see* inverse emulsion polymerization
minimum crystallization charge 184
miscibility of polymer chains 76
mode coupling theory (MCT) 300
molecular imprinting 369
monodisperse thermosensitive microgel 53 f
– with core-shell structures containing functional materials 63 f
– with multiphase complex structures 65 f
– preparation in a capillary microfluidic device 57 ff
– preparation in a PDMS microfluidic chip 54 ff
– with tunable volume-phase transition kinetics 62 f
monodispersity 53, 348
Mooney-Rivlin model 434
MSD. *see* mean square displacement
mud 408
multiple emulsion 66
multispecle scheme 289

n

N, N'-methylenebisacrylamide (MBA or BIS) 8, 10, 20, 55, 58, 97
N, N', N'', N-tetramethylethylenediamine (TEMED) 54, 59, 64
nanocapsule 375
nanogel 4, 35, 352, 365, 375, 399
NaPSS. *see* sodium poly(styrene sulfonate)
NEMAM. *see* N-ethylmethylacrylamide
nematic nanotube gel 232
N-ethylmethylacrylamide (NEMAM) 36
NIPA polymer 230 ff. *see also* poly(N-isopropylacrylamide) microgel
NIPAM. *see* N-isopropylacrylamide
NIPMAM. *see* N-isopropylmethacrylamide
N-isopropylacrylamide (NIPAM) 36, 55, 58
N-isopropylmethacrylamide (NIPMAM) 36
nucleation within an emulsion drop 15 f

o

oil well treatment
– casing 408
– cementing 408
– drilling 408
– perforating 408 f
– stimulation 409
oil-in-water emulsion 15
optical switching device 365
orientational order parameter 256
Ornstein-Zernike relation 176, 182
osmotic compressibility 91, 102, 157
osmotic compression 313
osmotic deswelling 102, 292, 341
osmotic modulus 108
osmotic pressure 74 f, 78, 86, 134, 296
– compressive measurements varying 312 f
– contribution from counterion correlations to 84 f
– contribution of polymer-solvent mixing to 89 f
– effect of inhomogeneous field-charge distribution 86 f
– effect of polymer additives 102 ff
– elastic 89 f
– electrostatic effects on 83 f
– external 102 f, 139, 312
– ideal gas contribution to 81 f
– ionic contribution to 81 ff, 98 f, 314 f
– relation to polymer volume fraction 92 f
– at thermodynamic equilibrium 88
– total 91
Oswald ripening 15

p

pairwise interaction free energy 133
particle aggregation 133
particle mobility in microgels 430 ff
particle tracking algorithm 235 ff
particle tracking velocimetry 291
particulate gels in concentrated biopolymer solutions 436
paste formation 291 f
PBG. *see* photonic bandgap
PCCA. *see* polymerized crystalline array
PDMS. *see* poly(dimethylsiloxane)
Percus-Yevick (PY) approximation 242
persistence length 80, 288, 426, 434
PGB material 362 f
PGPR. *see* polyglycerol polyricinoleate
phase behaviour
– influence of microgel particle size on 187
– influence of temperature on 213 f
– of ionic microgels 182 ff

– of monodisperse hard colloidal spheres   244
– of PNIPAM dispersions   210 ff, 215 ff
– polymer-concentration dependence of   213 ff
– of two-dimensional colloidal crystals   264 f
phase diagram   188, 190
– of ionic microgels   184
– of microgel formation   17
phase separation   439
phase transition
– order of   265
– temperature   53
photon correlation spectroscopy   288, 292
photonic bandgap   363
photonic crystal   362 f
– formation by self-assembly of microgels   364 f
– sensor   366
photonic glucose sensor   366
physical gel   37
Pickering emulsion   18
PNIPAM microgel. see poly(N-isopropylacrylamide) microgel
PNIPAM-co-acrylic acid microgel   218 f
PNIPAM-co-allylamine microgel dispersion   213 ff
PNIPMAM microgel. see poly(N-isopropylmethacrylamide) microgel
PNVF microgel. see poly(N-vinylformamide) microgel
Poisson ratio   167, 296, 315 f, 336
polyampholyte nanogel   35
poly(dimethylsiloxane) PDMS   54
polydispersity   239
polyelectrolyte   16
polyelectrolyte complex formation   17
polyelectrolyte gel   165, 389
polyelectrolyte multilayer capsule   393 f
polyelectrolyte (colloid) titration   23
polyester microgel   340
polyethylene glycol dimethacrylate   8
poly(GEMA) hydrogel   368
polyglycerol polyricinoleate (PGPR)   55, 58
polymer complexation   8, 16 ff
polymer gel   165
– for drug delivery   376 ff
polymer gelation   45
polymer network collapse   376 f
polymer network solubility   94
polymeric micelle   4
polymeric microsphere   34 f, 53 f, 57, 62
polymerized crystalline colloid array (PCCA)   366
– for glucose sensing   366

– for nerve agent sensing   367
polymer-solvent mixing   77 ff
poly(N-isopropylacrylamide) (PNIPAM) microgel   4 f, 8, 10, 53 f, 207, 230
– characterization of   208 f
– depletion flocculation of   151
– dispersion of   210 f
– in drug delivery   380
– effect of electrolyte on CFT of   142 f
– interparticle potential of   219 f
– melting behaviour of   231
– particles with added fluorophore   234
– particles with negative surface charge   234
– phase behaviour of PNIPAM dispersions   210 ff, 215 ff
– SANS experiment   124, 212
– structure of   121
– swelling behaviour of   74 f, 94 f, 105
– synthesis of   208 f, 232 ff
– UV-VIS spectra of PNIPAM dispersions   212 f
– viscosity of aqueous microgel dispersion of   158 ff
poly(N-isopropylmethacrylamide) (PNIPMAM) microgel   124 ff
– flocculation behaviour of   145
poly(N-vinylformamide) (PNVF) microgel   20
polystyrene latex   4, 11, 22
polyvinylamine (PVAm)   8, 16 f, 67 f
poly(2-vinylpyridine) (P2VP)   145, 151
positional fluctuation   246 f
– near defects   251 f
potentiometric titration   23
poylmer volume fraction   89 ff, 154
precipitation polymerization   42 f, 208
pregel emulsification   15
premelting   243. see also "first" melting
preparation
– of cationic microgels   17
– of charged microgels   20 f
– of commercial latex dispersion   11
– of core-shell microgels   14 f, 63
– of functionalized microgels   43, 64
– of hollow microgels   18 f
– of IPN microlgels   18
– of Janus microgels   18
– of microcapsules with multiphase structures   66
– microfluidic technique   54
– of microgels from linear polymers   14
– of microgels via capillary microfluidic techniques   35
– of microgels with internal voids   62

- of monodisperse microgels 35, 65
- of monodisperse polystyrene latex 11
- of monodisperse thermosensitive microgels 57
- of nanoparticle-filled microgels 21 f, 65
- of PDMS microfluidic chip 54
- of PNIPAM microgel 11 ff, 208 f
- of PVAm-CMC microgel 16 f
- of spherical microgels 63
- of thermally sensitive WSP 36 ff
- of two-dimensional colloidal crystal samples 258 f

procaine-hydrochloride drug release 383
proppant 409, 411 f
- settling 416 f
- suspending fluid 409
- transport 415 f
PVAm 8. see polyvinylamine
P2VP. see poly(2-vinylpyridine)
PY approximation. see Percus-Yevick approximation

r

radial distribution function 240, 247, 261
radical polymerization initiator 37
- decomposition rate of 40
- influence on polymerization rate 39 f
radical polymerization technique 35 ff
radical segregation 45
redox reaction approach 54 ff
redox reaction initiator 55
refractive index 75, 108, 208, 243, 359
rejuvenation 301
reverse electrolyte effect 149
reversed temperature solubility 376
rigidity criterion 197
rubber elasticity 79 f
RY closure 185

s

SANS. see small angle neutron scattering
SAXS. see small angle X-ray scattering
scaling behaviour of emulsions 195
scattering experiment 118 ff
- instrumental smearing in 123
- quantitative analysis of scattering curves 126
scattering intensity distribution 122
scattering intensity-segment density relation 119
scattering length density 119
screened electrostatic potential 171
SDS. see sodium dodecyl sulfate
self-degrading multilayer capsule 398

SEP. see surfactant-free emulsion polymerization
serum replacement 22
shear gel 348 ff
shear gelation 296, 350
shear modulus 108 f, 196, 202 ff, 224, 290, 294 ff, 306, 315 f, 318 f, 336, 436
- of concentrated microgel suspensions 336 f
shear rheology 286 f
- of agar microgels 337 f
- of concentrated microgel suspensions 332 ff
shear stress 305 ff, 319
shear thickening 345
shear thinning (pseudoplastic) 415
shearing surface 304 f
shear-thinning behaviour 329, 332, 340
Shockley partial dislocation 250
single-particle mechanics 311 ff
size distribution of microgels 35
slip velocity 304 ff
- direct measurement of 304
- elastohydrodynamic theory 306
- relation to shear stress 306 f
SLS. see static light scattering
small angle neutron scattering (SANS) 117 ff
- analysis of aggregation of PNIPAM microgel 149 f, 160
small angle X-ray scattering (SAXS) 117
sodium dodecyl sulfate (SDS) 12 f
sodium poly(styrene sulfonate) (NaPSS) 151 f
soft depletion 139
soft matter 176, 229, 268, 274, 299
soft particle elasticity 195 ff
solid-to-fluid transition 332
sol-to-gel transition 158
sol-to-gel/gel-to-sol transition 369 f
spatial density autocorrelation function 261
spin flipping 271
starch-based drilling fluid 419
static friction 319
static light scattering (SLS) 209, 217
steam stripping 22
steric interaction 187
Stokes-Einstein relation 118, 239, 289, 314, 429 f
storage modulus 156, 219, 286, 289 f, 293 f, 297 f, 301, 337, 414, 418, 426, 433 ff
strain-rate frequency superposition 298
strong screening regime 88
structure determination 117 ff, 129 ff
structure factor 87, 119 ff, 157, 160, 175 ff, 184, 212 ff, 220, 255, 261, 265
surface charge 19, 23, 145, 233 f, 257, 401

surface charge density 11, 24 f
surface functionalization of microgels 42 f
surfactant
– effect on microgel swelling behaviour 106 f
– influence on microgel particle size 12
– micelle 4
– in microchannel emulsification 55
surfactant-free emulsion polymerization (SEP) 11 ff
susceptibility 265
swelling behaviour 94 f
– effect of charge density of ionic microgels 98 f
– effect of network homogeneity on 400
– effect of polymer addition on 102 ff
– effect on drug activity 381
– influence of cross-linking density 97 f
– of ionic microgels 100 f
– response to enzymatic activity 369 f
– salt effects on 100
– scaling relation 98 ff
– in solvent mixtures 104 f
– surfactant effects on 107
swelling kinetics 108 f, 399 f
swelling process
– effect on microgel properties 75
– intermolecular forces in 73
– ionic effects on 81 ff
swelling thermodynamics 76 ff
swelling time 109 f, 321
swelling-deswelling transition 380 f
symmetry reduction 271

t
TEMED. see N,N',N'',N-tetramethylethylenediamine
temporal intensity autocorrelation function 239
thermodynamic symbol 111 ff
thermodynamics
– of elastic deformation 79 f
– of fluid microgel phase 177 f
– of ionic microgels 167 ff
– of polymer-solvent mixing 77 ff
throat channel 56 f
time-temperature superposition (TTS) 433 f
translational order parameter 256
transport mean path. 288
trigonal lattice 183
two-dimensional colloidal crystal 258 f
– correlation functions of 260 f
– geometric frustration of 266

– order of phase transitions in 265
– phase behaviour of 264 f
two-stage melting 254 f

u
ultrasoft interaction 178, 189
ultrasoft potential 172, 178, 181, 185
ultrasoft system 176 ff 187 ff
UV-VIS spectroscopy 212 f, 223

v
vaccine therapy 389, 392
van der Waals interaction 73 f, 136 f, 150, 158
video microscopy measurement of interparticle potentials 240
vinyl-based microgel monomer 8 ff
virial compressibility 175 f
virtual force field 200 f
viscoelastic modulus 287, 290 f, 301, 426, 433 ff
viscoelastic scaling algorithm 436, 438
viscosity
– of fracturing fluids 411, 413, 415
– influence of phase volume on 330
– intrinsic 329
– of microgel dispersions 158 ff, 292
– of microgel suspensions 330 ff
– stress dependence of 332 f, 335
– zero-shear 328 f
volume fraction
– of colloidal microgel suspensions 334
– of microgel dispersions 217, 292
– of microgel particles 154
volume swelling ratio 136, 155
volume term 170 ff
volume-phase transition 57, 62, 73 f
– temperature (VPTT) 36, 124, 130
– temperature induced 95
VPTT. see volume-phase transition temperature

w
wall slip 286, 303 f
Wannier model 271
WASP. see water-soluble polymer
water flooding 409
water-in-oil emulsion 15
water-soluble polymer (WSP) 36 ff
weak screening regime 88

x
xanthan
– consumer-product applications of 425

– industrial applications of   425
– in oil recovery   416 ff
– structure of   425 f
– viscoelastic properties of
  426 ff
xanthan microgel suspension   432

y

yield stress   286, 340, 348, 444
yielding   297 f
Young's modulus   316, 321
Yukawa potential   171. *see also* screened
  electrostatic potential